S. Namba · C. Hamaguchi · T. Ando (Eds.)

Science and Technology of Mesoscopic Structures

With 272 Illustrations

Springer Japan KK

Susumu Namba
Professor Emeritus of Osaka University
Professor of Nagasaki Institute of Applied Science, 536 Aba-machi, Nagasaki, 851-01 Japan

Chihiro Hamaguchi
Professor
Department of Electronic Engineering, Osaka University, Yamadaoka 2-1, Suita City, Osaka, 565 Japan

Tsuneya Ando
Professor
Institute for Solid State Physics, University of Tokyo, 7-22-1 Roppongi, Minato-ku, Tokyo, 106 Japan

ISBN 978-4-431-66924-1

Printed on acid-free paper

Library of Congress Cataloging-in-Publication Data
Science and technology of mesoscopic structures/S. Namba, C. Hamaguchi, T. Ando, eds. p. cm.
Includes bibliographical references.
ISBN 978-4-431-66924-1 ISBN 978-4-431-66922-7 (eBook)
DOI 10.1007/978-4-431-66922-7
1. Mesoscopic phenomena (Physics)—Congresses. 2. Tunneling (Physics)—Congresses, 3. Semiconductors—Design and construction—Congresses. 4. Nonlinear optics—Congresses. I. Namba, Susumu, 1928- . II. Hamaguchi, Chihiro, 1937- . III. Andō, Tsuneya, 1945- . QC176.8.M46S25 1992 530.4′1—dc20 92-30098

Typesetting: Asco Trade Typesetting Ltd., Hong Kong

Preface

The International Symposium on the Science and Technology of Mesoscopic Structures was held at Shin-Kohkaido in Nara from November 6–8, 1991. The symposium was sponsored by the International Institute for Advanced Study and partly by Nara Prefecture, Nara City, Nara Convention Bureau, and the Ministry of Education, Science and Culture of Japan, as well as industrial organizations. We would like to acknowledge the support of the symposium by these organizations.

The scope of the symposium was planned by the organizing committee to cover outstanding contributors in the fields of (1) ballistic transport, (2) electron waveguides and interference effects, (3) quantum confinement effects, (4) tunneling phenomena, (5) optical nonlinearity, and (6) fabrication technology of mesoscopic structures. Twenty-six invited speakers were selected from the United States, Europe, and Japan. In addition twenty-four contributed papers were accepted for presentation at the poster session. These papers are included in the proceedings.

We are grateful to the organizing committee, Ms. Yoshiko Kusaki of the International Institute for Advanced Study for the secretarial service, and Dr. Nobuya Mori, Osaka University, for his scientific cooperation. Thanks are also due to the authors and the participants for their contributions to a successful symposium.

SUSUMU NAMBA
CHIHIRO HAMAGUCHI
TSUNEYA ANDO

Opening Address

Ladies and Gentlemen,

On behalf of the organizing committee, I am happy to welcome you to the International Symposium on the Science and Technology of Mesoscopic Structures, sponsored by the International Institute for Advanced Study. First of all, I would like to express my gratitude to all of you who have contributed to this symposium. In particular I would like to thank the invited speakers who kindly agreed to present their interesting work at this symposium. It is a great honor for the organizing committee to have two winners of the Nobel Prize here as invited speakers. They are Dr. Leo Esaki and Professor Dr. Klaus von Klitzing. They have contributed a great deal to the initial development in the field of mesoscopic physics in semiconductors and are now playing a very important role in this area.

The technical program of the symposium covers a wide range of mesoscopic science. The topics covered and emphasized by the present symposium include: (1) ballistic transport, (2) electron waveguides and interference effects, (3) quantum confinement effects, (4) tunneling phenomena, (5) optical nonlinearity, and (6) fabrication technology of mesoscopic structures. Twenty-six invited speakers, who are known to be most active in each of these fields, were selected from all over the world. In addition to these invited talks we received twenty-four contributed papers which were all original, covering very recent progress in the field of mesoscopic science. These contributed papers will be presented at the poster session.

Finally I would like to talk about Nara in ancient times. We say that Nara is our cradle, because the oldest capital of Japan was built in Nara. Japanese culture started to grow in the 7th century when Buddhism was introduced to Japan. In addition to Buddhism many kinds of goods were brought to Nara through the Silk Road and Nara is known to be the east end of the Silk Road. When you visit museums and temples in Nara you will see many of these goods and how the Japanese culture was affected not only by Chinese culture but also by the Persian culture. In the year 607 Prince Shotoku built Horyuji Temple in Nara, and this still exists today as the oldest wooden building in the world. We decided to hold this symposium in November in Nara because a most famous exhibition is held in Nara during this period. The exhibition is called Shohsoin-ten and it displays the Emperor's treasures, including much of the Silk Road collection. It is a great pleasure for us to provide you with a chance to see the Shohsoin-ten at the National Museum of Nara. The museum is close to our conference hall, being only five minutes walk from here.

I would like to conclude by wishing us all an exciting and fruitful symposium which will contribute to further developments in the field of mesoscopic science.

SUSUMU NAMBA (Chairman)
Professor Emeritus
Osaka University

Foreword

The international symposium, "Science and Technology of Mesoscopic Structures", was held in the city of Nara, the ancient capital of Japan, under the auspices of the International Institute for Advanced Studies (IIAS).

IIAS was established in Japan in 1984 as a non-profit organization, a legal entity recognized by Japanese law, with the strong support of Japanese industry, academia, and the national and prefectural governments. Its prime objective is to create and develop new interdisciplinary scientific approaches that can deal with problems which could threaten human existence in the 21st century.

Scientific and technological advances have given many of the nations of the world rich material cultures, but they also have produced grave problems that are closely related to the crises currently facing human existence in this last decade of the 20th century, and that will continue into the next century. These problems include the drastic depletion of natural resources, pollution of the global environment, and a decline in ethics. It is imperative that we now pursue research that will lead to the solution of these problems and to the benefit and well-being of future humanity.

The fundamental concept on which all IIAS activities are based determines what should be studied and striven for to ensure the well-being of future human society. Not only will topics indispensable for the enrichment of human life be selected for research, but research strategies will be also developed that will be directed toward solving the pervasive problems that face humankind. From these strategies will come new directions in scientific research and novel concepts that will contribute to the development of world culture.

The IIAS research-residential complex, *Scholars' Village*, will be completed in the summer of 1994 in the Kansai Culture and Science City, which is under construction in the Keihanna area. This area was the center of historic Japanese culture, and is now a new urban zone where the modern prefectures of Kyoto, Osaka, and Nara meet. The Kansai Culture and Science City is a unique center designed to facilitate research on aspects of culture and in the sciences that will produce benefits for human society world-wide. The IIAS complex will be the central research facility of this new international cultural and scientific capital of Japan. IIAS *Scholars' Village* is to be a place for free and far ranging creative thinking; it is not designed to be an experimental facility. Its main objectives are:

- To establish new systems of knowledge and concepts of basic research that go beyond existing systems and concepts.

- o To respond with flexibility to scientific developments and to changes in the direction of world society.
- o To determine what forms partnerships between industry and science should take, and to collaborate with industry in the realization of these partnerships.

In keeping with the basic purpose of IIAS and the goals of its *Scholars' Village*, the research undertaken will be varied, and will be planned to incorporate parallel activities which will include the education of outstanding young researchers and collaboration between the natural and the social sciences. The kinds of research to be done at the *Scholars' Village* will be *thematic, basic, independent*, and *special research*. In addition, there will be related activities including periodic meetings (symposia and workshops), study groups (advanced research seminars and training courses), public lectures and publications (proceedings, monographs, annual reports, etc.).

For *thematic research*, projects will be chosen from a large array of possible research themes and will be supported for a specified period of time. As a rule, a project will last 3 years (including preparation time). The first research theme undertaken when the *Scholars' Village* opens will be in the area of *theoretical life sciences*. Topics include: (1) Brain science, (2) genes and evolution, and (3) biodiversity. The *basic research* will be characterized by broadly based long-term research projects in the fundamental natural and social sciences, which will complement and support the *thematic research* and other research activities of IIAS, and will include mathematical and philosophical investigations. The *independent research* is designed for eminent scholars world-wide, who will be invited to work at the *Village* for specific periods. Their interaction with other researchers in residence is expected to greatly stimulate all the institute's research activities. *Special research* is to be undertaken when topics calling for further investigation emerge from the results of *thematic, basic*, and *independent* research projects.

As mentioned above, IIAS organizes various international symposia and scientific exchange-programs. In the past six years, more than ten international symposia, covering topics in newly developing areas in the biological sciences, basic physics, and information science, have been held under IIAS auspices. International meetings on fundamental problems in human sciences and interdisciplinary areas in the natural and social sciences have also been held under IIAS auspices. In 1988, an international symposium in the area of basic physics, "Fluctuation and Relaxation in Condensed Phase—Disordered Systems, Fractals and Biological Materials", was held. In 1992, an international workshop, "Mathematical Approaches to Fluctuation in Astronomy, Physics and Biology" and an international seminar, "Frontiers of Nuclear Spectroscopy" will be held under IIAS, auspices.

MICHIO OKAMOTO
Director
International Institute
for Advanced Studies

Contents

Tunneling

Fabrication Technology

Nonlinear Optical Properties

List of Contributors

Historical View of Mesoscopic Physics

The Evolution of Semiconductor Superlattices and Other Quantum Structures

1

LEO ESAKI[1]

Summary. Following the past two decades' evolutionary path in the interdisciplinary research of semiconductor superlattices and quantum wells, significant milestones are presented, with emphasis on electric field-induced effects in the frontier of semiconductor physics associated with technological advances.

Introduction

In the early twentieth century, the encounter with physical phenomena for which Newtonian mechanics could not possibly provide an adequate explanation prompted the advent of quantum mechanics. Quantum mechanics, indeed, has been playing an indispensable role in the modern physics world, including our understanding of the behavior of semiconductors. Toward the late twentieth century, however, quantum mechanics has been expanding its role, that is, its principles have been utilized in the design of novel semiconductor structures which exhibit unprecedented transport and optical properties. The advanced techniques of thin-film growth and microfabrication have facilitated engineering for *designed quantum structures.*

In 1969, research on such structures was initiated with a proposal of an *engineered* semiconductor superlattice by Esaki and Tsu [1, 2]. In anticipation of advancements in epitaxy, two types of superlattices were envisioned, with alternative deposition of ultrathin layers: Doping and compositional, as shown at the top and bottom of Fig. 1, respectively.

Since the one-dimensional potential is introduced along with the superlattice (SL) axis (perpendicular to the deposited plane layers), we thought that, if our attempt was successful, elegantly simple examples in one-dimensional mathematical physics, for instance, resonant electron tunneling [3], Kronig-Penney bands [4], Tamm surface states [5], Zener band-to-band tunneling [6], and Stark ladders including Bloch oscillations [7, 8], all of which had remained more or less textbook exercises, could, for the first time, be practiced in a laboratory: *Do-it-yourself quantum mechanics* would be possible, since the principles of quantum theory dictate the design of semiconductor structures or devices.

[1] IBM Thomas J. Watson Research Center, Yorktown Heights, NY 10598, USA
Present address: University of Tsukuba, Tsukuba, Ibaraki, 305 Japan

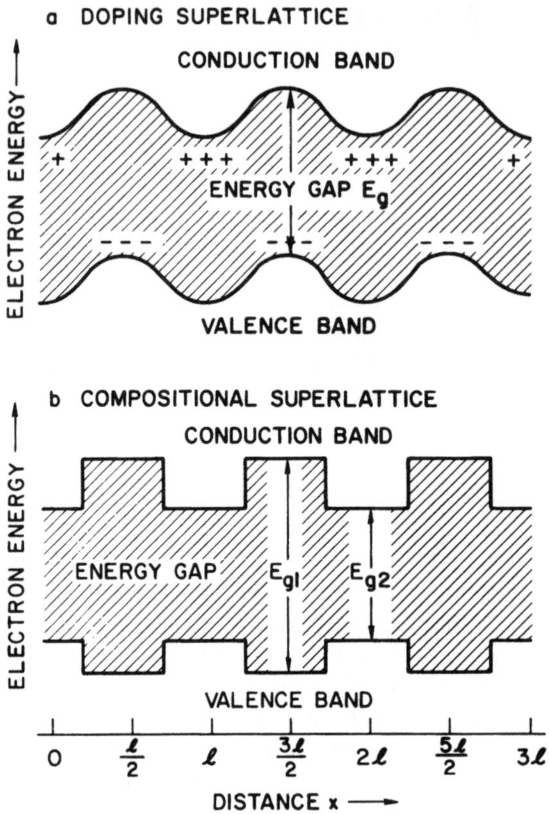

Fig. 1a, b. Spatial variations of the conduction and valence bandedges in two types of superlattices: **a** doping, **b** compositional

At the inception of the SL idea, it was recognized that the long, tailored lattice period provided the unique opportunity to exploit electric field-induced effects. Our early analysis of electron dynamics in a modestly high electric field, along with the SL axis, led to the prediction of the occurrence of a negative differential resistance which could be a precursor of the Bloch oscillation. The introduction of the SL, apparently, allows us to explore the regime of electric field-induced quantization. The formation of Stark ladders can be proved in an SL with a one-dimensional periodic potential [9]. In natural (three-dimensional) crystals, however, the existence and real nature of these localized states in a high electric field have been in controversy [10, 11].

Before reaching the SL concept, we had explored the feasibility of structural formation by epitaxy of ultra-thin barriers and wells which might exhibit resonant electron tunneling. It was thought that semiconductors, and technologies developed with them, might be suitable for demonstration of the quantum wave nature of electrons associated with the interference phenomena, since their small Fermi energies due to low carrier densities help make the de Broglie wavelength relatively large. Namely, the Fermi wavelength $\lambda_f = 2\pi/k_f$, where k_f is the magnitude of the Fermi wavevector, is given as a function of the carrier density n, as follows: $\lambda_f =$

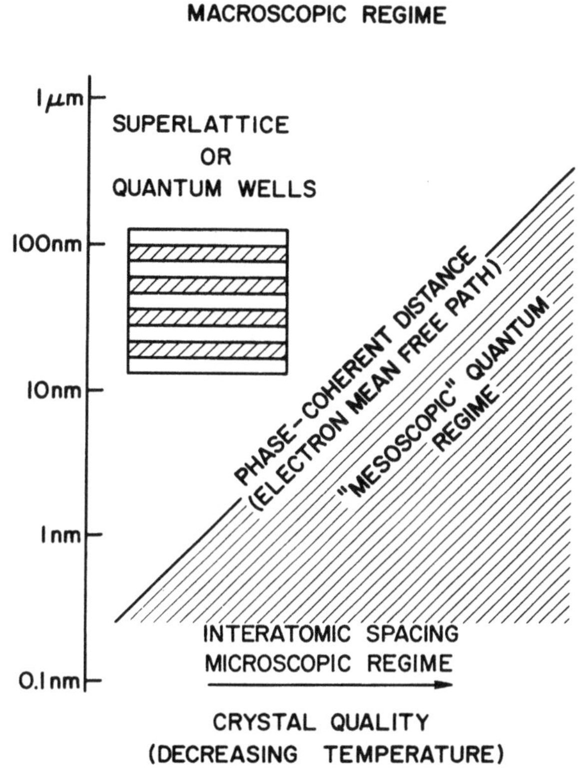

Fig. 2. Schematic illustration of a "mesoscopic" quantum regime (*hatched*) with a superlattice or quantum wells in the *inset*

$(8\pi/3n)^{1/3}$ for a three-dimensional (3D) system; $(2\pi/n)^{1/2}$ for a two-dimensional (2D) system; (4/n) for a one-dimensional (1D) system.

The idea of the SL occurred to us as a natural extension of double- and multi-barrier structures: Namely, the SL is a series of quantum wells (QWs) coupled by resonant tunneling, where quantum effects are expected to prevail. An important parameter relevant to the observation of such effects is the phase-coherent length or, roughly, the electron inelastic mean free path, which depends heavily on bulk, as well as on interface quality of crystals, and also on temperatures and values of the effective mass. As schematically illustrated in Fig. 2, if characteristic dimensions such as SL periods and QW widths are reduced to less than the phase-coherent length, the entire electron system will enter a *mesoscopic* quantum regime of reduced dimensionality, being placed in the scale between the macroscopic and the microscopic.

It was theoretically shown that the introduction of the SL potential perturbs the band structure of the host materials, yielding unprecedented electronic properties of quasi-two-dimensional character [1, 2]. Figure 3 shows the density of states $\rho(E)$ for electrons with $m^* = 0.067m_0$ in an SL with a well width of 100Å and the same barrier width, where the first three subbands are indicated with dashed curves. The figure also includes, for comparison, a parabolic curve $E^{1/2}$ for 3D, a steplike

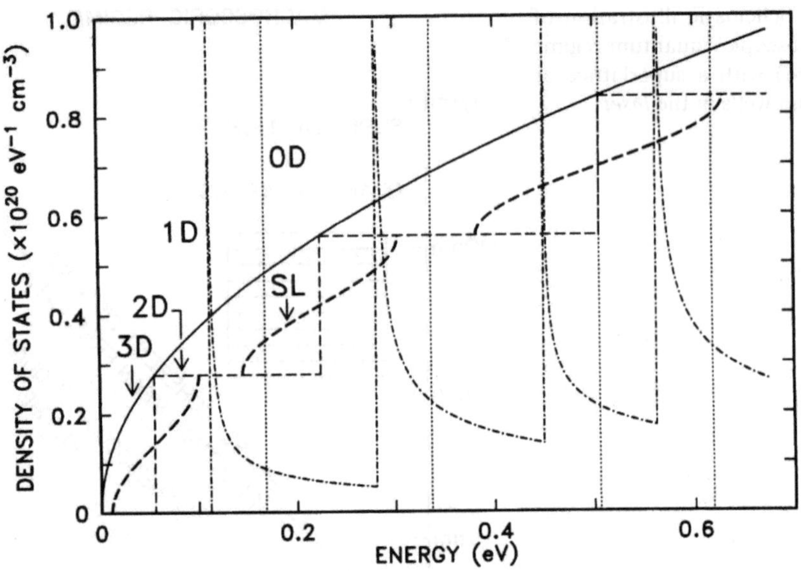

Fig. 3. Comparison of density of states in a three-dimensional (*3D*) electron system with those of a superlattice, and two-dimensional (*2D*), one-dimensional (*1D*), and zero-dimensional (*0D*) electron systems

density of states for 2D (quantum well), a curve $\sum_{mn} > (E - E_m - E_n)^{-1/2}$ for 1D (quantum wire), and a delta function $\sum \delta(E - E_l - E_m - E_n)$ for 0D system (quantum dot) where the quantum unit is taken to be 100Å for all cases and the barrier height is assumed to be infinite in obtaining the quantized energy levels, E_l, E_m, and E_n. Notice that the ground state energy increases with decrease in dimensionality if the quantum unit is kept constant. Each quantized energy level in 2D, 1D, and 0D is identified with one, two, and three quantum numbers, respectively. The unit for the density of states here is normalized to $eV^{-1}cm^{-3}$ for all the dimensions, although $eV^{-1}cm^{-2}$, $eV^{-1}cm^{-1}$ and eV^{-1} may be commonly used for 2D, 1D, and 0D, respectively.

Epitaxy and Superlattice Growth

In the early 1970s, we initiated the seemingly formidable task of engineering nanostructures in the search for novel quantum phenomena [12]. Heteroepitaxy is of fundamental interest for SL and QW growth. Innovations and improvements in growth techniques such as molecular beam epitaxy (MBE), metallo organic chemical vapor deposition (MOCVD), and metallo organic molecular beam epitaxy (MOMBE), or chemical beam epitaxy (CBE) during the last decade have made possible high-quality heterostructures. Such structures possess predesigned potential profiles and impurity distributions with dimensional control close to interatomic spacing and with virtually defect-free interfaces, particularly in a lattice-

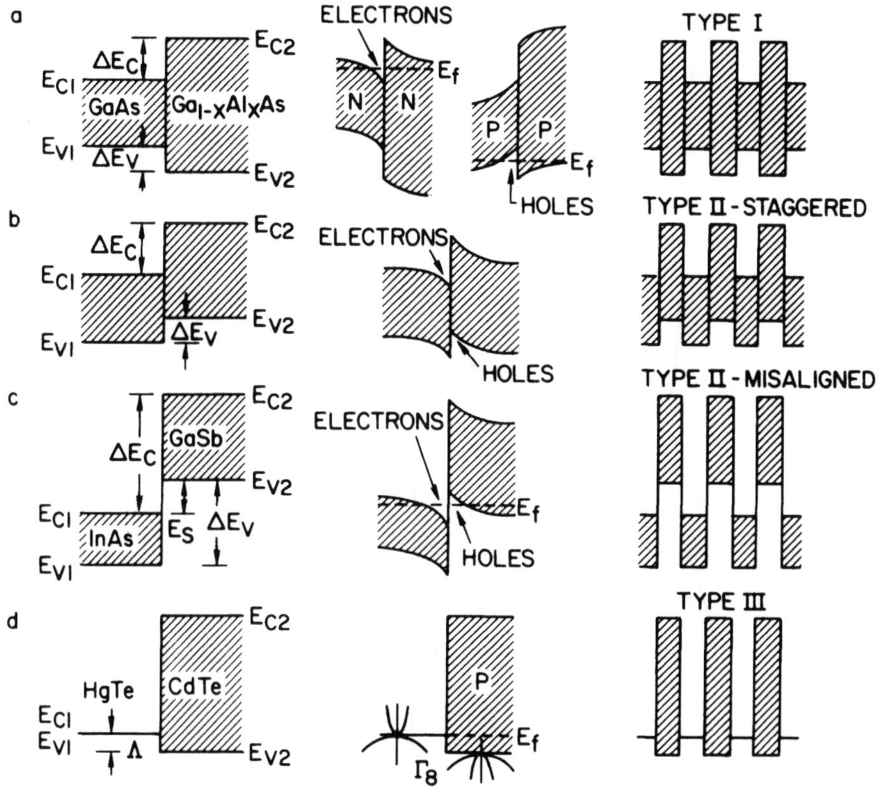

Fig. 4a–d. Discontinuities of bandedge energies at four types of hetero-interfaces: band offsets (*left*), band bending and carrier confinement (*middle*), and superlattice (*right*)

matched case such as GaAs – $Ga_{1-x}Al_xAs$. This great precision has cleared access to a *mesoscopic* quantum regime.

The semiconductor SL structures have been grown with III-V, II-VI, and IV-VI compounds, as well as elemental semiconductors. Semiconductor hetero-interfaces exhibit a band offset, usually associated with a gradual band-bending in its neighborhood which reflects space-charge effects. According to the character of such discontinuity, known hetero-interfaces can be classified into four kinds: Type I, type II-staggered, type II-misaligned, and type III, as illustrated in Fig. 4a, b, c, d, in which band offsets are shown on the left, band bending and carrier confinement in the middle, and SLs on the right.

Resonant Tunneling in an Electric Field

Resonant tunneling in multi-barrier structures can be observed with the application of an electric field perpendicular to the layer planes. In this context, Tsu and Esaki [13] derived the voltage dependence of the tunneling current in such struc-

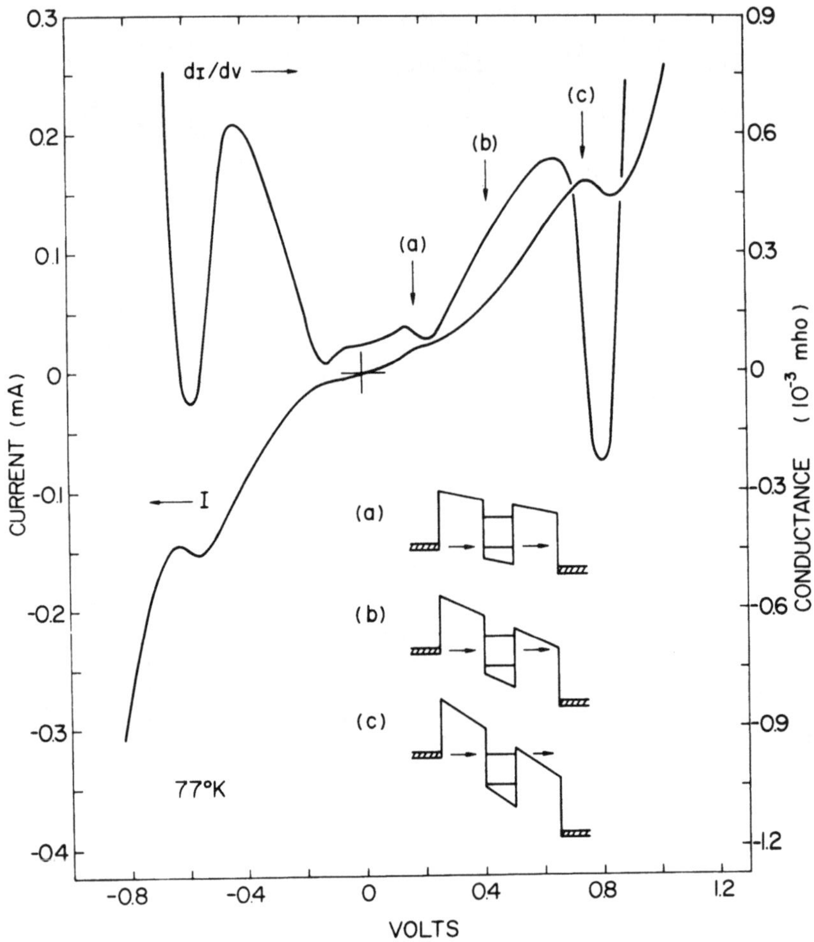

Fig. 5. Current- and conductance- voltage characteristics of a GaAs-GaAlAs double-barrier structure. Conditions at resonance (*a*) and (*c*), and at off-resonance (*b*), are indicated by *arrows* in the *inset*

tures with the transmission coefficient computed as a function of electron energy in assuming the conservation of the total electron energy, as well as its momentum transverse to the direction of tunneling. Chang, Esaki, and Tsu [14] measured the tunneling current and conductance as a function of applied voltage in GaAs-GaAlAs double barriers, as shown in Fig. 5, which exhibited a large component of the excess current primarily arising from imperfections in the structures, but agreed qualitatively with the Tsu-Esaki formula [13]. The energy diagram is shown in the inset where resonance occurs at such applied voltages as to align the Fermi level of the electrode with the bound states, as shown in cases (a) and (c) of Fig. 5. This experimental result probably constitutes the first observation of man-made bound states in single QWs.

The technological advance in MBE has resulted in remarkably improved characteristics in resonant tunneling, which spurred renewed interest in such structures. Recently, high-quality resonant-tunneling double-barrier junctions, exhibiting not only large values of the peak-to-valley ratio but also fine structure, have been the focus of intense experimental and theoretical investigations. The observation of resonant tunneling of holes reveals structure corresponding to respective bound states of both heavy and light holes [15]. The existence of optical phonon-assisted resonant tunneling was found in the valley region [16] and analyzed [17]. Resonant magneto-tunneling measurements manifested the energy levels quantized into Landau levels of index n [18]. Later, in magneto-tunneling, the longitudinal-optical (LO) phonon (magnetopolaron)—assisted process was clearly resolved [19, 20]. Charge accumulation in a double-barrier resonant tunneling structure was studied by photoluminescence [21] and photoluminescence-excitation spectroscopy [22].

A variety of tunneling heterostructures have been developed with a combination of materials other than GaAs/AlAs, such as InGaAs/InAlAs [23] and InAs/AlSb [24]. Research in polytype heterostructures of InAs/AlSb/GaSb [25, 26], has been rejuvenated with the observation of a large negative resistance based on resonant interband tunneling in GaSb/AlSb/InAs/AlSb/GaSb [27–29] and InAs/AlSb/GaSb/AlSb/InAs [30]. Those multi-heterojunctions, indeed, correspond to a portion of the proposed polytype superlattices [25].

A great deal of engineering interest in resonant tunneling, obviously, has arisen from its potential application to high-speed semiconductor devices. Picosecond bistable operation has been observed in a double-barrier resonant tunneling diode, where a rise time of 2 ps is said to be the fastest switching yet measured for an electronic device [31]. Oscillations up to frequencies of 420 GHz were reported in a GaAs resonant tunneling diode at room temperature [32].

Transport and Optical Properties in Superlattices and Quantum Wells Subject to an Applied Electric Field

Theoretical Background

The introduction of the SL potential, as shown in Fig. 1, perturbs the band structure of the host materials: The degree of perturbation obviously depends on its potential profile. With the SL period, l, of the order of 50–100Å, the Brillouin zone is divided into a series of minizones, as shown schematically in the bottom of Fig. 6, giving rise to the first and second subbands, E_1 and E_2, separated by forbidden regions, as shown in the top.

With application of an electric field in the SL direction, x, the electron dynamics in a narrow subband has been analyzed in the framework of a simplified path integration method to obtain a relationship between the applied field F and the average drift velocity v_d. The equations of electron motion are

$$\hbar \partial k_x/\partial t = eF \quad \text{and} \quad v_x = \hbar^{-1}\partial E_x/\partial k_x \tag{1}$$

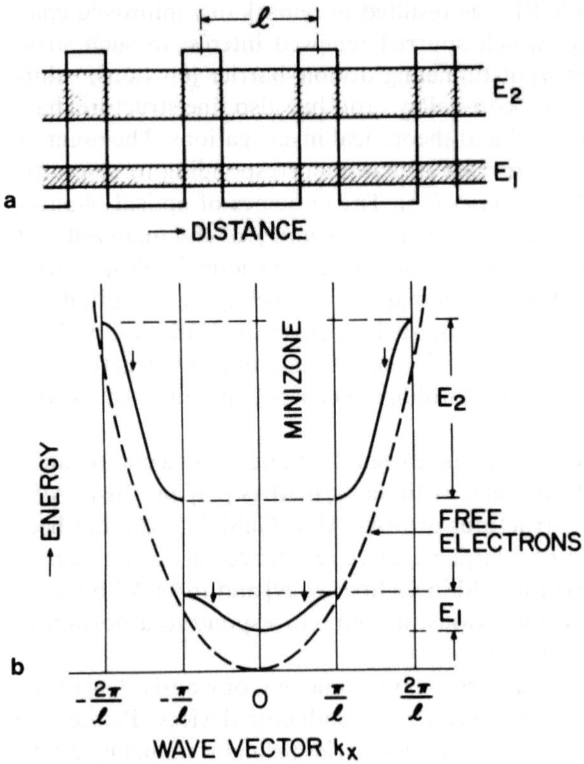

Fig. 6. a Potential profile of a superlattice, and **b** electron energy vs wave vector in the minizones

then, the velocity increment in a time interval dt is

$$dv_x = eF\hbar^{-2}(\partial^2 E_x/\partial k_x^2)dt \qquad (2)$$

The average drift velocity, taking into account the electron scattering time τ, is written as

$$v_d = \int_0^\infty \exp(-t/\tau)dv_x = eF\hbar^{-2}\int_0^\infty (\partial^2 E_x/\partial k_x^2)\exp(-t/\tau)dt \qquad (3)$$

If a sinusoidal relationship, $E_x = \frac{E_1}{2}(1 - \cos k_x l)$, is assumed, we obtain

$$v_d = eF\tau/m^*(0)[1 + (eFl\tau/\hbar)^2]^{-1} \qquad (4)$$

where $m^*(0)$ is determined by $\hbar^2(\partial^2 E_x/\partial k_x^2)^{-1}$ at $k_x = 0$. Thus, the current—field curve is given by

$$J = env_d \qquad (5)$$

where n is the electron concentration. The current J, plotted as a function of F in Fig. 7, has the maximum at $eFl\tau/\hbar = 1$ (indicated by an arrow) and thereafter decreases, giving rise to a differential negative resistance. The threshold field for the negative resistance is 10^3 V/cm if $l = 100$Å and $\tau = 0.66$ ps. This result indi-

Fig. 7. Current versus electric field curve

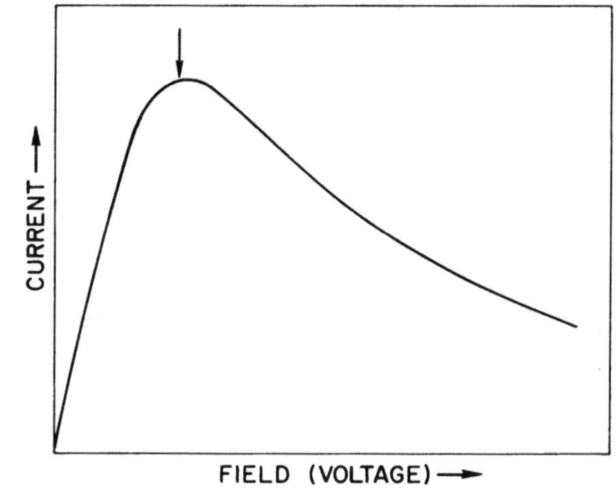

cates that, if, in applying modestly high electric fields, electrons gain enough energy to reach beyond the inflection points (indicated by arrows in Fig. 6), they will be decelerated rather than accelerated by such electric fields.

If, in further increasing electric fields, electrons reach the mini-Brillouin zone boundary ($k_x = \pi/l$ in Fig. 6), an umklapp process (or Bragg scattering) will occur, that is, the electron wave vector, k_x, is changed by a reciprocal lattice vector, $2\pi/l$. If this happens, electrons reappear at the opposite point of the Brillouin zone, $k_x = -\pi/l$, and the process will be repeated. This is the physical picture of the so-called Bloch oscillation (the centroid of the electron wave packet moves back and forth, steadily, in the k-space), whose frequency is determined by the time period T required to cross the Brillouin zone, namely, $T = 2\pi\hbar/eFl$. Thus, the Bloch angular frequency ω_B is given by $\omega_B = 2\pi/T = eFl/\hbar$. In order to realize this oscillation, the scattering time τ should be greater than T, that is, $\omega_B\tau > 2\pi$. A precursor of this oscillation is the above-mentioned differential negative resistance, which starts at $\omega_B\tau > 1$. The frequency of the Bloch oscillator is as high as 1 THz for $F = 4 \times 10^3$ V/cm and $l = 100$Å.

The Bloch oscillation involves resonating transitions between equally spaced and localized energy levels, called Stark ladders, which are the consequence of electric field-induced quantization. Figure 8 shows, schematically, electronic wave functions of Stark ladders formed in a superlattice under low and high electric fields; at the low field, optical transitions may be possible because of sufficient overlap of wave functions between adjacent Stark states; the energy separation of these levels ΔE is equal to $\hbar\omega_B = eFl$. At the high field, field-induced electron hopping to non-overlapping adjacent Stark states may occur, leading again to a differential negative resistance [33].

Before the introduction by Wannier [8] of the term Stark ladder, James [7] heuristically reached the basic concept of Stark localization. Experimental evidence supporting their existence is rather tenuous [34–36], because the observation, firstly, requires enough high fields to produce the energy separation greater than the collision broadening, that is,

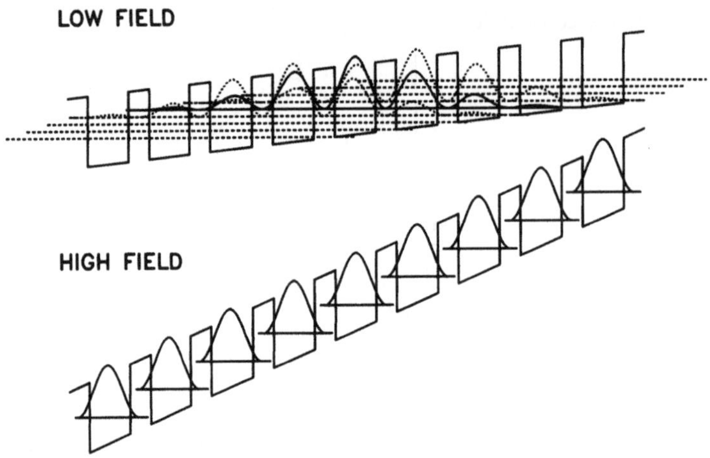

Fig. 8. Schematic potential profiles with electronic wave functions of Stark ladders formed in a superlattice under low and high electric fields

$$\Delta E = eFa > \hbar/\tau, \quad \text{then} \quad \omega_B \tau > 1 \tag{6}$$

where a is the lattice constant which is typically 2 to 3Å. If $\tau = 0.66$ ps, $\hbar/\tau = 1$ meV. Therefore, F should be of the order of 10^5 V/cm. On the other hand, the introduction of SL structures with large lattice constants l provides easy access to the electric field-induced quantization regime with a few tens of kV/cm.

Stark ladder states are localized over the length λ given by E_1/eF, where E_1 is the subband width. The k-space approach presented here is justified when $\Delta E = eFl \ll E_1$ or $l \ll \lambda$. If, however, the value of eFl approaches to $E_1 (\lambda \to l)$ because of high fields or narrow bandwidths, the hopping model of Tsu and Döhler [37] becomes valid. In this case, although a negative differential resistance may arise, the Bloch oscillation is unlikely to occur because of insufficient spacial overlap between adjacent Stark states.

Artaki and Hess [38] carried out Monte Carlo simulations for electron transport in SLs, taking into account scattering by acoustic and optical phonons and impurities; the collision broadening effect on SL transport associated with a negative resistance was assessed.

Incidentally, it should be appreciated that the quantum theory for the electric field effects in an SL presented here is quite different from the traditional approach to a bulk semiconductor with E_c and E_v bandwidths of a few electron-volts: The electro-optical properties (a tail below the energy gap and an oscillatory behavior above it) have been treated with the effective mass approach, where the envelope eigenfunctions under an external electric field are Airy functions: The observation is consistent with the resultant Franz-Keldish effect.

Early Attempts in the Search for the Quantum Effects

Esaki et al. [39] found a negative resistance, resembling Figure 7, starting at $F \sim 2 \times 10^4$ V/cm in an MBE-grown GaAs-GaAlAs SL with $E_1 \sim 40$ meV. Later,

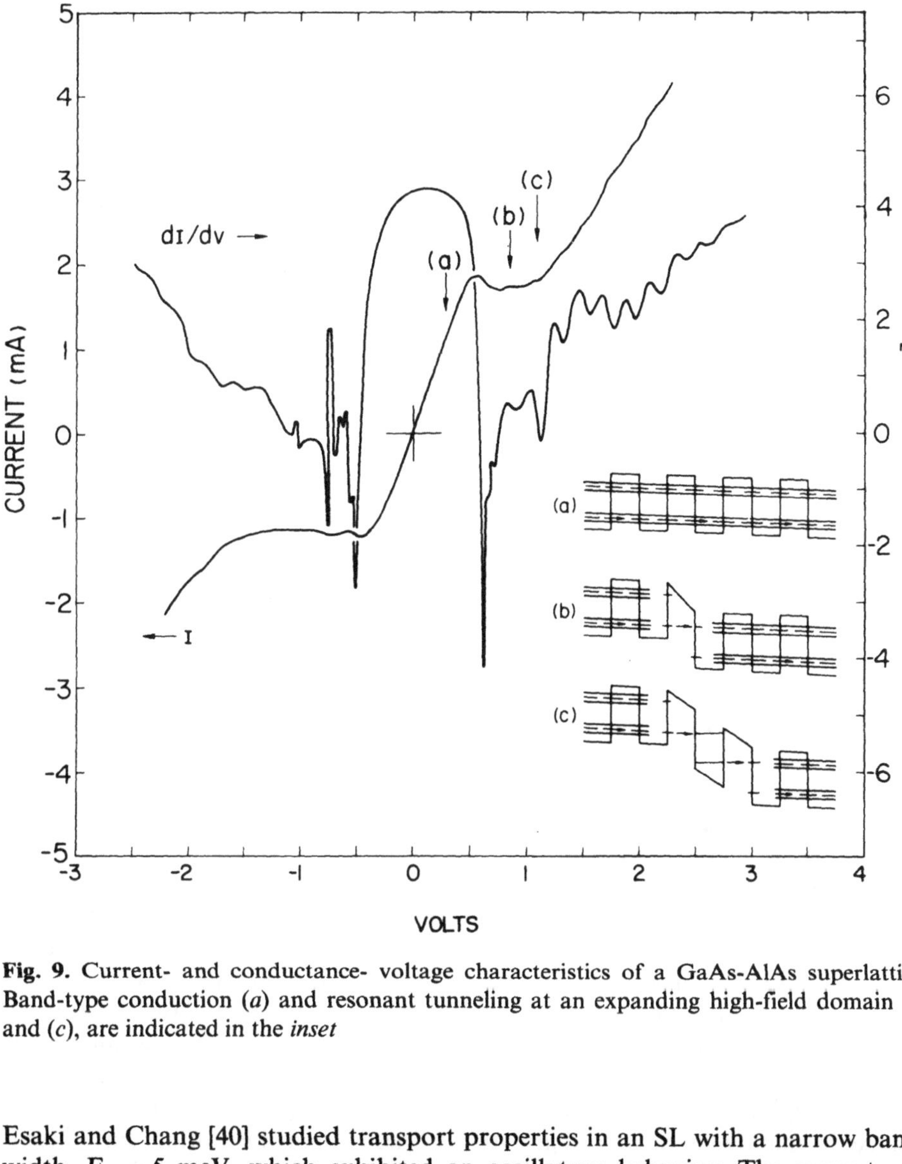

Fig. 9. Current- and conductance- voltage characteristics of a GaAs-AlAs superlattice. Band-type conduction (*a*) and resonant tunneling at an expanding high-field domain (*b*) and (*c*), are indicated in the *inset*

Esaki and Chang [40] studied transport properties in an SL with a narrow bandwidth, $E_1 = 5$ meV, which exhibited an oscillatory behavior: The current and conductance as a function of applied voltage are shown in Fig. 9. A marginal band-conduction mechanism [case (a) in the inset] is broken down, as eFl approaches to E_1, at relatively low voltages, resulting in a negative resistance with creation of a localized high-field domain. The current recovers due to resonant Zener tunneling between the first miniband and the second miniband in the adjacent QWs [case (b) in the inset] when their energies are aligned with each other. The oscillatory behavior in the conductance reflects the repetition of this resonance with increase in applied voltage, whose period, 0.24 V, agreed with the calculated value. Tsu et al. [41] performed photocurrent measurements on GaAs-GaAlAs SLs subject to an electric field perpendicular to the plane layers with the

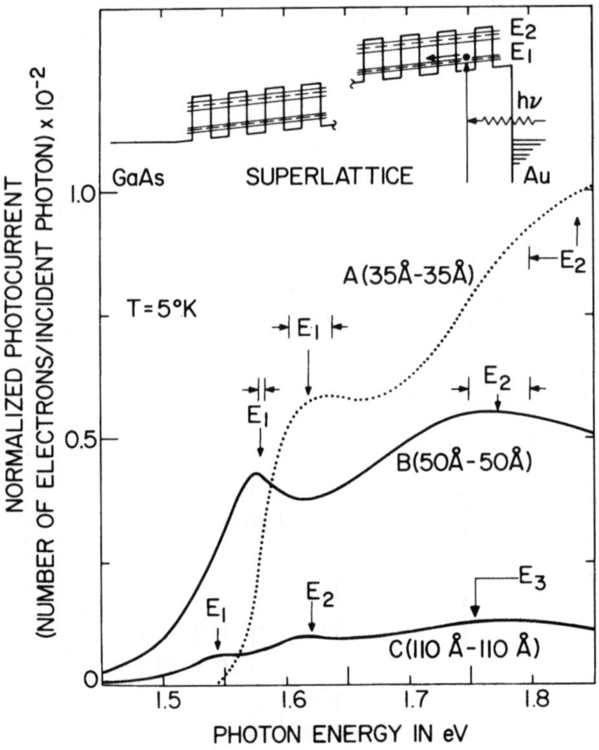

Fig. 10. Photocurrent versus photon energy for three superlattice samples A, B, and C. Calculated energies and bandwidths are indicated. The energy diagram of a Schottky-barrier structure is shown in the *upper part*

use of a semitransparent Schottky contact. The photocurrents as a function of incident photon energy are shown in Fig. 10 for samples of three different configurations, designated A (35Å GaAs–35Å GaAlAs), B (50Å GaAs–50Å GaAlAs), and C (110Å GaAs–110Å GaAlAs). The energy diagram is shown schematically in the upper part of the figure, where minibands created by the SL potential are labeled E_1 and E_2. In sample C, the minibands are essentially discrete states. Calculated energy locations and bandwidths shown in the figure are found to be in satisfactory agreement with the observation. The photocurrent, as a function of applied voltage, exhibited a pronounced negative resistance, which was interpreted with the hopping model [37], since it occurred in the range of $eFl > E_1$.

Since the mid-1980s, advanced growth facilities at many institutions have begun to provide high-quality SLs and QWs, meeting stringent requirements in the structural design. As a result, investigations into such quantum structures have been intensified and have proliferated extensively. We present here only a limited number of reports, selected rather arbitrarily by the author.

Recent Transport Experiments

Deveaud et al. [42] carried out transport measurements of 2D carriers in minibands using subpicosecond luminescence spectroscopy. Schneider et al. [43] studied the dynamics of electron transport along the growth direction of tight-binding

GaAs-AlAs SLs by electrical and optical time-of-flight experiments. Helm et al. [44] observed infrared light (intersubband) emission from SLs excited by sequential resonant tunneling. Tatham et al. [45] made time-resolved Raman measurements of intersubband relaxation in GaAs QWs. Sibille et al. [46–48] studied negative differential conductance in the transport properties of GaAs-AlAs SLs having relatively wide miniband-widths: The data were analyzed in terms of Esaki and Tsu's theory [1, 2]. Hadjazi et al. [49] obtained preliminary results for microwave applications of superlattices. Grahn et al. [50] reported the observation of photoluminescence from the second and third subbands in GaAs-AlAs SLs with an applied electric field: The occupation of the higher subbands was achieved by sequential resonant tunneling of electrons. Brozak et al. [51] observed thermal saturation of transport in a narrow band of an SL: The conductivity was quenched as the temperature was raised. Beltram et al. [52] studied transport properties in a transistor with an SL structure built between the base and the collector, observing negative differential conductance and scattering-controlled transmission resonances. Vengurlekar et al. [53] reported miniband transport of minority electrons in a superlattice transistor. Lei et al. [54] provided an advanced theory of negative differential conductance in a superlattice miniband. Zaslavsky et al. [55] observed negative differential resistance in an edge-regrown superlattice (a modulated 2D electron system) at low temperatures.

Stark Shifts in QWs

Mendez et al. [56] studied the field-induced effect on the photoluminescence in GaAs-GaAlAs QWs, where the main peak is attributed to the excitonic transition between 2D electrons and heavy holes: When the electric field was applied perpendicular to the plane layers, pronounced field-effects, Stark shifts, were found. More recently, Vina et al. [57] studied excitonic coupling, as indicated in Fig. 11, which shows Stark shifts of the light-hole exciton (open circles) and excited states of the heavy-hole exciton (open triangles), exhibiting strong coupling (solid circles) between them. Chemla et al. [58] observed Stark shifts of the excitonic absorption peak, even at room temperature; Miller et al. [59] analyzed such electroabsorption spectra and demonstrated optical bistability, level shifting, and modulation.

Formation of Stark Ladders in SLs with the Application of an Electric Field

Bleuse et al. [60] calculated field-induced Stark localization of the eigenstates in an SL, predicting a blue shift of the optical-absorption edge and the presence of oscillations in F^{-1}. Mendez et al. [61] observed the formation of Stark ladders in an SL by field-induced movement of the peak positions in photoluminescence and photocurrent spectra with applied electric fields. Figure 12 illustrates the conduction- and valence-band potential profiles, including the Stark ladder formation, for a 40Å–20Å GaAs-GaAlAs SL in the absence, (a), and in the presence of an electric field, (b) and (c). At $F = 0$, the electron and hole wavefunctions are delocalized and interband transitions take place between SL minibands. With the

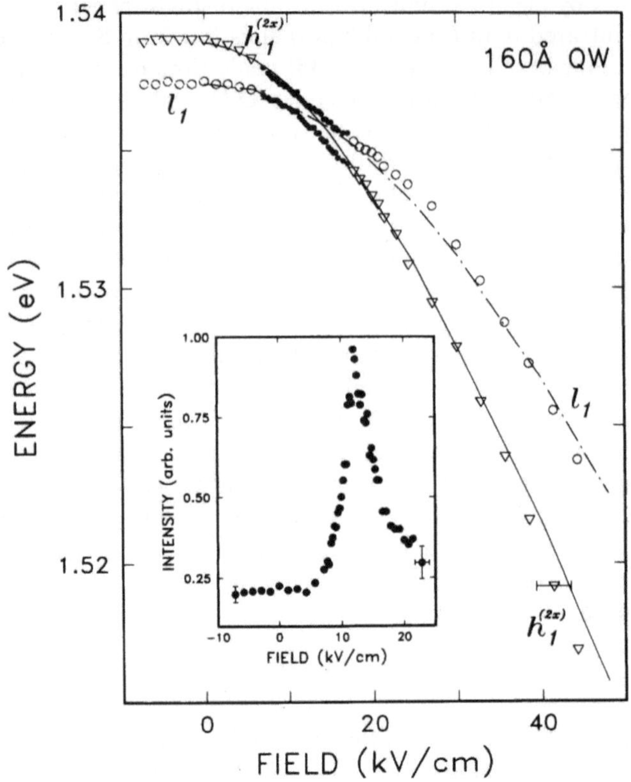

Fig. 11. Stark shifts of the light-hole exciton (*open circles*) and excited states of the heavy-hole exciton (*open triangles*), exhibiting strong coupling (*solid circles*) between them. The integrated intensity of $h_1^{(2x)}$ normalized to l_1 is shown as a function of electric field in the *inset*

application of an electric field, the localization of wavefunctions begins, forming the ladder whose energy spectrum is given by Eo + neFl (n = 0, ± 1, ± 2, $\cdots$), where Eo is the eigenenergy of an isolated QW. At a moderate field (2×10^4 V/cm), (b), electron and light-hole states are still extended over several SL periods, whereas heavy holes are fully localized. Interband transitions (indicated by -2, -1, 0, $+1$, $+2$) involve Stark-ladder states like the ones represented here for the conduction band. For simplicity, the light-hole ladder states are not included. At a high field (10^5 V/cm), (c), all states are completely localized and transitions take place between levels corresponding to isolated QWs. Figure 13 shows the interband transition energies as a function of field for the SL in Fig. 12, determined from the peak positions in photocurrent spectra. Each branch corresponds to a member of the Stark ladder, labeled by an index, n. With the same technique, Agullo-Rueda et al. [62] determined the coherence length of electrons from the number of observed interband transitions between the valence- and conduction-band Stark ladders, which was found to be at least ten periods when l = 60Å and

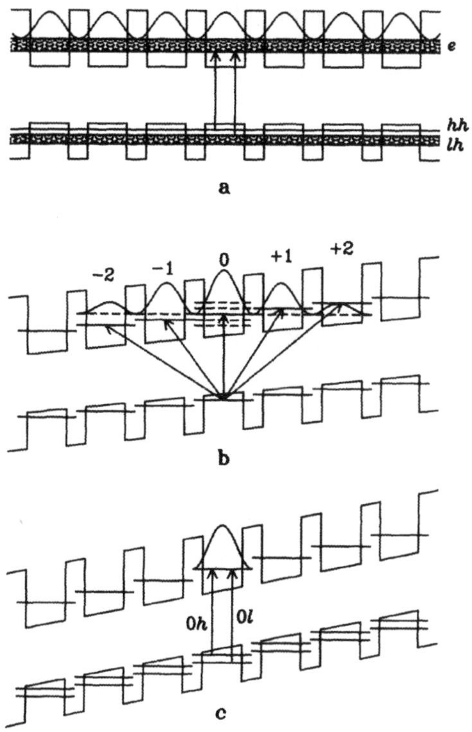

Fig. 12a–c. Conduction- and valence-band potential profiles, including Stark ladder formation, for a 40Å–20Å GaAs-GaAlAs superlattice at zero field, (**a**), at a moderate field (2×10^4 V/cm), (**b**), and at a high field (10^5 V/cm), (**c**)

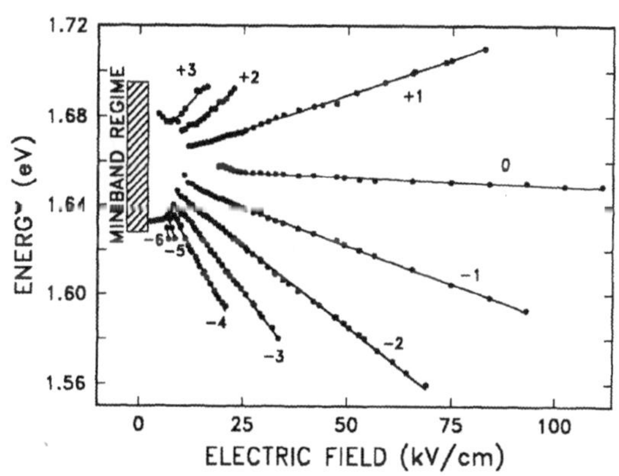

Fig. 13. Interband transition energies as a function of electric field for the superlattice shown in Fig. 12. Each branch corresponds to a member of the Stark ladder, labeled by an index, *n*

increased with decreasing *l* (or increasing miniband width). Later, it was shown that the coherence length did not depend strongly on temperature in the range 5–292K [63]. Voisin et al. [64] confirmed the predicted electro-optical effect [60] by an electroreflectance study for an SL. Dignam and Sipe [65] presented calculations of the energies and absorption spectra of GaAs-GaAlAs SL excitons in the presence of an applied electric field, which were in good agreement with the experimental photocurrent spectra [63]. Ohno et al. [66] observed localized surface states (Tamm states) intentionally introduced in GaAs-GaAlAs SLs by excitonic interband transitions in photoluminescence excitation spectra. Photocurrent experiments under an electric field showed additional transitions, as well as anticrossing interactions, between the Tamm states and the Stark-ladder states associated with the SL.

Conclusion

Our original proposal and early works apparently had universal appeal, which triggered a wide spectrum of experimental and theoretical investigations in the new domain of semiconductor science and engineering. A variety of *engineered quantum structures in low dimension* exhibited extraordinary transport and optical

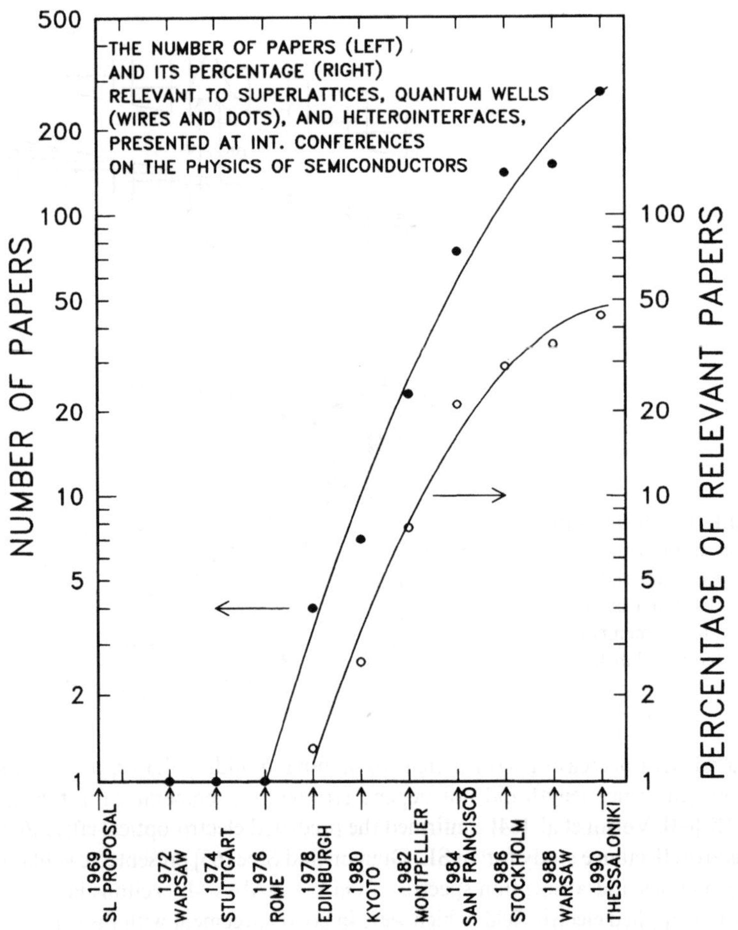

Fig. 14. Growth in the number of papers at ICPS

properties which may not even exist in any *natural* crystal. Thus, this new degree of freedom offered in semiconductor research *through advanced material engineering* has inspired many ingenious experiments, resulting in observations not only of predicted effects but also of totally unknown phenomena. The growth in the number of papers published on the subject over the last two decades is indeed phenomenal. For instance, the number of papers on the subject and the percentage of those to the total presented at the International Conference on the Physics of Semiconductors (ICPS), which is held every other year, has shown an extraordinary increase, as shown in Fig. 14. After an incubation period of several years, this field really took off, with a rapid annual growth, around ten years ago. It says a great deal to see that, currently, nearly one half of semiconductor physicists in the world are working in this field. Activity in this new frontier of semiconductor physics has, in turn, given immeasurable stimulus to device physics, provoking new ideas for applications. Thus, unprecedented transport and opto-electronic devices have emerged. In this interdisciplinary research, beneficial cross-fertilizations are prevalent.

In this review, we have primarily focused on intriguing phenomena which occur with the application of an electric field in the direction of the SL axis. The significance of such electric field-induced effects was recognized at the inception. The observation of Stark ladders probably consitutes one of the milestones in quantum structure research. It should be appreciated that an energy separation of 10 meV between adjacent Stark ladders was achieved only with 20 kV/cm, while bulk semiconductors would require far greater electric fields for such electric quantization. It is interesting to compare this with magnetic quantization of the orbital motion of electrons, in which an energy separation, $\hbar\omega_c$, of 10 meV between Landau levels for electrons with a small effective mass, is readily obtainable with a magnetic flux of 10 T or so. The electric counterparts of the Landau level and cyclotron resonance are considered to be the Stark ladder and Bloch oscillation, respectively. "Whether the Bloch oscillation in superlattices is really observable or not" is, undoubtedly, one of the most fascinating unanswered questions.

References

1. Esaki L, Tsu R (1969) IBM Research Note RC-2418
2. Esaki L, Tsu R (1970) IBM J Res Develop 14:61
3. Bohm D (1951) Quantum Theory: Prentice Hall, Englewood Cliffs, p 283
4. de L Kronig R, Penney WG (1931) Proc Roy Soc (London) A130:499
5. Tamm I (1932) (in Russian) Phys Z Sowjetunion 1:733
6. Zener C (1934) Proc Roy Soc (London) 145:523
7. James HM (1949) Phys Rev 76:1611
8. Wannier GH (1959) Elements of solid state theory. Cambridge University Press, Cambridge, pp 190–193; (1960) Phys Rev 117:432
9. Shockley W (1972) Phys Rev Lett 28:349
10. Zak J (1968) Phys Rev Lett 20:1477; (1991) Phys Rev B43:4519
11. Rabinovitch A, Zak J (1971) Phys Rev B4:2358
12. Esaki L (1986) IEEE J Quantum Electron, QE-22:1611

13. Tsu R, Esaki T (1973) Appl Phys Lett 22:562
14. Chang LL, Esaki L, Tsu R (1974) Appl Phys Lett 24:593
15. Mendez EE, Wang WI, Ricco B, Esaki L (1985) Appl Phys Lett 47:415
16. Goldman VJ, Tsui DC, Cunningham JE (1987) Phys Rev B36:7635
17. Cai W, Zheng TF, Hu P, Yudanin B, Lax M (1989) Phys Rev Lett 63:418
18. Mendez EE, Esaki L, Wang WI (1986) Phys Rev B33:2893
19. Leadbeater ML, Alves ES, Eaves L, Henini M, Hughes OH, Celeste A, Portal JC, Hill G, Pate MA (1989) Phys Rev B39:3438
20. Boelinger GS, Levi AF, Schmitt-Rink S, Passner A, Pfeiffer LN, West KW (1990) Phys Rev Lett 65:235
21. Young JF, Wood BM, Aers GC, Devine RL, Liu HC, Landheer D, Buchanan M (1988) Phys Rev Lett 60:2085
22. Yoshimura H, Schulman JN, Sakaki H (1990) Phys Rev Lett 64:2422
23. Takeuchi A, Inata T, Muto S, Miyauchi E (1989) Jpn J Appl Phys 28:L750
24. Luo LF, Beresford R, Wang WI (1988) Appl Phys Lett 53:2320
25. Esaki L, Chang LL, Mendez EE (1981) Jpn A Appl Phys 20:L529
26. Takaoka H, Chang CA, Mendez EE, Chang LL, Esaki L (1983) Physica 117B and 118B:741
27. Luo LF, Beresford R, Wang WI (1989) App Phys Lett 55:2023
28. Beresford R, Luo LF, Longenbach KF, Wang WI (1990) Appl Phys Lett 56:551; (1990) Appl Phys Lett 56:952
29. Mendez EE, Ohno H, Esaki L, Wang WI (1991) Phys Rev B43:5196
30. Soderstrom JR, Chow DH, McGill TC (1989) App Phys Lett 55:1094
31. Whitaker JF, Mourou GA, Sollner TCLG, Goodhue WD (1988) Phys Lett 53:385
32. Brown ER, Sollner TCLG, Parker CD, Goodhue WD, Chen CL (1989) Appl Phys Lett 55:1777
33. Tsu R, Esaki L (1991) Phys Rev B43:5204
34. Chynoweth AG, Wannier GH, Logan RA, Thomas DE (1960) Phys Rev Lett 5:57
35. Maekawa S (1970) Phys Rev Lett 24:1175
36. Koss RW (1972) Phys Rev B5:1479
37. Tsu R, Döhler G (1975) Phys Rev B12:680
38. Artaki M, Hess K (1985) Superlatt Microstruct 1:489
39. Esaki L, Chang LL, Howard WE, Rideout VL (1972) Proc 11th int conf on the Physics of Semiconductors, Warsaw, 1972. PWN-Polish Scientific Warsaw, p 431
40. Esaki L, Chang LL (1974) Phys Rev Lett 33:495
41. Tsu R, Chang LL, Sai-Halasz GA, Esaki L (1975) Phys Rev Lett 34:1509
42. Deveaud B, Shah J, Damen TC, Lambert B, Regreny A (1987) Phys Rev Lett 58:2582
43. Schneider H, Rühle WW, Klitzing K v, Ploog K (1989) Appl Phys Lett 54:2656
44. Helm M, England P, Colas E, DeRosa F, Allen SJ Jr (1989) Phys Rev Lett 63:74
45. Tatham MC, Ryan JF (1989) Phys Rev Lett 63:1637
46. Sibille A, Palmier JF, Wang H, Mollot F (1990) Phys Rev Lett 64:52
47. Sibille A, Palmier JF, Wang H, Esnaul JC, Mollot F (1990) Appl Phys Lett 56:256
48. Sibille A, Palmier JF, Mott F, Wang H, Esnault JC (1989) Phys Rev B39:6272
49. Hadjazi M, Sibille A, Palmier JF, Mollot F (1991) Electronics Lett 27:1101
50. Grahn HT, Schneider H, Rühle WW, von Klitzing K, Ploog K (1990) Phys Rev Lett 64:2426
51. Brozak G, Helm M, DeRosa F, Perry CH, Koza M, Bhat R, Allen SJ Jr (1990) Phys Rev Lett 64:3163
52. Beltram F, Capasso F, Sivco DL, Hutchinson AL, Chu SNG, Cho A (1990) Phys Rev Lett 64:3167

53. Vengurlekar AS, Capasso F, Hutchinson AL, Tsang WT (1990) Appl Phys Lett 56.262
54. Lei XL, Horing NJM, Cui HL (1991) Phys Rev Lett 64:3277
55. Zaslavsky A, Tsui DC, Santos M, Shayegan M (1991) Appl Phys Lett 58:1440
56. Mendez EE, Bastard G, Chang LL, Esaki L (1982) Phys Rev B26:7101
57. Vina L, Collins RT, Mendez EE, Wang WI (1987) Phys Rev Lett 58:832
58. Chemla DS, Damen TC, Miller DAB, Gossard AC, Wiegmann W (1983) Appl Phys Lett 42:864
59. Miller DAB, Weiner JS, Chemla DS (1987) IEEE J Quantum Electron QE-22:1816
60. Bleuse J, Bastard G, Voison P (1988) Phys Rev Lett 60:220
61. Mendez EE, Agullo-Rueda F, Hong JM (1988) Phys Rev Lett 60:2426
62. Agullo-Rueda F, Mendez EE, Hong JM (1989) Phys Rev B40:1357
63. Mendez EE, Agullo-Rueda F, Hong JM (1990) Appl Phys Lett 56:2545
64. Voisin P, Bleuse J (1988) Phys Rev Lett 61:1639
65. Dignam MM, Sipe JE (1990) Phys Rev Lett 64:1797
66. Ohno H, Mendez EE, Brum JA, Hong JM, Agullo-Rueda F, Chang LL, Esaki L (1990) Phys Rev Lett 64:2555

Mesoscopic Physics of Small Metallic Particles 2

Ryogo Kubo[1]

Summary. Some of the most fundamental points of mesoscopic physics of electrons in small metallic particles of submicron size are revisited emphasizing the quantum discreteness of electronic states and the effect of coulomb interaction.

Introduction

Mesoscopic properties of metals and semiconductors with nanometer extensions and very short time scales have attracted much attention in recent years, opening a new rich field of physics. Thirty years ago, the author published a paper [1] treating some peculiar properties of small metallic particles of submicron size. This was a theoretical prediction of quantum effects to be seen in such small systems and was probably the very earliest work of mesoscopic physics (for review papers, see [2–5]).

In the present paper, I would like to revisit some of the most fundamental points raised in my old paper, since they still retain their importance for the future developments of mesoscopic physics in general.

Metallic Electrons in a Particle

One-Particle Picture of Metallic Electrons

We consider a particle composed of metallic atoms. Our main concern is the states of metallic electrons confined in such a particle. They are those electrons which occupy the Bloch states in a bulk of the material. The energy of a Bloch state with a crystal momentum $\hbar k$ is denoted by $E(\vec{k}, \vec{\sigma})$. This may also depend on the spin operator $\vec{\sigma}$ if the spin orbit interaction is strong. We assume the atoms in the particle form essentially the same crystal lattice but are bounded by the boundary defining the particle. There may also be defects and distortions of the lattice. The effective one-electron Hamiltonian for the electrons in the particle is then assumed to be given by

[1] Frontier Research System, RIKEN, 2-1 Hirosawa, Wako City, Saitama, 351-01 Japan

$$H = E\left(\frac{1}{i}\vec{\nabla}, \vec{\sigma}\right) + V \tag{1}$$

where V represents the potential due to the boundary and defects. The boundary condition for the one-electron states is considered in V.

We adopt the one-electron picture and assume that each quantum state of the electrons as a whole is defined by a configuration of occupied one-electron quantum states. Thus, the ground state of the electrons is that in which the states are occupied from the bottom to the Fermi level E_F following the Pauli principle. It should be remembered that this simple picture is somewhat beyond the Sommerfeld free electron model, since excitation of the electron states corresponds to the Landau quasiparticles.

Statistical Considerations

In practice, it is difficult to prepare small particles with a precise control of size, shape and defects. Therefore we regard these factors as random parameters when we deal with an assembly of particles. Then the potential V in Eq. (1) is considered as a random variable, and accordingly the set of the eigenfunctions and eigenvalues of (1) is random.

The problem associated with a statistical ensemble of Hamiltonian is known as the theory of random matrices [6]. This concept was first introduced by Wigner in treating the quantum states of atomic nuclei and has been well developed since then by many authors. Here we mention only a few important points.

If the eigenvalues of (1) are the set $\{\varepsilon_j\}$, then the probability W for realization of the set is shown to have the form [7]

$$W(\varepsilon_{j-1}, \varepsilon_j, \varepsilon_{j+1}, \ldots) = C \exp\left(-\frac{n}{2}\sum \varepsilon_j^2\right) \Pi |\varepsilon_j - \varepsilon_k|^n \tag{2}$$

where Gaussian distributions are assumed for the matrix elements of the random Hamiltonian and n is an integer to be discussed later. In particular, the level spacing Δ_j of neighboring eigenvalues

$$\Delta_j = \varepsilon_j - \varepsilon_{j-1} \tag{3}$$

is a random object. Its average is denoted by δ, namely,

$$\delta = \overline{\Delta}_j \tag{4}$$

Since the total number of quantum states for the Hamiltonian (1) is equal to the number of atoms, N, in a particle without counting the spin degeneracy, and the Fermi energy is typically a few electron volts, the average spacing δ is in the order of magnitude,

$$\delta = E_F/N \tag{5}$$

This amounts to 10^{-4} eV, if N is of the order of 10^4, which corresponds to the thermal energy at 1°K.

The probability distribution for small values of Δ is given in the form,

$$P(\Delta) = C(\Delta/\delta)^n, \quad \text{for} \quad \Delta \ll \delta \tag{6}$$

The number n is an important parameter appearing in Eq. (2) and is shown to be determined by the basic symmetry [7] of the Hamiltonian as follows:

1. *Orthogonal ensemble*: $n = 1$
   If the spin orbit interaction is weak and there is no magnetic field present, the Hamiltonian is invariant under orthogonal transformation and its matrix elements are all real numbers.
2. *Unitary ensemble*: $n = 2$
   If the spin-orbit interaction is weak and there is a magnetic field present, the Hamiltonian is invariant under unitary transformation and its matrix elements are complex numbers.
3. *Symplectic ensemble*: $n = 4$
   If the spin orbit interaction is strong, each quantum level is a Kramers doublet and the matrix elements are all quartanions.

In the orthogonal case, two eigenvalues come close when the relevant part of the Hamiltonian matrix has the form

$$H = \begin{pmatrix} x & y \\ y & -x \end{pmatrix}$$

which yields two eigenvalues $\varepsilon = \pm\sqrt{x^2 + y^2}$ with random real numbers x and y, thus leading to distribution (2) with $n = 1$. Similar consideration for cases (2) and (3) gives $n = 2$ and 4, respectively.

Before we proceed further to mesoscopic considerations of small particles, we must note an essential point of the problem.

Charge Neutrality Condition

When a metallic particle of radius a is charged by Q, the work done in the process is

$$W = Q^2/2a \tag{7}$$

If we put $Q = e =$ the electron charge and $a = 0.57$Å, the Bohr radius, W is the ionization energy of a hydrogen atom, 13.7eV. Making a as large as 57Å, W still amounts to 0.14eV, much larger than the thermal energy at room temperature.

The above estimate is too simple and is not quite correct, since it must be done by quantum mechanics, for example, using a jellium model and the density functional method. But it is enough to support the idea that each small metallic particles very strictly retains charge neutrality, charge fluctuation being totally suppressed. It has been found that a thin metallic film consisting of granular islands exhibits tunneling conduction of the activation type [8]. Recent experiments on metallic clusters also give some support to the neutrality. Also in recent studies of mesoscopic tunneling the phenomenon of Coulomb blockade, which is another indication of charge neutrality, has been described.

Thermodynamics at Low Temperatures

The Fermi Distribution

It is an interesting and important feature of the mesoscopic physics of metallic small particles that the familiar Fermi distribution function loses its validity at low temperatures, $T \ll \delta/k$. By the neutrality condition, the number of electrons in a given particle is fixed, so that in making the statistical mechanics one must use a canonical ensemble with a fixed number of electrons in each particle, not a grand canonical ensemble. The asymptotic equivalence of the two ensembles is guaranteed only if the number of quantum states is large in the energy interval of the order of the thermal energy, kT, namely if $\delta \ll kT$. Therefore the statistical theory of electrons in mesoscopic particles at low temperatures is a hard task permitting no simple analytical treatment. Using a constant spacing instead of random spacing [9], the occupation probability is in fact found to be discontinuous at the Fermi level, unlike the Fermi distribution function.

Heat Capacity at Low Temperatures

Figures 1 and 2 illustrate the configurations of electrons occupying the one-electron states in a given particle. Here we assume the spin-orbit interaction is sufficiently weak so that each of the discrete orbital states can accommodate two electrons with opposite spins.

Obviously there is a remarkable difference between the two cases where the total number of electrons is even or odd.

If the electron number is even, the ground state of the electrons as a whole is that in which each of the states up to ε_0 in Fig. 2 are occupied by a pair of electrons. The excited states are created by bringing an electron from one of the occupied states up to an empty state above ε_0.

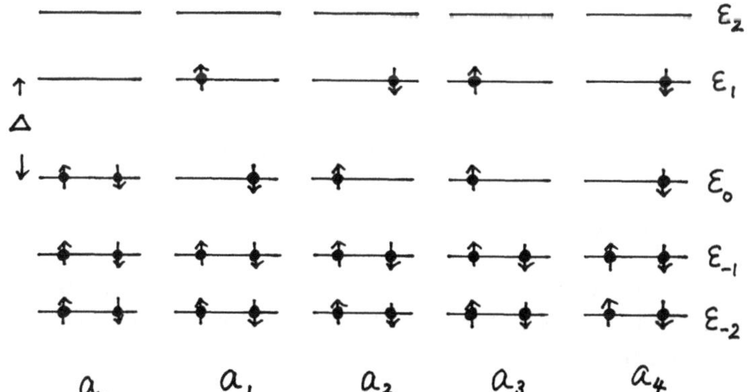

Fig. 1. Electron configurations of a metallic particle containing an even number of electrons; a_0 is the ground state and $a_1, a_2, \ldots$ are the excited states

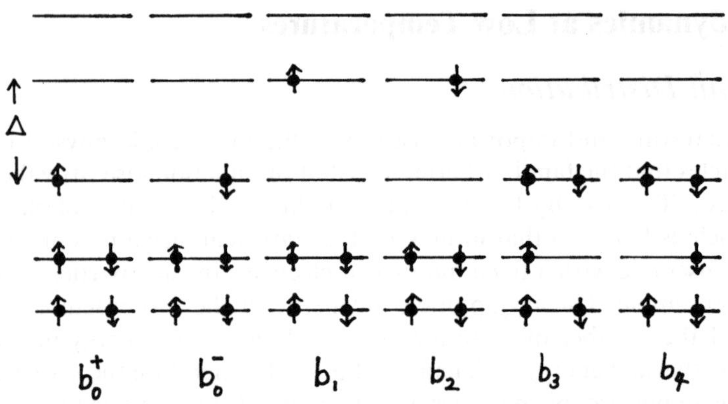

Fig. 2. Electron configuration of a metallic particle containing an odd number of electrons; b_0^+ and b_0^- are the ground states and b_1, b_2, b_3, ... are the excited states

If the electron number is odd, the higher occupied state has a free spin. As seen easily from the above considerations, the minimum energy of excitation is the level spacing Δ.

Therefore the average energy of excitation at a temperature T is given by

$$\bar{E} = \int P(\varepsilon)\varepsilon e^{-\varepsilon/kT} d\varepsilon \tag{8}$$

and accordingly the heat capacity is given by

$$C = d\bar{E}/dT = k \int P(\varepsilon)(\varepsilon/kT)^2 e^{-\varepsilon/kT} \quad d\varepsilon = k(kT/\delta)^{n+1} \tag{9}$$

where P is the asymptotic law (4).

In the limit of a large particle, we can show that the heat capacity per particle approaches the Sommerfeld value

$$C = k^2 T/\delta \tag{10}$$

Thus the mesoscopic feature of the electronic heat capacity is a reduction by the factor, $(kT/\delta)^n$. Experimentally, a decrease of heat capacity has been reported for platinum particles [10], but determination of the index n has not yet been done.

Spin Paramagnetism

As was already pointed out, if the electron number is odd, the highest occupied state in the ground state is a free spin. Therefore it shows a Curie susceptibility near the absolute zero,

$$\chi_{odd} = \mu^2/kT \tag{11}$$

where μ is the Bohr magneton.

If the electron number is even, the spin is quenched in the ground state so that the magnetic susceptibility is zero near the absolute zero.

At higher temperatures, excitation creates spin triplets and the susceptibility increases as (n = 2)

$$\chi_{\mathrm{even}} = (2\mu^2/\mathrm{kT}) \int \mathrm{P}(\varepsilon)\mathrm{e}^{-\varepsilon/\mathrm{kT}} \quad \mathrm{d}\varepsilon = \mu^2\mathrm{kT}/\delta^2 \tag{12}$$

As temperature still increases, the susceptibility approaches the Pauli value,

$$\chi_{\mathrm{bulk}} \simeq 2\mu^2/\delta \tag{13}$$

Thus the mesoscopic effect results in the reduction of magnetic susceptibility by the factor kT/δ for particles with an even number of electrons.

Spin-Orbit Interaction

The above consideration applies when the spin-orbit interaction is weak. If it is not weak, it mixes states such as those shown schematically in Fig. 1 and 2, so that the anomalies, discussed in the previous section, are greatly reduced. Since the spin orbit interaction is the cause of g-shift and the relaxation of conduction electron spin polarization in bulk metal, the measure of the effectiveness of spin-orbit interaction is the parameter,

$$\rho = \pi\hbar/(\tau_{\mathrm{so}}\delta) \tag{14}$$

where τ_{so} is the free electron spin relaxation time given by

$$\hbar/\tau_{\mathrm{so}} = 2\pi|(\mathrm{V}_{\mathrm{so}})_{\mathrm{mn}}|^2/\delta \tag{15}$$

Here $(\mathrm{V}_{\mathrm{so}})_{\mathrm{mn}}$ represents the off-diagonal matrix elements of spin-orbit interaction containing the imperfections or boundary potential. For small values of ρ, the previous consideration applies and this results in anomalous spin susceptibility behavior. For large values of ρ, the anomalies disappear and the Pauli susceptibility is recovered [11], even at low temperatures.

The spin susceptibility of conduction electrons can be observed by nuclear magnetic resonance through the Knight shift. The first observation of Knight shift anomaly was done for very small particles of lithium [12]. Later, line shape observations of Cu and Al particles [13–15] gave evidence of the anomalies.

Relaxation Processes

Discreteness of the one-electron energy spectrum in small particles causes significant effects on the relaxation processes in which metallic electrons participate. In bulk, the energy spectrum is continuous and so electrons can exchange any amount of energy, whereas in small particles the discreteness hinders the matching of energy exchange.

Korringa Mechanism

When nuclear spins are polarized in a metal the Korringa mechanism causes relaxation of polarization. Through contact interaction, a metallic electron is scattered by a nucleus with spin flip, the Zeeman energy being compensated by the change of orbital energies of the electron. This mechnism is quenched in small particles by the discreteness of electron energies [14].

Elliott Mechanism

In bulk, conduction electron spins achieve relaxation through spin orbit interaction, exchanging orbital energies with the Zeeman energy. Phonons or other lattice defects are the agency of such inelastic scattering. In the mesoscopic conditions of small particles, such defects are no longer agents for scattering, but play their role in determining one-electron quantum states, and, even if there is a remainder of the effect of such defects, the discreteness causes a mismatch of the energy balance. Thus, the Elliott mechanism is expected to be quenched in small particles.

The expected elongation is, in some cases, really observed in nuclear magnetic and electron spin relaxation. However, the effect is not so clear cut, probably because small particles are supported by some other medium, which may interact with electrons in some complicated manner.

Conductance

In a bulk metal, lattice defects and phonons are the agents which scatter free electrons producing the electric resistance. In a particle, lattice defects work to determine the quantum states rather than working as scatterers. Phonons also have a discrete spectrum, and the effects of electron-phonon interaction can be very different from that in bulk cases.

Infra-Red Absorption

In an aggregate of small particles, if the particles are not in contact, the dc conductance is zero, but the ac conductance is observed as the response to an external electromagnetic radiation with a finite frequency. If the size of particles under observation is smaller than the wavelength of the radiation, the absorption is the dipole transition between one-electron quantum states. Thus, the absorption is expected to reflect the level spacing distribution. Simply, it will be proportional to ω, ω^2, and ω^4, respectively, according to the symmetry of the electrons. A more careful analysis (K. Frahm, B. Muehlschegel, R. Nemeth, unpublished note, 1989) shows that it is proportional to ω^{2+n} when $h\omega \ll \delta$. The infra-red absorption spectra of thin granular film composed of particles of aluminum, tin, and gold showed more subtle oscillatory behaviors [16], which may be attributed to the level distributions. The data are in good accord with theoretical predictions. On the other hand, the anomalous enhancement of polarizability predicted by Gorkov and Eliashberg [17] has not been detected.

Damping of Plasma Oscillation

Colloidal solutions of metallic small particles exhibit beautiful colors. This is explained by classical electromagnetic theory, but in the modern picture it is explained by resonance scattering by the plasma oscillation of electrons. As the particle size gets smaller, the resonance width increases in inverse proportion to the size [17]. Naiively, it is interpreted as the shortening of the mean free time of conduction electrons by the boundaries. This is not quite consistent with the mesoscopic picture. The following picture is more appropriate [19]: The plasma oscillation produces an oscillating electric field inside a particle. This induces dipolar transitions between electronic quantum states, dissipating the plasma oscillation. Namely, the energy of collective motion is transferred to the individual modes of electrons. It has been shown that this picture is indeed equivalent to the classical picture and that both coincide in the bulk limit.

Conclusion

I have summarized the important features of the mesoscopic physics of small metal particles. In recent years remarkable progress has been made in the physics and chemistry of clusters; I have been impressed by the cleverness of experimentalists who have been successful in preparing materials in nanometer scales with very fine structures to make nature speak of a new world of mesoscopic physics. Now theorists can feel free to use their imagination in speculating on the secrets of nature without being afraid of making experimentalists waste their energy. In conclusion, I would like to quote from one of my old essays [3] on small particles. "Probably one can go on speculating in this way. Although many of the speculations will just be enjoyable exercises unrealizable in laboratories, there can be quite a large number of realizable effects which may also cast some light on the very basic principles of physics."

References

1. Kubo R (1962) J Phys Soc Jpn 17:975
2. Kubo R (1977) J de Physique Colloq 38C 2–70
3. Kubo R (1969) In: Polarisation, matiere et rayonnement, volume jubilaire en l' honneur d' Alfred Kastler. Pres Univ France Paris, p 325
4. Kubo R, Kawabata A, Kobayashi S (1984) Ann Rev Mater Sci 14:49
5. Halperin WP (1986) Rev Mod Phys 58:533
6. Mehta ML (1967) Random matrices and statistical theory of energy levels Academic, New York. See also, Efetov KB (1983) Adv in Phys 32:53
7. Dyson FJ (1962) J Math Phys 3:140
8. Neugebauer CA, Webb MB (1962) J Appl Phys 33:74
9. Denton R, Muehlschlegel B, Scalapino DJ (1973) Phys Rev B7:3589
10. Stewart GR (1977) Phys Rev B15:15

11. Sone J (1977) J Phys Soc Jpn 43:1457
12. Taupin C (1967) J Phys Chem Solids 28:41
13. Yee P, Knight W (1975) Phys Rev B11:3261
14. Kobayashi S (1977) J Phys C2:121
15. Nomura K, Kobayashi S, Sasaki S (1980) J Phys Soc Jpn 48:37
16. Tanner DB, Sievers DB, Buhrman AJ (1975) Phys Rev B11:1330
17. Gor'kov LP, Eliashberg GM (1965) (in Russian) Zh Eksp Teor Fiz 48:1407
18. Kreibig V (1974) J Phys F4:999
19. Kawabata A, Kubo R (1968) J Phys Soc Jpn 21:1765

Quantum Fluctuations and Interferences

Probing Spin-Glass Configurations with Mesoscopic Conductance Fluctuations

3

PAUL G.N. DE VEGVAR, LAURENT P. LÉVY, AND TED A. FULTON[1]

Summary. The natural sensitivity of the conductance of mesoscopic-sized samples to the specific scattering potential has been theoretically proposed as a method to explore the complex frozen magnetic order in systems such as spin-glasses. Recent magneto-transport measurements on 1000 ppm *Cu*Mn films and wires demonstrate phase coherent electron propagation over 0.4-μm distances. Exploiting the broken time-reversal symmetry of the spin-glass state, a unique magneto-fingerprint of the *specific* frozen spin-configuration is extracted from a pair of resistance measurements. This technique is then applied to study how the spin configuration changes along different paths in the $H - T$ plane, reaching some surprising conclusions not obtainable by other experimental means.

Introduction

When an electron propagates phase coherently through a disordered metal the quantum interferences between Feynman paths are sensitive to the specific microscopic arrangement of impurities imbedded in the sample. An experimental signature for this is found in magneto-transport measurements performed on samples with sizes comparable to the dephasing length l_ϕ. In such experiments conductance fluctuations are observed [1], arising from the change of interference as a flux quantum $\Phi_0 = h/e$ is threaded through a phase coherent volume. This sensitivity to the particular scattering potential has recently been proposed as a tool to investigate sample specific phenomena in systems possessing *frozen* magnetic order, such as spin-glasses [2]. The idea is that different spin-configurations may be distinguished by their magneto-conductances, possibly yielding information about spin-glass states and some measure of their overlap. Additionally, as the sample dimensions are reduced, any characteristic length scale of the spin-glass phase, which could be some correlation length or a size for droplet excitations [3], might emerge were a measurable phase coherent fingerprint to be demonstrated. In order for this to work it is essential that local fields be sufficiently large to freeze out spin-flip dephasing. Some experimental studies, however, have cast some doubt on whether this is possible [4] (A. Benoit and V. Chardrasekhar, 1989, private communication).

[1]AT & T Bell Laboratories, Murray Hill, NJ 07974, USA

We have recently performed experiments on 1000 ppm mesoscopic *Cu*Mn structures that surmount these hurdles, and we discuss observations of phase coherent transport in this spin-glass system. We also demonstrate how to extract from the observed conductance fluctuations only that part arising from the frozen spins. This "spin-fingerprint" or snap-shot of a spin-glass configuration is then studied as the temperature and magnetic field are varied. As the system is warmed through T_g the spin fingerprint disappears ("melts"), and upon refreezing a new one arises as the spin-glass settles into a new state. Yet, most surprisingly, these spin fingerprints are remarkably insensitive to field cycling to many times the exchange field $h_g = k_B T_g/\mu$. We also present a simple theoretical interpretation of which spin properties are probed by these spin-fingerprints.

Materials: Choice and Characterization

To obtain a long phase breaking length l_ϕ, care was taken in selecting a suitable material. A dilute *Cu*Mn alloy (1000 atomic ppm, with inter-Mn spacing $d \approx 23$Å, bulk $T_g \approx 1.6K$, and a bulk residual resistivity $\rho \approx 4.31 \mu\Omega \cdot cm/at.$ %; Ref. [5]), was chosen for its large diffusion constant D, weak spin-orbit scattering, and small out-of-phase susceptibility $\chi''(\omega)$ implying a more rigidly frozen spin system. *Cu*Mn films, wires, and rings (with linewidths ≤ 850Å) were deposited to depths of 420–500Å using two different techniques: low-pressure argon ion sputtering and flash-evaporation. Bi-layer Cr-PMMA resist and e-beam lithography were used. The material was deposited at a rate of 0.2Å/s in a UHV sputtering system ($p \approx 1 - 2 \times 10^{-9}$ Torr) to avoid Mn oxidation [6]. The sputtered samples gave

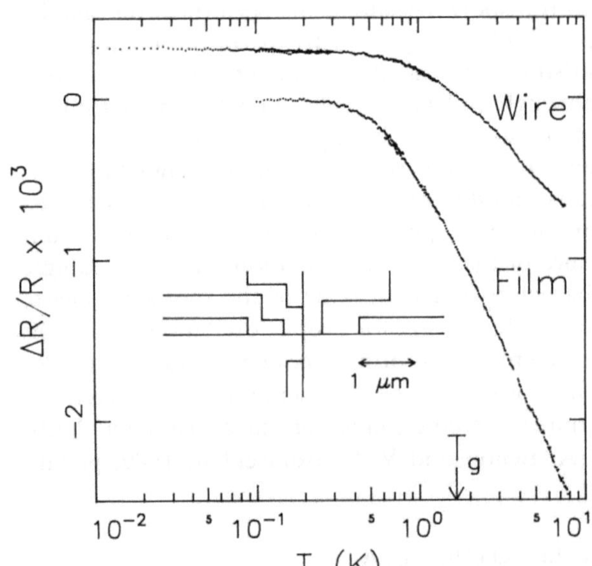

Fig. 1. Temperature dependence of resistance for a 420-Å-thick by 2-μm-wide 1000 ppm *Cu*Mn film and a codeposited 850–900-Å-wide by 2-μm-long wire in zero applied field. The bulk T_g is indicated for comparison. *Inset*: A scaled schematic of a wire sample. (From [13], with permission)

the longest elastic mean free path $\lambda_e = 310$Å, and these will be discussed below. Qualitatively similar findings were obtained for the flash-evaporated material.

The temperature dependence of the resistivity ρ in zero applied magnetic field for sputtered wires and co-deposited films is given in Fig. 1. It was found to obey $\Delta\rho(T) = -A \cdot \ln(T)$ over the range $2 < T < 8$ K. Comparing to measured bulk values [5] of A, we find these $R(T)$ are in accord with Kondo-like behavior for T above T_g. The measured As are nearly two orders of magnitude too large to be accounted for by weak localization or electron-electron interaction effects. Below T_g, however, the resistance of both sputtered and flash evaporated films and wires gradually saturates. This contrasts markedly with the decrease in resistance observed in bulk samples as T drops below T_g [5]. The bulk behavior is frequently attributed to the slow freeze-out of single ion spin-flip scattering as the moment fully polarizes in its local field. Therefore the saturation of the film and wire resistances below $0.6 \times T_g$ implies a total polarization of the Mn impurities and an unexpectedly large suppression of spin-flip scattering relative to the bulk. Because this saturation effect was also observed for films as thick as 2000Å, it is most likely not a finite size phenomenon. It might, however, be due to some kind of local order arising during the deposition process. Although we presently lack a complete understanding of this suppression, the key to observing the mesoscopic effects discussed below lay in reducing the spin-flip dephasing for $T < T_g$ from its estimated bulk value.

The magneto-resistance of a film, shown in Fig. 2, can be used to confirm that

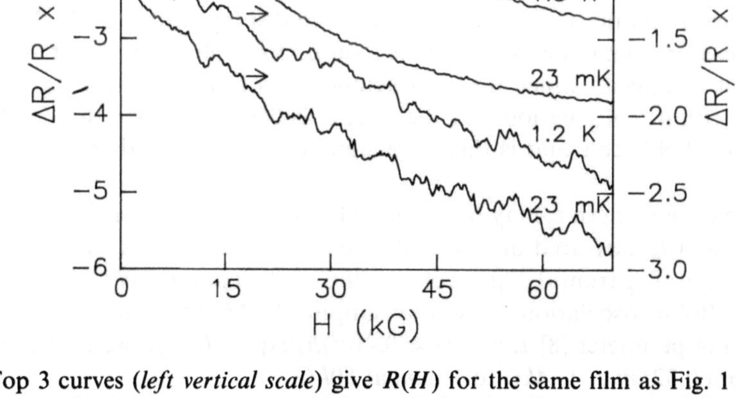

Fig. 2. Top 3 curves (*left vertical scale*) give $R(H)$ for the same film as Fig. 1 at various temperatures. The bottom two traces (right vertical scale) display $R(H)$ of the 2-μm-long wire. (From [13], with permission)

these samples' magneto-transport properties are similar to those of bulk spin-glasses of identical concentration. The magnitude of the large negative magneto-resistance is in accord with that observed in bulk samples after scaling by the ratio of magnetic to non-magnetic scattering, but is too large ($8 \times e^2/h$) to be explained by weak localization (which is destroyed in vector spin-glasses [7]). At low temperatures, the inflection point of the magneto-resistance provides a crude indication for a characteristic field of order $9-11kG$, comparable to the typical bulk spin-glass exchange field $h_g \approx k_B T_g/\mu \approx 8kG$.

Mesoscopic Transport Measurements

Figure 2 also displays the magneto-resistance of a 2-μm-long wire segment. The observed quantum fluctuations (a magneto-fingerprint) arise from two sources: (a) The magnetic field induces an orbital phase shift in the electronic wavefunctions, giving rise to fluctuations on a field scale $h_\phi = \Phi_0/(wl_\phi)$, where w is the wire width. (b) A spin configuration has a finite susceptibility and is distorted by the external field. This produces conductance fluctuations on a field scale h_s, determined below to be larger than h_ϕ. The magneto-fingerprints are therefore determined primarily by the *frozen* initial configuration (a), and the spin stiffness (b) plays only a small role. At $23mK$ the high field conductance fluctuations have a peak to peak amplitude $\Delta G = 0.20e^2/h$. Their magnitude does not change up to about $1.0K$ and decreases gradually thereafter. The fluctuations possess a correlation coefficient $C = 0.94$ when cycling the magnetic field to 70 kG: the spin-configurations are therefore quite robust upon field cycling to $9 \times h_g$. The conductance fluctuations, which are expected in the fully polarized limit ($h \gg h_g$) [8], persist into the spin-glass regime existing for fields below h_g, where the larger spin-flip dephasing rate could destroy the quantum effects. This is apparent from Fig. 3a, where we plot the magneto-resistance of a $0.35\mu m$ wire section measured between $58-70kG$ and $\pm 6kG$. These data are highly repeatable, two successive low field scans having $C = 0.992$. Aharonov-Bohm resistance oscillations have also been observed in a square-shaped annulus $0.4\mu m$ on a side and are displayed in Fig. 3b. The oscillations decrease in amplitude as the field is reduced from $50-52kG$ down to $1-3.5kG$, but do not disappear. Since the conductance was found to have no additional structure within a Φ_0 period, the field $h_\phi \approx 260G$ determines the conductance fluctuations' field scale and is therefore significantly smaller than h_s for the ring sample.

The dephasing length l_ϕ may be estimated in two ways. From the decorrelation field $h_\phi = \Phi_0/wl_\phi$ measured on a $L = 0.35\mu m$ segment at $22mK$, we find a monotonic increase of l_ϕ from $0.37\mu m$ below $10kG$ to $0.96\mu m$ at $60kG$. If we take the Aharonov-Bohm oscillation peak-peak amplitude ΔG to decrease exponentially with the ring perimeter [8] L as $\Delta G \approx 0.4(e^2/h)\exp(-L/2l_\phi)$, we also find that l_ϕ ranges from $0.32\mu m$ at $1-4kG$ to $1.2\mu m$ at $50kG$.

Since both magnetic and non-magnetic scattering contribute to the conductance fluctuations, information regarding the specific spin-configuration in the sample

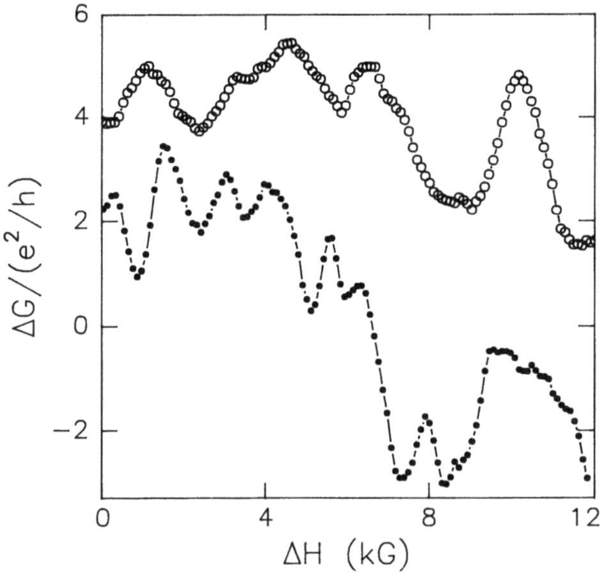

a

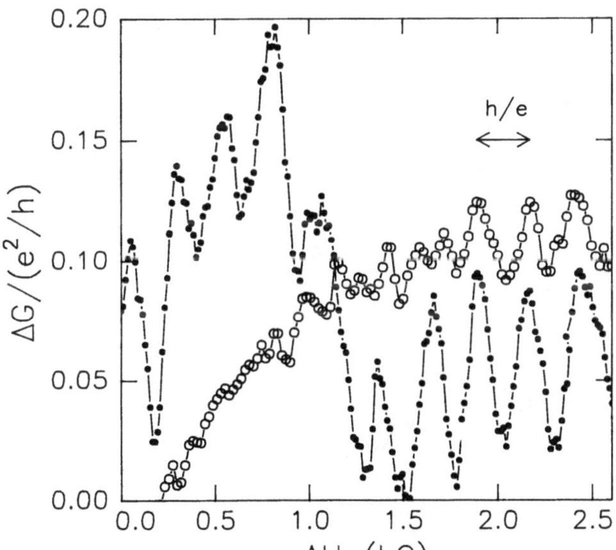

b

Fig. 3. a Conductance fluctuations in a wire with a 0.35μm voltage probe spacing at 23 mK. *Open* (*solid*) symbols refer to a −6 to 6 kG (58–70 kG) field window. **b** Aharonov-Bohm conductance oscillations measured in a square annulus 0.4 μm on a side at 23 mK. The *open* (*solid*) data were collected from 1 to 3.5 kG (50–52.5 kG). (From [13], with permission) can only be obtained by separating the spin-dependent and spin-independent contributions to the magneto-resistance. This is particularly important here, since the fraction of scattering (potential and spin dependent) that occurs from the Mn spins is only about 15% due to scattering from grain boundaries and surfaces. To this end, we recall the Onsager-Büttiker relation: $R_{i,v}(H, M) = R_{v,i}(-H, -M)$, which is valid only for a system invariant under time-reversal [9]. Here i, v refers to a set of current and voltage leads used to perform the measurement and M is the magnetization. Since a frozen magnetic state breaks time reversal symmetry: $M(-H) \neq -M(H)$, the Onsager-Büttiker anti-symmetrized resistance $R_a = (1/2)(R_{i,v}(H) - R_{v,i}(-H))$ no longer vanishes and represents the contribution from the magnetic order. Actually, this magnetic contribution can be further decomposed into one component $R_{a,s}$ symmetric in H and another one $R_{a,a}$ anti-symmetric in the field. Here we will focus on the former, defined as:

$$R_{a,s}(H) = \tfrac{1}{4}(R_{i,v}(H) - R_{v,i}(-H) + R_{i,v}(-H) - R_{v,i}(H))$$

This quantity is depicted in Fig. 4 for a 0.65-μm-long segment as a conductance fluctuation, where the data were obtained on warming from 23mK. The 23mK trace is very similar to the 0.9K ones with an average $C = 0.89$. This implies that there is little change in the magnetic state on this length scale below 0.9K. As the sample is warmed further through the bulk T_g to 2.2K (not shown), $\Delta G_{a,s}$ is only partially correlated with the 0.9K curves, having $C = 0.49$. By $T = 3.9K$, this com-

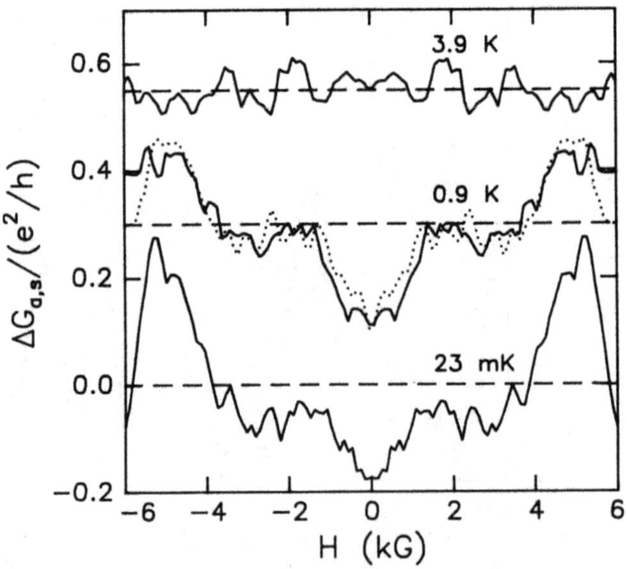

Fig. 4. $\Delta G_{a,s}$ component of the magneto-conductance fluctuations as extracted from measurements on a 0.65-μm-long segment. The sample was cooled from 4 K to 12 mK in 7 kG and subsequently warmed to the indicated temperatures. The curves are offset vertically for clarity. The two traces at 0.9K indicate the experimental reproducibility and have $C = 0.90$. (From [13], with permission)

ponent of the conductance contrasts vividly with the lower temperatures data: it has dropped into the noise floor of the measurement. This behavior is expected if the spins start to lose their rigidity ("melt") at T_g. On the other hand, the Onsager-Büttiker symmetric components of the conductance fluctuations did not display such a sharp decrease in amplitude and were highly correlated across T_g with a mean $C = 0.85$.

How sensitive is the conductance to a specific spin configuration? That is, how many spins must reorient before $\Delta G_{a,s} = e^2/h$? Following the arguments first put forward by B.L. Al'tshuler and P.A. Lee [10], any given Feynman path visits a large number of impurities, occupying a volume $v \sim (4\pi\lambda_e/k_F^2)(L/\lambda_e)^2$. If the scattering potential of a single impurity changes, it can induce a conductance change $\Delta G \sim (e^2/h)\sqrt{f}$, where $f = v/V$ is the fraction of Feynman paths visiting the impurity and V is the sample volume. Taking a wire of length $L = l_\phi \approx 0.35\mu m$ and correcting for the fraction of scattering due to the Mn, we find that $\sim$28% of the 1.1×10^5 spins in V must reorient to produce $\Delta G = e^2/h$, provided each reoriented spin produces a relative phase shift of order π. For $L = l_\phi$ we can resolve $\Delta G = 0.07e^2/h$. For lengths $L > l_\phi$, ΔG will be reduced by a factor $(L/l_\phi)^{-3/2}$ due to the random addition of each l_ϕ segment.

This technique may then be applied to measure the extent to which the spin configuration changes with T or H. To investigate the former we measured the low field magneto-resistances of a $L = 2\mu m$ wire at $T/T_g \approx 0.4$ separated by an anneal to $T/T_g = 2.2$ in $H = 0$ kG (see Fig. 5) and found an rms $(L/l_\phi)^{3/2}\Delta G_{a,s} \approx 0.33(e^2/h)$. (We also carried out a high field (70 kG) control measurement to demonstrate that

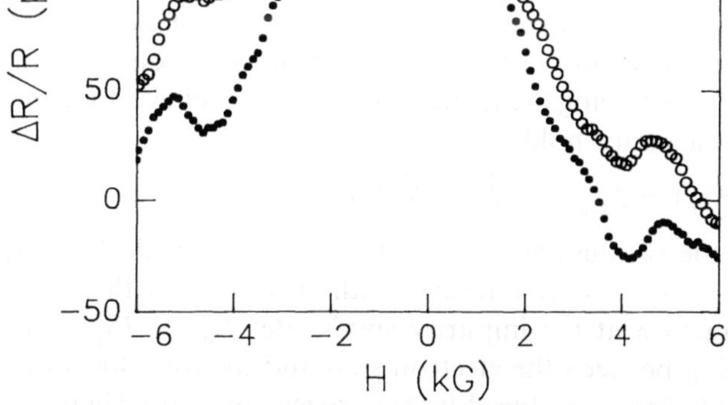

Fig. 5. $R(H)$ for a zero field cooled $2\mu m$ section before (*open*) and after (*solid*) an anneal to 3.3 K in 0. kG. Due to an instrumental vertical offset of unknown but constant size, the curves have been forced to coincide near $H = 0$

this change was not due to impurity diffusion.) This value implies a large fraction of the spins in an l_ϕ long volume reoriented due to the anneal. Such a reorientation is reflected in the two $\Delta G_{a,s}(H)$ spin fingerprints, which are typically only weakly correlated across the anneal: $C = 0.34 - 0.68$. This behavior stands in stark contrast to cycling H starting from a zero-field cooled configuration to $75kG \approx 9h_g$ and back at fixed $T = 0.8K$. Then the low field fingerprints had $C = 0.90$, implying a *nearly reversible* distortion of the spin configuration as the field was cycled; i.e., the magnetic state of the system remained in nearly the same energy valley even though it had been polarized by H [11]. We find this result rather startling, since we have no idea how the spin-configuration finds its way back to its initial low field state after being polarized by such a large applied field.

Theoretical Analysis

Although the above combinations of 4-point resistances are sensitive only to the scattering off the frozen spins, it leaves unanswered precisely what *spin* property is probed by $R_{a,s}(H)$. In this section we investigate this problem theoretically. Because of the well-known Landauer-Büttiker formula relating the 4-point resistance to the transmission probabilities for scattering from lead to lead [9, 12], we focus on these latter quantities. To this end, consider the transmission probability from point 1 to point 2 expressed as a sum over paths p and p',

$$T^{1\to 2} = \left|\sum_p t_p^{1\to 2}\right|^2 = \sum_{p,p'} t_p^{1\to 2} t_{p'}^{1\to 2*} \equiv \sum_{p,p'} T_{p,p'}^{1\to 2}(\vec{H}, \{\vec{S}_i(\vec{H})\})$$

Here $T_{p,p'}^{1\to 2}(\vec{H}, \{\vec{S}_i(\vec{H})\})$ is that part of the transmission probability generated by interfering path p with p' in the presence of the external field $\vec{H}$ and the frozen configuration $\{\vec{S}_i(\vec{H})\}$. We note three facts about $T_{p,p'}^{1\to 2}(\vec{H}, \{\vec{S}_i(\vec{H})\})$. First, it is rotationally invariant in spin space (a spin scalar), so it is made of combinations of spins such as 1, $\vec{S}_1(p)\cdot\vec{S}_1(p')$, $\vec{S}_1(p')\cdot(\vec{S}_1(p)\times\vec{S}_2(p))$, $\vec{H}_{loc}(p')\cdot\vec{S}_1(p)$, $J_{\alpha\beta}(\vec{S}_1(p))_\alpha(\vec{S}_1(p'))_\beta$, and so on. Here $\vec{S}_i(p)$ is the ith spin on path p. (We are treating the spins as classical vectors, the generalization to quantum spins is straightforward.) Also, we inject a spin unpolarized mixture of electron states and we do not detect the polarization of the outgoing electron in our measurement. Lastly, since we are not studying weak interactions, we believe that microscopic time reversal invariance must hold:

$$T_{p,p'}^{1\to 2}(\vec{H}, \{\vec{S}_i(\vec{H})\}) = T_{p,p'}^{2\to 1}(-\vec{H}, \{-\vec{S}_i(\vec{H})\})$$

Now we list the various spin and field interactions an electron experiences in its journey from 1 to 2. (a) It interacts with frozen spins through the δ-function exchange interaction at the impurity spin's site: $\hat{H}_{exc} \sim J\vec{S}_p\cdot\hat{\sigma}$. (b) There is the Zeeman coupling between the electron spin and the total local field produced by $\vec{H}$ and the dipole fields produced by the frozen spins. (c) There is the Aharonov-Bohm coupling of the electron's charge to the local field on the path. (d) There is the spin-orbit interaction. Next, we estimate the relative importance of these contributions. At the concentrations we are studying, the dipole fields are only 40 G and are negligible relative to the external fields (a few kG) and the exchange fields (8 kG). The Zeeman interaction can be neglected with respect to the exchange effects provided $g\mu_B H/2 \ll \hbar/\tau_s$, or $H \ll 910$ kG. The spin orbit interaction is also unimportant here, because $1/\tau_{so} \ll 1/\tau_s$, where $1/\tau_s$ is the elastic scattering rate of the magnetic impurities. Therefore the exchange interaction dominates the situation. In examining the list of spin scalars above we can then make some simplifications. The trivial invariant 1 does not enter $R_{a,s}$ or $R_{a,a}$ since it is totally insensitive to the spin system. Any positive power of the field H^n represents an n-fold Zeeman interaction, which we neglect with respect to the exchange fields. Therefore the invariants entering $T^{1\to 2}_{p,p'}(\vec{H}, \{\vec{S}_i(\vec{H})\})$ and then $R_{a,s}$ are composed solely of products of spins such as $\vec{S}_1(p)\cdot\vec{S}_1(p')$, $\vec{S}_1(p')\cdot(\vec{S}_1(p)\times\vec{S}_2(p))$, and so on. The first invariant enters as J^2 to leading order in perturbation theory, the second invariant as J^3. We then restrict ourselves to considering these two lowest order products. The only roles of H_{ext} are to thread flux through the area between p and p', and to elastically distort the spin configuration $\{\vec{S}_i(H)\}$.

The 2-spin invariant corresponds to one spin on each of the paths p and p'. Using microscopic T-invariance it is easy to show that the Onsager-Büttiker anti-symmetrized resistance is

$$R_a(2) \propto \sum_{p\neq p'} \tilde{T}^{1\to 2}_{p,p'}(H)[\vec{S}_p(H)\cdot\vec{S}_{p'}(H) - \vec{S}_p(-H)\cdot\vec{S}_{p'}(-H)]$$

Here $\tilde{T}^{1\to 2}_{p,p'}(H)$ is the usual (spin-independent) Aharonov-Bohm type interference intensity arising from paths p and p'. Notice that this quantity not only vanishes for paramagnetic spins $\vec{S}(-H) = -\vec{S}(H)$ as it should, but also for totally rigid spins $\vec{S}(-H) = \vec{S}(H)$. Therefore $R_a(2)$ measures the change of relative orientation of pairs of spins upon $H \to -H$; that is, it corresponds to a spin elasticity. In that case $R_a(2)$ remains small until $|H| \geq h_s$, the field at which the spins start to rearrange.

The 3-spin invariant corresponds to interferences involving two spins on one path (the vector product) and one spin on the other path (the scalar product). Again we find that

$$R_a(3) \propto \sum_{p\neq p'} \tilde{T}^{1\to 2}_{p,p'}(H)[\vec{S}_{p'}(H)\cdot(\vec{S}_{p1}(H)\times\vec{S}_{p2}(H)) + \vec{S}_{p'}(-H)\cdot(\vec{S}_{p1}(-H)\times\vec{S}_{p2}(-H))]$$

This vanishes as expected for paramagnetic spins but is non-zero for totally rigid spins. $R_a(3)$ then measures the configuration "volume" of spin triplets and can be non-negligible even for $|H| < h_s$. Experimentally we observe that the auto-correlation function of the Onsager-Büttiker anti-symmetrized wire resistance falls off at about the same field as that of the corresponding symmetrized resistance, namely 1-2 kG. Also, the film magneto-resistance at 23 mK already has $|\Delta R/R|$ of a few hundred ppm for fields of this order. We may then infer that $h_s \approx 1$–2 kG, so that $R_{a,s}$ and $R_{a,a}$ as measured from -6 to $+6$ kG include contributions from both pairs of spins, as in $R_a(2)$, as well as from spin triplets, as in $R_a(3)$ above.

Conclusions and Future Directions

We have demonstrated that phase-coherent transport is experimentally observable in a spin-glass and that information about the spin configuration on sub-micron length scales can be obtained from a set of magneto-resistance measurements. A simple theoretical analysis showed that R_a couples primarily to pairs and triplets of spins through the invariants $S_1 \cdot S_2$ and $S_1 \cdot (S_2 \times S_3)$, respectively. The method has been experimentally illustrated by showing that warming through T_g (melting) changes the magnetic state far more effectively than field cycling (polarization). Another application of this new tool would be to try to measure the field scale h_s over which *reversible* spin distortion occurs. As indicated above this field is roughly the same or larger than that produced in the wire or ring conductances by the orbital Aharonov-Bohm effect (h_ϕ); i.e., a few kG for wires. To separate these two scales we are carrying out an experiment with the wire oriented along H. This increases the orbital field scale by roughly a factor 10, but leaves h_s unchanged. This powerful technique can, in principle, provide quantitative information on the overlaps between spin-glass configurations when the theoretical link between magneto-conductance traces and configuration overlaps becomes more quantitatively described.

Acknowledgments. We have benefited from conversations with S. Hershfield and B.L. Al'tshuler. We also gratefully acknowledge L.N. Dunkleberger, G. Espinosa, G. Kammlott, R. Miller, and P. Trevor for their contributions to the processing and C.A. Murray for the UHV system.

References

1. Umbach CP, Washburn S, Laibowitz RB, Webb RA (1984) Phys Rev B30:4048; Lee PA, Stone AD (1985) Phys Rev Lett 55:1622; Altshuler BL (1985) Pis'ma Zh Eksp Teor Fiz (in Russian) 41:530 [JETP Lett (1985) 41:648]
2. Al'tshuler BL, Spivak BZ (1985) Pis'ma Zh Eksp Teor Fiz (in Russian) 42:363 [JETP Lett., 42:447 (1985)]; Feng S, Bray AJ, Lee P, Moore MA (1987) Phys Rev B36:5624; Israeloff NE, Weissman MB, Nieuwenhuys GJ, Kosiorowska J (1989) Phys Rev Lett 63:794
3. Lévy LP, Ogielski AT (1986) Phys Rev Lett 57:3288; Kenning GG, et al (1990) Phys Rev B42:2393; Fisher D, Huse D (1988) Phys Rev B38:386
4. Van Hasendonck C, Vloeberghs H, Bruynseraede Y (1989) In: Reed N, Kirk W (eds) Nanostructure Physics and Fabrication.
5. Bass J (1982) In: Hellwege K.-H. (ed) Landolt-Börnstein Numerical Data and Functional Relationships in Science and Technology, series III, vol 15a. Springer, New York p 166; Laborde O, Radhakrishna P (1973) J Phys, F3:1731; Ford PJ, Mydosh JA, (1976) Phys Rev B14:2057; Nigam AK, Majumdar AK (1983) Phys Rev B27:495
6. Dolan GJ, Dunsmuir J (1980) Physica, B152:7
7. Wei Wei, Bergmann G (1988) Phys Rev B38:11751; Lévy LP, de Vegvar PGN (1989) Bull Am Phys Soc 34:415

8. Washburn S, Webb RA (1986) Adv Phys 35:375
9. Büttiker M (1988) IBM J Res Dev 32:317
10. Lee PA (1987) Proc 18th Conf on Low Temp Phys, Kyoto 1987, JJAP 26:1934 Al'tshuler BL (1987) ibid 26:1938
11. Binder K, Young AP, Rev Mod Phys 58:802; Parisi G, Mezard M, Virasoro M (1987) Spin-Glasses and beyond. World Scientific, Singapore
12. Landauer R (1988) IBM J Res Dev 32:306
13. de Vegvar PGN, Lévy LP, Fulton TA (1991) Phys Rev Lett 66:2380

Coherent Ballistic Transport in Micro-Junctions: Quenching, Fluctuations, and Chaos

4

HAROLD U. BARANGER,[1] RODOLFO A. JALABERT,[2*] AND A. DOUGLAS STONE[2]

Summary. We compare the quantum and classical transport properties of ballistic microstructures and find two important differences. First, there are large fluctuations of the quantum conductance in experimentally realizable structures caused by coherent scattering from the geometric features. These fluctuations arise from interference between long paths trapped in a junction or cavity. A semiclassical argument shows that the scale of the magnetic field correlation function for the *quantum* fluctuations is determined by the underlying chaotic *classical* scattering, a result which is confirmed numerically. Second, we extract the average quantum behavior using an ensemble of very weak impurities and some thermal broadening. While the classical and quantum results are in good *qualitative* agreement, we find substantial *quantitative* differences that persist well into the many-channel (classical) limit. The first effect should be observable in low-temperature resistance measurements on high-mobility heterostructure devices; the second is relevant to recent experiments on quenching of the Hall resistance.

Introduction

Transport in microstructures has been intensively investigated in the last decade, (for a review, see Ref. [1].) Several length scales are important in defining the character of transport in these systems at low temperature: the elastic mean-free-path caused by impurity scattering, the Fermi wavelength controlled by the density, and the phase-coherence length—the distance over which electrons retain phase information and therefore the distance over which quantum interference effects can occur. By using the high-mobility two-dimensional electron gas formed at the interface between GaAs and AlGaAs as a basis for making wires, one can now make junctions between wires whose size is less than the mean-free-path [1]. Transport through such junctions is then ballistic. In addition, at low temperature, the phase coherence length is larger than the size of the junction which can be of the order of the Fermi wavelength. In this regime, then, the electrons are coherent

[1] AT&T Bell Laboratories 1D-230, Murray Hill, NJ 07974-0636, USA
[2] Applied Physics, Yale University, New Haven, CT 06520, USA
* Present Address: Service de Physique de l'Etat Condensé, CEA Saclay, 91191 Gif-sur-Yvette Cédex, France

and ballistic, and the micro-junctions act like electron waveguides with a rather complex shape. Many novel magnetoresistance effects have been observed in various structures in the ballistic regime [1].

In this paper, we concentrate on the conductance of two-probe structures and the Hall resistance of single junctions. We consider several models and, in particular, compare classical [2] and quantum [3] calculations. Typically the zero temperature resistance fluctuates strongly as a function of either magnetic field or Fermi energy [3, 4], and this necessitates the definition of an *average* resistance in order to compare to experiments at $T \sim 1K$ *even in the ballistic regime*. The classical model itself, introduced in this context by Beenakker and van Houten [2], provides one natural definition of the average resistance. In particular, although the classical model fails completely to reproduce the low temperature fluctuations (which come from interference), it does yield the correct qualitative behavior of the high-temperature resistance. Nonetheless, we expect effects due to mode quantization and diffraction of the electron waves to be substantial in this few-mode regime, and this leads us to attempt to define a quantum average resistance for these systems.

Figure 1 summarizes our main results by showing four calculations [5] of $R_H(B)$ (at a particular Fermi energy) described in detail below. The Hall resistance, R_H, for a junction is proportional to the voltage difference between the two side probes when current is passed straight through (inset to Fig. 1). R_H is, of course, produced by the Lorentz force in the junction region, and in a macroscopic sample it is linear in the magnetic field: $R_H = (1/nec)B$ in two dimensions where n is the electron density. A surprising suppression, or "quenching", of the Hall resistance at low

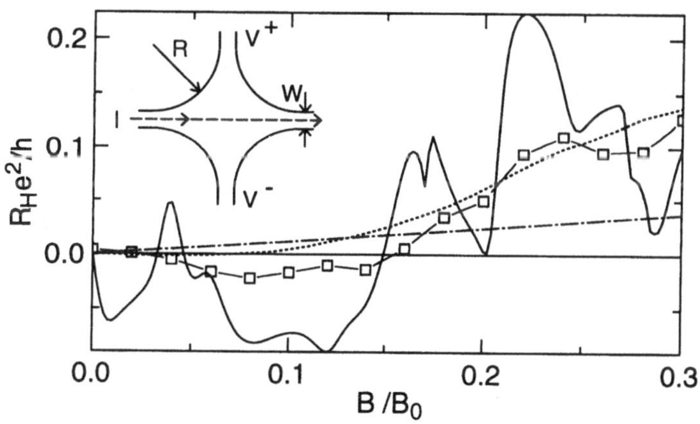

Fig. 1. Hall resistance for four-disk junction with R/W = 4. The quantum calculation for a ballistic structure at T = 0 (*solid line* for $k_F W/\pi = 3.5$) shows fluctuations not present in either the classical (*dotted*) or quantum square-corner junction (R/W = 0, *dash-dotted*) cases. The quantum fluctuations are removed by a weak impurity and energy average; the resulting average quantum trace (*squares*) shows enhanced quenching compared to the classical result. The *inset* shows both the structure and the lead configuration which defines the Hall resistance. $B_0 = mcv_F/eW$. (From [5] with permission)

magnetic field has been observed in a variety of microjunctions [6–13]. The nearly straight line in Fig. 1 (dash-dotted) is the result for a simple square-corner junction and is nearly equal to the two-dimensional linear behavior. The smooth dotted line is the classical behavior of a four-disk junction (inset to Fig. 1), which shows quenching at small magnetic fields and enhancement of R_H at larger fields. The remaining two calculations are the full quantum behavior (solid line) and the average quantum behavior (squares). The deviations of all three calculations for the four-disk structure from that for the square-corner structure shows the strong influence of geometry in these ballistic structures. Both the large fluctuations in the full quantum behavior [3, 4] and the fact that the average quantum behavior shows enhanced quenching compared to the classical result [5] are evident from these results.

In order to describe transport in the ballistic regime, it seems reasonable to neglect scattering from impurities completely and to view the transport properties in terms of scattering from the geometry of the junction. Thus it is natural to use the approach advocated and developed by Landauer [14] and Büttiker [15], in which one expresses the resistance of the system in terms of its transmission properties between reservoirs attached to leads. In particular, Büttiker showed [15] that, in a multiprobe structure, the current in any lead m, I_m, is related to the voltages applied by reservoirs to the other leads n, V_n, through the transmission probability to go from lead n to lead m, T_{mn},

$$I_m = \frac{e^2}{h} \sum_{n=1}^{N} T_{mn}(V_n - V_m) \tag{1}$$

This equation has formed the basis for nearly all of the theoretical discussion of ballistic magnetotransport anomalies. It can also be derived from the more familiar Kubo-type linear response theory in an arbitrary magnetic field [16]. After applying the appropriate boundary conditions—one lead used to supply the current, another used to sink the current, and zero current in the voltage leads—Eq. (1) can be solved for the four-point resistance, defined as the voltage difference between two leads divided by the current. In this paper we restrict ourselves to either the two-probe case or the case of four identical leads, so that the junction has four-fold symmetry. In the symmetric four lead case, one finds that the Hall resistance is

$$R_H = (h/e^2)(T_R^2 - T_L^2)/D \tag{2}$$

where

$$D = (T_R + T_L)[2T_F(T_F + T_R + T_L) + T_L^2 + T_R^2] \tag{3}$$

T_F is the forward transmission, T_R is the probability to turn right, and T_L is the probability to turn left. We emphasize that this expression for the resistance is that appropriate to a transport measurement in which currents are applied and the chemical potential of reservoirs is measured [15].

The first observation of quenching of the Hall resistance was made by Roukes et al. [6] and it generated a considerable amount of experimental [7–13] and theoretical [2, 3, 5, 17–25] interest. The experimental work by several groups established that quenching exists in a variety of differently fabricated structures and persists over a sizable range of density and magnetic field ($\sim 100mT$). Quenching at a wide range of energies was obtained theoretically by Baranger and Stone [3] who emphasized the role of collimation caused by rounded corners in producing quenching.

Collimation is closely related to the idea of adiabatic transport [26], and the importance of collimation in ballistic structures was first discussed by Beenakker and van Houten [27] and by Imry [28]. Classically [27], a particle injected into a rounded-corner structure with a large transverse momentum has its k-vector rotated by the gradual widening of the wire so that it arrives at the junction region with a large component of its k-vector parallel to the injecting lead. Thus, graded wires produce a collimated electron beam. Quantum-mechanically [3], suppose one injects a wave-packet from the left in the horizontal arm; what is the behavior of this wave-packet at later times? If the widening of the wire is gradual, the electrons will travel adiabatically and will conserve their transverse mode number [26]. Thus, particles injected from a reservoir such that the current is evenly distributed [14, 15, 29] across all the subbands below E_F end up in the low-lying transverse subbands as they approach the junction. Thus, arguing either classically or quantum-mechanically, we find that a non-equilibrium momentum distribution is created in the junction region. It is this structure in the momentum distribution that we refer to as collimation.

Collimation influences the magnetotransport properties because the scattering properties of the junction [the T_{mn} in Eq. (1)] depend on the momentum distribution in the junction. Near the junction, the adiabatic approximation breaks down, and the wave-packet scatters into different modes and probes. Because the scattering properties of a junction are very different for the different modes [8, 30], the result of injecting a non-equilibrium momentum distribution could be quite different from injecting the current in all the modes. Collimation may suppress R_H through, for instance, a decrease in the total turning probability or through equalizing the probability to turn left and right. Indeed, by analyzing the transmission properties of a square-corner junction, Baranger and Stone found that low-lying modes make a much smaller contribution to R_H than high-lying modes and that both suppression mechanisms may be present [3]. Thus, the non-equilibrium momentum distribution caused by the graded wires generates quenching.

Since collimation is clearly a classical concept, a classical approach ought to give qualitatively similar results. Indeed, Beenakker and van Houten [2] performed a classical calculation of the transmission coefficients through a junction (a "billiard ball" approach described in more detail below). They obtained both quenching and inversion of R_H in appropriate geometries. Because of the nature of the classical calculation, the authors were able to look at the distribution of particles close to the junction and show that the injected particles are collimated.

In this paper, we investigate the relationship between the classical and quantum results. The most apparent difference, evident from Fig. 1, is that the classical resistance is much smoother as a function of field than the quantum resistance: quantum fluctuations in $R_H(B)$ are large. These fluctuations result from the quantum interference among the many complicated trajectories trapped in the junction region [4]. This raises the question of whether one can make a connection between the quantum interference properties and the nature of the classical dynamics. In fact, for the structures considered here (and for typical ballistic structures), the classical motion is chaotic. Our first goal in this paper is to connect the conductance fluctuations in these structures to the chaotic classical mechanics through a semiclassical theory [4].

In comparing with experiments on the quenching of the Hall resistance, we are concerned with measurements at $T \approx 4K$. Fluctuations are not evident in these experimental results, presumably because the phase-breaking rate is sufficient at these temperatures to remove the interference effects of the relatively long trajectories involved. The experiments done at low temperature ($T \approx 300mK$) in the ballistic regime [7, 12] (L. Kouwenhoven, 1990 private communication; S. Washburn, 1990 private communication) do show fluctuations which may well be caused by junction scattering as in our calculations. Our second goal in this paper is to define an appropriate quantum "average" resistance and to show that it deviates substantially from the classical results in the regime of several subbands (≈ 6).

The organization of the rest of the paper is as follows. In the next section, we review the methods of calculation used for the numerical work in this paper [5]. The semiclassical theory of the transmission amplitude is used in the following section to connect the correlation function of conductance fluctuations to the chaotic classical scattering in the structure [4]. The predictions of this argument are verified numerically [4]. In the next section, the average properties of the quantum system are studied. We define an average resistance appropriate for the few mode regime and show that quenching of the Hall resistance is enhanced compared to the classical behavior [5]. This is followed by our conclusions.

Methods of Calculation

Both our quantum and classical calculations of resistance are based on the relation between resistance and transmission coefficients, Eq. (1). The problem, then, is to find the transmission intensities for a particular ballistic structure. There are two essential ingredients to our quantum mechanical calculation: the relation of transmission coefficients to the Green function, and the recursive calculation of the Green function for the discretized problem.

Scattering theory shows that there is a very simple relation between the transmission amplitude for a quantum scattering problem and the Green function in the absence of a magnetic field: the transmission amplitude is simply the projection of the Green function onto the transverse wave functions in the asymptotic part of the lead, $\chi_a(y)$ [16],

$$|t_{mn,ba}| = \hbar\sqrt{v_b v_a}\left|\int dy_m \int dy'_n \chi_b(y_m)\chi_a(y'_n) G^+_\varepsilon(\mathbf{x}_m, \mathbf{x}'_n)\right| \tag{4}$$

where the transmission is from mode a in lead n to mode b in lead m and v_a is the longitudinal velocity of mode a (at the Fermi energy). In fact this projection can be done at any point in the lead outside the scattering region, since the population of the various transverse states will not change in the ideal lead. In all of the calculations performed here, the projections are done in zero magnetic field; the magnetic field is graded from the value in the junction to zero in the leads. For the low field properties discussed here, we expect that this approximation should be accurate, and we checked in several cases that the rate of the grading did not affect the results. For high field calculations, the full relation between the amplitudes and the Green function [16] should be used.

In order to calculate the Green function numerically, one needs to discretize the continuum problem. The lowest-order finite difference approximation to the derivatives in the Schrödinger equation produces the standard nearest-neighbor tight-binding Hamiltonian on a square lattice (see Ref. [5] and references therein for detailed explanation of our numerical method) for which the Green function must be evaluated. In a magnetic field, we use the Peierls substitution (see Ref. [5] and references therein) $V_{i,j}^{x} = \exp(ieA_{i,j}^{x}/\hbar c)$ and $V_{i,j}^{y} = \exp(ieA_{i,j}^{y}/\hbar c)$ to relate the hopping matrix element to the vector potential. In all of our calculations, the flux per unit cell of the lattice is small, so we expect this to be a good approximation to the continuum problem. The single-site energy $\varepsilon_{i,j}$ in the Hamiltonian can be chosen to follow the potential confining the particles to the microstructure. The ability to choose the single-site energies is central to our work, since it allows one to study different junction geometries.

Finally, the Green function for the discrete problem must be evaluated. We use a standard approach (see Ref [5] and references therein) based on Dyson's equation [31], which relates the Green function, G, of a system in the presence of a perturbation, U, to the Green function, G_0, in the absence of U: $G = G_0 + G_0 UG$. By judiciously choosing the unperturbed system and the perturbation, one can derive recursive relations for the Green functions which correspond to building up the system slice by slice from left to right. For the multi-lead systems which are of interest here, somewhat more complicated recursive relations must be developed using the same basic idea [5]. An important advantage of this recursive technique is its numerical stability.

The classical calculations also use Eq. (1) to relate resistance to transmission coefficients, but the transmission coefficients are evaluated for classical ballistic particles, as introduced by Beenakker and van Houten [2]. Tracing the ballistic trajectories between the leads allows one to evaluate the probability of transmission from one lead to another, a quantity exactly comparable to the quantum result. This intuitively appealing result has been derived from a Boltzmann equation approach in which Fermi-Dirac statistics is assumed but the particles are otherwise treated classically [30]. The distribution function (in the Boltzmann equation sense) of particles going into the junction region is continuous in k-space and uniform in angle, which corresponds to the current being equally distributed among the modes in Landauer-type arguments for the quantum case. An alternative derivation of the classical transmission method can be given starting from the semiclassical WKB expression for the transmission amplitude; this approach makes the relationship to the quantum amplitude somewhat clearer [5].

Fluctuations: Semiclassical Approach

We start by discussing the striking difference between the fluctuating quantum resistance and the smooth classical resistance in Fig. 1, closely following the treatment of Ref. [4]. These fluctuations arise due to the complex scattering dynamics from geometric features which can trap the electrons for times much longer than the ballistic transit time. Long trapping times are essential because, e.g., the Hall resistance of a square junction, which cannot trap particles, shows no such behavior (Fig. 1). We will show below that the characteristic scale in B for these fluctuations in the *quantum* transport coefficients can be predicted from knowledge of the irregular *classical* scattering dynamics of the same system [4]. The results in Fig. 1 show that fluctuation effects in the classical transport coefficients are completely absent in these structures at these fields. The fluctuations of Fig. 1 are a quantum interference effect, not obtainable from the classical transmission matrix.

The connection between the quantum resistance fluctuations and the classical dynamics suggests a connection between this mesoscopic effect and "quantum chaos" (for a recent treatment of issues in "quantum chaos" see [32]). A signature of chaos in a classical dynamical system is "random" (unpredictable) behavior due to exponential sensitivity to initial conditions. It is now understood that quantum systems whose classical analogs are chaotic also exhibit "randomness" in a more subtle manner. For example, the energy levels of a Hamiltonian begin to exhibit [33] the long-range level repulsion characteristic of random matrices [34] at the classical transition to chaos. Recent studies of "quantum chaos" have turned attention to open systems [4, 35–38], whose classical scattering dynamics are irregular or chaotic [39–44] in the sense that the final scattering angle varies strongly with the incident angle on an arbitrarily fine scale. In these systems, the signature of quantum chaos is fluctuations of the quantum scattering matrix, with properties described by random-matrix theory [38]. Since the transport properties of small phase-coherent conductors may be expressed completely in terms of the transmission matrix [Eq. (1)], this raises the possibility of an *experimental* probe of this random behavior characteristic of quantum chaos.

We outline a semiclassical derivation of the magnetic field correlation function, $C(\Delta B) \equiv \langle \delta g(B + \Delta B) \delta g(B) \rangle$, since B is the most convenient experimental control parameter, focusing on the two-probe case for simplicity [4]. In the exact expression for the transmission amplitude Eq. (4) we approximate the Green function by its semiclassical path-integral expression (see Ref. [32], pp 184–190; 283–287),

$$G(y', y, E) = \frac{2\pi}{(2\pi i\hbar)^{(3/2)}} \sum_{s(y,y')} \sqrt{D_s} \exp\left(\frac{i}{\hbar} S_s(y', y, E) - i\frac{\pi}{2}\mu_s\right) \tag{5}$$

where S_s is the action integral along *classical* path s, at energy E, $D_s = (v \cos\theta'/m)^{-1} \, |(\partial\theta/\partial y')_y|$, θ and θ' are the incoming and outgoing angles, and μ is the Maslov (see Ref [32], pp 184–190; 283–287) index given by the number of constant-energy conjugate points. The restriction to paths at energy E arises from a stationary-phase approximation in the time Fourier transform of the WKB propagator (see Ref [32], pp 184–190; 283–287). We evaluate t_{mn} from Eqs. (4) and (5) in an analogous manner. Assuming hard walls in the leads, the integration over the transverse (sine) wavefunctions can be performed by stationary phase approximation. This is valid in the semiclassical $\hbar \to 0$ limit, that is, for large mode index m and n. For example, the stationary-phase condition on the y-integration is $(\partial S/\partial y)_{y'} = -p_y = -\bar{m}\hbar\pi/W$ for $y = y_0$ ($\bar{m} = -m, m$); i.e., the classical trajectories that contribute are those for which the incident transverse momentum is equal to the appropriate quantized value. We obtain

$$t_{nm} = -\frac{(2\pi i\hbar)^{(3/2)}}{2W} \sum_{s(\bar{m},\bar{n})} sgn(\bar{m}) sgn(\bar{n}) \sqrt{\tilde{D}_s} \exp\left(\frac{i}{\hbar}\tilde{S}_s(\bar{n},\bar{m},E) - i\frac{\pi}{2}v_s\right) \tag{6}$$

The sum is now over trajectories between the cross sections x and x' at angles $\sin\theta = \bar{m}\pi/kW$ and $\sin\theta' = \bar{n}\pi/kW$. The action is $\tilde{S}(\bar{n}, \bar{m}, E) = S(y_0', y_0, E) + \hbar\pi\bar{m}y_0/W - \hbar\pi\bar{n}y_0'/W$ and the prefactor is $\tilde{D}_s = (mv\cos\theta')^{-1}|(\partial y/\partial\theta')_\theta)|$, $v = \mu + H(-(\partial\theta/\partial y)_{y'}) + H(-(\partial\theta'/\partial y')_\theta)$ (H is the Heaviside step function). (The direct paths are not properly treated in Eq. (6), since $\tilde{D}_s$ becomes infinite in the second stationary phase integration; a correct treatment will be given in Ref. [45].

The magnetic field correlation function $C(\Delta B)$ can be calculated using Eq. (6) by arguments similar to those used in Ref. [36] for $C(\Delta E)$. First, in the multiple sum over paths, only the interference of a path s at field **B** with the same path at $B + \Delta B$ is retained. Second, the difference of the action is expanded to find the relative phase accumulated:

$$\frac{1}{\hbar}\left(\frac{\partial\tilde{S}_s}{\partial B}\right)\Delta B = \frac{e}{\hbar c}\int \Delta\vec{A}\cdot d\vec{l} \equiv \frac{\Theta_s \Delta B}{\phi_0} \tag{7}$$

where $|\nabla \times \Delta\vec{A}| = \Delta B$, $\phi_0 = hc/e$. Θ can be positive or negative (see Fig. 3), and we find that for weak fields (except near the origin) the distribution of Θ over classical trajectories, $N(\Theta)$, has an approximately symmetric exponential form, $N(\Theta) \propto \exp(-\alpha_{cl}|\Theta|)$. We emphasize that the entire analysis is only valid for weak fields, where the cyclotron radius is larger than the device dimensions. For closed orbits Θ is the area enclosed (times 2π). For typical open orbits this is true to a good approximation, and the inverse decay constant α_{cl}^{-1} will then give the typical area enclosed by trajectories traversing the structure (although Θ is not itself gauge-invariant, we have checked that α_{cl} is independent of gauge [45]). Finally, making the approximation $N(\Theta) \propto \exp(-\alpha_{cl}|\Theta|)$ for *all* Θ, we can replace the sum over trajectories by an integral over Θ, giving

$$C(\Delta B) = \frac{C(0)}{[1 + (\Delta B/\alpha_{cl}\phi_0)^2]^2} \tag{8}$$

We have calculated by classical simulations α_{cl} for the structures shown, and hence can compare the predictions of Eq. (8) to exact numerical results for the correlation function with no free parameters. We consider systems with 4-15 modes at E_f. It is important to note that the semiclassical approximation leading to Eqs. (5) or (6) is *not* well-justified in the few-mode limit [38] typical of experiments, since the difference between the actions of classical trajectories with the same end-points is not always much greater than $\hbar$. Thus our numerical results provide a crucial test of the accuracy of the semiclassical approximation in such systems.

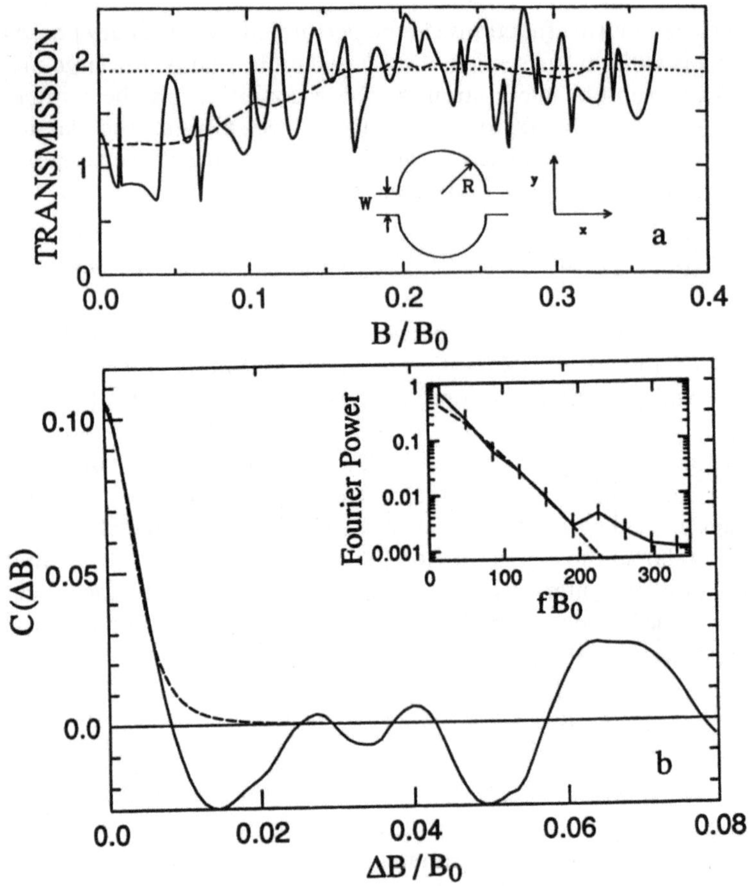

Fig. 2. **a** T(B) for open stadium with R/W = 2(*inset*). *Solid line* is quantum result ($k_f W/\pi = 4.5$), *dashed line* is smoothed $\langle T(B) \rangle$ used in computing $C(\Delta B)$, *dotted line* is classical result. $B_0 = mcv_f/eW$. **b** Magnetic field correlation function (for $W = 1000Å$, half-width $\Delta B_c = 40G$). *Dashed line* is the semiclassical prediction of Eq. (8). *Inset*:smoothed power spectrum of T(B). *Error bars* indicate rms variation of raw data. *Dashed line* is best fit to Fourier transform of Eq. (8) in interval [50,200]. (From [4] with permission)

First we show (Fig. 2) a typical correlation function $C(\Delta B)$ and the semiclassical prediction of Eq. (8). The agreement is excellent, except in the tail which corresponds to the non-universal short trajectory behavior. Because the behavior in the tail affects the half-width, it is necessary to calculate δT from a smoothed $\langle T(B) \rangle$ curve (Fig. 2); which introduces some arbitrariness in the determination of α_{qm}. To eliminate this freedom, we instead extract α_{qm} from fitting the Fourier power spectrum of the data to the Fourier transform of Eq. (8) (inset to Fig. 2); the value thus extracted is independent of the non-universal "low-frequency" behavior.

Figure 3 shows that α_{qm} is indeed given by the classical quantity α_{cl} to high accuracy while it is varied over roughly two orders of magnitude by changing R/W in the two- and four-probe structures. *Thus it is possible to predict quantitatively*

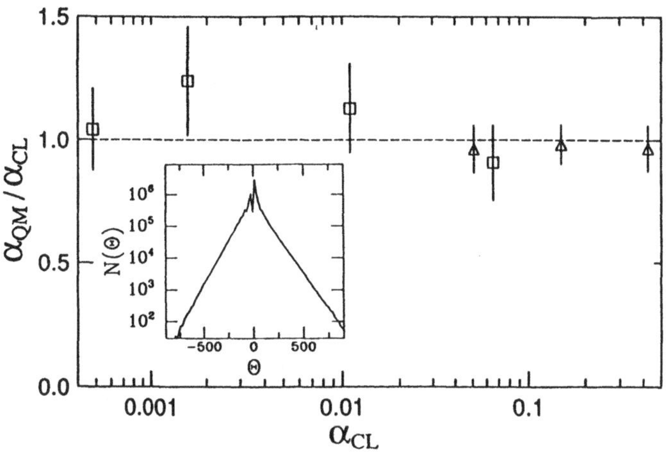

Fig. 3. Ratio of magnetic field correlation length (obtained as in inset of Fig. 2) to α_{cl}, the exponent of the distribution of effective areas, as a function of α_{cl} for both structures. Four-disk structure (*triangles*) with $R/W = 1, 2, 4$, and open stadium (*squares*) with $R/W = 1, 2, 4, 6$. α_{cl} (in units W^{-2}) is obtained from classical $N(\Theta)$ curve. *Inset*:$N(\Theta)$ for structure of Fig. 2. The correlation lengths of the quantum fluctuations agree with the semi-classical prediction over two decades. (From [4] with permission)

measurable properties of these ballistic quantum conductors from a knowledge of the chaotic classical scattering dynamics.

We briefly comment on two important issues to be treated elsewhere ([45]; H.U. Baranger, R.A. Jalabert, and A.D. Stone, 1992, unpublished work) First, the non-universal behavior due to short trajectories does cause very low frequency peaks in the Fourier spectra whose intensity depend on the mode index. Second, there is the fundamental question of the extent to which these fluctuation effects are a *unique* signature of quantum chaos. We have performed exactly the same analysis of scattering from non-chaotic cavities, e.g., a rectangular box, and still find substantial aperiodic structure (see related work by R.B.S. Oakeshott and A. MacKinnon, in [46]). However a careful analysis of the Fourier spectra indicates that they have a qualitatively different and non-universal shape in the many-mode limit, which is clearly distinguishable given two or three orders of magnitude sensitivity.

Average Quantum Behavior: Enhanced Quenching

We now turn to the question of the average quantum resistance and how it differs from the classical resistance, particularly with regard to quenching of the Hall resistance [5]. First, we must remove the fluctuations due to quantum interference discussed in the last section by defining an appropriate "average" quantum resistance. The most straightforward way to perform this average would be to include a phase-breaking process as in the experiments, as well as some thermal broadening. However, though partially phase-coherent transport has received considerable attention recently, calculations for only the simplest structures have been

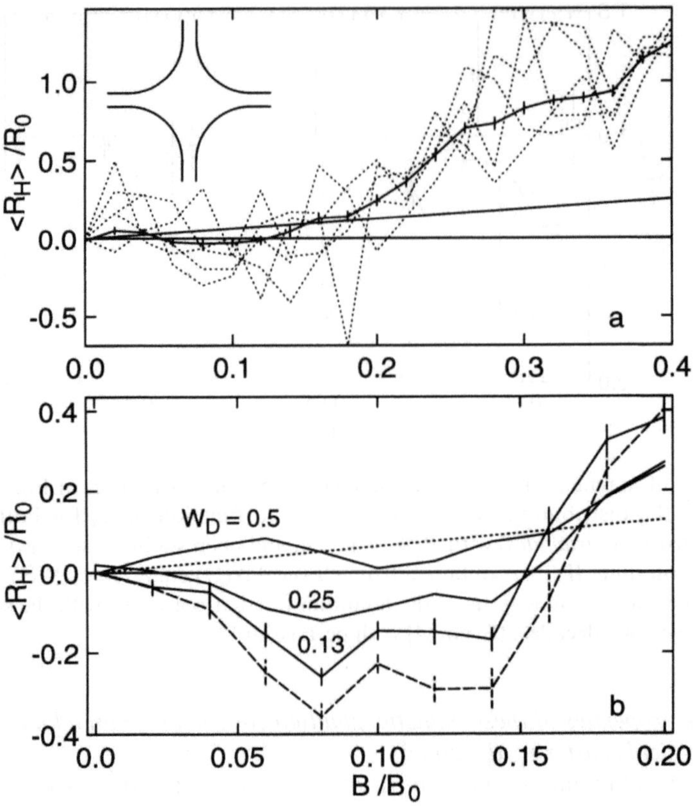

Fig. 4a, b. Quantum calculation of Hall resistance as a function of magnetic field in a four-disk junction with **R/W** = 4 (*inset*). **a** The *solid line* results from an average over both weak disorder ($W_D = 0.13$) and $k_F W/\pi$ in the range [4.44,4.56]; the *error bars* represent the statistical standard deviation. The *dotted lines* are results without disorder at specific energies in the same range, and the *straight solid line* is the 2D result. The averaging suppresses the fluctuations in the ballistic traces and produces a trace showing the typical behavior of the set of ballistic results. **b** Impurity-energy averaged results (*solid lines*) using three different strengths of disorder ($W_D = 0.13$, 0.25, 0.5 bottom to top) are compared to an energy average without disorder (*dashed*). For clarity, statistical error bars are shown on only one solid line; the others have comparable errors. The range of energies is $k_F W/\pi \in [4.44, 4.56]$ and the *dotted line* is the 2D result. Stronger impurity scattering suppresses the quenching. $R_0 = (h/e^2)(\pi/k_F W)(mv_F/\hbar k_F)$, $B_0 = mcv_F/eW$. (From [5] with permission)

carried out (see, for example [47]). Thus, we prefer to use the approach developed for diffusive systems of averaging over the energy and the impurity configuration and we ask the reader to recall the ergodic hypothesis for such systems [48]: the statistical properties of the resistance fluctuations are assumed to be independent of the control parameter producing the fluctuations, whether it be energy, magnetic field, or impurity configuration. Thus, averaging over one quantity (impurity configuration, say) reveals the smooth average behavior as a function of another quantity (field, say).

In our ballistic structures, it is clear that the ergodic hypothesis does not hold on large energy scales or for strong impurity averaging. In the case of energy, the discrete modal structure of the leads affects the resistance in ways not consistent with the ergodic hypothesis. However, the scale of the fluctuations in transport coefficients as a function of energy found in graded structures is one to two orders of magnitude smaller and is related to the trapping time, not to modal quantization. Thus we believe that the ergodic hypothesis holds approximately on this more restricted energy scale. In the case of impurity averaging, strong impurity scattering will obviously degrade the ballistic effects of interest here; yet we empirically show below that a weak impurity average is possible.

We define the average quantum resistance by averaging over a distribution of very weak random impurities as well as a small energy window. From a semiclassical point of view, the weak impurity average (the mean free path is kept larger than the size of the junction) "dephases" the long trajectories while retaining the coherence of the short paths: the small phase shifts from individual impurities accrue over the trajectory, and the subsequent average over an ensemble of impurities removes interference effects of those trajectories whose total phase shift is $\sim\pi$ while retaining interference effects of short trajectories. The average over impurities is a crude way of including some aspects of the phase randomization occurring in real structures. Of course, the structures are no longer completely ballistic, and the average resistance of the structure will be modified by an impurity strength that is too large. It is important to establish, then, if an impurity strength exists which will average the fluctuations but not change the typical ballistic behavior. In addition, a small amount of energy averaging serves to smooth the traces further and is physically sensible in terms of incorporating some of the effects of a non-zero temperature. A second method of averaging, involving an average over magnetic field and energy, is discussed elsewhere [5] and gives good agreement with the results presented here.

An example of the results of this "impurity-energy" averaging technique is given in Fig. 4a [5], which compares ballistic traces at five closely spaced energies to an impurity-energy averaged trace for the $R/W = 4$ four-disk structure[1]. Since our calculation is always performed on a lattice, it is trivial to include impurity scatter-

[1] The parameter values for the quantum calculations are as follows. For the linearly-widened and four-disk structures, $W = 18$ sites for which $E_1/V = 0.03$. For the soft-wall structures, $W = 39$ in the junction at the maximum energy which corresponds to the threshold for the sixth or seventh mode. Writing $U = a_n x^n y^n / W^{2n}$, we use $a_4 = 430$ and match onto a y^n cross-section at $x = W$; $E_1(n = 4) = 0.062$. The widening of the wires is about a factor of two for the maximum energy used. Because the calculations are performed on a lattice, the dispersion relation is neither parabolic nor isotropic. The non-parabolicity is easily accounted for by distinguishing between $\hbar k_F$ and $m v_F$ in scaling the data. In contrast, distortion of the Fermi circle leads to preferential propagation in certain directions and must be avoided; for instance, a ballistic bend resistance of the wrong sign can occur. Our calculations are done for $k_x a < 1.4$ and $E_F/V < -2.4$ for which the maximum deviation of k_F (v_F) from the mean is 3% (7%).

ing within the Anderson model of disorder: the onsite energies are chosen uniformly from the interval $[-W_D, W_D]$. Such impurities are placed throughout the entire junction, i.e., the region where the walls of the leads are not parallel. These impurities are certainly not meant to be a realistic representation of the impurity potential in these structures, but rather a tool for extracting the average quantum behavior. The energy averaging was done using the five energies for which ballistic traces are shown; $W_D = 0.13$ for which the mean free path [30] in the junction in terms of the direct transit length is $\approx 17 L_D$ (averaged over energy)[2]. The average trace lies in the middle of the five ballistic traces at all fields; this confirms that the averaging suppresses the large quantum fluctuations while yielding a faithful representation of the typical behavior.

Figure 4b shows the effect of different impurity strengths: impurity-energy averaged traces for three different strengths are compared to a pure energy averaged trace. Stronger impurity scattering eliminates the transport anomalies [13]: the quenching of the Hall resistance and the magnitude of the bend resistance are both suppressed. The mean free paths for the three cases shown are ≈ 1.2, 4.4, and $17 L_D$ for $W_D = 0.5$, 0.25, and 0.13, respectively[2]. For the weakest impurity strength, the impurity-energy averaged traces are in qualitative agreement with the purely energy averaged ones. However, similarly smooth traces from only energy averaging require a larger energy window and hence the sacrifice of energy resolution. The impurity-energy averaging that we use provides a smooth magnetic field trace at a relatively well-defined energy; we average over about 1/6 of the subband spacing in the results shown here.

Figures 5–7 show impurity-energy averaged quantum traces, $\langle R_H(B) \rangle$, and classical traces for three different structures (see[1] on p. 55): the $R/W = 4$ four-disk structure (Fig. 5), the gradually-widened structure (Fig. 6), and the $n = 4$ soft-wall structure[3] (Fig. 7). Thc quantum traces mostly show the same qualitative behavior as the classical traces. However, these traces show explicitly the large quantitative differences between the classical and quantum results: the quenching width is

[2] The mean free path is calculated numerically for a strip with a narrow disordered region through the Born approximation, as in Ref. [30]. For the values given here, the strip width is 36 sites, which corresponds to $3W$ for the $R/W = 4$ structure, $2W$ for the gradually-widened structure, and twice the classical width at the highest energy for the soft-wall structure. (Note that L_D, the direct length, is approximately the same in these three structures.) The mean free path varies substantially with energy, by as much as a factor of two; the values given result from averaging over energies for which 6–13 modes are contributing [(−3.72, −2.72) in units of the hopping matrix element]. In narrower strips 18 sites wide, the averaged mean free path is nearly the same (within 5%), but the energy variation is stronger.

[3] The scaling used in a soft-wall potential $U(y)$ is as follows. First, we retain the relation between our resistance scale and the experimental scale, namely, that one multiplies our scaled value by h/e^2 times the number of modes. Defining the effective width of the wire by $W_{eff} \equiv \int_{y_1(E)}^{y_2(E)} dy \sqrt{1 - U(y)/E}$ where $y_1(E)$ and $y_2(E)$ are the classical turning points, the resistance is scaled by $R_0 = (h/e^2)(\pi/k_F(0) W_{eff})$ (times $mv_F/\hbar k_F$ in the lattice case). Second, we require that the slope in the two-dimensional case be $2/\pi$. This is accomplished by using W_{eff} in the magnetic field scale, $B_0 = mcv_F(0)/eW_{eff}$.

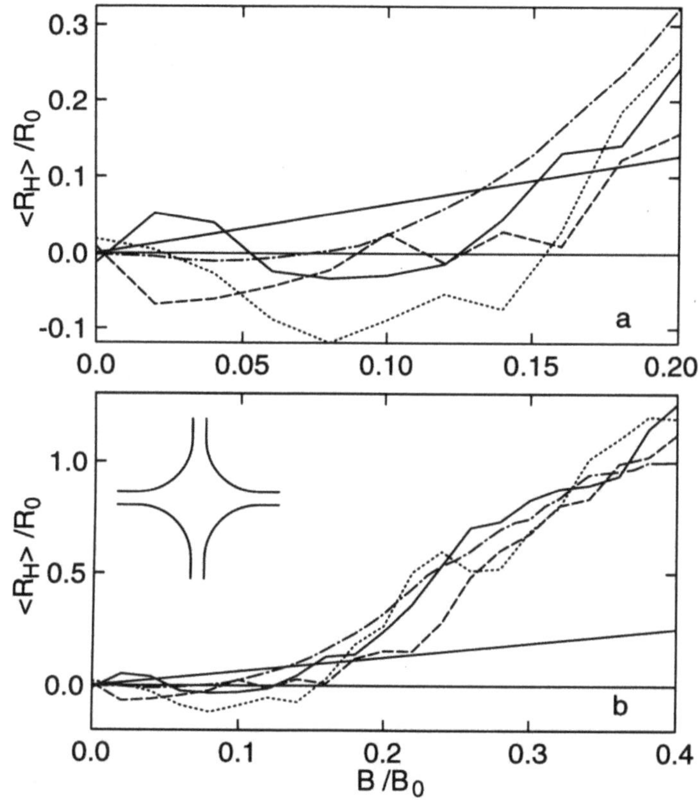

Fig. 5a, b. Quantum calculation of Hall resistance (on two different scales **a** and **b**) as a function of magnetic field in a four-disk junction with R/W = 4 (*inset*). Impurity-energy averaged results ($W_D = 0.13$) at three energies ($k_F W/\pi = 3.5$ (*dotted*), 4.5 (*solid*), and 6.5 (*dashed*)) are compared to the classical result (*dash-dotted*) and the 2D result (*solid straight line*). The region of anomalous behavior is larger in the quantum results than in the classical trace. There is little variation with energy after using reduced units. (From [5] with permission)

substantially larger in both the four-disk structure and the soft-wall structure, and the traces are very different for the gradually-widened structure.

In these three figures, quantum traces at three different energies show how the quenching and bend resistance change as the number of modes contributing changes from three to four to six. For the four-disk structure (Fig. 5), there is a slight narrowing of the quenching region. The linearly-widened structure (Fig. 6) is very sensitive to energy, showing inversion, quenching, and normal linear behavior of the Hall resistance at these three energies. Finally, for the soft-wall structure, Fig. 7 shows that the quenching region in the Hall resistance becomes narrower (in reduced units) as the energy increases. This change in the width of the ballistic anomaly in R_H as energy changes is observed in some experimental traces [12].

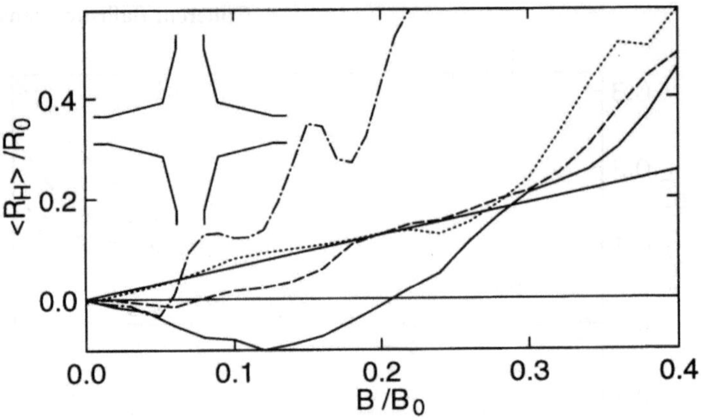

Fig. 6. Quantum calculation of Hall resistance as a function of magnetic field in the gradually-widened structure. Impurity-energy averaged results ($W_D = 0.13$) at three energies [$k_F W/\pi = 3.5$ (*dotted*), 4.5 (*solid*), and 6.5 (*dashed*)] are compared to the classical result (*dash-dotted*) and the 2D result (*solid straight line*). The quantum results are strikingly different from the classical trace and vary substantially with energy. (From [5] with permission)

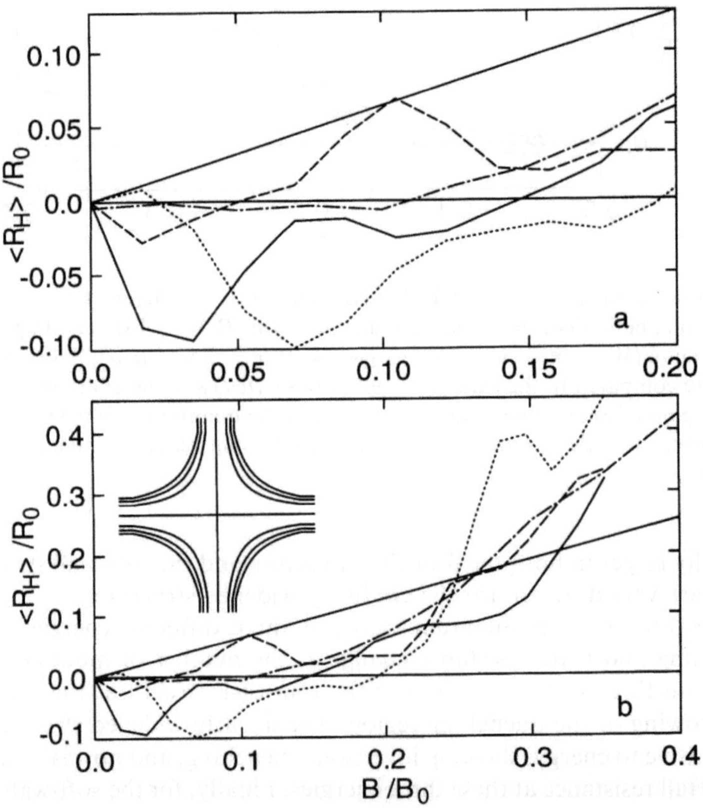

Fig. 7a, b. Quantum calculation of Hall resistance (on two different scales **a** and **b**) as a function of magnetic field in the $n = 4$ soft-wall structure (*inset*). Impurity-energy averaged results ($W_D = 0.13$) at three energies [$k_F W/\pi = 3.5$ (*dotted*), 4.5 (*solid*), and 6.5 (*dashed*)] are compared to the classical result (*dash-dotted*) and the 2D result (*solid straight line*). The region of quenching in R_H is substantially larger than in the classical trace and is enhanced at lower energy. (From [5] with permission)

Conclusions

In this paper we have summarized our recent work [4, 5] concerning the differences between classical and quantum transport properties of ballistic microstructures. The most important difference is the existence of quantum fluctuations in the conductance of structures which trap the electrons. These fluctuations are caused by interference between the long trapped trajectories. For structures in which the classical dynamics is chaotic, our semiclassical and exact numerical calculations show that the scale of the magnetic-field correlation function is determined by the underlying chaotic classical scattering. This scale can be varied over a large range by varying the degree of openness of the structure. A clear experimental signature of fluctuations from geometric scattering would be, then, a large change in the scale of the fluctuations as the degree of trapping is varied. Such a clear experimental demonstration has not yet been achieved.

In addition to the quantum fluctuations, the average quantum resistance differs from the classical resistance. We remove the quantum fluctuations and define an average quantum resistance using both an ensemble of very weak impurities and thermal broadening. The resulting average quantum magnetic field traces show good qualitative agreement with the classical result. However, quantitatively, the ballistic anomalies in R_H, such as the quenching, are enhanced in the quantum regime, and the quantum traces vary with energy, quite dramatically for the gradually-widened structure, while no such variation (in the scaled units) occurs classically.

Acknowledgments. We thank O.L. Alerhand, M. Büttiker, D.P. DiVincenzo, M. Gutzwiller, R. Jensen, M.L. Roukes, and U. Smilansky for many helpful discussions. Research at Yale University was supported, in part, by NSF Grant No. DMR-8658135, an IBM postdoctoral fellowship (R.A.J.), and a Sloan Foundation fellowship (A.D.S.).

References

1. Beenakker CWJ, van Houten H (1991) In: Ehrenreich H, Turnbull D (eds) Solid state physics, vol. 44. Academic, New York, pp 1–228
2. Beenakker CWJ, van Houten H (1989) Phys Rev Lett 63:1857
3. Baranger HU, Stone AD (1989) Phys Rev Lett 63:414
4. Jalabert RA, Baranger HU, Stone AD (1990) Phys Rev Lett 65:2242
5. Baranger HU, DiVincenzo DP, Jalabert RA, Stone AD (1991) Phys Rev B44:10637
6. Roukes ML, Scherer A, Allen SJ, Jr, Craighead HG, Ruthen RM, Beebe ED, Harbison JP (1987) Phys Rev Lett 59:3011
7. Ford CJB, Thornton TJ, Newbury R, Pepper M, Ahmed H, Peacock DC, Ritchie DA, Frost JEF, Jones GAC (1988) Phys Rev B38:8518
8. Timp G, Baranger HU, deVegvar P, Cunningham JE, Howard RE, Behringer R, Mankiewich PM (1988) Phys Rev Lett 60:2081

9. Ford CJB, Washburn S, Büttiker M, Knoedler CM, Hong JM (1989) Phys Rev Lett 62:2724
10. Chang AM, Chang TY, Baranger HU (1989) Phys Rev Lett 63:1860
11. Takagaki Y, Gamo Y, Namba S, Ishida S, Takaoka S, Murase K, Ishibashi K, Aoyagi Y (1988) Solid State Commun 68:1051; Takagaki Y, Gamo K, Namba S, Takaoka S, Murase K (1990) Solid State Commun 75:873
12. Timp G, Behringer R, Sampere S, Cunningham JE, Howard RE (1989) In: Reed MA, Kirk WP (eds) Nanostructure physics and fabrication. Academic New York, p 331
13. Roukes ML, Scherer A, Van der Gaag BP (1990) Phys Rev Lett 64:1154
14. Landauer R (1957) IBM J Res Dev 1:233; (1987) Z Phys B68:217
15. Büttiker M (1986) Phys Rev Lett 57:1761
16. Baranger HU, Stone AD (1989) Phys Rev B40:8169 (and references therein)
17. Beenakker CWJ, van Houten H (1988) Phys Rev Lett 60:2406
18. Peeters FM (1988) Phys Rev Lett 61:589
19. Kirczenow G (1988) Phys Rev B38:10958
20. Akera H, Ando T (1989) Phys Rev B39:5508
21. Ravenhall DG, Wyld HW, Schult RK (1989) Phys Rev Lett 62:1780; Schult RL, Wyld HW, Ravenhall DG, (1990) Phys Rev B41:12760
22. Kirczenow G (1989) Solid State Commun 71:469; (1989) Phys Rev Lett 62:2993
23. Baranger HU, Stone AD (1990) Surf Sci 299:212
24. Li Q, Thouless DJ (1990) Phys Rev Lett 65:767
25. Akera H, Ando T (1990) Phys Rev B41:11967
26. Glazman LI, Lesovick GB, Khmelnitskii DE, Shekhter RI (1988) Pis'ma Ah Teor Fiz (in Russian) 48:218 [JETP Lett 48:238 (1988)]
27. Beenakker CWJ, van Houten H (1989) Phys Rev B39:10445
28. Imry Y (1989) In: Reed MA, Kirk WP (eds) Nanostructure physics and fabrication. Academic, New York, p 379; Yacoby A, Imry Y, (1990) Phys Rev B41:5341
29. Imry Y (1986) In: Grinstein G, Mazenko E (eds) Perspectives on condensed matter physics. World Publishing, Singapore
30. Baranger HU (1990) Phys Rev B42:11479
31. Messiah A, Quantum mechanics (1985) Wiley, New York, pp 825–828
32. Gutzwiller M (1991) Chaos in classical and quantum mechanics. Springer, New York
33. Bohigas O, Giannoni MJ, Schmit C (1984) Phys Rev Lett 52:1 Berry MV (1985) Proc Roy Soc Lond A400:229
34. Brody TA, Flores J, French JB, Mello PA, Pandey A, Wong SSM (1981) Rev Mod Phys 53:385 (and refs. therein)
35. Gutzwiller MC (1983) Physica D7:341
36. Blümel R, Smilansky U (1988) Phys Rev Lett 60:477; (1990) Phys Rev Lett 64:241; Physica (1989) D36:111
37. Doron E, Smilansky U, Frenkel A (1990) Phys Rev Lett 65:3072; (1991) Physica D50:367
38. Lewenkopf CH, Weidenmüller HA (1991) Ann Phys 212:53
39. Jung C, Scholz HJ (1987) J Phys A20:3607
40. Eckhardt B (1987) J Phys A20:5971
41. Tél T (1989) J Phys A22:L691
42. Gaspard P, Rice SA (1989) J Chem Phys 90:2225, 2242, 2255
43. Bleher S, Ott E, Grebogi C (1989) Phys Rev Lett 63:919
44. Roukes ML, Alerhand OL (1990) Phys Rev Lett 65:1651
45. Jalabert RA (1992) Bull Am Phys Soc 36:743

46. Oakeshott RBS, MacKinnon A (1992) In:Kirk W Reed MA (eds) Nanostructure physics and fabrication. Academic, New York
47. Büttiker M (1986) Phys Rev B33:3020; Datta S (1989) Phys Rev B40:5830; Feng S (1990) Phys Lett A143:400; Stern A, Aharonov Y, Imryy (1990) Phys Rev 41:3436; Kirczenow G (1990) Solid State Commun 74:1051; D'Amato JL, Pastawski HM (1990) Phys Rev 41:7441; Hershfield S (1991) Phys Rev B43:11586
48. Lee PA, Stone AD, Fukuyama H (1987) Phys Rev B35:1039 (and references therein)

5

Quasi-Particle Interferometer Controlled by Andreev Reflection

HAYATO NAKANO AND HIDEAKI TAKAYANAGI[1]

Summary. We propose a quasiparticle interferometer for experimentally confirming the predicted interaction between the microscopic phase of a quasiparticle and the macroscopic phase of a superconducting state. Andreev reflection at the superconductor- normal-metal (S-N) interface causes phase interaction. The proposed interferometer consists of a Josephson junction and a Y-junction composed of normal electron waveguides. In this setup, the phase interaction due to Andreev reflection affects the quasiparticle interference. Thus the amount of supercurrent flowing through the Josephson junction controls the resistance across the end of a waveguide branch and one electrode of the Josephson junction.

Introduction

In recent years, the phase coherence phenomena of normal electrons in mesoscopic systems have been widely observed [1–4]. In addition, phase coherence phenomena in superconducting states have already been applied to devices such as superconducting quantum interference device (SQUID) [5, 6]. What is, however, the relationship between the microscopic 'phase' of a normal electron and the macroscopic 'phase' of a superconducting state? The answer is not trivial. Andreev reflection [7], which occurs at the superconductor-normal-metal(S-N) interface, will give an answer. The phase of a quasiparticle (electron or hole) that is Andreev-reflected is expected to be shifted by the phase of the superconductor [8]. Here, we propose a quasiparticle interferometer that should enable us to directly check this predicted phase interaction between a quasiparticle and a superconducting state.

Acquisition of Macroscopic Phase Through Andreev Reflection

We assume step-like pair potentials as:

$$\Delta(\boldsymbol{r}) = \begin{cases} 0, & \text{for } x < 0\text{: normal-metal} \\ \Delta_0 e^{i\theta}, & \text{for } x > 0\text{: superconductor} \end{cases}$$

[1] NTT Basic Research Laboratories, Midori-cho 3-9-11, Musashino, Tokyo, 180 Japan

where θ is the macroscopic phase of the superconductor. Using the Bogoliubov-de Gennes equation [9], the wavefunctions can be described conveniently in terms of two-component representation. For simplicity, hereafter we treat the case where there is no *intrinsic* normal reflection at the S-N interface. The effect of the intrinsic normal reflections will be discussed elsewhere (H Nakano, H Takayanagi, manuscript in preparation). In that case, the wavefunction in the normal-metal region is given by [10] $\psi = \psi_{\rm inc} + \psi_{\rm ref}$, where

$$\psi_{\rm inc} = \begin{bmatrix} 1 \\ 0 \end{bmatrix} \exp[i(k^{+}x + \boldsymbol{k}_{\perp}\cdot\boldsymbol{r}_{\perp})]$$

$$\psi_{\rm ref} = a\begin{bmatrix} 0 \\ 1 \end{bmatrix} \exp[i(k^{-}x + \boldsymbol{k}_{\perp}\cdot\boldsymbol{r}_{\perp})]$$

These represent the incident electron and the Andreev-reflected hole, respectively. Here, $k^{\pm}$ is the wave-number for the direction perpendicular to the interface given by

$$k^{\pm} = (2m/\hbar^2)^{1/2}(E_{\rm F} - E_{\perp} \pm E)^{1/2}$$

where m and E_F are the electron effective mass and the Fermi energy in the normal-metal, $\boldsymbol{k}_{\perp}$ is the transversal wave-number, which gives $E_{\perp} = (\hbar^2/2m)\boldsymbol{k}_{\perp}{}^2$, and E is the excitation energy measured from the Fermi energy.

When the energy is within the superconducting gap energy Δ_0, the electron is Andreev-reflected, changing into a hole. A simple calculation gives the Andreev reflection coefficient a as

$$a = \mathrm{e}^{-i(\phi+\theta)}$$

where $\phi = \arctan\sqrt{(\Delta_0/E)^2 - 1}$ is the extra energy-dependent phase shift. When a hole is incident, it is phase-shifted in the same way by the Andreev reflection, changing into an electron.

It should be emphasized that the phase shift $\phi + \theta$ by Andreev reflection does not depend upon $\boldsymbol{k}_{\perp}$; that is, it is independent of the incident angle for the interface.

Phase Shift and Interference in the Setup

This phase shift cannot be confirmed in a single S-N geometry. Figure 1 shows our apparatus for observing the phase shift. It consists of a Josephson junction (JJ) and a Y-type junction composed of normal electron waveguides. The first branch of the Y-junction is the entrance for the incident electron. The second branch is in contact with one electrode of the JJ, and the third branch with the other electrode; these contacts constitute the S-N interfaces where Andreev reflections occur. The length L of these two branches must be shorter than the phase coherence length L_{ϕ} of the waveguide, so that an electron or hole keeps itself phase-coherent during the round trip between the Y-junction point and the S-N interface.

Let us examine the resistance between the first branch of the Y-junction (Q) and

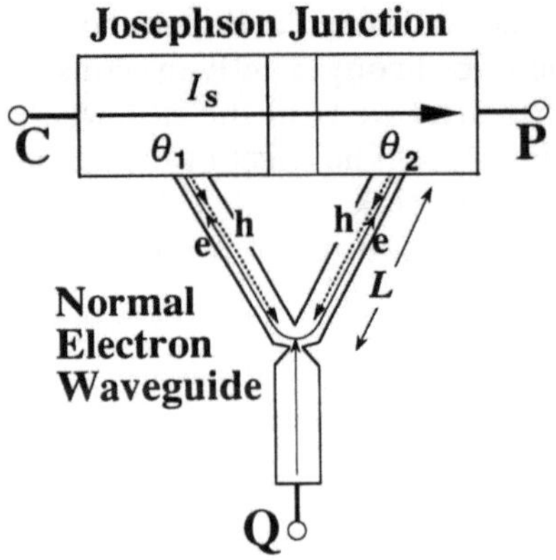

Fig. 1. The proposed quasiparticle interferometer for confirming the predicted phase interaction between a quasiparticle and a superconducting state. We concentrate on the resistance across Q-P, which varies as the function of the supercurrent flowing in the Josephson junction

an electrode of the JJ (P). The narrow constriction on the first branch functions as a Sharvin resistance. The potential difference across Q-P concentrates at the constriction. This gives the ordinary resistance R_N across Q-P.

An electron wave incident through the narrow constriction is divided between the second and third branches. Each partial electron wave propagates along a waveguide branch, reaches the S-N interface, and is then Andreev-reflected. The Andreev reflection phase-shifts the partial hole wave by the macroscopic phase of the electrode, θ_1 or θ_2. Andreev-reflected holes return on the original partial electron's paths and meet again at the Y-junction point. The phase advances during the round trip cancel out because of the retroactive property of Andreev-reflected holes. Thus, the phase difference when two partial hole waves meet is determined by the phase shifts they undergo when they are Andreev-reflected, that is, $\theta_1 - \theta_2$. As a result of the interference, a proportion $\frac{1}{2}[1 + \cos(\theta_1 - \theta_2)]$ of the holes is received by the first branch, and this amount contributes as the excess current, resulting in a decrease in the resistance across Q-P.

When the Andreev-reflected hole (electron) cannot be received by the waveguide as the result of the interference, the original incident electron (hole) suffers from *normal* reflection, and makes no current contribution. The amount of normal-reflected electrons is $\frac{1}{2}[1 - \cos(\theta_1 - \theta_2)]$. This is different from the *intrinsic* normal reflection and sharpens the interference effect, whereas the intrinsic normal reflection smears the effect. As the result, the resistance across Q-P is given by

$$R_N/[1 + \cos(\theta_1 - \theta_2)]$$

Therefore, the phase difference of the JJ changes the resistance across Q-P, and thus the setup works as an interferometer.

Modification of Interference by Supercurrent

The relationship between the phase difference $\theta_1 - \theta_2$ of a JJ and the supercurrent I_s flowing through the JJ is given by [5] $I_s = I_0 \sin(\theta_1 - \theta_2)$, where I_0 is the critical current of the JJ. Therefore, the phase difference can be controlled from $-\pi/2$ to $\pi/2$ by changing the bias-current from $-I_0$ to I_0.

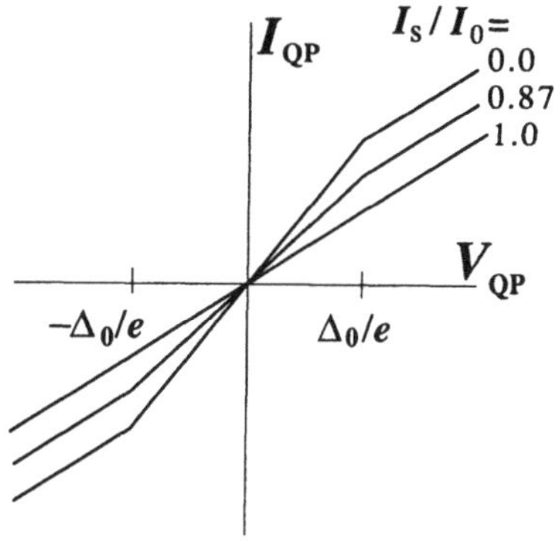

Fig. 2. Schematic $I_{QP} - V_{QP}$ characteristics for various supercurrents flowing in the Josephson junction. The amount of the supercurrent modifies the quasiparticle interference and changes the resistance

Consequently, the differential resistance across Q-P varies from $\frac{1}{2}R_N$ to R_N as a function of the supercurrent of the JJ. Schematic *I-V* curves are shown in Fig. 2. Observing this phenomenon experimentally would confirm the phase interaction between a quasiparticle and a superconducting state.

Points to Note in the Interferometer

Schrödinger's Cat Between Two Electrodes

The Josephson coupling between two superconducting electrodes is crucial for the quasiparticle interference. Andreev reflection on an electrode adds one Cooper-pair in the electrode. Unless the two electrodes are quantum mechanically coupled, this Cooper-pair addition classically determines on which electrode a quasiparticle was Andreev-reflected, and the interferenece never occurs. In our setup, a superconducting electrode does not act merely as a phase-bath. Owing to the Josephson coupling, the Andreev reflections on both electrodes are superimposed quantum mechanically with the phase difference $\theta_1 - \theta_2$. You cannot know on which electrode the Andreev reflection occurred. This is an example of *Schrödinger's cat* [11]. This aspect of our interferometer will be discussed in detail elsewhere (H Nakano, N Hatakenaka, manuscript in preparation).

Effect of Multimodes in the Waveguide

In order to observe the Aharonov-Bohm effect [12], the width of the waveguide should be on the order of or shorter than the Fermi wavelength, since a wide waveguide has many transmission modes which experience different phase shifts and smear the interference effect [13]. In our interferometer, (if limited to the intrinsic normal reflection free case,) the phase shift by Andreev reflection is independent of the incident angle of the quasiparticle. In other words, all modes in the waveguide are phase-shifted by the same amount, and the average on the modes does not smear the interference effect. Therefore, the restriction on the waveguide width is not so severe as for observations of the Aharonov-Bohm effect.

Feasible Materials and Dimensions

Now we consider materials and dimensions necessary for this interferometer. The electron waveguide must be made of a material with high mobility and long phase coherence length L_ϕ. A degenerate semiconductor is suitable for the material of the waveguide from the fabrication viewpoint. Taking into account the limitations of fabrication techniques, a length of a few μm for the waveguide branch L is reasonable. Since a low temperature is already required to get a long-L_ϕ waveguide, the Josephson junction need not to be composed of high-T_c superconductors. The most important point is to prepare a good S-N interface, since the intrinsic normal reflection strongly degrades the Andreev reflection and prevents us from observing the interference effect. For example, the combination of n-doped InAs (as the normal-metal) and Nb (as the superconductor) is promising. The InAs-Nb interface is Schottky barrier free and it gives a large Andreev reflection probability [14], that is, small intrinsic normal reflection. Moreover, n-doped InAs has a rather long L_ϕ. At a few K, the L_ϕ is more than a few μm [15] for a sample with a carrier density $n \sim 10^{18}\text{cm}^{-3}$.

Negligible Phase Shift by Magnetic Flux

Note that the supercurrent I_s generates a magnetic field. The magnetic flux piercing the area enclosed by the JJ and the Y-branches causes a shift in the phase of normal electrons or holes in the waveguide, as happens in the Aharonov-Bohm effect. This effect might hinder the observation of the phase shift by Andreev reflection. The magnetic flux is estimated to be on the order of $\mu_0 I_s G L$, where μ_0 is the permeability of vacuum, and G is a geometrical factor on the order of unity. When the length of the branch L is a few μm, and the supercurrent is a few μA, the flux is about $10^{-4}h/e$. Therefore, the phase shift caused by the magnetic flux is negligible.

Acknowledgments. We are indebted to N. Hatakenaka of NTT for his helpful suggestions. We also wish to thank Dr. S. Kurihara and Dr. J. Nitta of NTT for their fruitful discussions.

References

1. Umbach CP, Washburn S, Laibowitz RB, Webb RA (1984) Phys Rev B30:4048–4051
2. Webb RA, Washburn S, Umbach CP, Laibowitz RB (1985) Phys Rev Lett 54:2696–2699
3. Ford CJB, Fowler AB, Hong JM, Knoedeler CM, Laux SE, Wainer JJ, Washburn S (1990) Surface Sci 229:307–311
4. Tokura Y, Tsubaki K (1987) Appl Phys Lett 51:1807–1808
5. Josephson BD (1962) Phys Lett 1:251–253
6. Van Duzer T, Turner CW (1981) Principles of Superconductive Devices and Circuits. Elsevier North Holland, New York (and references therein)

7. Andreev AF (1964) Sov Phys JETP 19:1228–1231
8. Spivak BZ, Khmel'nitskiĭ DE (1982) JETP Lett 35:412–416
9. de Gennes PG (1966) Superconductivity of Metals and Alloys. Benjamin, New York
10. Blonder GE, Tinkham M, Klapwijk TM (1982) Phys Rev B25:4515–4532
11. Schrödinger E (1935) Naturwiss 23:844–849
12. Aharonov Y, Bohm D (1959) Phys Rev 115:485–491
13. Feng S, Hu Y (1988) Phys Rev B38:11871–11873
14. Nitta J, Nakano H, Akazaki T, Takayanagi H (1991) In: Koch H, Lübbig H (eds) Proceedings SQUID 1991. Springer, Heidelberg
15. Nakano H, Akazaki T, Nitta J, Takayanagi H (1991) Solid State Commun 80:251–253

Conductance Oscillations in Quantum Wires with a Stub Structure

6

KIMIHISA AIHARA, MASAFUMI YAMAMOTO, AND TAKASHI MIZUTANI[1]

Summary. Quantum interference effects of electron waves in a Y-shaped structure of a quantum wire with a stub fabricated in a split gate configuration were experimentally studied. Periodic conductance oscillations as a function of gate voltage, which were characterized by a single peak in a Fourier spectrum, were observed. In contrast, aperiodic conductance oscillations were observed in a Y-shaped branch in which the reflection wall in the stub had been removed. The clear difference of the conductance characteristics in these structures suggests that the observed periodic oscillations were caused by the quantum interference in the stub. In addition, the calculated gate voltage period based on a model that the conductance oscillations are due to quantum interference in the stub was in good agreement with the experimental period. From these experimental results, it was concluded that the observed periodic conductance oscillations for the Y-shaped structures were due to the quantum interference effect of electron waves in the stub.

Introduction

A quantum wire with a stub in the form of a T-shaped quantum waveguide structure [1–3] has the potentiality for three terminal quantum interference devices. In a previous experimental study [4] carried out using T-shaped structures fabricated by reactive ion etching, conductance oscillations as a function of the electrostatic potential of a remote gate placed on the edge of the stub were observed. However, the possible origins of conductance oscillations, i.e., quantum interference in a stub or universal conductance fluctuations, were not clearly separated.

The purpose of the present study was to investigate quantum interference effects in a quantum wire with a stub fabricated in a split gate configuration. This configuration allows the effective channel width to be decreased continuously to a pinch-off by applying negative gate voltage. Intuitively, it was expected that the degree of coupling of the stub to the quantum wire in Y-shaped structures would be increased compared with the degree of coupling in T-shaped structures, and that the quantum interference effect would be enhanced. In this study, conductance, G,

[1] NTT LSI Laboratories, 3-1, Morinosato Wakamiya, Atsugi, 243-01 Japan

versus gate voltage, V_g, was measured for a Y-shaped structure of a quantum wire with a stub. Periodic conductance oscillations as a function of gate voltage, which were characterized by a single peak in the Fourier spectrum, were observed. In order to clarify the origin of periodic conductance oscillations, G-V_g characteristics were also measured for a Y-shaped branch in which a reflection wall in the stub had been removed, and the results were compared. In addition, the gate voltage period was calculated, based on a model that the conductance oscillations are due to the quantum interference in the stub, and the calculated period was compared with the experimental one.

Fabrication

The devices were made on high mobility AlGaAs/GaAs heterostructures grown by molecular-beam epitaxy. The carrier concentration and mobility of the 2DEG were 5.2×10^{11} cm^{-2} and 3.0×10^5 cm^2/Vs at 4.2 K, respectively, corresponding to an electron mean free path of 3.6μm. The split gates were fabricated from Ti/Au by electron beam lithography and lift-off techniques. The insets in Figs. 1 and 3 show SEM micrographs of a fabricated Y-shaped structure and a Y-shaped branch. The lithographic gap width of the split gates and the device length from entrance to exit along the V-shaped path in the two devices were 0.4 μm and 1.2μm, respectively.

Results and Discussion

Resistance was measured using a low-frequency four-terminal AC technique with a lock-in amplifier. In Fig. 1, conductances, G, obtained by inverting the measured resistance versus gate voltage, V_g, at 0.4 K for a Y-shaped structure are shown for several current levels. Clear periodic conductance oscillations as a function of gate voltage were observed for low current levels of 5 nA and 50 nA. Figure 2 shows the Fourier power spectrum of the second differential of G(V_g) for the lowest current level of 5 nA. A single peak, with a frequency of $(25.6 \text{ mV})^{-1}$, was clearly observed. The amplitude of this peak decreased as the current level or temperature was increased, which suggests that these periodic oscillations are caused by the quantum interference effect of electron waves. The clear peak of the Fourier spectrum corresponding to a period of 25.6 mV at 0.4 K was reproducible after repeated thermal cycles between 300 K and 4.2 K.

In order to clarify the origin of the periodic conductance oscillations observed in Y-shaped structures, Y-shaped branches were fabricated, and G-V_g characteristics were measured. The measured conductances for several current levels are shown in Fig. 3. Oscillations were observed for low current levels of 5 nA and 50 nA. Figure 4 shows the Fourier power spectrum of the second differential of G(V_g) for the lowest current level of 5 nA. Several frequency peaks whose amplitudes were nearly equal were observed in the Fourier power spectrum, which contrasted with the single frequency peak observed in the Y-shaped structure. The

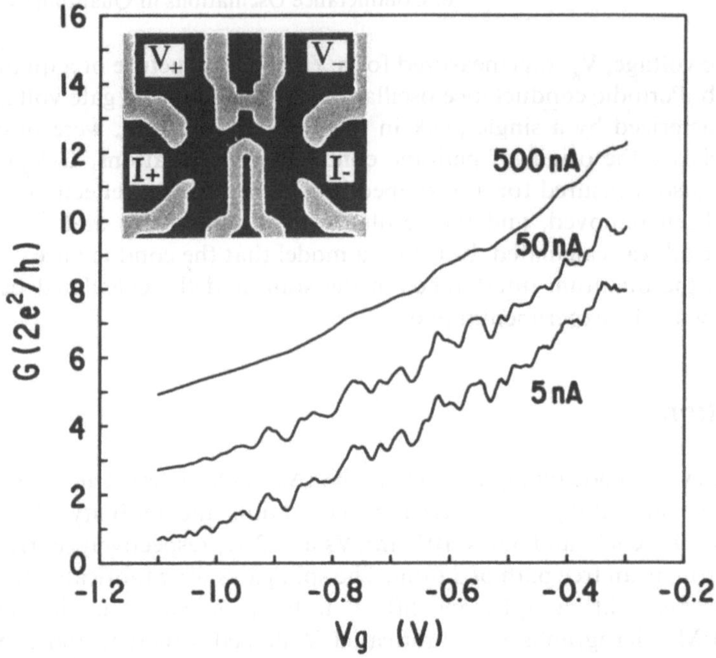

Fig. 1. Conductance of Y-shaped structure as a function of voltage applied to the split gate for several current levels. Subsequent curves from the bottom are offset by 2. An electron micrograph of a fabricated device is shown in the *inset*

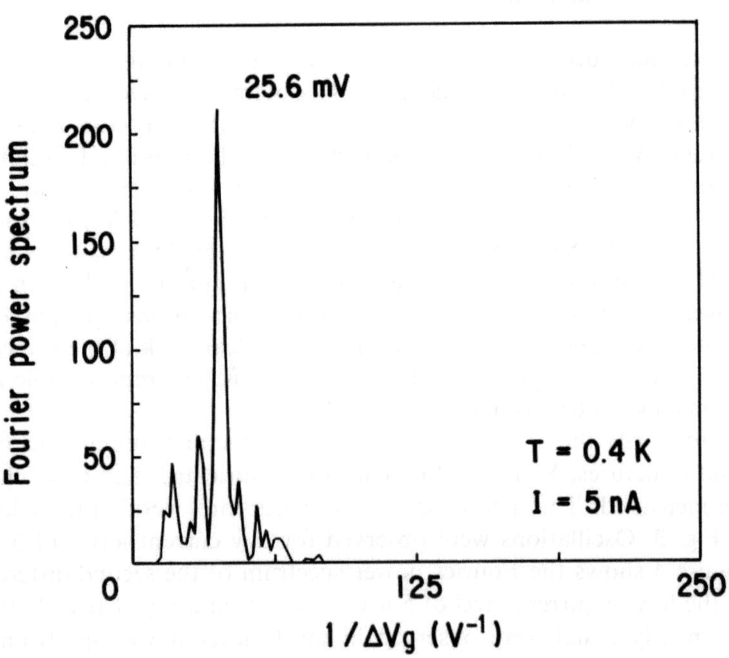

Fig. 2. Fourier power spectrum of conductance in Y-shaped structure

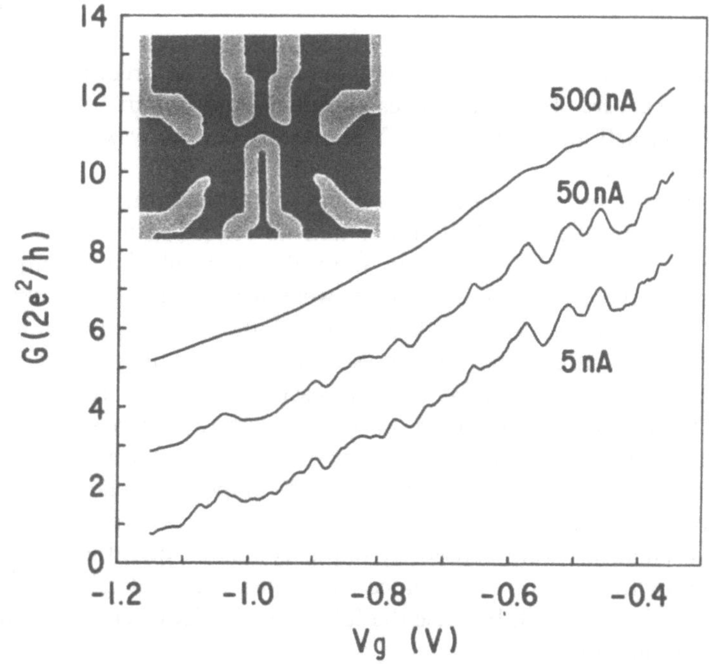

Fig. 3. Conductance of Y-shaped branch as a function of voltage applied to the split gate for several current levels. Subsequent curves from the bottom are offset by 2. An electron micrograph of a fabricated device is shown in the *inset*

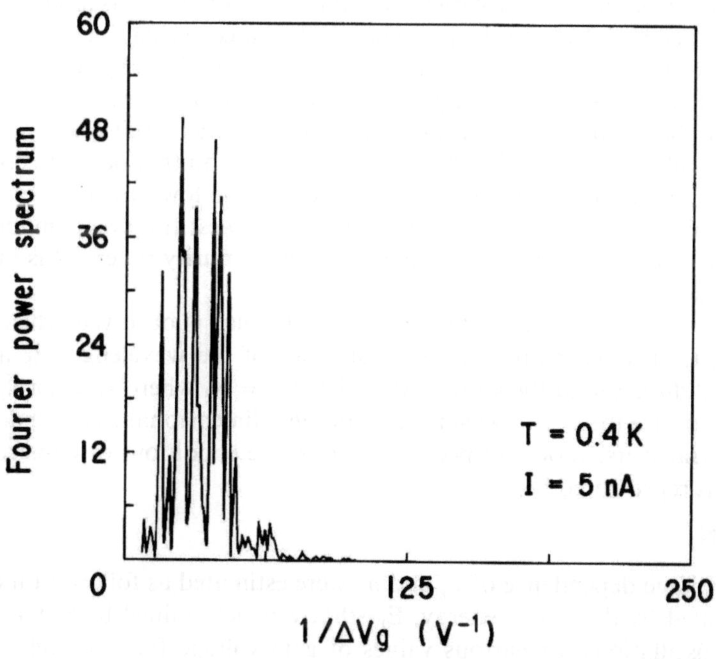

Fig. 4. Fourier power spectrum of conductance in Y-shaped branch

amplitudes of these peaks were about one-fifth of the predominant peak in the Y-shaped structure. The peak corresponding to the 25.6 mV periodic oscillation, which was clearly observed in the Y-shaped structure, was not observed in the Y-shaped branch.

The difference between these two devices is that the former had a reflection wall in the a stub structure, and the latter had no reflection wall. The characteristic feature that the periodic oscillations were observed only for Y-shaped structures indicates that these periodic oscillations were aroused from the quantum interference in the stub. If periodic oscillations are a) due to the quantum interference of electron waves scattered by impurities or unintentionally formed structural disorders in the channel; b) caused by transmission resonances associated with reflections at the entrance and exit [5–6]; or c) associated with successive quantized channel opening under the influence of impurity scattering [7] they should be observed for both Y-shaped structures and Y-shaped branches.

Moreover, the reproducibility after the thermal cycle is not explained if the periodic oscillations were caused by impurity related quantum interference. The possibility of transmission resonance associated with reflections at the entrance and exit can be excluded, because the fabricated devices were smoothly tapered at the entrance and exit [6].

Comparison of the experimental results for the Y-shaped structures and Y-shaped branches strongly indicates that the periodic oscillations observed in Y-shaped structures are due to the quantum interference in the stub. In order to further confirm this point, we calculated the period of gate voltage corresponding to the conductance oscillation and compared it with the experimental one. The calculation was based on the following assumptions: First, it was assumed that the predominant contribution to the periodic conductance oscillation comes from the quantum interference of electron waves traveling along the longitudinal direction in a stub. This assumption is appropriate, since the difference in the structure appeared in the longitudinal direction whether there was a reflection wall or not. Second, the calculation was simplified by the assumption that periodic oscillations through the quantum interference were caused by the lowest transverse mode in the stub. This is reasonable, considering that the lowest transverse mode has the highest kinetic energy and so the influence of the impurity potential is minimized for this mode.

For this simplified model, constructive interference is observed when the stub length, L_s, is an integer multiple N of one-half of the wavelength λ_F along the longitudinal direction in the stub, that is, $2\,L_s/\lambda_F = N$, where $\lambda_F = 2\pi/k_F = 4/n_1$, and k_F and n_1 are the Fermi wavelength and one-dimensional electron density for the lowest transverse mode, respectively. Thus, the condition for constructive interference is represented by

$$L_s n_1/2 = N \tag{1}$$

The gate voltage dependence of L_s and n_1 were estimated as follows: First, $n_1(V_g)$ was calculated by the Fermi energy, E_F, that was determined by S-dH magnetoresistance oscillations for various values of gate voltage for a straight quantum wire which has an identical nominal split gate gap width. The obtained experimental dependence of n_1 on V_g is expressed as

$$n_1 = 1.18 \times 10^8(V_g + 1.35)(m^{-1}) \tag{2}$$

Second, L_s was calculated using the experimental dependence of the effective channel width W_{eff} on V_g. $W_{eff}(V_g)$ was estimated from the value of the magnetoresistance, R, of the quantum wire mentioned above for a magnetic field B_0, at which boundary scattering is sufficiently suppressed, using the relation

$$R(B_0) = (L/W_{eff})(1/n_2 e\mu) \tag{3}$$

where L is the length of the quantum wire, n_2 is the two-dimensional electron gas (2DEG) electron density determined by S-dH oscillations, and μ is the mobility. Here, it was assumed that μ for the magnetic field B_0 was the same as that of the as grown two-dimensional electron gas (2DEG). The obtained experimental dependence of L_s on V_g is expressed as

$$L_s = 6.06 \times 10^{-7} + 1.32 \times 10^{-7} V_g(m) \tag{4}$$

The calculated dependence of N on V_g was almost linear, and the calculated period of V_g was 28.5 mV, which was in good agreement with the experimental one of 25.6 mV.

Conclusions

A Y-shaped structure of a quantum wire with a stub was fabricated in a split gate configuration and conductance versus gate voltage characteristics were measured. Periodic conductance oscillations as a function of gate voltage, characterized by a single peak in the Fourier spectrum, were observed. In contrast, aperiodic conductance oscillations were found in a Y-shaped branch in which a reflection wall in the stub structure had been removed. The fact that periodic oscillations were observed only in the Y-shaped structures suggests that these periodic oscillations were caused by the quantum interference in the stub. The calculated gate voltage period, based on a model in which the conductance oscillations are due to the quantum interference in a stub, was in good agreement with the experimental period.

From these experimental results, we conclude that the observed periodic conductance oscillations for Y-shaped structures are due to the quantum interference effect of electron waves in the stub.

Acknowledgments. The authors wish to thank K. Arai and T. Hasuike for the molecular–beam epitaxy (MBE) growth of heterostructure wafers. They are also very much indebted to K. Hirata for his continuous encouragement.

References

1. Hohkawa K, Yamamoto M (1988) (in Japanese) The Institute of Electronics, Information and Communication Engineers (IEICE) technical report ED88-92:33–40
2. Sols F, Macucci M, Ravaioli U, Hess K (1989) J Appl Phys 66:3892–3906
3. Datta S (1989) Superlattices and Microstructures 6:83–93

4. Miller DC, Lake RK, Datta S, Lundstorm MS, Melloch MR, Reifenger R (1989) In: Reed M, Kirk W (eds) Nanostructure physics and fabrication. Academic, New York pp 165–174
5. Szafer A, Stone AD (1989) Phys Rev Lett 62:300–303
6. Tekman E, Ciraci S (1989) Phys Rev B39:8772–8775
7. Masek J, Lipavsky P, Kramer B (1989) J Phys: Condens Matter 1:6395–6401

Sub-Quantum-Limit Electron Interferometers 7

MASAHIRO KITAGAWA AND MASAHITO UEDA[1]

Summary. We propose a scheme for surpassing the standard quantum limit (SQL) of phase sensitivity in electron interferometry. A nonlinear interferometer with Coulomb coupling between two arms is shown to generate phase-correlated electron states. The minimum detectable phase is improved from the SQL (total electron number)$^{-1/2}$ to (total electron number)$^{-5/6}$ by replacing an input linear coupler of a phase-sensing interferometer with the nonlinear interferometer. A possible experimental scheme in a mesoscopic system is discussed.

Introduction

The ultimate phase sensitivity of an interferometer is limited by quantum uncertainty of the phase difference between the waves fed to both arms of the interferometer. For conventional interferometers, this is given by $\delta\phi = N_0^{-1/2}$—the standard quantum limit (SQL) of the phase correlation with respect to partition of N_0 quanta by linear beamsplitter. For optical interferometers, the phase correlation can be enhanced up to $\delta\phi = 2N_0^{-1}$ by feeding the *squeezed state* to the open input port [1]. (For recent special issues on squeezed light, see [2] and [3]). Yurke has pointed out that the phase sensitivity of fermion interferometers can also reach the same limit if phase-correlated states are used [4, 5]. However, it has been an open question as to how to prepare such states, since a fermion state analogous to the boson squeezed state is not known. In this paper we propose a scheme for generating phase-correlated electron states which achieve sub-quantum-limit phase sensitivity up to $\delta\phi \approx N_0^{-5/6}$ [6].

We consider a nonlinear Mach-Zehnder interferometer (Fig. 1) to generate a pair of states with enhanced phase-correlation. Two quasi-monochromatic electron fields

$$A(z,t) = \frac{1}{\sqrt{L}} \sum_k a_k e^{i(kz-\omega_k t)}$$

$$B(z,t) = \frac{1}{\sqrt{L}} \sum_k b_k e^{i(kz-\omega_k t)}$$

[1]NTT Basic Research Laboratories, Midori-cho 3-9-11, Musashino, Tokyo, 180 Japan

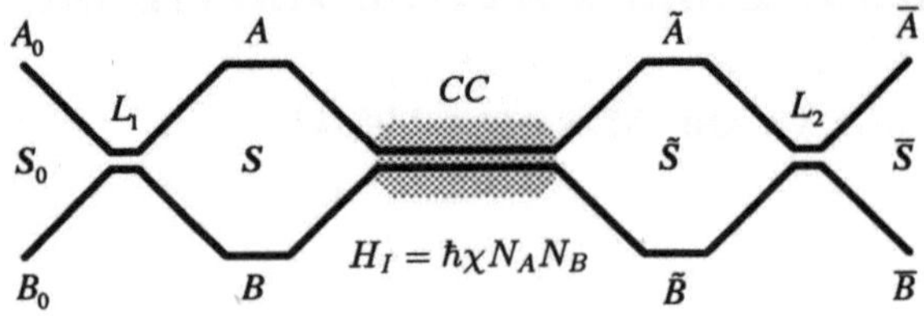

Fig. 1. Schematic illustration of a nonlinear Mach-Zehnder interferometer with a Coulomb coupler as a correlated electron generator. The *solid lines* denote paths for electrons. L_1 is an input linear coupler which splits an input electron field A_0 into A and B. CC is a Coulomb coupler which establishes a quantum-mechanical correlation between A and B. L_2 is an output linear coupler which combines $\tilde{A}$ and $\tilde{B}$ to make suitably correlated outputs $\bar{A}$ and $\bar{B}$. (From [6] with permission)

are propagating and interacting through the interferometer, where L denotes the quantization length. Electron creation and annihilation operators $a_k^\dagger$ and a_k for system A, and $b_k^\dagger$ and b_k for system B obey the anticommutation relations: $\{a_k^\dagger, a_{k'}\} = \delta_{kk'}$, $\{a_k, a_{k'}\} = 0$, $\{b_k^\dagger, b_{k'}\} = \delta_{kk'}$, $\{b_k, b_{k'}\} = 0$, $\{a_k^\dagger, b_{k'}\} = 0$, and $\{a_k, b_{k'}\} = 0$. The various accents in Fig. 1 distinguish the fields and the associated operators at different points. Electrons are fed only into an input port A_0. L_1 is an input linear coupler which divides input electron field A_0 into A and B. CC is a nonlinear coupler which establishes a quantum-mechanical correlation between A and B. The output fields of this coupler, $\tilde{A}$ and $\tilde{B}$, are combined at another linear coupler L_2, yielding phase-correlated outputs $\bar{A}$ and $\bar{B}$.

Spin Representation

We make extensive use of spin notation [4, 5, 7] because it best describes quantum mechanical motion in interferometers. The problem of improving phase sensitivity is then reduced to the squeezing of a spin [8].

The two interacting electron systems can be expressed in terms of abstract spin operators $S = (S_x, S_y, S_z)$,

$$S_+ \equiv \sum_k a_k^\dagger b_k, \quad S_z \equiv \frac{1}{2}(N_A - N_B) \tag{1}$$

where $S_\pm = S_x \pm iS_y$, $N_A = \sum_k a_k^\dagger a_k$, and $N_B = \sum_k b_k^\dagger b_k$. If A_0 has N_0 electrons and B_0 is in the vacuum state, the combined system S_0 is in the simultaneous eigenstate of S^2 and S_z,

$$S^2|S,S\rangle = S(S+1)|S,S\rangle, \quad S_z|S,S\rangle = S|S,S\rangle$$

where $S = \dfrac{N_0}{2}$ [4, 5].

Linear couplers L_1 and L_2 are described by the interaction Hamiltonian,

$$H_I = \hbar g\left(e^{-i\phi}\sum_k a_k^\dagger b_k + e^{i\phi}\sum_k b_k^\dagger a_k\right) = 2\hbar g\boldsymbol{n}\cdot\boldsymbol{S} \tag{2}$$

where $g > 0$ and $n \equiv (\cos\phi, \sin\phi, 0)$. This couples two electron systems as

$$\begin{bmatrix} a'_k \\ b'_k \end{bmatrix} = \begin{bmatrix} \cos\frac{\Omega}{2} & -ie^{-i\phi}\sin\frac{\Omega}{2} \\ -ie^{i\phi}\sin\frac{\Omega}{2} & \cos\frac{\Omega}{2} \end{bmatrix} \begin{bmatrix} a_k \\ b_k \end{bmatrix}$$

where $\Omega = 2g\tau$, with τ being the interaction time. In spin notation, this can be expressed as a rotation of the spin through the angle Ω about the axis n lying in the x–y plane,

$$\boldsymbol{S}' = \exp(i\Omega\boldsymbol{n}\cdot\boldsymbol{S})\boldsymbol{S}\exp(-i\Omega\boldsymbol{n}\cdot\boldsymbol{S}) \tag{3}$$

Standard Quantum Limit

Let L_1 be a 50–50 beamsplitter with $\Omega = \frac{\pi}{2}$ and $n = y$ be a unit vector pointing in the $+y$ direction. Then, L_1 rotates the spin about the y-axis through $\frac{\pi}{2}$, giving an eigenstate of S_x described by the coherent spin state $|\theta, \phi\rangle = \left|\frac{\pi}{2}, 0\right\rangle$, as shown in Fig. 2 a. (For a review of coherent spin states, see [9]). Since $\langle S_z\rangle = 0$, A and B contain the same average number of electrons $\langle N_A\rangle = \langle N_B\rangle = \frac{N_0}{2}$. This is a conventional equal partition of an electron source by a linear coupler. For future reference, let us find correlations in particle number and in wave phase between these two electron systems. The particle number correlation is measured by the standard deviation of the number difference,

$$\delta N \equiv \langle[\Delta(N_A - N_B)]^2\rangle^{1/2} = 2\langle \Delta S_z{}^2\rangle^{1/2} = \sqrt{N_0} \tag{4}$$

while the phase correlation is measured by that of the phase difference,

$$\delta\phi \equiv \langle[\Delta(\phi_A - \phi_B)]^2\rangle^{1/2} \approx \frac{\langle \Delta S_y{}^2\rangle^{1/2}}{|\langle S_x\rangle|} = \frac{1}{\sqrt{N_0}} \tag{5}$$

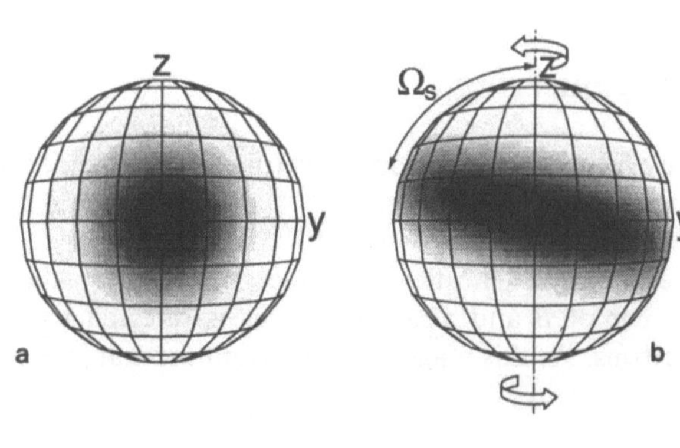

Fig. 2a, b. State evolution in terms of the quasi-probability distribution $Q(\theta, \phi) = |\langle\theta, \phi|\Psi\rangle|^2$. **a** Coherent spin state, and **b** one-axis twisting by $U = \exp[i\mu S_z^2]$ ($\mu = 0.1$), where $S = 20$. The density in each part of the figure is normalized by its maximum value

The minimum detectable phase difference of a phase-sensing interferometer is limited by $\delta\phi$, which is $N_0^{-1/2}$ for a conventional interferometer using a linear coupler as an input partition device. Equations (4) and (5) are referred to as the standard quantum limit of the number-partition noise and that of the interferometric phase sensitivity, respectively. Note that they satisfy the minimum uncertainty relationship

$$\delta N \delta\phi = 1 \tag{6}$$

We will show that $\delta\phi$ can be reduced at the expense of δN by squeezing the spin vector [8] with some kind of nonlinear interaction.

Nonlinear Coupler

A nonlinear coupler *CC* is described by the interaction Hamiltonian,

$$H_I = \hbar\chi N_A N_B = \hbar\chi\left(\frac{N^2}{4} - S_z{}^2\right) \tag{7}$$

This can be obtained through a Coulomb interaction between electrons. Consider two quasi-monochromatic electron fields $A(z, t)$ and $B(z', t)$ along one-dimensional paths—waveguides. If two paths are parallel, then the Coulomb interaction energy between the two fields can be written as

$$H_C = \int_{-L/2}^{L/2} dz \int_{-L/2}^{L/2} dz' \frac{e^2 A^\dagger(z, t) A(z, t) B^\dagger(z', t) B(z', t)}{4\pi\varepsilon\sqrt{(z \mp z')^2 + d^2}}$$

where d denotes the distance between the paths, ε the dielectric constant, and e the electron charge. The interaction length is chosen to be identical to the quantization length L for correlation of the entire fields. The $\mp$ signs in the denominator correspond to co-propagating and counter-propagating schemes, where the coordinates z and z' are taken in the opposite directions for the latter case. H_C consists of H_I and residual terms. H_I is dominant for $d > L$; the coefficient χ approaches $e^2/4\pi\hbar\varepsilon d$ as d/L increases. The coefficient itself can be enhanced by decreasing d, while H_I becomes less dominant as d/L decreases. The counter-propagating scheme is useful for suppressing the residual terms. We will refer to the device expressed by Eq. (7) as a *Coulomb coupler*.

Spin Squeezing

The evolution or electron operators by the interaction H_I is

$$\tilde{a}_k = a_k \exp(-i\mu N_B), \quad \tilde{b}_k = b_k \exp(-i\mu N_A)$$

where $\mu = \chi\tau$. The field A undergoes a phase shift proportional to the particle number of B and *vice versa*. This is referred to as *mutual phase modulation* of electrons. At this stage, the quantum-mechanical correlations are established between the two systems. To see how they can be converted into desirable phase correlations, we again use the spin representation.

The unitary transformation by the Coulomb coupler *CC* can be written in terms of spin as

$$\tilde{\boldsymbol{S}} = \exp(-i\mu S_z^2)\boldsymbol{S}\exp(i\mu S_z^2) \tag{8}$$

The quadratic term S_z^2 in Eq. (7) gives $\tilde{S}_+ = S_+ e^{-i2\mu(S_z+1/2)}$—a rotation proportional to S_z. It *twists* the noise distribution around the z-axis as shown in Fig. 2 b and generates a *squeezed spin state* (SSS) [8]. This is analogous to the *self phase modulation* in photon systems [10]. The magnitude of squeezing is characterized by the parameters

$$\alpha = \frac{N_0}{2}\mu, \quad \beta = \frac{N_0}{2}\mu^2$$

assuming $\alpha \gg 1$ and $\beta \ll 1$. The uncertainties in the y–z plane are now increased by a factor of 2α in the direction angled $\Omega_s = \frac{\pi}{4} + \frac{1}{2}$ arctan α from the z-axis and are reduced by the same factor in the orthogonal direction.

Correlated Outputs

To obtain phase-correlated electrons, the spin must be rotated about the x-axis by $\Omega = -\Omega_S$. This is accomplished at the output linear coupler L_2 by choosing $n = x$ and reflectivity

$$R = \frac{1}{2}\left[1 - \sqrt{\frac{1}{2}\left(1 - \frac{\alpha}{\sqrt{\alpha^2 + 1}}\right)}\right]$$

Then $\langle \Delta \bar{S}_y^2 \rangle$ is reduced and phase-correlated electron states are generated as

$$\delta\phi \approx N_0^{-3/2}\mu^{-1}(1 + N_0^4\mu^6/4)^{1/2} \geqslant \left(\frac{4}{3}\right)^{1/6} N_0^{-5/6} \tag{9}$$

$$\delta N \approx N_0^{3/2}\mu$$

$\delta\phi$ is reduced by a factor of μN_0 from the SQL and δN is increased by the same factor, with the minimum uncertainty relationship in Eq. (6) almost satisfied. The term $N_0^4\mu^6$ in Eq. (9) arises from the *swirliness*—the deviations from a geodesic—of a one-axis squeezed spin state and imposes a limit [8] on the enhancement of the phase-correlation. The minimum $\delta\phi$ is achieved at $\mu = 6^{1/6}N_0^{-2/3}$, where the last equality in Eq. (9) holds.

Discussion and Conclusions

We have shown that a nonlinear Mach-Zehnder interferometer in which two arms are coupled via Coulomb interaction generates phase-correlated electron states.

The interferometric phase sensitivity $\delta\phi$ surpasses the standard quantum limit $N_0^{-1/2}$ and can reach $(\frac{4}{3})^{1/6} N_0^{-5/6}$, if we replace the input linear coupler of a conventional Mach-Zehnder interferometer with the phase-correlated electron generator. Although the sensitivity does not reach the limit $2N_0^{-1}$ exemplified by Yurke [4, 5], further improvement can be expected if we use a more sophisticated nonlinear interaction such as two-axis twisting [8].

In a mesoscopic electron interferometer fabricated with solid waveguides, the Coulomb coupling can be enhanced by placing the two electron waveguides very close to each other, while avoiding linear coupling by separating them with an oxide insulator. This will be a promising structure for sub-quantum-limit phase-sensing interferometers.

References

1. Caves CM (1981) Phys Rev D23:1693
2. Kimble HJ, Walls DF (eds) (1987) J Opt Soc Am B4:1450
3. Loudon R, Knight PL (1987) J Mod Opt 34:709
4. Yurke B (1986) Phys Rev Lett 56:1515
5. Yurke B (1988) Physica (Amsterdam) 151B:286
6. Kitagawa M, Ueda M (1991) Phys Rev Lett 67:1852
7. Yurke B, McCall SL, Klauder JR (1986) Phys Rev A33:4033
8. Kitagawa M, Ueda M In: Musha T, Sato S, Yamamoto M (eds) Proceedings of the international conference on noise in physical systems and 1/f fluctuations (ICNF-91). Ohmsha, Tokyo, pp 355
9. Klauder JR, Skagerstam BS (1985) Coherent states. World Scientific, Singapore
10. Kitagawa M, Yamamoto Y (1986) Phys Rev A34:3974

Phase Velocity Tuning in Quantum Point Contacts

8

SHINGO KATSUMOTO,[1] NAOKATSU SANO,[2] AND SHUN-ICHI KOBAYASHI[1]

Summary. A simple method is proposed to tune the interference by gate voltage in a ring-shaped 2-dimensional electron system at the AlGaAs-GaAs interface. Experimental evidence of the effect is also presented.

Introduction

Quantum interference effects have been widely investigated in micro-fabricated metallic rings [1]. (For review see [2, 3].) Magnetic field is most frequently used to shift the phase of the electrons in these experiments. Another possible parameter for tuning the interference is the electrostatic potential. Washburn et al. [4] observed the phase shift of the interference pattern of an Aharonov-Bohm (AB) metallic ring [5] due to the external electrostatic field. However, the origin of the shift was not clear, since the screening length in the metallic ring was very short. De Vegvar et al. [6] attempted to find the electrostatic AB effect in a ring-shaped 2-dimensional electron gas (2DEG) at the AlGaAs-GaAs interface. In their experiment, a metallic short gate was fabricated over one of the two electron paths. They succeeded in turning the AB effect on and off, although clear oscillatory behavior for the gate voltage was not found because of re-population of the electrons between subbands.

In this article we propose a novel mechanism for tuning the interference in a simple device by applying external voltage, and we present experimental evidence of the effect.

Principles of the Experiment

It is not so easy to pick up the electric field effect on the electron phase. When there are few conductance channels, the applied field changes the current path and the local conductance [7]. When the number of conductance channels is so large

[1]Department of Physics, Faculty of Science, University of Tokyo, 7-3-1 Hongo, Bunkyo-ku, Tokyo, 113 Japan
[2]Department of Physics, Faculty of Science, Kwansei-Gakuin University, Uegahara, Nishinomiya, 662 Japan

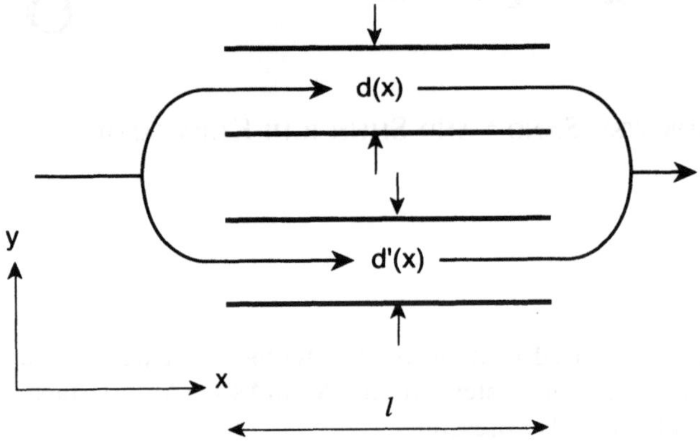

Fig. 1. Schematic layout of an ideal system with two parallel quantum point contacts

that the current path effect is negligible, the screening length should be short, and the voltage effect on the conductance is then very small.

We use quantum point contact (QPC) [8], to avoid the difficulty, or at least to see this change in the number of conductance channels. We adopt here the simplest adiabatic approximation of QPC [9], i.e., we solve the Schrodinger equation in QPCs separately for x- and y-axes for the configuration given in Fig. 1. The transverse (y) solution gives the energies

$$E_n(x) = \frac{h^2}{2m}\left(\frac{n\pi}{2d(x)}\right)^2 \tag{1}$$

where the integer n denotes the mode and d(x) is the local width of the constriction. This means that narrowing of the contacts results in the increase of "effective potential" (EP) for the x-direction. The total kinetic energy does not change with the EP in the adiabatic approximation, but the direction of the k-vector changes. Equation (1) also suggests that the conductance of QPC is quantized by the unit of $2e^2/h$. It is important to note that the conductance of the QPC does not change when the gate voltage, i.e., the EP is varied within a quantized conductance step if the reflection and the tunneling are negligible.

Suppose two QPCs are placed closely and in parallel as illustrated in Fig. 1. Their widths are designed to be slightly different for the same gate voltage. The incoming electron wave with Fermi wave vector is divided into two parts, goes through the QPCs and interferes. For simplicity, we assume each QPC has only one conductance channel. In the EP region, the x-component of the wave vector decreases and the phase velocity increases because the total kinetic energy does not change. Thus the difference in two EPs results in a phase shift at the rejoining point. The phase shift is estimated as

$$\frac{1}{h}\sqrt{\frac{m}{2E_F}}\int_0^1 (E(x) - E'(x))\,dx. \tag{2}$$

Here E(x) and E'(x) are the effective potential energies for two QPCs and l is the length of the QPCs. This shift can be modified through the gate voltage V_G because the EP depends non-linearly on d(x), as can be seen in (1). Hence we can hopefully tune the interference by varying the gate voltage. The important point here is that the conductance of QPC is quantized and constant in a certain range of the gate voltage. Thus we do not have to care about the subband re-population effect or the reflection by the effective potential in the voltage range. The tuning effect should be most distinct when the mode number is 1 for both QPCs, because the period for the oscillation is different for different modes.

Here we have no electrostatic AB effect, because the effective potential is only for the longitudinal motion and the total kinetic energy is not changed by the potential. The voltage tuning effect is somewhat similar to the vector potential AB effect in the sense that the wavevector part of the electron phase is modified.

Experimental Setup

The experimental setup was as follows. A 2DEG at AlGaAs-GaAs interface was prepared by molecular beam epitaxy. The Hall mobility was 9×10^4 cm^2/Vs and the concentration was 2×10^{11} cm^{-2} in the dark. These values were increased to 1.2×10^5 cm^2/Vs and to 5×10^{11} cm^{-2}, respectively, by photo-exposure. The metallic gate configuration is illustrated in Fig. 2. Using electron beam lithography, we removed the regions (shown by hatching in Fig. 2) of the PMMA film with a thickness of about 2000A coated over the surface of the sample. An Au film 4000A thick was deposited over it to make the Schottky gates and also to make a bridge to the central circle-shaped gate [10]. We did not create an intentional imbalance between the two paths, but we expected that error in the fabrication would naturally provide it. The specimen was cooled down to 0.5K in liquid ^{3}He.

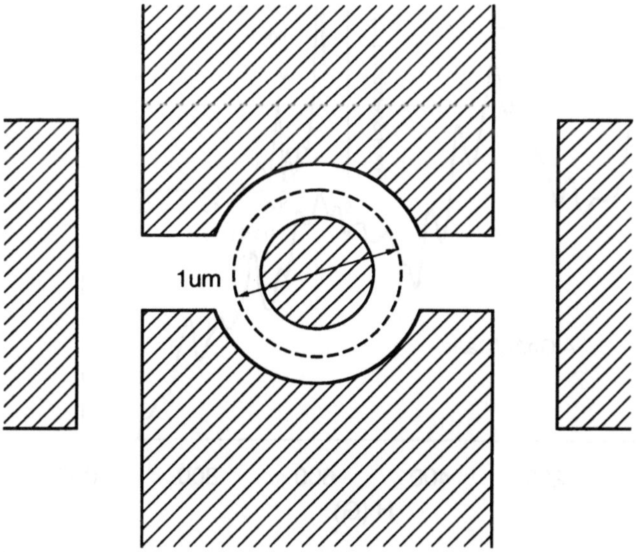

Fig. 2. Metallic gate pattern used in the experiment. The metallic overlayer is attached to the semiconductor surface in the *hatched regions*, which worked as the gates

The resistance was measured by a simple 4-terminal AC bridge at a frequency of 400Hz.

Results and Discussion

Figure 3 shows the gate voltage variation of the resistance. Well-defined conductance quantization was observed only at a range of conductance higher than $7 \times (2e^2/h)$. The appearance of conductance quantization means that the motion of

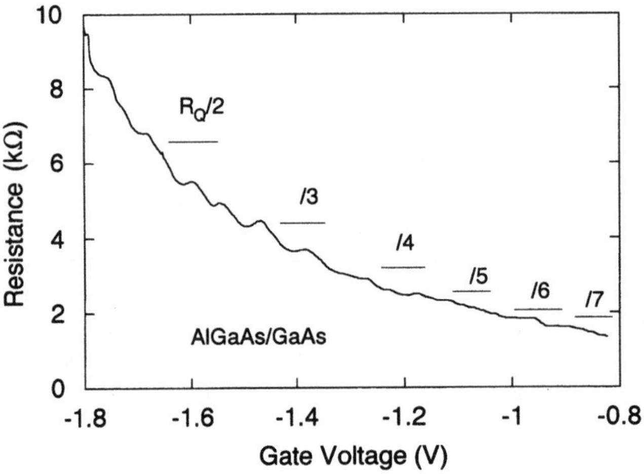

Fig. 3. Gate voltage dependence of the sample resistance. $R_Q = h/2e^2 = 12.95k\Omega$

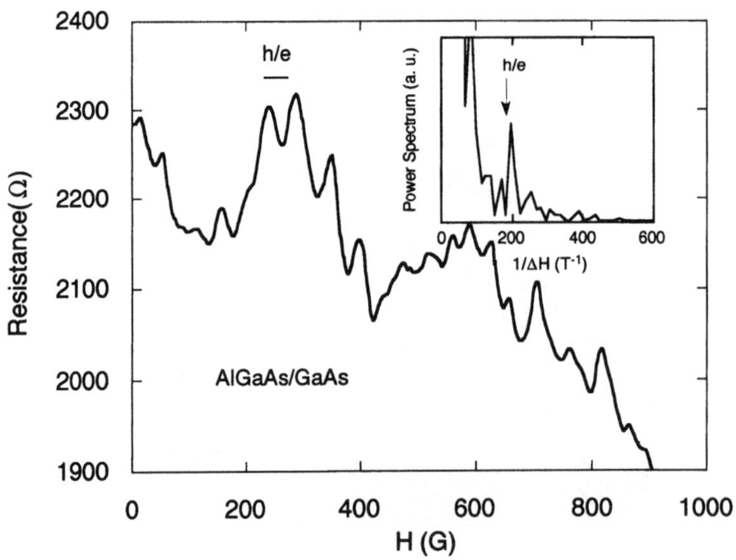

Fig. 4. Weak field magnetoresistance of the 2DEG constricted to the ring. The *inset* shows the Fourier power spectrum in which a peak appeared at the position corresponding to the magnetic flux h/e. *a.u.*, arbitrary units

the electrons through the split gates is nearly ballistic, despite comparatively low mobility [11]. For a lower number of modes the conductance showed a wavy structure, which was probably due to the reflection at the entrance of the QPCs and to the effect of the impurity scattering [12, 13]. The wavy structure appeared near $2e^2/h$. This means that the first conductance channels for two paths opened for different voltages, i.e., there was some imbalance between the two paths. The voltage tuning effect should be distinct when the number of the modes is low. This is because the EP is different for different modes, as can be seen in (1), and the effect should be canceled out when the number of the modes is large. For these reasons we investigated the voltage effect at a conductance range lower than $\sigma = 7 \times (2e^2/h)$.

The magnetoresistance exhibited AB oscillation, a part of which is shown in Fig. 4. The inset shows the Fourier spectrum which has a peak at h/e. This con-

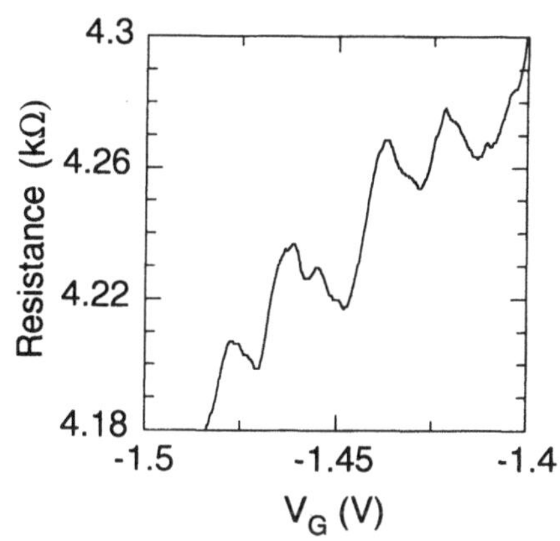

Fig. 5. An example of the resistance oscillation to the gate voltage. The resistance is about $R_Q/3$. The pattern was reproducible

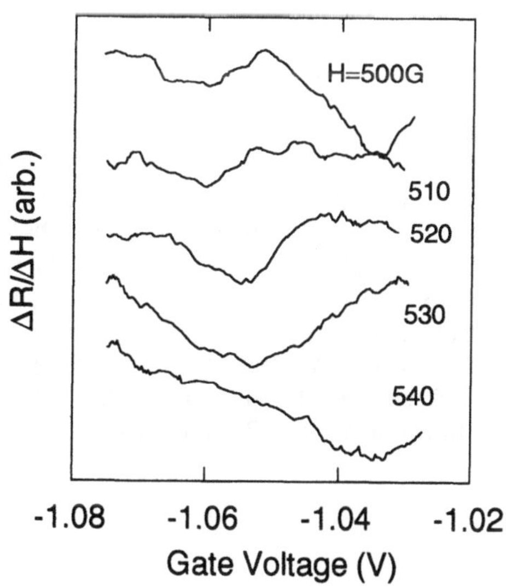

Fig. 6. Gate voltage variance of the differential magnetoresistance for a center magnetic field series. The resistance is about $R_Q/7$

firms that the system is a ring-shaped 2D electron gas. Non-regular structures are probably due to the reflection effect.

We then examined the voltage effect. Figure 5 shows an example of the gate voltage dependence of the conductance. Some oscillation can be seen. Such oscillation was also observed in a single QPC and was attributed to the effect of reflection at the entrance. Although the reflection effect is inevitable in these systems, we can distinguish it from the interference effect by applying a magnetic field.

The shape of the oscillation changed with a small magnetic field comparable with the period of the AB oscillation. However, because of the oscillation due to reflection, the voltage tuned AB effect cannot be seen clearly. In order to eliminate the reflection effect, we measured the differential magnetoresistance by using field modulation. This process emphasizes the interference of higher modes. Reproducible results are shown in Fig. 6. We can see that the pattern shifts with the magnetic field. It needs about 40 G to shift one period, and this is very close to the period of the magnetic AB effect.

Though the oscillation is not very clear, probably because the mode number is not 1 and the oscillation contains several frequencies, we may conclude from the agreement of the periods that we have observed voltage tuning of the interference. However, the result is not distinct enough for further quantitative analysis. In order to observe clearer oscillation, we need to improve the figure and the alignment of the gates to reduce the reflection and we need to perform the experiment with a lower mode number. We are now attempting such experiments.

In conclusion, we propose interference modification by gate voltage in a parallel QPC system and we have presented experimental evidence of this effect.

References

1. Webb RA, Washburn S, Umbach CP, Laibowitz RB (1985) Phys Rev Lett 54:2696
2. Washburn S, Webb RA (1985) Adv Phys 35:375
3. Aronov AG, Sharvin YV (1987) Rev Mod Phys 59:755
4. Washburn S, Schmid H, Kern D, Webb RA (1987) Phys Rev Lett 59:1791
5. Aharonov Y, Bohm D (1959) Phys Rev 115:485
6. De Vegvar PG, Timp G, Mankiewich PM, Behringer R, Cunningham J (1989) Phys Rev B40:391
7. Sivan U, Heiblum M, Umbach CP, Shtrikman H (1990) Phys Rev B41:7939
8. Van Wees BJ, Van Houten H, Beenakker CW, Williamson JG, Kouwenhoven LP, Van der Martel D, Foxon CT (1988) Phys Rev Lett 60:848
9. Yacoby A, Imry Y (1990) Phys Rev B41:5341
10. Ford CJB, Thornton TJ, Newbury R, Pepper M, Ahmed H, Foxon CT, Harris JJ, Roberts CC (1988) J Phys C21:L325
11. Beenakker CWJ, Van Houten H (1991) Solid State Physics 44:1
12. Kirkzenow G (1989) Phys Rev B39:10452
13. Hirayama Y, Saku T, Horikoshi Y (1989) Phys Rev B39:5535

Analysis of Hall Resistance Anomalies with Wave Mechanics

9

Kiyoshi Kawamura, Hiroyuki Sawano, and Tsuyoshi Ueta[1]

Summary. Electrons injected into a semi-infinite plane from a narrow wave guide form two beams as a consequence of zigzag paths in the latter. With this property and the contour maps of electron wave functions, the Hall resistance anomalies are analyzed from the point of view of electron wave mechanics.

Introduction

The Hall resistance anomalies [1, 2] exhibited by devices fabricated on interfaces of GaAs/AlGaAs have been analyzed in terms of the billiard model, that is, in terms of classical electrons pictures [3, 4]. Here, the Hall resistance anomalies mean the quenching of the Hall resistance, the negative Hall resistance, and the enhancement of the Hall resistance, respectively, observed on devices of type 1, 2, and 3, shown in Fig. 1.

In the present paper we discuss the phenomena from the point of view of wave mechanics. Since diffraction is a characteristic property exhibited by waves we first discuss the Fraunhoffer diffraction, in a two-dimensional anisotropic medium on the type 4 device shown in Fig. 1. Electrons, which propagate to the right in the narrow wave guide, form diffracted beams in the semi-infinite plane. The anisotropy is caused by the crystal potential of solids on which these devices are fabricated. It will be pointed out that the boundary condition within the wave guide plays an important role in determining the behavior of electrons after they are emitted in an open space.

The plan of this paper is as follows: The Kirchhoff-Huygens theorem will be derived for tight binding electrons in a magnetic field. Some aspects of the Fraunhoffer diffraction of tight binding electrons will then be reviewed and the Hall resistance anomalies will be discussed.

[1]Department of Physics, Keio University, 3-14-1 Hiyoshi, Kohoku-ku, Yokohama, 223 Japan

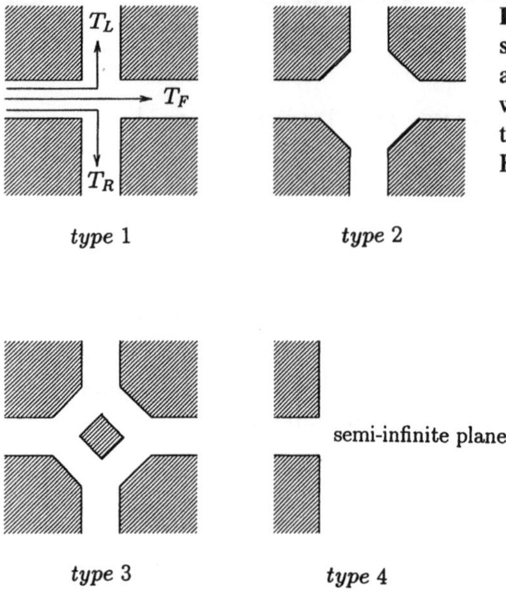

Fig. 1. The four types of devices considered in the present paper. Electrons are injected into the junction from the wave guide on the left-hand side. The type 4 device is considered to represent Fraunhoffer diffraction

Kirchhoff-Huygens Formula

In the present section a model of our electron system and the basic formulation which is used in the following sections will be presented. It is assumed that a wave function of an electron obeys the tight binding equation for a square lattice

$$E\psi(\mathbf{n}) = \sum_i T_{\mathbf{n},\mathbf{n}+\mathbf{a}_i}\psi(\mathbf{n}+\mathbf{a}_i) \tag{1}$$

with appropriate boundary conditions which describe the confinement of electrons within channels (shown in Fig. 1) by potential barriers of infinite height. In the above equation, $\mathbf{n}$ represents the lattice sites and $\mathbf{a}_i$ represents displacements to the nearest neighbors. Throughout the present paper, the lattice constant is employed as the unit of length. The hopping parameter, which is T in the absence of a magnetic field, is modified by it as follows:

$$T_{\mathbf{n},\mathbf{n}+\mathbf{a}_i} = T\exp(2\pi i\tilde{\mathbf{A}}\cdot\mathbf{a}_i) \tag{2}$$

where the vector potential has been scaled by the flux quantum $\phi_o = h/e$ as $\tilde{\mathbf{A}} = \mathbf{A}/\phi_0$. This expression holds when a magnetic flux through a unit cell is much smaller than the flux quanta [5].

To solve Eq. 1 with an arbitrary boundary condition, the Green function is defined by the outgoing solution to the equation

$$\sum_{i=1}^{4} T_{\mathbf{n}'+\mathbf{a}_i,\mathbf{n}'}G(\mathbf{n},\mathbf{n}'+\mathbf{a}_i) + \delta_{\mathbf{n},\mathbf{n}'} = EG(\mathbf{n},\mathbf{n}') \tag{3}$$

Elimination of E from Eqs. 1 and 3 and summation over $\mathbf{n}'$ within a restricted region Ω on the square lattice yield

$$\psi(\mathbf{n}) = \sum_{i=1}^{4} \sum_{\substack{\mathbf{n}' \in \Omega \\ (\mathbf{n}'+\mathbf{a}_i \notin \Omega)}} [G(\mathbf{n},\mathbf{n}')T_{\mathbf{n}',\mathbf{n}'+\mathbf{a}_i}\psi(\mathbf{n}'+\mathbf{a}_i) - G(\mathbf{n},\mathbf{n}'+\mathbf{a}_i)T_{\mathbf{n}'+\mathbf{a}_i,\mathbf{n}'}\psi(\mathbf{n}')] \tag{4}$$

if the site labelled by $\mathbf{n}$ is involved within the region Ω. Since the Taylor expansion of the right-hand side of Eq. 4 with respect to $\mathbf{a}_i$ yields the Kirchhoff-Huygens formula in the presence of a magnetic field [6], eq. 4 is its modified version in the tight-binding approximation [7].

Although one of the present authors (TU) has derived a rigorous expression for the Green function (see T. Ueta, *Green's function in magnetic fields*, Chap. 10 of this volume) for electrons in a two-dimensional free space in the presence of a magnetic field, the semiclassical formula will be employed in the following studies.

Throughout the present paper, we take the boundary condition so that an incident wave is a standing wave in the transverse direction and the mode number is denoted by α, which is equal to the number of loops of the standing wave.

Fraunhoffer Diffraction of Anisotropic Electrons

The diffraction of electron waves, studied by applying Eq. 4 to the type 4 device shown in Fig. 1 d in the absence of a magnetic field, is briefly described in the present section in comparison with the Fraunhoffer diffraction of ordinary light.

When the x-axis is defined by the axis of the wave guide, the ordinary diffraction pattern is represented by a function of y/r which expresses a profile of the electron intensity on a wall perpendicular to the x-axis where the origin of the coordinates is defined by the center of the orifice. In the ordinary theory of Fraunhoffer diffraction, y/r is equal to k_y/k (k represents the wave vector), but the latter is prior to the former in describing the diffraction of electrons whose dispersion relation is anisotropic. The diffraction pattern is given by the formula

$$f_\alpha(k_y) = \frac{1}{2i}\left[\exp\left(i\frac{\alpha\pi}{2}\right)F(k_y - p_{y\alpha}) - \exp\left(-i\frac{\alpha\pi}{2}\right)F(k_y + p_{y\alpha})\right] \tag{5}$$

In this formula, $F(k_y)$ represents the Fraunhoffer diffraction pattern of ordinary light:

$$F(k_y) \simeq \frac{1}{\sqrt{d}}\frac{\sin(k_y d/2)}{k\sin(k_y/2k)} = \frac{1}{\sqrt{d}}\frac{\sin(yd/2)}{k\sin(y/2r)} \tag{6}$$

where d is the width of the slit.

The important feature of this expression is that the right-hand side consists of two terms. This implies that electrons emitted from a narrow wave guide are separated into two beams and that the electron intensity profile in each beam is the same as the Fraunhoffer diffraction pattern as a function of k_y.

The separation of beams is attributed to the zigzag paths of electrons in the wave guide. In the wave guide, electrons move in the longitudinal direction with a constant momentum, while the transverse component of the momentum changes alternately due to collision with two walls. The quantities $p_{y\alpha}$ and $-p_{y\alpha}$ in Eq. 5

are the transverse momenta of electrons which belong to the subband α in the waveguide.

Although we have emphasized that emitted electrons form two beams, there is an exception. More precisely, all the diffraction patterns do have double peaks, except when incident electrons belong to the fundamental transverse mode. The degeneracy for $\alpha = 1$ is the result of the fusion of two broad peaks. In fact, each beam has the width $\delta k = \pi/d$ which is of the same order as the separation $2p_{1y} = 2\pi/d$ of the centers of those two lines, where d is the width of the wave guide.

Transmission Coefficients and the Hall Resistance

The transmission coefficients T_R, T_L, and T_F, which express the transmission probability to the wave guides in the rightward, leftward, and forward directions (see Fig. 1 a) and the Hall resistance, have been evaluated by using the Landauer-Bütticker formula

$$R_H = \frac{h}{e^2} \frac{T_R - T_L}{2T_F(T_F + T_L + T_R) + T_R^2 + T_L^2} \tag{7}$$

The Hall resistance was computed and plotted against the magnetic induction for three types of crosses (Fig. 2). It is seen that the Hall resistance anomalies are well realized.

These behaviors of the Hall resistance are analyzed in terms of wave mechanics in comparison with analysis in terms of the billiard model.

In the billiard model, we think of classical particles injected from a wave guide into a junction from which they cannot turn sidewards easily in a weak magnetic field. The quench of the Hall resistance is attributed to this. However, electrons of high transverse mode have initially large transverse velocity and, consequently,

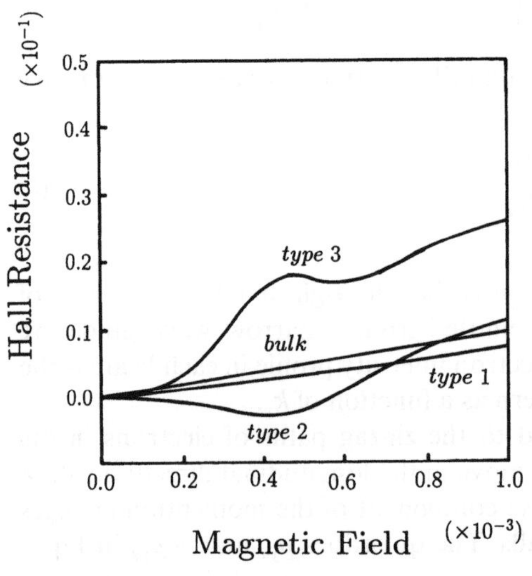

Fig. 2. The Hall resistance of the devices shown in Fig. 1. The magnetic induction is scaled by the flux quantum per area of a unit cell

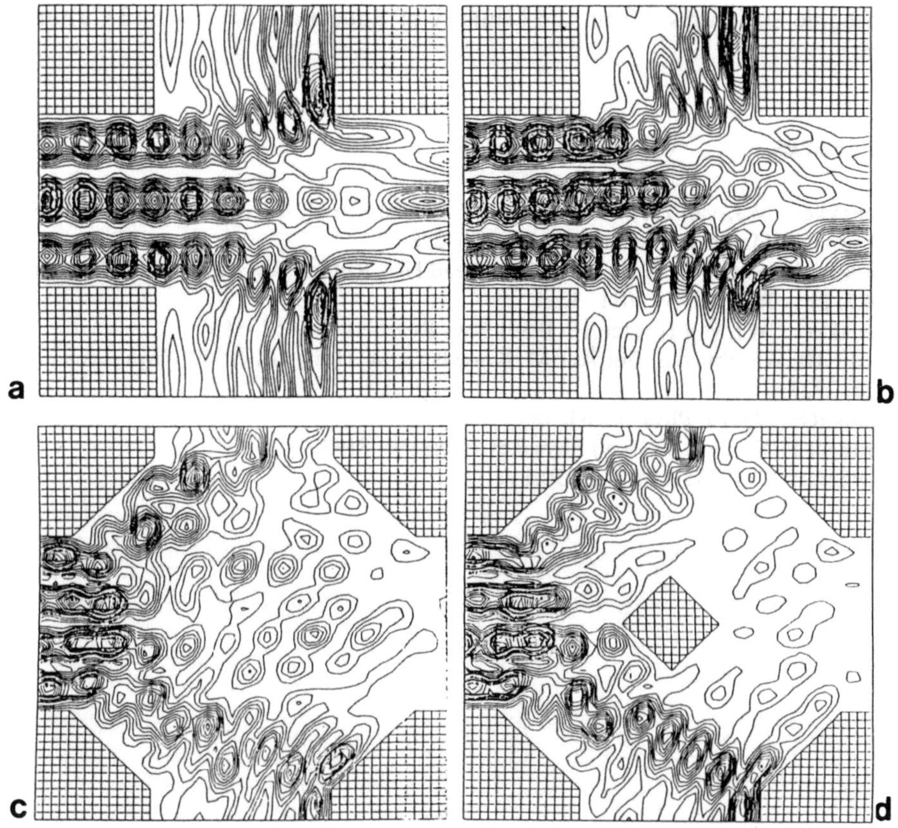

Fig. 3a–d. Example of contours of electron waves for $E/T = 2.0$

they enter into sideward wave guides directly, not being helped by the Lorentz force see Fig. 3 a. Therefore, electrons with large mode numbers do contribute to the Hall resistance, even in type 1 devices (Fig. 3 b).

In the billiard model, we further imagine electrons which enter into the "wrong" probe after reflection at the rounded corner in the cross of a type 2 device, or those which enter into the "right" probe after scattering off the object at the center of the cross of a type 3 device.

All these behaviors imagined in the billiard model are exhibited by electrons of low transverse modes. When high wave modes enter the junction, however, reflection at the corner in the wrong direction does not take place (Fig. 3 c). In such a case the object at the center of the junction does not efficiently scatter electrons to sideward probes (Fig. 3 d).

Acknowledgment. The present work was partially supported by a Grant-in-Aid for Scientific Research on the Priority Area, "Electron Wave Interference Effects in

Mesoscopic Structures," from the Ministry of Education, Science and Culture of Japan.

References

1. Beenakker CWJ, van Houten H (1991) In: Ehrenreich H, Turnbull K (eds) Solid State Physics, vol 44. Academic San Diego, pp 1–228
2. Roukes ML, Sherer A, Allen SJ Jr, Graighead HG, Ruthen RM, Beeke ED, Harbison JP (1987) Phys Rev Lett 59:3011–3014
3. Beenakker CWJ, van Houten H (1989) Phys Rev Lett 63:1857–1860
4. Ford CJB, Washburn S, Büttiker M, Knoedler CM, Hong JM (1989) Phys Rev Lett 62: 2724–2727
5. Peierls RE (1933) Z Phys 80:763–791
6. Baranger H, Stone AD (1989) Phys Rev B40:8169–8193
7. Kawamura K (1978) Z Physik B30:1–11

Green's Function in Magnetic Fields 10

TSUYOSHI UETA[1]

Summary. It is shown that Green's function for a charged particle in a uniform magnetic field is resolved into a gauge-dependent exponential factor breaking the translational symmetry and the translationally symmetric function which is independent of gauge, but influenced by the magnetic field, even if the system contains periodic potential. In the absence of the periodic potential, the exact Green's function is derived. This solution is calculated numerically. For the tight-binding approximation, Green's function is represented by a Fourier transform of the infinite continued fraction. By making use of this Green's function and the Kirchhoff-Huygens theorem, we can apply the "boundary element method" to analyze the quantum transport of a two-dimensional electron gas confined in an arbitrary structured boundary in the presence of a uniform magnetic field.

Introduction

Various quantum phenomena have been observed in mesoscopic systems due to recent rapid progress in semiconductor fabricating techniques. For a ballistic transporting regime in particular, electron focusing [1] and quenching of the Hall resistance in weak magnetic fields [2] have been observed. These experimental results have fundamentally been explained in terms of the classical "billiard model". However, the boundaries of the structure confining the electron gas determine the eigenstates taking part in transport, since in mesoscopic systems the wave functions of an electron are spread throughout the systems. Therefore, it is important to deal with these wave functions by quantum theory.

It is very advantageous to represent the wave functions in terms of boundary integration involving Green's function of a charged particle in a uniform magnetic field [3].

Translational Symmetry Breaking

This section deals with the investigation of the way in which the translational symmetry of the single particle Green's function $G(\mathbf{r}_2, \mathbf{r}_1)$ is broken in the presence of a magnetic field. We consider a charged particle moving under the influence of

[1] Department of Physics, Faculty of Science and Technology, Keio University, 3-14-1 Hiyoshi, Kohoku-ku, Yokohama, 223 Japan

the periodic potential $V(\mathbf{r}) = V(\mathbf{r} + \mathbf{R})$ and a homogeneous magnetic field $\mathbf{B} = \nabla \times \mathbf{A}$, where $\mathbf{R}$ represents the translation vector.

The Hamiltonians for a particle of electric charge q subjected to the magnetic field and magnetic translation operators [4] are taken to be

$$\mathscr{H} = \frac{1}{2m}(\mathbf{P} - q\mathbf{A})^2 + V(\mathbf{r}) \quad \text{and} \quad T(\mathbf{R}) \equiv \exp\left(\frac{i}{\hbar}\mathbf{R}\cdot(\mathbf{P} + q\mathbf{A})\right) \tag{1}$$

Then, we require it to satisfy

$$T(\mathbf{R})f(\mathbf{r}) = \exp\left(\frac{i}{2\hbar}q\mathbf{R}\cdot(\mathbf{B}\times\mathbf{r})\right)f(\mathbf{r}+\mathbf{R}), \quad \text{for any } \mathbf{R} \tag{2}$$

where $f(\mathbf{r})$ is an arbitrary function of $\mathbf{r}$. This condition is satisfied if and only if the vector potential is given by $\mathbf{A} = \frac{1}{2}\mathbf{B} \times \mathbf{r} + \mathbf{A}_0$, where $\mathbf{A}_0$ is an arbitrary constant vector. From now on, we assume this gauge by taking $\mathbf{A}_0 = 0$.

A direct calculation yields an arbitrary function of $\mathbf{P} + q\mathbf{A}$ that commutes with one of $\mathbf{P} - q\mathbf{A}$ and $[\mathscr{H}, T(\mathbf{R})] = 0$ for arbitrary $\mathbf{R}$ [4].

This means that when $|n\rangle$ is an eigenket of $\mathscr{H}$ which has the eigenvalue E_n, $T(\mathbf{R})|n\rangle$ is also the eigenstate of $\mathscr{H}$ which has the same eigenvalue.

The retarded Green's function $G = (E - \mathscr{H} + i\delta)^{-1}$ can then be expanded as

$$\langle\mathbf{r}_2|G|\mathbf{r}_1\rangle = \sum_n \frac{\langle\mathbf{r}_2|T(\mathbf{R})|n\rangle\langle n|T^\dagger(\mathbf{R})|\mathbf{r}_1\rangle}{E - E_n + i\delta} \tag{3}$$

On making use of the fact that

$$\langle\mathbf{r}|T(\mathbf{R})|n\rangle = e^{-i(q/2\hbar)\mathbf{B}\cdot(\mathbf{R}\times\mathbf{r})}\langle\mathbf{r}+\mathbf{R}|n\rangle \tag{4}$$

we find that

$$\langle\mathbf{r}_2|G|\mathbf{r}_1\rangle = e^{i(q/2\hbar)\mathbf{B}\cdot\{\mathbf{R}\times(\mathbf{r}_1-\mathbf{r}_2)\}}\langle\mathbf{r}_2+\mathbf{R}|G|\mathbf{r}_1+\mathbf{R}\rangle \tag{5}$$

Putting $\mathbf{R} = -\mathbf{r}_1$, we obtain

$$\langle\mathbf{r}_2|G|\mathbf{r}_1\rangle = e^{i(q/2\hbar)\mathbf{B}\cdot(\mathbf{r}_1\times\mathbf{r}_2)}\langle\mathbf{r}_2-\mathbf{r}_1|G|\mathbf{0}\rangle \tag{6}$$

where $\mathbf{r}_2$ *is always arbitrary*, whereas $\mathbf{r}_1$ *has to be taken as an arbitrary lattice vector in the presence of a periodic potential.*

Performing the gauge transformation $\mathbf{A} \to \tilde{\mathbf{A}} = \mathbf{A} + \nabla\Lambda$ on Eq. (6) gives

$$\langle\mathbf{r}_2|\tilde{G}|\mathbf{r}_1\rangle = \exp\left(i\frac{q}{\hbar}\int_{\mathbf{r}_1}^{\mathbf{r}_2}\tilde{\mathbf{A}}\cdot d\mathbf{r}\right)\langle\mathbf{r}_2-\mathbf{r}_1|G|\mathbf{0}\rangle \tag{7}$$

where $\tilde{G}$ denotes the Green's function when $\tilde{\mathbf{A}}$ is used in place of $\mathbf{A}$; for convenience we have used the fact that the integration of $\mathbf{A}$ from $\mathbf{r}_1$ to $\mathbf{r}_2$ along the straight path gives $\frac{1}{2}\mathbf{B}\cdot(\mathbf{r}_1 \times \mathbf{r}_2)$. The arrow in the left-hand side denotes that the integration is performed along a straight path.

Thus, it is clear that the factor breaking the translational symmetry of the Green's function of a charged particle under a uniform magnetic field is only the phase factor.

If $\langle\mathbf{r}_2 - \mathbf{r}_1|G|\mathbf{0}\rangle$ in Eq. (7) is replaced by the Green's function in the absence of a magnetic field, we obtain the well-known expressions in the semi-classical approximation [5].

If, further, we put $\mathbf{r}_2$ on a lattice point, we obtain the expression in the tight-binding approximation in the presence of a homogeneous magnetic field.

Exact Green's Function for the Free Case

We calculate the Green's function of a charged particle in a homogeneous magnetic field in the absence of a periodic potential. Here, we choose the vector potential as $\mathbf{A} = \frac{B}{2}(-y, x, 0)$.

The Green's function, then, satisfies the differential equation

$$\left[-\frac{\hbar^2}{2m}\nabla_2{}^2 + \frac{i\hbar qB}{2m}\left(-y_2\frac{\partial}{\partial x_2} + x_2\frac{\partial}{\partial y_2}\right) + \frac{q^2B^2}{2m}\frac{1}{4}(x_2{}^2 + y_2{}^2) - E\right]G(\mathbf{r}_2, \mathbf{r}_1)$$
$$= -\delta(\mathbf{r}_2 - \mathbf{r}_1) \tag{8}$$

where $\mathbf{r}_i = (x_i, y_i)$ $(i = 1, 2)$.

We introduce the dimensionless variables $\mathbf{r} = \mathbf{r}_2/\ell$, $\mathbf{r}' = \mathbf{r}_1/\ell$, $\varepsilon = \dfrac{E + i\delta}{\hbar\omega}$, and $z = \frac{1}{2}(\mathbf{r} - \mathbf{r}')^2$, where ω is the cyclotron frequency and ℓ represents the magnetic length $\sqrt{\hbar/m\omega}$.

In the present unit, the relation of Eq. 6 is written as

$$G(\mathbf{r}, \mathbf{r}') = e^{i(1/2)(x'y - y'x)}G_0(\mathbf{r} - \mathbf{r}')$$

and $G_0(\mathbf{r} - \mathbf{r}')$ obeys the following equation,

$$\left[\frac{\partial}{\partial z}\left(z\frac{\partial}{\partial z}\right) - \frac{1}{4}z + \varepsilon\right]G_0(z) = \frac{m}{2\pi\hbar^2}\delta(z) \tag{9}$$

We here note that the differential operator on the left-hand side of Eq. 9 has the Sturm-Liouville form. The Green's function, therefore, may be calculated systematically [6], if we can get two independent solutions to

$$\left[\frac{\partial}{\partial z}\left(z\frac{\partial}{\partial z}\right) + \left(\varepsilon - \frac{1}{4}z\right)\right]\varphi(z) = 0 \tag{10}$$

The physical condition $\varphi(z)$ vanish as $z \to \infty$. Then, let us consider the boundary condition as $z \to 0$. The limit $z \to 0$ means $(\mathbf{r}_2 - \mathbf{r}_1) \to 0$ or/and $B \to 0$. Therefore, $\varphi(z)$ for $z \to 0$ approaches the Green's function in the absence of a magnetic field, that is the Hankel function, for $(\mathbf{r}_2 - \mathbf{r}_1) \to 0$, so that $\varphi(z)$ diverge logarithmically as $z \to 0$.

Two such independent solutions of Eq. 10 are

$$\varphi^0(z) = e^{-z/2}M(\tfrac{1}{2} - \varepsilon, 1, z) \quad \text{and} \quad \varphi^\infty(z) = e^{-z/2}U(\tfrac{1}{2} - \varepsilon, 1, z)$$

where $U(z)$ and $M(z)$ are defined as follows [7],

$$U(\tfrac{1}{2}-\varepsilon,1,z)=\frac{-1}{\Gamma(\frac{1}{2}-\varepsilon)}\Bigg[M(\tfrac{1}{2}-\varepsilon,1,z)\ln z + \sum_{r=0}^{\infty}\frac{(\frac{1}{2}-\varepsilon)_r z^r}{(r!)^2}\{\psi(\tfrac{1}{2}-\epsilon+r)-2\psi(1+r\}\Bigg] \tag{11}$$

and

$$M(\tfrac{1}{2}-\varepsilon,1,z)=\sum_{n=0}^{\infty}\frac{(\frac{1}{2}-\varepsilon)_n}{(n!)^2}z^n$$

$$(a)_n=a(a+1)(a+2)\cdots(a+n-1),\quad (a)_0=1,\quad \psi(a)=\frac{\Gamma'(a)}{\Gamma(a)} \tag{12}$$

The Wronskian of them is $\Delta(\varphi^0(z),\,\varphi^\infty(z))=-\dfrac{\Gamma(1)}{\Gamma(\frac{1}{2}-\varepsilon)}\dfrac{1}{z}$. On making use of the general procedure [6] to obtain the Green's function corresponding to the Sturm-Liouville equation, our Green's function is written

$$\begin{aligned} G_0(z) &= -\frac{m}{2\pi\hbar^2}\Gamma(\tfrac{1}{2}-\varepsilon)\varphi^\infty(z) \\ &= -\frac{m}{2\pi\hbar^2}\Gamma(\tfrac{1}{2}-\varepsilon)e^{-z/2}U(\tfrac{1}{2}-\varepsilon,1,z) \end{aligned} \tag{13}$$

By means of [7] $W_{\kappa,\mu}(z)=e^{-z/2}z^{\mu+1/2}U(\frac{1}{2}+\mu-\kappa,\ 1+2\mu,\ z)$, the exact Green's function of a free charged particle in a uniform magnetic field is obtained

$$G(\mathbf{r},\mathbf{r}')=e^{i(x'y-y'x)/2}\left(-\frac{m}{2\pi\hbar^2}\right)\frac{\pi}{\cos\pi\varepsilon}\frac{1}{\Gamma(\varepsilon+\frac{1}{2})}\frac{W_{\varepsilon,0}(z)}{\sqrt{z}} \tag{14}$$

where $W_{\kappa,\mu}(z)$ is Whittaker's function.

This expression is also derived by taking the Fourier transform of the Feynman kernel, derived by Glasser in 1964 [8], with respect to relative time.

Several Properties of the Green Function

Symmetry with Respect of Magnetic Field

From the geometrical symmetry, we obtain $G(x_2, y_2, x_1, y_1;\ -B)=G(x_2, -y_2, x_1, -y_1; B)$. Therefore $G_0(z; B)=G_0(z; B)$, since G_0 is a function of $z=\frac{1}{2}(\mathbf{r}-\mathbf{r}')^2$. This means that G_0 does not include information regarding the direction of the magnetic field, and that whether the motion is clockwise or counterclockwise is determined only by the phase factor.

The $B \to 0$ Limit

Making use of the properties 13.3.4 and 13.3.5 in Ref. [6], we can prove

$$\lim_{B\to +0} G(\mathbf{r}_2, \mathbf{r}_1; E) = \begin{cases} -i\dfrac{m}{2\hbar^2} H_0^{(1)}\left(\dfrac{\sqrt{2mE}}{\hbar}|\mathbf{r}_2 - \mathbf{r}_1|\right) & \operatorname{Im} E > 0 \\ i\dfrac{m}{2\hbar^2} H_0^{(2)}\left(\dfrac{\sqrt{2mE}}{\hbar}|\mathbf{r}_2 - \mathbf{r}_1|\right) & \operatorname{Im} E < 0 \end{cases} \tag{15}$$

Asymptotic Expansion and Limiting Forms

Asymptotic expansion and limiting form of the Green function is given as
1) for large z

$$G_0(z;\varepsilon) = -\frac{m}{2\pi\hbar^2}\frac{\pi}{\cos\pi\varepsilon}\frac{1}{\Gamma(\varepsilon+\frac{1}{2})}e^{-z/2}z^{\varepsilon-(1/2)}\left\{\sum_{n=0}^{R-1}\frac{(\frac{1}{2}-\varepsilon)_n{}^2}{n!}(-z)^{-n} + \mathcal{O}(|z|^{-R})\right\} \tag{16}$$

2) for small z

$$G_0(z;\varepsilon) = \frac{m}{2\pi\hbar^2}\{\ln z + \psi(\tfrac{1}{2}-\varepsilon)\} + \mathcal{O}(|z\ln z|) \tag{17}$$

on making use of Eq. 13 and the equations 13.5.2 and 13.5.9 in Ref. [6].

Numerical Calculation

Although the expression in Eq. 14 is useful for investigating the analytical properties of the Green function, it is not practical to use, for instance, the integral representation of the Kummer's function [6] to obtain the numerical values of the Green function. This is because, in this case, we would have to perform numerical integration for each value of z, wasting long computation time to obtain the Green function for changing values of z with sufficiently small intervals.

Another method is to solve Eq. 10 numerically. Necessary boundary conditions are fortunately given if we evaluate Eq. 16 at sufficiently large z. This method gives the Green function with enough accuracy for sufficiently small intervals in z at little calculation cost. This method has another advantage, in that calculation for the differentiation of $G_0(z;\varepsilon)$ is not necessary, since it is simultaneously evaluated in the intermediate step of the calculation of $G_0(z;\varepsilon)$. It has been found that the 4th order Runge-Kutta method is sufficient to calculate this, even for large ε.

Bloch Electrons in Magnetic Fields

The result in the section on translation symmetry breaking is valid for Bloch electrons. It also enables us to derive the expression of the Green's function of a Bloch electron in a magnetic field. From now on, we assume the Landau gauge,

$\mathbf{A} = B(0, x, 0)$. Under the tight-binding approximation, the Green's function in two dimension is defined by

$$G(x+a,y;x',y') + G(x-a,y;x',y') + e^{-i(2\pi\alpha/a)x}G(x,y+a;x',y') + e^{i(2\pi\alpha/a)x}G(x,y-a;x',y') = \varepsilon G(x,y;x',y') - \delta_{x,x'}\delta_{y,y'} \quad (18)$$

, where $\alpha = \dfrac{qBa^2}{\hbar}$, $\varepsilon = \dfrac{E - 4E_0}{T}$, a is the lattice constant, and E_0 and T are site energy and the nearest-neighbor transfer integral, respectively.

By substituting

$$G(x,y;x',y') = e^{i(2\pi\alpha/a)x'(y-y')}G_0(\xi,\eta) \quad \text{and} \quad G_0(\xi,\eta) = \int_{-\pi}^{\pi} e^{ik_y\eta}g(\xi,k_y)\mathrm{d}k_y$$

where $\xi = (x - x')/a$, $\eta = (y - y')/a$, our equation turns into a one-dimensional difference equation:

$$g(\xi+1,k_y) + g(\xi-1,k_y) + [2\cos(k_y - 2\pi\alpha\xi) - \varepsilon]g(\xi,k_y) = -\delta_{\xi,0} \quad (19)$$

We can write all $g(\xi, k_y)$s with continued fractions as follows:

$$\text{for } \xi = 0 \quad g(0) = \frac{1}{\varepsilon - 2\cos k_y - C_{-1} - B_0} \quad (20)$$

$$\text{for } \xi > 0 \quad g(\xi) = \left(\prod_{i=0}^{\xi-1} B_i\right)g(0) \quad (21)$$

$$\text{for } \xi < 0 \quad g(\xi) = \left(\prod_{i=\xi}^{-1} C_i\right)g(0) \quad (22)$$

where

$$C_\xi = \frac{g(\xi,k_y)}{g(\xi+1,k_y)} = \frac{1}{\varepsilon - 2\cos(k_y - 2\pi\alpha\xi) - C_{\xi-1}} \quad (\xi < 0) \quad (23)$$

and

$$B_\xi = \frac{g(\xi+1,k_y)}{g(\xi,k_y)} = \frac{1}{\varepsilon - 2\cos(k_y - 2\pi\alpha(\xi+1)) - B_{\xi+1}} \quad (\xi \geqslant 0) \quad (24)$$

The property of the Green's function is quite different depending on whether α, being the ratio of flux through a lattice cell to one flux quantum, is rational or irrational. When $\alpha = p/q$, the infinite continued fractions C_ξ and B_ξ are indeed periodic in ξ, with period q.

For example, if $\alpha = \frac{1}{2}$,

$$C_\xi = \cfrac{1}{\varepsilon - 2\cos(k_y - \pi\xi) - \cfrac{1}{\varepsilon + 2\cos(k_y - \pi\xi) - C_\xi}} \quad (25)$$

so that C_ξ is obtained by solving the quadratic equation,

$$C_\xi^{\,2} - [\varepsilon + 2\cos(k_y - \pi\xi)]C_\xi + \frac{\varepsilon + 2\cos(k_y - \pi\xi)}{\varepsilon - 2\cos(k_y - \pi\xi)} = 0.$$

For other $\alpha = p/q$ also, C_ξ and B_ξ are determined by quadratic equations, which are equations of degree of q with respect to ε or $\cos k_y$. This fact predicts the splitting of a Bloch band into q subbands which occasionally happen to touch each other [9, 10].

On the other hand, when α is irrational, C_ξ and B_ξ do not have periodicity, so that the quadratic equations which determine C_ξ and B_ξ are equations of infinite degree with respect to ε or $\cos k_y$. Therefore the band structure is infinitely complex. The Green's function has never actually been calculated; this is a subject for future resolution.

Application

The Green's function in a free space is a useful tool for studying wave propagation in a charged particle gas with arbitrary boundary geometry, if it is combined with the Kirchhoff-Huygens theorem

$$\phi(\mathbf{r}) = -\frac{\hbar^2}{2m}\int [G(\mathbf{r},\mathbf{r}')\nabla'\phi(\mathbf{r}') - \phi(\mathbf{r}')\nabla' G(\mathbf{r},\mathbf{r}')]\cdot \mathbf{n}' dS' + \frac{i\hbar q}{m}\int [G(\mathbf{r},\mathbf{r}')A(\mathbf{r}')\phi(\mathbf{r}')]\cdot \mathbf{n}' dS'$$

in the presence of a magnetic field. Here the integral is performed on an arbitrary closed boundary surface which encloses the point $\mathbf{r}$.

The method for the numerical calculation of such an equation is well-known as the boundary element method (BEM) [3]. This method is advantageous in that that it conveniently treats boundary value problems which have arbitrary structured boundaries. It enables us to investigate the influence of geometry in ballistic wires on conductance with full quantum effects.

Acknowledgments. The author would especially like to thank Professor K. Kawamura for his discussions, encouragement, and continual interest in this work.

This work was supported, in part, by a Grant-in-Aid for Scientific Research on the Priority Area, "Electron Wave Interference Effects in Mesoscopic Structures," from the Ministry of Education, Science, and Culture of Japan.

References

1. van Houten H, Beenakker CWJ, Williamson JG, Broekaart MEI, van Loosdrecht PHM (1989) Phys Rev B39:8556–8575
2. Ford CJB, Washburn S, Büttiker M, Knoedler CM, Hong JM (1989) Phys Rev Lett 62:2724–2727
3. Brebbia CA (1978) The boundary element method for engineers. Pentech, London

4. Brown E (1964) Phys Rev 133:A1038–A1044
5. Gor'kov LP (1959) (in Russian) Soviet Phys JETP 9:1364–1367
7. Abramowitz M, Stegun IA (1970) Handbook of mathematical functions. Dover New York, pp 504–510
6. Courant R, Hilbert D (1961) Methods of mathematical physics, vol 2. Interscience, New York
8. Glasser ML (1964) Phys Rev 133:B831–B834
9. Hofstadter DR (1976) Phys Rev B14:2239–2249
10. Rauh A (1975) Phys Stat Sol (b) 69:K9–K13

Electronic Transport in a Quasi-Ballistic Narrow Wire Confined by Split Metal Gates

11

KOJI ISHIBASHI, YOSHINOBU AOYAGI, SUSUMU NAMBA,[1] YUICHI OCHIAI, and MITSUO KAWABE[2]

Summary. A quasi-ballistic narrow channel was confined in a 2 dimensional electron gas (2DEG) formed at the heterointerface between GaAs and AlGaAs in terms of split metallic gates defined by electron beam lithography. Magnetoresistance (MR) measurements were made at 1.2 K and under a magnetic field of up to 8 T. Shubnikov-de Haas (SdH) oscillations and conductance fluctuations were observed. The effect of side wall scatterings on the relaxation process and the width dependence of conductance fluctuations were studied. It was found that (1) in high magnetic fields where edge states begin to be important, scattering with side walls was a dominant relaxation mechanism, and (2) the amplitude of conductance fluctuations began to decrease from a constant value with decreasing width. The observed effects were specific to the quasi-ballistic transport regime.

Introduction

Recent development in crystal growth techniques and microfabrication techniques have made it possible to study electronic transport on a micron scale. Various new effects which are not observable on a macroscopic scale, have been observed in this new field. So far, broad studies have been done in a diffusive regime and in a ballistic regime [1, 2]. In the former regime, quantum interference effects such as universal conductance fluctuations and Aharanov-Bohm magnetoresistance oscillations were observed, while in the latter, ballistic effects such as quantized conductance and quenching of Hall resistance were observed. However, few studies have been reported in the intermediate transport regime, where several scatterings are likely to occur.

In this paper, we address this intermediate regime and investigate how the transition occurs from the diffusive to the ballistic regime. Shubnikov-de Haas (SdH) oscillations were measured in a narrow channel to study the transition of the carrier relaxation mechanism. We also studied the transition of universal conductance fluctuations.

[1] Frontier Research Program, RIKEN, 2-1 Hirosawa, Wako City, Saitama, 351-01 Japan
[2] Institute of Materials Science, University of Tsukuba, Tsukuba, Ibaraki, 351-01 Japan

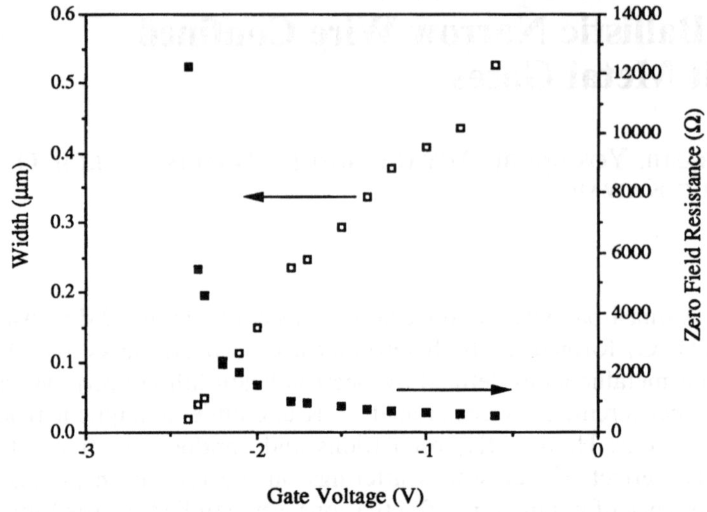

Fig. 1. Gate voltage dependence of the resistance and the width at B = 0 T at 1.2 K

Experimental Technique

The starting wafer of GaAs/AlGaAs heterostructure was grown by molecular beam epitaxy. A narrow channel was confined by metallic split gates formed by the combination of electron beam lithography and lift-off technique. The length of the channel and width between the split gates were about 2 μm and 0.8 μm, respectively. The carrier concentration and the mobility at zero gate voltage were 7.2×10^{15} m^{-2} and 8.1 m^2/Vs, respectively, at 1.2 K, and the corresponding mean free path was 1.1μm. The magnetoresistance (MR) was measured at 1.2 K in a magnetic field of up to 8 T. Figure 1 shows the gate voltage dependence of the resistance and the width of the channel. The 2DEG underneath the gate was depleted at a gate voltage of -0.6 V and the narrow channel was completely depleted at -2.4 V. The width can be roughly estimated from Ohm's law and a zero field resistance by assuming that the resistivity does not change with reductions of the channel width [3]. It should be noted that the carrier density estimated from the period of SdH oscillations does not change above a gate voltage of -2.2 V. Therefore, the width for a gate voltage smaller than -2.2 V is unclear.

Results and Discussion

SdH Oscillations

The amplitude analysis of SdH oscillation gives information about single particle relaxation time. The amplitude can be expressed by the following equation [4],

$$\Delta R \propto \exp(-\pi/\omega_c \tau_f) \tag{1}$$

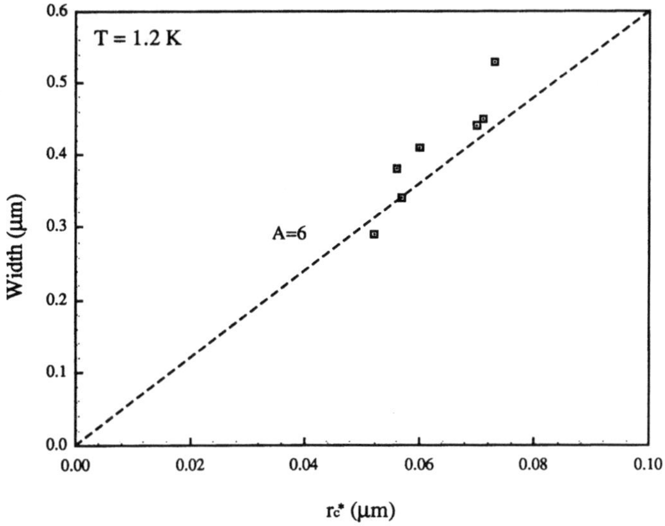

Fig. 2. Relationship between r_c* and the width (*W*)

where ω_c is the cyclotron angular frequency and τ_f is the combination of the single particle relaxation time (τ_s) and the temperature equivalent relaxation time (τ_T). In the present temperature range, τ_f is equal to τ_s. From the plot of ΔR as a function of inverse magnetic field, τ_s can be estimated. As already reported for an etched wire [5], τ_s has two different values in low (τ_{sl}) and high magnetic fields (τ_{sh}), while τ_s for a wide film has a single value in the entire magnetic field range. The boundary field (B_c*) which distinguishes τ_{sl} and τ_{sh} is a function of the channel width. Figure 2 shows the relation between the width (W) and the r_c*, which is the corresponding cyclotron radius of B_c* and is given by $\hbar k_F/eB_c^*$. From the figure, it is noted that r_c* and W have a linear relationship given by

$$W = A\, r_c^* \tag{2}$$

where A is about 6.

We believe these experimental observations are due to changes of electron motion in a wire in a magnetic field. Suppose there are no impurity scatterings in a wire. In a low magnetic field, both a skipping orbit along the edge of the wire and a traversing motion across the wire are possible. But, once the cyclotron orbit becomes smaller than the width, the traversing orbit never occurs and the cyclotron motion is possible in the bulk of the wire. In this case, the transition occurs at the magnetic field given by Eq. 2 and A = 2. When some impurities are present inside a wire, the transition is expected to occur at a higher magnetic field, because interaction between two edge states (left-ward and right-ward) is possible through impurities [6]. Therefore, more field is necessary to separate edge states. In this case, A is larger than 2. "A" is expected to be a measure of the degree of impurity scatterings in a channel, but the mobility dependence should be investigated to confirm this.

Figure 3 shows the width dependence of relaxation times in low (sl) and high

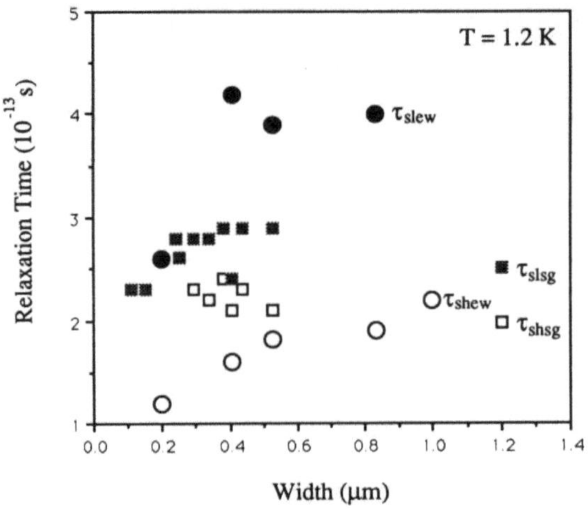

Fig. 3. Width dependence of various relaxation times for etched and split gate wire

(sh) fields determined by using Eq. 1 for the split gate wire (sg) together with those for the etched wire (ew) [7]. In both wires, the relaxation time in the high field is shorter than that in the low field and the difference between τ_{sl} and τ_{sh} in the same wire is larger for the etched wire than for the split gate wire. We might explain these results as follows: The relaxation can occur by scatterings with impurities in the bulk and by diffusive scatterings with edges of the wire. It can be said that the scattering time in the bulk is independent of the magnetic field, as indicated by the experimentally obtained single relaxation time for the wide film in the present magnetic field range. On the other hand, the scattering time in an edge depends on the field, since the interaction between electrons and an edge increases with increases in the magnetic field. In a low magnetic field, relaxation in the bulk is dominant, while relaxation in the edge is dominant in a high magnetic field. Therefore, the shorter relaxation time in a high field indicates frequent diffusive scatterings in the edge of a wire. Because the specularity of the edge scattering in a split gate wire is larger than that in an etched wire [8], the difference between τ_{sl} and τ_{sh} for the split gate wire is smaller than that for the etched wire. If we follow the above considerations, the relaxation time in a high field could have a field dependence, which was not seen in the experiment. Further study is necessary for quantitative agreement.

Conductance Fluctuations

Conductance fluctuations are observed in a low magnetic field where SdH oscillations are small and in a high field where SdH oscillations have maximum values. Figure 4 shows the amplitude of conductance fluctuations (δG) as a function of the channel width for $B = 0$ and $B = 6$ T. The amplitude of the fluctuations is defined by averaged peak to peak value. As seen from the figure, δG is nearly constant for widths larger than 0.3 μm. However, it decreases gradually for widths smaller than 0.3μm.

The fluctuation observed here is due to the quantum interference effect. Conductance fluctuations have been widely studied in metallic systems, where the

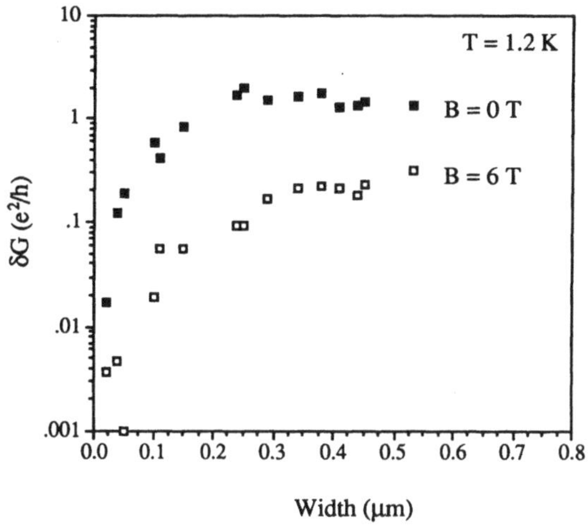

Fig. 4. Width dependence of the amplitude of conductance fluctuations

system size is much larger than the mean free path. In that case it is found, both theoretically and experimentally, that the amplitude of the fluctuation is of the order of e^2/h and does not depend on the sample size and the degree of disorder. In that sense, it is called universal conductance fluctuations (UCF). However, in the case of a quantum wire made from a high mobility semiconductor, it has been pointed out that the amplitude of the fluctuations has a size dependence [9, 10]. In a quantum wire with a small number of impurities, δG could depend upon the number of the transverse mode (N), since correlation between modes is unlikely to occur, which is in contrast to a diffusive transport regime. Clearly, the conductance never fluctuates, but is quantized, to $N x(e^2/h)$, for the ballistic wire. The observed decrease in δG might be due to a decrease in the number of modes.

It should be noted that for a gate voltage smaller than -2.2 V, not only the width but also the carrier concentration decreases, so that the transport regime moves from the metallic to non-metallic. For the non-metallic regime at zero temperature, the conductance is zero and fluctuations are also zero. Therefore, a decrease in δG below a gate voltage of -2.2 Volts could be due to the effect of the small number of modes and the transition to a non-metallic transport regime.

The amplitude of fluctuations at $B = 6$ T is smaller than that at zero field. In a metallic sample, it has been reported to be constant at very large magnetic fields ($B > 8$ T) [1]. However, in a high mobility system, a small magnetic field can change electron trajectories and form edge states, as mentioned before. In a high magnetic field, the conductance approaches the quantized value so that the fluctuations become small [10].

Conclusion

We studied the relaxation mechanism by analyzing SdH oscillations, and we also studied conductance fluctuations in a quasi-ballistic narrow channel. It was found that the main scattering for the relaxation changes with the magnetic field, and that the boundary magnetic field depended upon the width and the degree of disorder. The conductance fluctuations had a width dependence when the number of subbands became small.

Acknowledgment. We thank Dr.J.P.Bird and Mr.T.Onishi of the University of Tsukuba for their help and useful discussions. This work was partially supported by a Grant-in-Aid for Scientific Research on the Priority Area, "Electron Wave Interference Effects in Mesoscopic Structures," from the Ministry of Education, Science, and Culture of Japan.

References

1. Washburn S, Webb RA (1986) Adv Phys 35:375
2. Beenakker CWJ, van Hauten H (1991) Solid State Physics 44:1
3. Ochiai Y, Onishi T, Bird JP, Kawabe M, Ishibashi K, Aoyagi Y, Namba S (1991) Jpn J Appl Phys 30:3859
4. Ando T (1974) J Phys Soc Jpn 37:1233
5. Ochiai Y, Mizuno M, Kawabe M, Ishibashi K, Aoyagi Y, Namba S (1990) Jpn J Appl Phys 29:L739
6. Buttiker M, (1988) Phys Rev B38:9375
7. Ochiai Y, Abe S, Kawabe M, Ishibashi K, Aoyagi Y, Namba S (1991) Phys Rev B43:14750
8. Thornton TJ, Roukes ML, Scherer AL, Van de Gaag BP (1989) Phys Rev Lett 63:2128
9. Higurashi H, Iwabuchi S, Nagaoka Y (1992) Surf Sci 263:382
10. Tamura H, Ando T (1991) Phys Rev B44:1792

Conductance Fluctuations in Quantum Wires: Effects of Strong Spin-Orbit Interaction

12

TSUNEYA ANDO[1]

Summary. Conductance fluctuations of quantum wires are studied numerically in the presence of a strong spin-orbit interaction. When the wire length is much smaller than the localization length, the fluctuation is reduced by a factor close to 1/2 of that in the absence of spin-orbit interaction. When the length is comparable to the localization length, the fluctuation is considerably enhanced.

Introduction

It is well-known that both fluctuations and localization effects are quite sensitive to symmetry or universality classes. Perturbation calculations in the diffusive regime have shown that conductance fluctuations are reduced by $1/\sqrt{2}$ in the presence of magnetic fields (unitary ensemble) versus those in its absence (orthogonal ensemble) and by 1/2 in the presence of strong spin-orbit interaction (symplectic ensemble) [1–4].

In quantum wires, the wire width is much smaller than the mean free path and is comparable to the Fermi wavelength, in contrast to diffusive metallic wires where the width is much larger than both mean free path and Fermi wavelength. Therefore, we can expect quite different fluctuation behavior in quantum wires. The purpose of this paper is to present the results of a numerical study of fluctuations in quantum wires in the presence of strong spin-orbit interactions. We calculate the conductance using Landauer's conductance formula [5] and study its fluctuations as a function of the wire length.

Model

We consider a wire consisting of a square lattice with width M. The equation of motion is given by

$$(E - \mathscr{H}_0)\mathbf{C}_j - V^+\mathbf{C}_{j-1} - V\mathbf{C}_{j+1} = 0 \tag{1}$$

[1] Institute for Solid State Physics, University of Tokyo, 7-22-1 Roppongi, Minato-ku, Tokyo, 106 Japan

where $\mathbf{C}_j$ is a $2M$ component vector describing the amplitudes of the jth cell consisting of M sites and two spin directions. The Hamiltonian $\mathscr{H}_0$ is the $2M \times 2M$ matrix given by

$$\mathscr{H}_0 = \begin{bmatrix} 4V_1 & -V_y & 0 & \dots & 0 \\ -V_y^+ & 4V_1 & -V_y & \dots & 0 \\ 0 & -V_y^+ & 4V_1 & \dots & 0 \\ \vdots & \vdots & \vdots & \ddots & \vdots \\ 0 & 0 & 0 & \dots & 4V_1 \end{bmatrix} \tag{2}$$

and the inter-cell matrix element is given by $V = -V_x$, with V_x and V_y the 2×2 transfer integral, defined by

$$V_x = \begin{pmatrix} V_1 & V_2 \\ -V_2 & V_1 \end{pmatrix}, \quad V_y = \begin{pmatrix} V_1 & -iV_2 \\ -iV_2 & V_1 \end{pmatrix} \tag{3}$$

where V_1 is the spin diagonal and V_2 describes the spin-flip element [6].

In the bulk two-dimensional (2D) system ($M \to \infty$), the energy for wave vector (k_x, k_y) is obtained as an eigenvalue of the matrix:

$$\begin{pmatrix} 4V_1 - 2V_1(\cos k_x a + \cos k_y a) & -2V_2(i \sin k_x a + \sin k_y a) \\ -2V_2(-i \sin k_x a + \sin k_y a) & 4V_1 - 2V_1(\cos k_x a + \cos k_y a) \end{pmatrix} \tag{4}$$

For small $k_x a$ and $k_y a$ or sufficiently large λ_F/a with λ_F the Fermi wavelength in 2D, this becomes

$$\begin{pmatrix} \hbar^2 k^2/2m & -2\hbar^2 k_s(ik_x + k_y)/2m \\ -2\hbar^2 k_s(-ik_x + k_y)/2m & \hbar^2 k^2/2m \end{pmatrix} \tag{5}$$

with

$$V_1 = \frac{\hbar}{2ma^2} = E_F \left(\frac{1}{2\pi}\right)^2 \left(\frac{\lambda_F}{a}\right)^2, \quad V_2 = \frac{\hbar^2 k_s}{2ma} = E_F \frac{1}{2\pi} \frac{\lambda_F}{a} \frac{k_s}{k_F} \tag{6}$$

The energy takes a minimum along the circle defined by $k = k_s$ instead of $k = 0$, where k_F is the Fermi wave vector in 2D. The parameter characterizing the strength of spin-orbit interaction is k_s/k_F.

The present model simulates actual n-channel inversion layers on narrow-band-gap semiconductors or GaAs/AlGaAs heterostructures, where the lack of inversion symmetry leads to a k-linear term in the subband dispersion and then to spin-splitting for nonzero wave vectors [7]. Note that the spin splitting does not mean the removal of the Kramers degeneracy and that states are always doubly degenerate.

Effects of scatterings from impurities are introduced through the randomness of site energy ε_i distributed uniformly with width Q ($-Q/2 \le \varepsilon_i + 4V_1 \le Q/2$). This model corresponds to a wire containing a high concentration of scatterers with δ-potential. The scattering strength is characterized by Λ/λ_F, where Λ is the mean free path in 2D. The conductance is calculated using Landauer's formula and Green's function technique, as described elsewhere [8].

Results

We choose $\lambda_F/a = 8$ and $\Lambda/\lambda_F = 50$. Figure 1 shows a comparison of the results in the presence of strong spin-orbit interaction, $k_s/k_F = 0.2$, with those in its absence, $k_s = 0$, for $2W/\lambda_F = 5.25$ corresponding to channel number 5. Two different averages of the conductance, arithmetic $\langle G \rangle$ and geometric $\exp\langle \ln G \rangle$, begin to differ from each other around the localization length ξ estimated by fitting to $\xi^{-1} = -(1/2)\partial\langle \ln G \rangle/\partial L$, with L being the wire length. The spin-orbit interaction tends to reduce the localization effect drastically (see, for example, Ref. [6] and references cited therein).

In the absence of spin-orbit interaction, the fluctuation increases first, takes a maximum, and begins to decrease gradually with the wire length [9]. In the presence of spin-orbit interaction, the fluctuation is reduced by a factor close to 1/2 for a length region much shorter than the localization length, but with further increase of the length it gradually becomes larger and takes a maximum around the localization length. This feature becomes much clearer for wider wires, as is shown in Fig. 2, and is similar to that in weak magnetic fields [9].

For wires longer than the localization length, the conductance and its fluctuation are essentially the same if the length is scaled by the localization length. This can easily be understood, because the conductance in the strong localization regime consists of sharp peaks at energies of localized states which are distributed almost randomly and have broadening determined by the localization length [10, 11]. The present result suggests that this picture still holds even for lengths comparable to the localization length.

Figure 3 shows the dependence on the strength of spin-orbit interaction. The amount of the reduction in the fluctuations becomes smaller with the decrease of k_s/k_F. For smaller k_s the fluctuation for very short wires in the near-ballistic regime is enhanced. This shows that the spin-orbit interaction is not effective in the ballistic regime and is quite analogous to the effects of a weak magnetic field. The magnetic field does not play a role unless the flux passing through the system becomes much larger than the flux quantum hc/e.

The features appearing in short wires may also be related to the enhancement of fluctuations suggested by Higurashi et al. [12]. The present numerical calculations, however, give essentially no enhancement in the absence of spin-orbit interaction. A possible origin of the disagreement is a difference in the models: Higurashi et al. studied fluctuations at a "band center" where lattice effects may manifest themselves most strongly, in contrast to the present model, which simulates a continuum system quite well.

Acknowledgments. The author would like to thank H. Tamura for his valuable discussions. This work is supported, in part, by the Industry-University Joint Research Program "Mesoscopic Electronics" and by Grants-in-Aid for Scientific Research on the Priority Areas, "Electron Wave Interference Effects in Mesoscopic Structures" and "Computational Physics as a New Frontier in Condensed Matter Research," from the Ministry of Education, Science, and Culture of Japan.

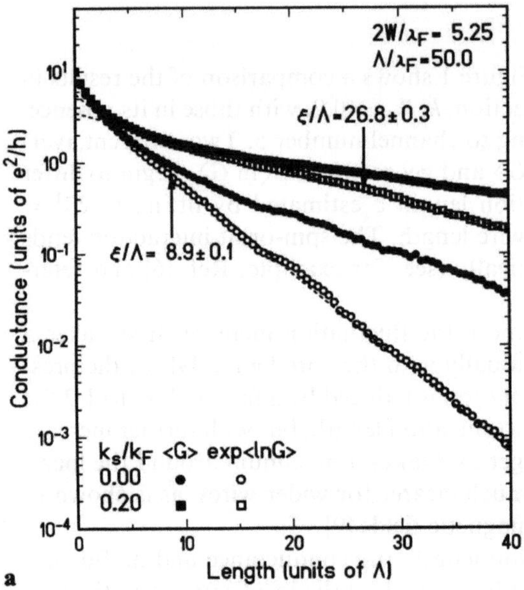

a

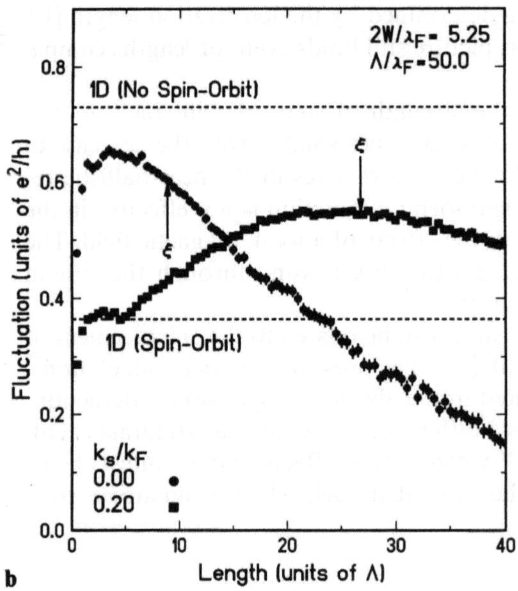

b

Fig. 1a, b. Calculated conductance: **a** and fluctuation; **b** as a function of length for wires with width $2W/\lambda_F = 5.25$ in the presence ($k_s/k_F = 0.2$) and absence ($k_s = 0$) of spin-orbit interaction. The *vertical arrows* indicate the localization length ξ

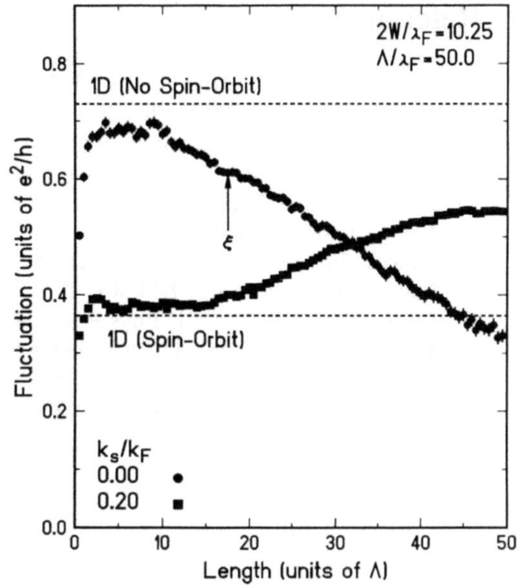

Fig. 2. Calculated conductance fluctuation as a function of length for wires with width $2W/\lambda_F = 10.25$ in the presence ($k_s/k_F = 0.2$) and absence ($k_s = 0$) of spin-orbit interaction. The *vertical arrows* indicate the localization length ξ

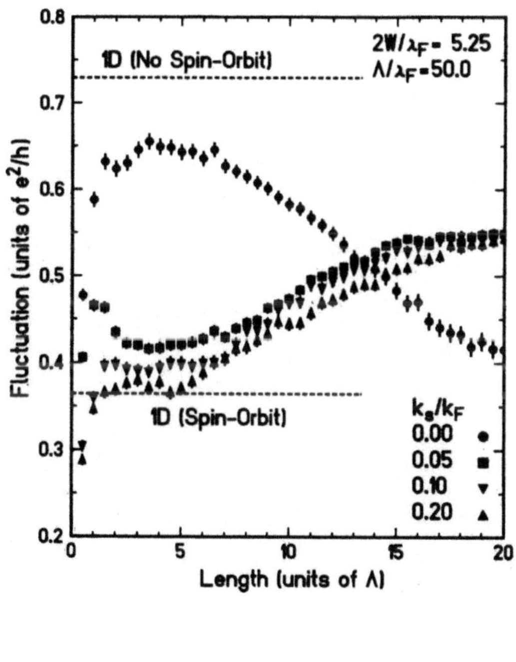

Fig. 3. Calculated conductance fluctuation as a function of length for wires with width $2W/\lambda_F = 5.25$ for different strengths of spin-orbit interaction

References

1. Altshuler BL (1985) JETP Lett 41:648
2. Lee PA, Stone AD (1985) Phys Rev Lett 55:1622
3. Lee PA, Stone AD, Fukuyama H (1987) Phys Rev B35:6701
4. Kane CL, Serota RA, Lee PA (1988) Phys Rev B37:6701
5. Landauer R (1957) IBM J Res Dev 1:223
6. Ando T (1989) Phys Rev B40:5325
7. Ando T, Fowler AB, Stern F (1982) Rev Mod Phys 54:437
8. Ando T (1991) Phys Rev B44:8017
9. Tamura H, Ando T (1991) Phys Rev B44:1792
10. Anderson PW, Thouless DJ, Abrahams E, Fisher DS (1980) Phys Rev B22:3519
11. Tankei K, Sawada A, Nagaoka Y (1989) J Phys Soc Jpn 58:368
12. Higurashi H, Iwabuchi S, Nagaoka Y (1992) Surf Sci 263:382

Quantum Transport

Mesoscopic Phenomena in Magnetotransport Measurements on Two-Dimensional Systems

13

Klaus von Klitzing[1]

Summary. Magnetotransport measurements on two-dimensional electronic systems demonstrate new physical phenomena when the sample size becomes smaller than certain characteristic lengthscales such as the electronic localization length in a disordered system or the equilibration length for the electrochemical potential in adjacent edge channels. The experimental results show that, in general, the measured voltages and currents cannot be directly related to bulk parameters, such as the resistivity or the conductivity tensor components, of the infinite homogeneous system. The measurements show that the finite size of the devices modifies the metal-insulator transition for energies close to the center of a Landau level, as well as the energy dissipation due to backscattering. The experimental data can be interpreted on the basis of the edge channel picture for adiabatic transport and the theory of localized states of a two-dimensional system in strong magnetic fields.

Introduction

Until the discovery of the quantum Hall effect (QHE), most magnetotransport measurements on two-dimensional electron systems were analyzed on the basis of the magnetic field dependent components σ_{xx}, ρ_{xx}, and ρ_{xy} of the conductivity and resistivity tensor usually determined from transport measurements on devices with Corbino- and Hall geometries [1]. The quantum oscillations $\sigma_{xx}(B)$ observed in silicon *metal-oxide semiconductor field-effect transistors* (MOSFETs) with relatively low electron mobilities, $\mu < 10^4 cm^2/Vs$, (short range scatterers) could be especially well described on the basis of the equation [2]

$$\sigma_{xx} \sim (N + \tfrac{1}{2}) \cdot D^2(E) \tag{1}$$

which depends exclusively on the Landau quantum number N and the density of states D(E) at the Fermi energy. A small amount of localized states in the tails of the Landau levels were assumed in order to explain the finite magnetic field ranges with $\sigma_{xx} = 0$ (quantum Hall regime). However, the main part of the σ_{xx}-curves, for devices with short range scatterers, could be interpreted on the basis of Eq. (1) with peak values $\sigma_{xx}^{max} = \dfrac{2e^2}{\pi \cdot h}(N + \tfrac{1}{2})$.

[1] Max-Planck-Institut für Festkörperforschung, Heisenbergstr. 1, W-7000 Stuttgart 80, Germany

The relative importance of localization has changed dramatically in the recent past. Measurements at very low temperatures show that not only a small amount of electrons in the tails of the Landau levels, but nearly all of them, are localized at zero temperature, except those at the center of a Landau level [3–5]. This result can be easily understood in the percolation picture with electronic states extended along equipotential lines in the limit of strong magnetic fields (magnetic length l_0 much smaller than the range of the impurity potential). In a disordered system, most of the equipotential lines form closed loops and therefore localized states. The localization length l of these states diverges at the percolation threshold, which is located energetically close to the center of each Landau level [6–8]. For an infinitely large system at $T = 0K$, δ-function like spikes in the $\sigma_{xx}(B)$- and $\rho_{xx}(B)$-curves are expected when the Fermi energy coincides with the extended states. Simultaneously, the Hall resisitivity $\rho_{xy}(B)$ increases steplike with an infinite slope $d\rho_{xy}/dB$ between the Hall plateaus. Such δ- and steplike behaviour smears out in systems of finite size, since electronic states with a localization length larger than the sample size are effectively delocalized. Conductivity measurements on a Corbino device give direct information about the extended states at the Fermi energy.

For measurements on devices with Hall geometry, two different types of extended equipotential lines at the Fermi energy and, therefore, current carrying states are involved. Close to the edge of the sample, the energy of the Landau levels is raised so that all occupied Landau levels cross the Fermi energy. The corresponding edge channels (the number of these channels is identical to the number of occupied Landau levels) are extended along the whole boundary and are used to discuss the quantum Hall effect within the Landauer-Büttiker formalism [9]. Figure 1 is a sketch of a rectangular two-dimensional device with the

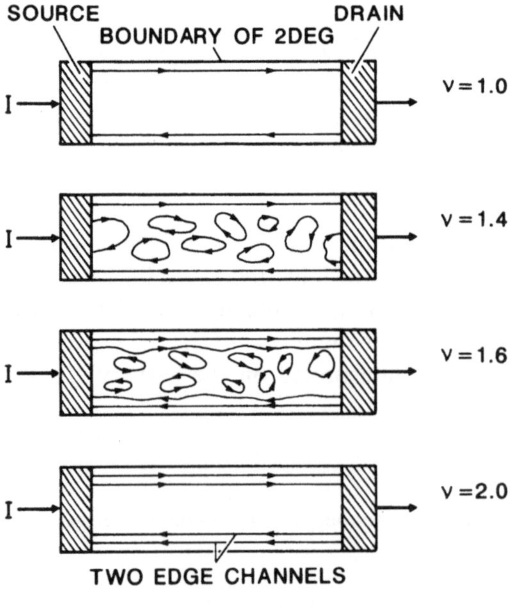

Fig. 1. Equipotential lines at the Fermi energy for different filling factors ν of the Landau levels. *2DEG*, two-dimensional electron gas

source and drain contacts as metallic reservoirs. The equipotential lines for electronic states at the Fermi energy are plotted in the range $1 \leq \nu \leq 2$ for the Landau level filling factor ν.

At integer filling factors, the two-terminal resistance is determined exclusively by the edge channels (quantized Hall resistance) [10, 11]. The experimental results are equivalent to ballistic transport measurements on an ideal one-dimensional system [12, 13]. The suppression of backscattering between the one-dimensional channels with opposite k-values is easily realized in QHE-experiments, since these channels are spatially separated. In this respect, the edge channels of a two-dimensional system in a strong magnetic field can be treated as ideal one-dimensional channels. Backscattering is only possible if the edge channels on one side of the sample communicate with the edge channels on the opposite side. This scattering process leads to increased resistance and energy dissipation, usually observed as a finite resistivity ρ_{xx}. The next section discusses the influence of edge channels on magnetotransport and the application of adiabatic transport for a new type of infrared photoconductor, while the following section examines the delocalization of electronic states due to the finite size of the system or as a result of the finite phase coherence length at nonzero temperatures.

Edge Channel Transport

Magnetotransport measurements on a two-dimensional system (Si-MOSFET) which showed an unexpected behavior of the Shubnikov-de Haas oscillations as a function of the sample size were published as early as 1984 [14]. A typical result is shown in Fig. 2. In this figure the four terminal resistance of a device with a large aspect ratio is plotted. The experimental results do not scale with the length and width of the device, so that the shape of the device not only between the potential probes but also outside this active region (nonlocal effects) must be known for an interpretation of the data (see Fig. 6). The most pronounced anomaly in Fig. 2

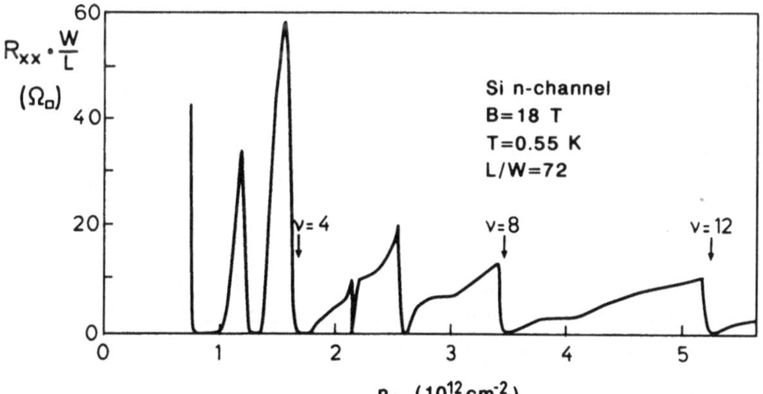

Fig. 2. Experimental curve for the four terminal resistance R_{xx} of a long silicon MOSFET ($L/W = 72$) at a magnetic field $B = 18T$ as a function of the carrier density n_s

[compared to the expected curve on the basis of Eq. (1)] is the triangular lineshape and the strongly reduced peak values of the measured resistance. This result can be understood on the basis of edge channels and adiabatic transport which leads to different local chemical potentials for the edge channels [15–19]. Especially, the uppermost occupied Landau level N^{max}, which is responsible for backscattering and the observed four terminal resistances, can easily be decoupled from the edge channels of the lower lying Landau levels $N \leq N^{max} - 1$. In the absence of sufficient interchannel scattering most of the current is carried (without energy dissipation) by these edge channels, whereas the uppermost level, which could contribute to the backscattering, is electrically decoupled from the current carrying edge channels. This leads to a reduction in the energy dissipation. The strength of the interchannel scattering rate can be expressed as an inverse equilibration length l_{eq} for the electrochemical potential in adjacent edge channels. An experimental arrangement for a quantitative determination of this characteristic length l_{eq}, which can influence transport measurements on Hall devices if the size of the sample is smaller than l_{eq}, will be discussed in the following paragraph.

Interchannel Scattering in the Quantum Hall Regime

The quantum Hall regime is characterized by the absence of backscattering (no extended states at the Fermi energy across the width of the device) and all experimental results can be quantitatively described within the edge channel picture [20]. Within this approach, each channel flowing out of a reservoir j (metallic contact at the boundary of the sample with an electrochemical μ_j) carries the current $I_{out}(j) = e \cdot \mu_j/h$. The direction of the edge currents depends exclusively on the direction of the magnetic field. Normally, each of the adjacent edge channels (ν channels for a filling factor ν) carries the same current, since all of them originate from the same reservoir with the same electrochemical potential. However, a gate across the sample, which reduces the carrier density to a filling factor $\nu_g < \nu$ (index g for gate), leads to a selective reflection of the uppermost $\nu - \nu_g$ channels, whereas ν_g channels are transmitted [21]. In this way the electrochemical potentials of adjacent channels can be fixed at *different* values as shown in Fig. 3a. The two edge channels between the two gates (upper edge) originate from different reservoirs with electrochemical potentials μ_1 and μ_2. Depending on the distance $\Delta x = d$ between the two gates, relative to the equilibration length l_{eq}, different values for the electrochemical potential μ_3 can be established. In the limit $l_{eq} \gg d$ (adiabatic transport) a value $\mu_3 = \mu_1$ is obtained, whereas for strong interchannel scattering ($l_{eq} \ll d$) the mean value $\mu_3 = \frac{\mu_1 + \mu_2}{2}$ is expected. In general, the resistance $R_{12,23}$ (where the subscripts 12, 23 indicate that the current source is connected to the contacts 1 and 2 and the voltage is measured between the contacts 2 and 3) can be expressed as a function of the equilibration length,:

$$R_{12,23} = \frac{h}{2e^2}\left(1 + exp\frac{-2d}{l_{eq}}\right) \tag{2}$$

The length-dependence of this resistance originates from the change in the proba-

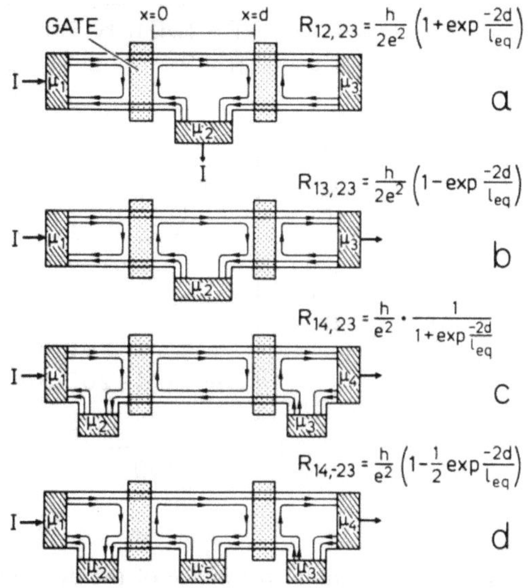

Fig. 3a–d. Edge channel picture for different devices with gate barriers. The reservoirs are characterized by the electrochemical potentials μ_i; the filling factors are $\nu = \nu_b = 2$ (bulk-regions) and $\nu_g = 1$ (gate-regions). The calculated four terminal resistances depend on the distance d between the gate barriers relative to the equilibration length l_{eq}

bility $P_1(x) = \frac{1}{2} + \frac{1}{2} exp - \frac{2x}{l_{eq}}$ to find an electron along one channel ($P_1(x = 0) = 1$). The corresponding probability P_2 to find the electron in the adjacent channel is therefore $P_2(x) = \frac{1}{2} - \frac{1}{2} exp - \frac{2x}{l_{eq}}$. Depending on the arrangement of the contacts, quite different resistances can be calculated as shown in Fig. 3b–3d. All these devices can be used to determine the equilibration length.

For a system with more than two channels such a direct determination of l_{eq} is not possible as long as no assumptions are made for the different interchannel scattering rates.

The situation of two edge channels in the bulk and one edge channel below the gates has been investigated in detail [22] and Fig. 4 shows a result obtained on the structure shown in Fig. 3c. For a distance $d = 20\mu m$ between the gates nearly ideal adiabatic transport is found $\left(R_{14,23} \approx \frac{h}{2e^2}\right)$, so that from the small deviation from the quantized value only a very inaccurate value for l_{eq} can be calculated ($300\mu m < l_{eq} < 1000\mu m$). The measurements on a device with a spacing $d = 200\mu m$ between the gates show a clear contribution of the interchannel scattering on the resistance, and the calculated equilibration length is $l_{eq} = 940\mu m$. The inset of Fig. 4 shows the temperature dependence of the inverse equilibration length which is proportional to the spin-flip rate. The equilibration length saturates at temperatures below T = 250 mK and the corresponding value is interpreted in terms of the spin-orbit interaction, which also leads to a local g-factor enhancement at the edge.

The equilibration lengths are much smaller for devices with smaller mobilities so that not only intrinsic properties of the two-dimensional system determine the

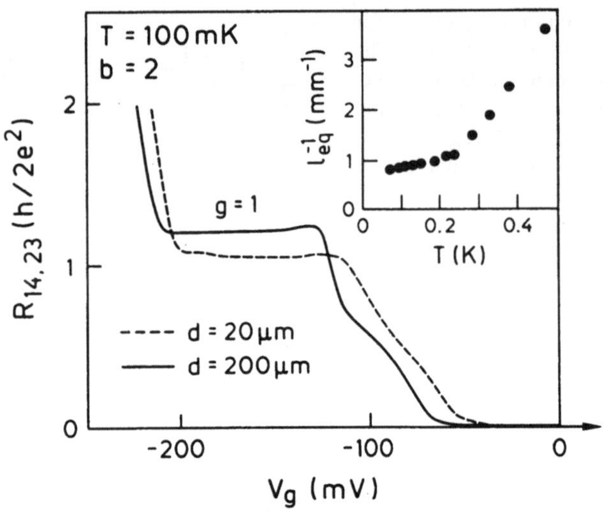

Fig. 4. Four terminal resistance of the device shown in Fig. 3c as a functon of the gate voltage V_g. The filling factor $\nu_g = 1$ is realized at $V_g \approx -170mV$. Results for two different spacings d between the gates are shown. The *inset* shows the temperature dependence of the inverse equilibration length for $d = 200\mu m$ [22]

interchannel scattering. However, a simple connection between the mobility of a device and the equilibration length is not possible.

Adiabatic transport can always be destroyed if metallic contacts are connected to the sample [19]. Measurements on a device such as that shown in Fig. 3d demonstrated that the resistance $R_{14,23}$ could be switched from the adiabatic value $\frac{h}{2e^2}(l_{eq} \gg d)$ to the equilibrated value $\frac{h}{e^2}$ if an additional reservoir was connected to the upper edge between the two gates. This means that additional contacts (=reservoirs which enforce equilibrium) between potential probes can be used to test whether adiabatic transport contributes to the measured resistance. Such adiabatic transport is not only observed in samples where edge channels are selectively reflected, as shown in Fig. 3, but it is also important in measurements on conventional devices with Hall bar geometry [23], especially if the length to width ratio of the device is large. The anomalous Shubnikov-de Haas oscillations in Fig. 2 can be explained on the basis of adiabatic transport.

Influence of Adiabatic Transport on Shubnikov-de Haas Oscillations

Magnetotransport measurements on standard Hall devices show that the measured resistance can be strongly reduced due to adiabatic transport. The experimental results in Fig. 5 demonstrate that the contacts 5 and 6 (which enforce equilibrium at these points between all edge channels for $V_{gb} = +0.3V$) drastically influence the experimental data [24]. A disconnection of these contacts (negative

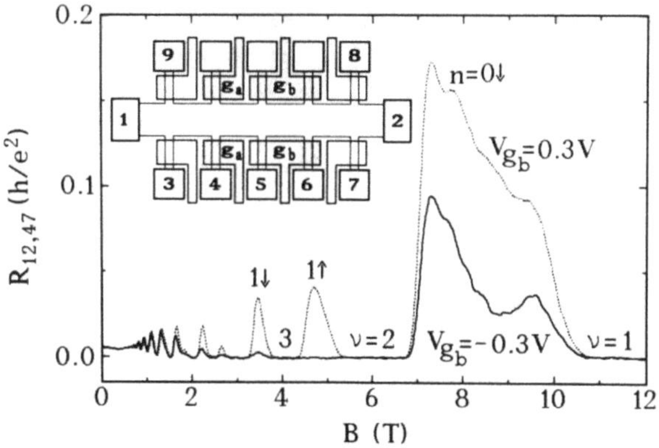

Fig. 5. Four terminal resistance $R_{12,47}$ of a GaAs/AlGaAs heterostructure for different gate voltages V_{gb}, which influence the coupling of the reservoirs to the two-dimensional system. The geometry of the device is sketched in the *inset*

gate voltage across the leads to the reservoirs) leads to much smaller peak values for the resistance in strong magnetic fields. It is not necessary to disconnect (electrically) the reservoirs completely by applying a negative gate voltage larger than the pinch-off voltage. Already a gate voltage $V_{gb} = -0.3V$, which disconnects the uppermost (partly occupied) Landau level N^{max} from the reservoir, avoids a breakdown of the adiabatic transport at these contacts. This means that all edge states originating from the fully occupied Landau levels $N < N^{max}$ are in equilibrium (characterized by the same electrochemical potential), whereas the uppermost Landau level N^{max} is only weakly coupled to these edge channels. Thus, strong backscattering within this Landau level electrically decouples this level from the edge channels.

The properties of backscattering can be simulated using a gate barrier with a finite transmission probability t for the uppermost Landau level. The transmission probability varies from $t = 0$, for an empty Landau level, up to $t = 1$ for a fully occupied level. These changes in t correspond to changing the resistivity ρ_0 for this level from infinite to zero [23]

$$\rho_0(N^{max}) \cdot \frac{L}{W} = \frac{h}{e^2}(1 - t)/t \qquad (3)$$

The resistivity ρ_0 is an intrinsic property of the two-dimensional system, whereas the transmission probability depends on the length to width ratio of each of the segments of the device. The segments are defined by the reservoirs (potential or current probes) which fix the equilibrium between all edge channels.

On the basis of this model (no interchannel scattering between the uppermost channel and the other channels), one can calculate the resistances for different aspect ratios L/W of a device. For simplicity, one assumes that the variation of the resistivity ρ_0 as a function of the filling factor $\Delta\nu$ of the uppermost Landau level is given by Eq. 3 with the relation $t = \Delta\nu$ for $L/W = 1$.

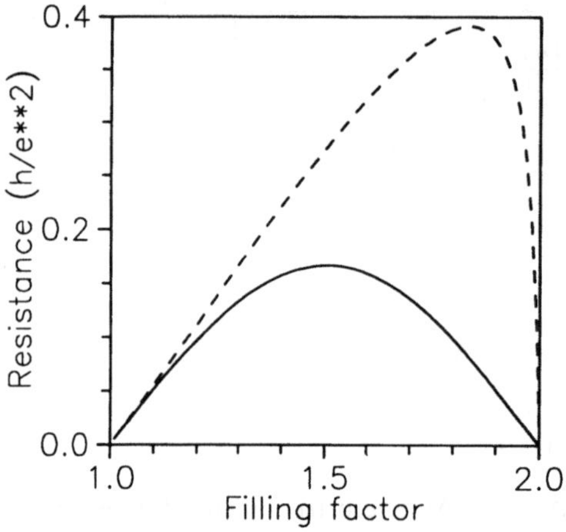

Fig. 6. Calculated four terminal resistance for a long device ($L/W = 72$, *dotted line*) and a short device ($L/W = 2$, *unbroken line*) as a function of the filling factor

Figure 6 shows the result for two samples. The dotted line corresponds to a device where the length between the potential probes is increased by a factor 36 relative to the short device (unbroken line) which had a length to width ratio of two. All other geometries are identical: $L/W = 4$ for the regions between the current and voltage probes and $L/W = 1$ for the arms of the potential probes. These parameters are the design-parameters of the sample used in Fig. 2. The triangular lineshape of the Shubnikov-de Haas oscillation for the long device is very similar to the experimental result shown in Fig. 2. Also, the fact that the resistance does not increase linearly with the length of the device is clearly demonstrated.

The analysis of the curves in Fig. 2 indicates that not only the uppermost energy level is decoupled from the rest of the edge channels, but that a group of four channels (in the region $8 < \nu < 12$) or two channels ($4 < \nu < 6$ or $6 < \nu < 8$) should be treated as one partly occupied uppermost level. This indicates that the interchannel scattering between valley- and spinsplit levels is relatively strong, which is a result of the relatively small energy gap between these levels.

It should be noted that a "normal" resistance (scaling with the length to width ratio) is observed at currents above, typically, 1 μA and temperatures above 4 K [25]. Under these conditions, the interchannel scattering between all of the edge channels becomes so strong that within the experimentally accessible length scale of some μm no nonlocal phenomena are observed. The strong sensitivity of the resistance in the adiabatic regime on both the variation in the interchannel scattering rate and the transmission probability opens the possibility of using the device as a far infrared (FIR) detector, as discussed in the following section [26].

Influence of FIR-Absoption on Adiabatic Transport

Magnetotransport measurements on devices as sketched in Fig. 3c have shown that the resistance increases by a factor of two if the transport changes from adiabatic transport $\left(l_{eq} = \infty,\ R_{12,34} = \dfrac{h}{2e^2}\right)$ to equilibrated transport $\left(l_{eq} = 0,\right.$

$R_{12,34} = \frac{h}{e^2}\Big)$. Cyclotron resonance absorption corresponds to electronic transitions between adjacent Landau levels, which can be interpreted as an interchannel scattering process. The experimental results in Fig. 7 demonstrate that only under the condition of adiabatic transport (negative gate voltage on *both* gates) is a large photosignal observed [26]. Since adiabatic transport can be observed not only on devices with two gate barriers across the current path, but also on devices with strong backscattering and a large length to width ratio, experiments similar to that shown in Fig. 7 have been repeated on long devices without gate barriers for backscattering. The curves in Fig. 8 show that the resistance R is influenced by adiabatic transport, which is visible as a strongly asymmetric lineshape of the spin-split Landau level $N = 2$ (curve b). Under FIR-illumination with a wavelength of $\lambda = 392 \mu m$, a drastic increase in the resistance is observed in the magnet-

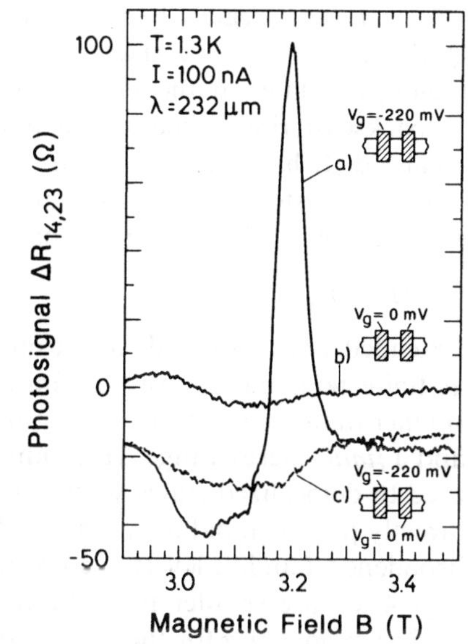

Fig. 7. Photosignal $\Delta R_{14,23}$ (device geometry see Fig. 3c) under FIR illumination in the magnetic field range of cyclotron resonance absorption. A large signal is only observed if a negative gate voltage of $V_g = -220 mV$ (corresponding to $\nu_g = 1$) is applied on *both* gates

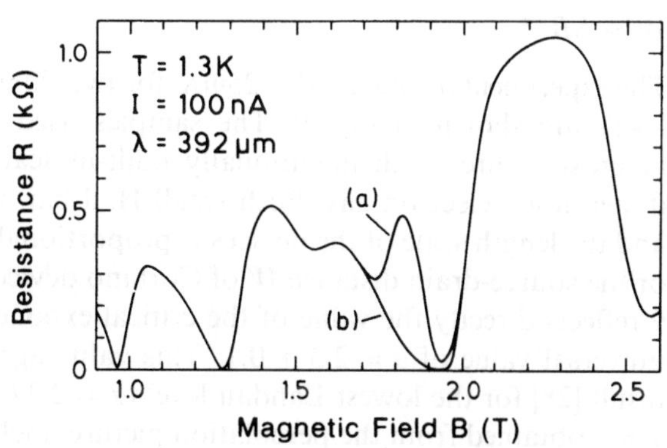

Fig. 8. Magnetotransport data of a GaAs/AlGaAs heterostructure with (*curve a*) and without (*curve b*) FIR absorption

ic field range of cyclotron resonance absorption (curve a). The photosignal is reduced by an order of magnitude if the device current is increased to $I = 1\mu A$ (equilibrated transport). All experiments show that the FIR-absorption process drastically changes the interchannel scattering rate. In principle, an increase in the electron temperature due to the absorption process or a change in the filling factor (quasi Fermi levels for each of the Landau levels involved in the resonance) can produce a photosignal, but for a quantitative interpretation of the data, a more direct influence of the FIR-radiation on the interchannel scattering rate is necessary. Even without a microscopic interpretation of the photosignal, it is clear that the presented data show one of the first applications of a mesoscopic system—application as a FIR-detector.

Metal-Insulator Transition in the Quantum Hall Regime

The transition region between Hall plateaus and the finite conductivity σ_{xx} in QHE-experiments is connected with a metal-insulator transition. In the plateau region, all states at the Fermi energy which can contribute to the conductivity (=backscattering in measurements on samples with Hall bar configuration) are localized. It has been shown that the localization length ξ of a two-dimensional electronic system in strong magnetic fields is always finite, except for energies E_N at the center of the impurity-broadened Landau levels [6–8]:

$$\xi \sim |E - E_N|^{-\alpha} \tag{4}$$

Extended states exist only at the energies E_N. This statement requires modification at finite temperatures or for finite system sizes. If the size W of the sample is smaller than the localization length, the electronic states behave as extended states and a *finite* energy range ΔE around the center of the Landau levels can be attributed to delocalized states with a finite σ_{xx}. The half width ΔB of the σ_{xx}-peaks is directly proportional to ΔE, since the density of states close to the maxima of the broadened Landau levels is constant. (The energy range ΔE and the thermal energy 4 kT are smaller than the halfwidth of the Landau levels.) Therefore, the critical exponent α for the localization length can be determined directly from the halfwidths ΔB of the σ_{xx}-peaks and the size W of the sample [27]

$$W \sim \Delta B^{-\alpha} \tag{5}$$

The experimental data, at $T = 25mK$, for two different devices and different Landau levels are shown in Fig. 9. The samples used are low-mobility GaAs/AlGaAs heterostructures with intentionally built-in scatterers in the region of the two-dimensional electron gas. Both small Hall bar and Corbino geometries are used and the length scale of the devices is proportional to the widths W of the Hall bars or the source-drain distance W of Corbino devices. The slope of the curves in Fig. 9 reflects directly the value of the critical exponent, and all devices give the same universal value of $\alpha = 2.3 \pm 0.1$. This value agrees very well with the numerical result [28] for the lowest Landau level ($\alpha = 2.34 \pm 0.04$), and the calculated value $\alpha = \frac{7}{3}$ obtained from the percolation picture, including quantum tunneling [29].

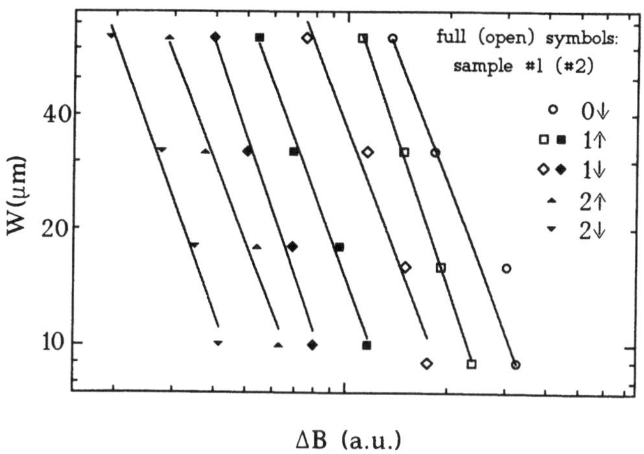

Fig. 9. Halfwidth ΔB of the Shubnikov-de Haas peaks in the low temperature limit $T < 30mK$ as a function of the width W of the device

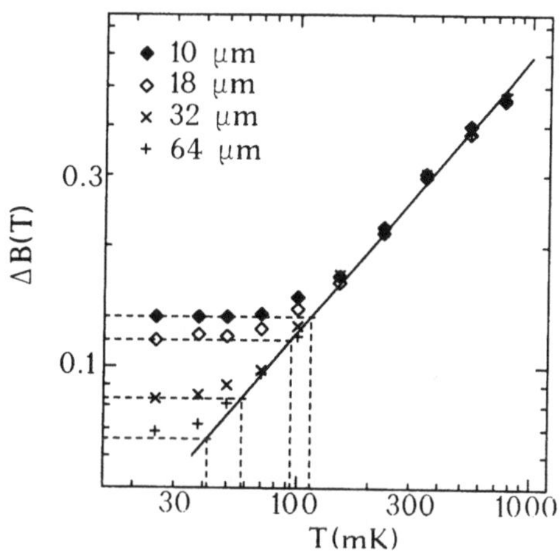

Fig. 10. Temperature dependence of the Shubnikov-de Haas peak at $\nu \approx 3.5$ for different sample sizes

An increase in the halfwidth ΔB for the conductivity peaks is not only observed if the size of the sample is reduced but also if the temperature is increased. Figure 10 shows the experimentally determined temperature-dependent halfwidths for different devices, of different sizes, from the same wafer (W = widths of the Hall bar, the distance between the potential probes changes proportional to the width). At temperatures $T > 100mK$ all data points follow a common curve

$$\Delta B \sim T^{\kappa} \tag{6}$$

The critical exponent κ is related to the exponents α and p via $\kappa = p/2\alpha$, with p the temperature exponent of the inelastic scattering rate [30]. The corresponding inelastic mean free path is determined by $l_{in} \sim T^{-p/2}$ and characterizes the size of the system. The analysis of the curves allows an independent determination of the exponents α and p. All measurements show that the localization exponent is universal, whereas the inelastic scattering rate exponent p is influenced by the specific form of the random potential and changes from $p = 1.2$ to $p = 3.8$ with increasing impurity concentration in the plane of the two-dimensional system.

Conclusion

Magnetotransport measurements on two-dimensional systems show that, very often, the measured resistances do not scale with the size of the device as expected from Ohm's law. At very low temperatures ($T < 100mK$), the inelastic mean free path of the electrons may exceed the size of the sample, resulting in a size-dependent energy range of delocalized states close to the center of each Landau level. This phenomenon changes the width of the Shubnikov-de Haas peaks, whereas the amplitude of these peaks can be strongly influenced by the adiabatic transport. This reduction of the energy dissipation is visible in measurements on devices with Hall bar geometry and can be explained within the edge channel picture.

Acknowledgments. This work has been partly supported by the BMFT under contract number TK 0366/5.

References

1. Kawaji S, Igarashi T, Wakabayashi J (1975) Progress of Theoretical Physics, Suppl 57:176
2. Ando T, Matsumoto Y, Uemura Y, Kobayashi M, Komatsubara K (1972) J Phys Soc 32:859
3. Ebert G, v Klitzing K, Probst C, Ploog K (1982) Solid State Commun 44:95
4. Wei HP, Tsui DC, Paalanen MA, Pruisken AMM (1988) Phys Rev Lett 62:1294
5. Koch S, Haug RJ, v Klitzing K, Ploog K (1991) Phys Rev B43:6828
6. Aoki H (1977) J Phys C10:2583
7. Ono Y (1982) J Phys Soc Jpn 51:2055, 3544
8. Trugman SA (1983) Phys Rev B27:7539
9. Landauer R (1957) IBM J Res Rev 1:223; Büttiker (1986) Phys Rev Lett 57:1761
10. Streda P, Kucera J, MacDonald AH (1987) Phys Rev Lett 59:1973
11. Büttiker M (1988) Phys Rev B38:9375
12. v Wees BJ, v Houten H, Beenakker CWJ, Williamson JG, Kouwenhoven LP, v d Marel D, Foxon CT (1988) Phys Rev Lett 60:848
13. Wharam DA, Thornton TJ, Newbury R, Pepper M, Ahmed H, Frost JEF, Hanko DG, Peacock DC, Ritchie DA, Jones GAC (1988) J Phys C21:L209
14. v Klitzing K, Ebert G, Kleinmichel N, Obloh H, Dorda G, Weimann G (1985) In: Chadi JD, Harrison WA (eds) Proc seventeenth int conf on physics of semiconductors, San Francisco. Springer, New York, p 271
15. Haug RJ, v Klitzing K (1989) Europhys Lett 10:489
16. v Wees BJ, Kouwenhoven LP, Willems EMM, Harmans CJPM, Mooij JE, v Houten H, Beenakker CWJ, Williamson JG, Foxon CT (1991) Phys Rev B43:12 431
17. Komiyama S, Hirai H, Sasa S, Hiyamizu S (1989) Phys Rev 40:12 566

18. Alphenaar BW, McEuen PL, Wheeler RG, Sacks RN (1990) Phys Rev Lett 64:677
19. Müller G, Weiss D, Koch S, v Klitzing K, Nickel H, Schlapp W, Lösch R (1990) Phys Rev B42:7633
20. v Klitzing K (1990) In: Rössler U (ed) Advances in solid state physics, vol 30. Vieweg, Braunschweig, p 25
21. Haug RJ, MacDonald AH, Streda P, v Klitzing K (1988) Phys Rev Lett 61:2797
22. Müller G, Weiss D, Khaetskii AV, v Klitzing K, Koch S, Nickel H, Schlapp W, Lösch R (1992) Phys Rev B45:3932
23. McEuen PL, Szafer A, Richter CA, Alphenaar BW, Jain JK, Stone AD, Wheeler RG, Sacks RN (1990) Phys Rev Lett 64:2062
24. v Klitzing K, Nieder J, Haug RJ, Müller G, Koch S, Weiss D, Ploog K (1992) Semicond Sci Technol 7:82
25. Müller G, Weiss D, Koch S, v Klitzing K, Nickel H, Schlapp W, Lösch R (1990) In: Anastassakis EM, Joannopoulos JD (eds) Proc of the 20th int conf on physics of semiconductors, Thessaloniki, Greece. World Scientific, Singapore, p 829
26. Dießel E, Müller G, Weiss D, v Klitzing K, Ploog K, Nickel H, Schlapp W, Lösch R (1991) Appl Phys Lett 58:2231
27. Koch S, Haug RJ, v Klitzing K, Ploog K (1991) Phys Rev Lett 67:883
28. Huckestein B, Kramer B (1990) Phys Rev Lett 64:1437
29. Mil'nikov GV, Sokolov IM (1988) JETP Lett 48:536
30. Wei NP, Tsui DC, Paalanen MA, Pruisken AMM (1988) Phys Rev Lett 61:1294

Effect of Scattering on Ballistic Transport in GaAs/GaAlAs Heterostructures: Magnetotransport in Anti-Dot Lattice

14

JUNNICHI TAKAHARA, KENJI GAMO, SUSUMU NAMBA[1]*, AND KAZUO MURASE[2]

Summary. Magnetoresistance of a hexagonal array of anti-dots is investigated. Several magnetoresistance peaks are observed and the results are in good agreement with simple geometrical calculation. However, the calculation cannot explain the relative peak intensities at lower carrier concentration.

Introduction

There has been an increasing interest in electron wave devices. The principle of electron wave devices is based on the wave nature of electrons and utilizes quantum interference effects. For such devices, scattering and device geometry have a decisive influence on the performance, as optical guided wave devices and device size are limited by elastic and inelastic scattering length. We have been investigating scattering effects in wires and crossed junctions [1]. For quantum wires, it was observed that both impurity and boundary scattering limit elastic mean free path and that boundary scattering depends on fabrication techniques and sample geometry.

Remote ionized donors, surface roughness, and defects distribute randomly in crystals, thus elastic scattering is considered to be a stochastic process [2]. However, such recent advances in crystal growth techniques as molecular beam epitaxy (MBE) have made it possible to realize two-dimensional electron gas (2DEG) systems with a very high mobility in GaAs-AlGaAs heterostructures [3]. The elastic mean free path is longer than 10μm and electrons travel without suffering scattering for a long distance. Recent nanostructure fabrication technology enables one to introduce artificial potential modulations or scatteres into high mobility samples, where modulation periodicity is shorter than the mean free path. Thus we are able to perform a systematic modification of the transport properties.

A novel low-field magnetoresistance oscillation was discovered by Weiss et al. in a 1D lateral superlattice [4]. They introduced weak potential modulation ($V \ll E_F$, where V is magnitude of potential and E_F Fermi energy) in 2DEG using holographic illumination. This experiment has attracted much interest in the trans-

[1] Faculty of Engineering Science and Research Center for Extreme Materials and
[2] Faculty of Science, Osaka University, Toyonaka, Osaka, 560 Japan
*Nagasaki Institute of Applied Science, 536 Amibacho, Nagasaki-shi, Nagasaki, 851-01 Japan

port phenomena of periodic structures in a magnetic field, and experiments for many kinds of potential configurations and potential profiles have been performed. In contrast to the weak modulation case, transport properties are affected differently by strongly modulated ($V > E_F$) potential, where a strong potential is called "anti-dot" [5–7]. The interplay of the two characteristic lengths, cyclotron radius $r_c = \hbar k_F/eB$ and periodicity of modulation potential a, is important to explain these phenomena.

In this paper, we report on our investigations of the magnetotransport of 2DEG in the presence of an array of hexagonal anti-dots [8, 9]. Anti-dots were introduced into a high mobility heterojunction by electron beam (EB) lithography and dry etching. We can control potential profile around anti-dots by light emitting diode (LED) illumination, which changes the carrier concentration. Several magnetoresistance peaks were observed in the low magnetic field. These were attributed to commensurate classical orbits in hexagonal lattices, but simple geometrical calculation cannot explain the relative peak intensities.

Fabrication and Experiment

The starting materials were high-mobility GaAs-AlGaAs heterostructures grown by the MBE technique, as shown in Fig. 1a. The mobility of the substrate was $130 m^2/Vs$ and carrier concentration was $3.2 \times 10^{15}/m^2$ at 4.2 K in the dark. The corresponding elastic mean free path is $13 \mu m$.

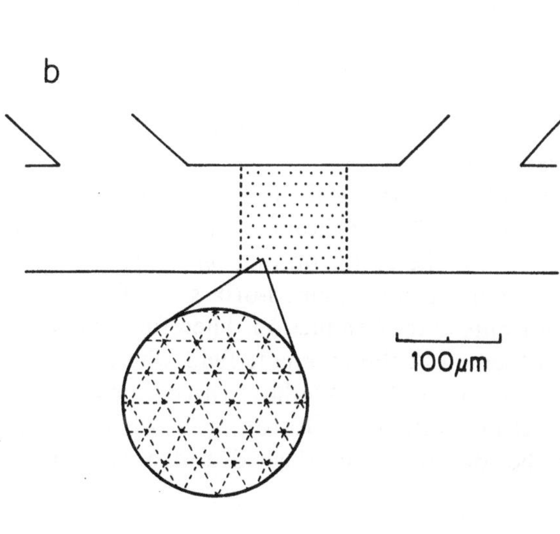

Fig. 1. a Structure of the heterojunction. **b** Schematic view of the sample

A schematic view of the sample is shown in Fig. 1b. The interval of anti-dots is 0.8 and 1.0μm, and the radius of an anti-dot is 0.1μm. The sample fabrication process was as follows: First, 100μm-wide four-terminal shaped mesa structures were defined by wet-etching. Au/Ge layers were evaporated and alloyed to form current and potential probe contacts. Dot patterns were delineated on EB resist polymethylmethacrylate (PMMA) by a 50keV EB lithography system, and the exposed resist was developed by a conventional wet process. Finally, the samples were shallow etched using 1keV Ar ion to remove a depth of about 30nm. We observe that the removal of the heterostructure surface by 30nm induces an almost completely depleted region [10], resulting in the scatterers.

Measurements were done at temperatures ranging from 1.4 to 50K in the dark, and after illumination using an ac bridge operating at 30Hz. LED (wavelength 920nm) or white-light was used to control carrier concentration by illumination.

Results and Discussion

Experimental Results

Figure 2 shows the low-field resistivity after white-light illumination. Shubnikov-de Haas (SdH) oscillations were observed at the higher field. At the lower field, resistivity increased and several peaks were observed. Although the SdH oscillations were quenched at 5K, the low-field peaks were clearly observed even above 10K. In particular, the largest peaks around 0.2T were still observed at 50K (not shown). From the zero-field resistivity, we obtained the mobility of 17 and 20$m^2V^{-1}s^{-1}$ for $a = 0.8$ (Fig. 2a) and 1μm (Fig. 2b), respectively. The decrease of the mobility at the lower field is due to the imposed scatterers. At the magnetic field of 0.4T, the resistivity was sharply decreased. Thus we obtained the mobility of 102 (Fig. 2a) and 122 (Fig. 2b) $m^2V^{-1}s^{-1}$, respectively, from the minimum resistivity. This suggests that the unpatterned region maintained the initial characteristics of an as-grown sample and that the imposed scatterers do not play an important role above the critical field. We may attribute this to the condition that the cyclotron diameter becomes less than the effective separation between the anti-dots [6] (see Fig. 2 inset). This yields a radius of anti-dot r_d to 0.15μm from the trace shown in Fig. 2a and 0.13μm from the trace shown in Fig. 2b. These values are in reasonable agreement with the values expected from the fabrication conditions.

We attribute the peaks to the commensurability problem on the hexagonal lattice. The Fig. 2 inset shows several commensurate classical cyclotron orbits. When electrons make cyclotron motion without being scattered, electrons move in several commensurate pinned-orbits [5]. Electrons moving in pinned-orbits cannot contribute to the conduction. Therefore, the maxima of the magnetoresistance will be observed at the magnetic field where the cyclotron radius coincides with the pinned-orbit radius. We have three kinds of commensurate orbits. The solid line circles illustrate the cyclotron orbit, where the center of the orbits is placed at one of the anti-dots. The dashed and dotted circles illustrate the other groups of

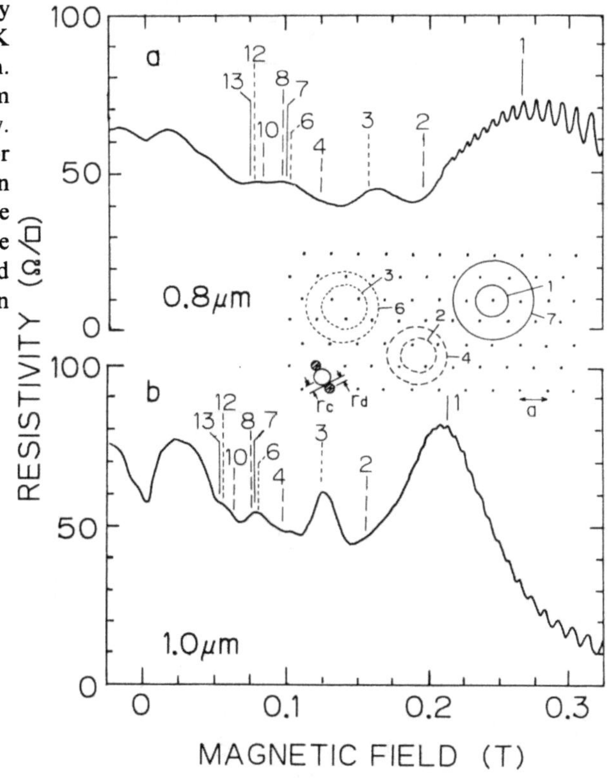

Fig. 2a, b. Low-field resistivity versus magnetic field at 1.4K after white-light illumination. The dot spacing *a* is **a** 0.8μm and **b** 1.0μm, respectively. *Arrows* indicate the field for pinned-orbits as illustrated in the *inset*. *Inset*: Commensurate orbits on hexagonal lattice. The number of anti-dots encircled by the cyclotron orbit is shown beside the orbit

pinned-orbits, where the center of the orbits is placed at the middle point of the side or the center of a triangle. The field positions for commensurate orbits are marked by arrows and the number of anti-dots encircled by cyclotron orbits is denoted beside the arrows in Fig. 2. This number has a close correlation with the maxima in the observed resistivity, except for the large peak at 0.02 T. However, the No. 2 peak was not observed. The absence of the No. 2 peak is probably due to larger scattering compared with other orbits which have wider open space for cyclotron motion such as No. 3 and No. 7.

The amplitude of the peaks becomes smaller with increasing index number, or with increasing cyclotron radius, but the near-zero field peaks are relatively large. We cannot explain these peaks by commensurate orbit.

Calculation

Figure 3 shows the calculated resistivity. The calculation was based on a simple geometrical model according to Weiss et al., where a fraction of the pinned, scattered, and drifting orbit was estimated [5]. The calculated result was in good agreement with the experimental results. Here we can see that the resistivity is sensitive to the aspect ratio a/r_d. For a of 0.8μm and r_d of 0.15μm, the peaks are significantly reduced. The reason that the amplitude of the low-field peaks for the

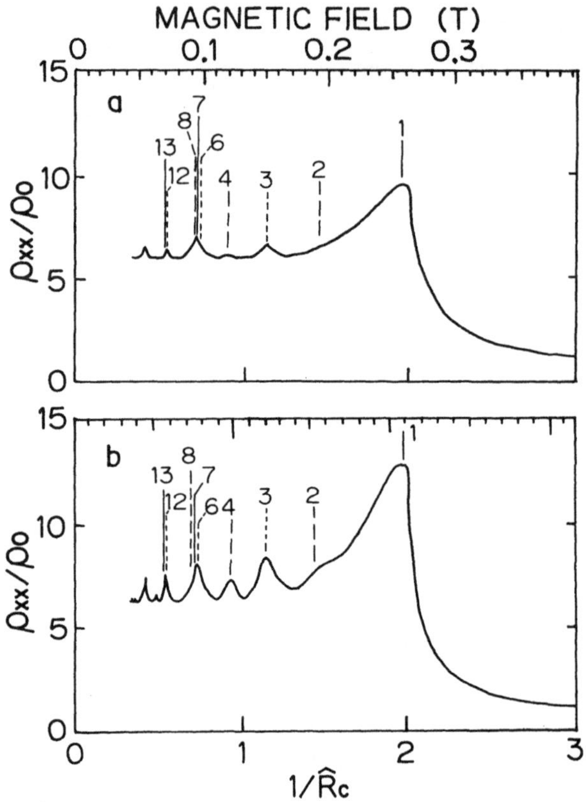

Fig. 3a, b. Calculated normalized resistivities versus normalized field $1/\acute{R}_c$ ($\acute{R}_c = r_c/a$). The parameters a and r_d are given from the experimental results. The field positions for commensurate orbits are depicted. **a** $a = 0.8$, $r_d = 0.15\mu$m. **b** $a = 1.0$, $r_d = 0.13\ \mu$m

trace shown in Fig. 2a was smaller than that for the trace shown in Fig. 2b is that the aspect ratio is smaller for the trace in Fig. 2a. Moreover, we can estimate the fraction of pinned-orbits f_p from the peak to valley ratio. From Fig. 2b we obtained $f_p \approx 0.5$. This is in good agreement with the prediction of Fig. 3b.

Carrier Concentration Dependence

Magnetoresistance for $a = 0.8\mu$m observed at various carrier concentrations is shown in Fig. 4. The uppermost trace was observed at 4.2K in the dark and all others were observed at 1.4K from dark to gradual LED illumination. After the illumination, a persistent photoconductivity (PPC) effect arose and an increased carrier concentration was maintained. The carrier concentration was varied by changing the intensity of illumination. We can see that the peaks shift with increasing carrier concentration, except for the peak observed at near zero field. Based on the classical commensurate orbit model, the peak shift depends upon the square of carrier concentration through Fermi wavenumber $k_F = \sqrt{2\pi n_s}$. The relative intensity of the peaks also changed with increasing carrier concentration. This is probably associated with the change in potential profiles around the anti-dots. The potential may be sloped at low carrier concentration. This may

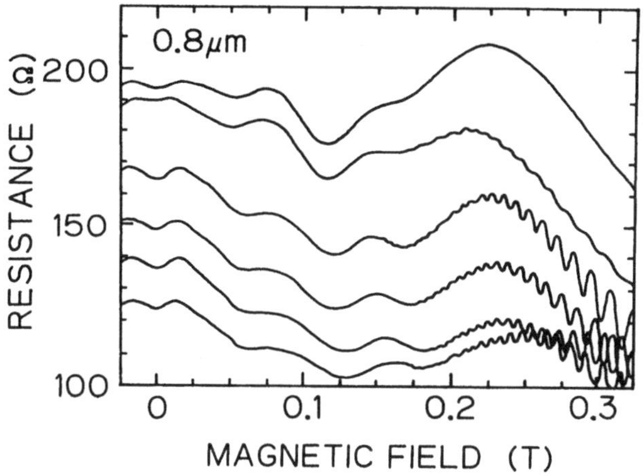

Fig. 4. Magnetoresistance for $a = 0.8\mu m$ at various carrier concentrations. The top trace was measured at 4.2K in the dark and all the others were measured at 1.4K from dark (the *second trace*) to gradual illumination. The carrier concentration, determined from SdH oscillations, is 3.16, 3.22, 3.36, 3.53, 3.75, 3.92 × $10^{15}m^{-2}$ from *top* to *bottom*

cause guiding of electrons and more electrons will enter in pinned orbits at low magnetic field or with large cyclotron radius.

Conclusions

Hexagonal anti-dot lattices were fabricated by EB lithography and Ar ion beam etching. Novel low field magnetoresistance peaks were observed. The peak position is attributed to classical commensurate orbits on a hexagonal array of anti-dots.

Acknowledgments. We would like to thank F. Wakaya and Y. Yuba for the high-mobility heterojunctions, and Y. Takagaki and J. Yanagisawa for useful discussions. This work was supported, in part, by a Grant-in-Aid for Scientific Research in the Priority Area, "Electron Wave Interference Effects in Mesoscopic Structures", from the Ministry of Education, Science, and Culture (MESC) of Japan. One of the authors (K.G.) is grateful for the partial support given by the Industry-University Joint Research Program, "Mesoscopic Electronics" through MESC.

References

1. Sakamoto T, Takagaki Y, Gamo K, Namba S, Takaoka S, Murase K (1991) Solid State Commun 80(7):535–539
2. Roukes ML, Thornton TJ, Scherer A, Van der Gaag BP (1990) In: Chamberlain JM, Eaves L, Portal JC (eds) Electron properties of multilayers and low-dimensional semiconductor structures. Plenum, New York, pp 95–116
3. Pfeiffer L, West KW, Stormer HL, Baldwin KW (1989) Appl Phys Lett 55:1888
4. Weiss D, von Klitzing K, Ploog K, Weimann G (1989) Europhys Lett 8:179

5. Weiss D, Roukes ML, Menschig A, Grambov P, von Klitzing K, Weimann G (1991) Phys Rev Lett 66:2790 (and references therein)
6. Ensslin K, Petroff PM (1990) Phys Rev B41:12307
7. Lorke A, Kotthaus JP, Ploog K (1991) Phys Rev B44:3447
8. Yamashiro T, Takahara J, Takagaki Y, Gamo K, Namba S, Takaoka S, Murase K (1991) Solid State Commun 79:885
9. Takahara J, Kakuta T, Yamashiro T, Takagaki Y, Shiokawa T, Gamo K, Namba S, Takaoka S, Murase K (1992) Jpn J Appl Phys 30(11B):3250–3255
10. Takagaki Y, Gamo K, Namba S, Ishida S, Takaoka S, Murase K (1990) J Appl Phys 67:340

Transport in Quasi-Ballistic Quantum Wires and Rings

15

KHALED ISMAIL [1]

Summary. We report on the fabrication and measurement results in relatively long (2–4 microns) quantum wires and rings. Despite the presence of elastic scattering in these geometries, quasi-ballistic transport is manifested in quantized conductance steps, similar to those observed in point contacts. In addition, the wires show a new resonance at higher temperature in the range of 50K–250K. This is attributed to a phonon inter-subband resonance. The rings show a variety of novel effects, among them classical electron focusing, and the observation of Aharonov-Bohm oscillations in the extremely high resistance limit.

Introduction

With the evolution in nanometer-scale fabrication techniques, combined with the progress in epitaxial growth of extremely high mobility layers, a new breed of structures and devices has become available. In such structures, a stunning variety of novel effects has been observed, which has contributed significantly to the improvement of our understanding of low-dimensional transport properties. Such effects are also potentially useful when applied to novel devices and circuits. Among such effects, the two most significant ones are the Aharonov-Bohm (AB) oscillations and the presence of distinct modes in quantum point contacts.

The former effect is observed in small conducting rings [1] where the diameter of the loops is smaller than, or comparable to, the inelastic diffusion length (coherence length) and is manifested in conductance oscillations periodic in the magnetic flux threading the ring. This is attributed to consecutive constructive and destructive interference due to a change in phase of the electron wave function in one branch relative to the other.

The latter effect is observed [2, 3] in narrow, short constrictions fabricated using high-mobility GaAs/AlGaAs modulation-doped layers. Due to the ballistic transport of electrons through the constriction, the conductance between the ohmic contacts is quantized in steps of multiples of $2e^2/h$. Each step corresponds to the transmission of an additional mode through the constriction.

[1] IBM, Watson Research Center, Yorktown Heights, NY 10598, USA, and Dept. of Electronics and Communications, Faculty of Engineering, Cairo University, Egypt

In this paper, we shall review some of the important fabrication techniques of quantum wires and rings. We will focus on transport in the quasi-ballistic regime, as opposed to electron transport in the fully-ballistic or diffusive regimes. We will present some novel experimental results which we have observed in quasi-ballistic quantum wires and rings, and we will discuss the significance of these results in terms of the underlying physics as well as in terms of possible future applications.

Fabrication

There has been a continuous effort during the past two decades to fabricate ultra-small structures with dimensions comparable to relevant physical length scales such as the electrons Fermi wavelength, their mean-free-path, their coherence length, or the magnetic cyclotron orbit diameter. Such structures have been fabricated in Si metal-oxide-semiconductor (MOS) configurations, as well as in GaAs/AlGaAs modulation-doped layers. The latter offer the advantage of providing extremely high mobilities in layers grown by molecular beam epitaxy (MBE). In such layers, the electrons' motion is confined to a plane parallel to the hetero-interface.

Further confinement can be achieved by various techniques, such as etching, selective ion implantation, using the field effect to create quasi one-dimensional conducting channels, or by pinching off a two-dimensional electron gas (2DEG) from the sides using a split gate.

High mobility samples have proven to be quite sensitive to the fabrication techniques. Upon using reactive ion etching, ion milling, or selective ion implantation, the electron mobility in the layers suffers a dramatic reduction. To our knowledge, there has been no clear quantitative explanation for this effect. The split-gate configuration does not involve the penetration of ions into the original material and therefore suffers the least from mobility reduction. However, ballistic transport in those geometries appeared only over length scales which are at least an order of magnitude shorter than the electron mean free path estimated from the Drude formula. This has been attributed to the small angle scattering by the fluctuating potential caused by the random location of the donor ions in the AlGaAs layer.

We have adopted an etching technique, which is based on a heavily diluted crystallographic wet etch, to define quantum wires and rings. Perfect control and uniformity has been achieved [4] over single wires and arrays of up to 10000 parallel wires. The uniformity is corroborated by the sharp photoluminescence

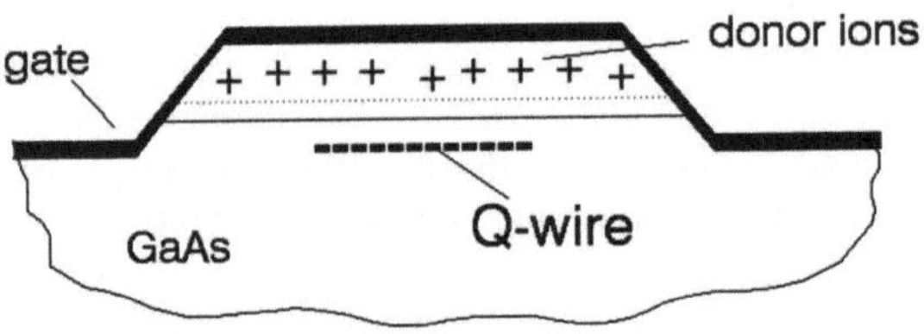

Fig. 1. Schematic cross-section of a quantum wire fabricated using wet chemical etching followed by gate metallization

peak and by the observation of distinct subband depopulation in the presence of a magnetic field.

To achieve external control of the electron density, the structures are gated. Right after the wet etching (which is typically ended by a passivation step in which the surface oxide is removed), the sample is introduced into an evaporation chamber which is pumped down instantaneously. Figure 1 shows a crossection of a wire after etching and gating. There are several advantages in this fabrication technique: (1) The wet etch results in no damage or strain to the material, in addition to being well controlled due to the slow etch rate (2 nm/s). (2) The side walls of the wires are almost perfectly flat in view of the crystallographic nature of the etching. (3) The gate metal being deposited on a freshly cleaned and passivated surface suffers the least from interface states or impurities residing at the metal/semiconductor interface. (4) The fact that the gate surrounds the structure (i.e., from the top and both sides) results in better control on the electron density, which is translated into a high transconductance. It also provides much better screening of the random electrostatic potential from the DX centers in the AlGaAs layer.

In terms of lithography, for isolated wires and rings, or for arrays with a small number of wires and rings, electron beam lithography is the most adequate. It is interesting to note that in our fabrication process, the electron beam irradiation is performed on the area to be etched. Thus, possible damage by the electron beam to the conduction path is avoided. To study optical properties, on the other hand, large arrays with up to 10000 wires have to be fabricated. In this case, one is forced to resort to either holography or x-ray lithography to ensure uniformity and a reasonable throughput.

Measurement Results

Various interesting effects have been observed in the temperature range of 0.1–200K in our quasi-ballistic quantum wires and rings. Starting with measurements at 0.1 K, which were conducted in the mixing chamber of a dilution refrigerator, we have observed very clear Aharonov-Bohm oscillations in one, two, and four parallel rings of 1.25 inner diameter and 1.65 outer diameter. The parallel rings were designed such that they would always share one arm, to investigate whether coupling between modes in the parallel loops is possible or not [5].

AB oscillations with up to 20% modulation were observed in isolated and parallel ring structures. The condition of strongest oscillations was the same in all cases, which is that the conductance in the structure corresponds to approximately $2Ne^2/h$, where N is the number of rings. Sakaki [6] has previously predicted that electron back-scattering in quantum wires would be suppressed when the electrons are confined to the lowest energy subband. In other words, the "purest" transport can be observed under single mode operation. This is consistent with our observation.

Figure 2 shows a magneto-resistance sweep containing AB oscillations. Two sweeps are shown for comparison, one at a small and one at a large resistance value. Two interesting effects are worth noting at this point. First, AB oscillations

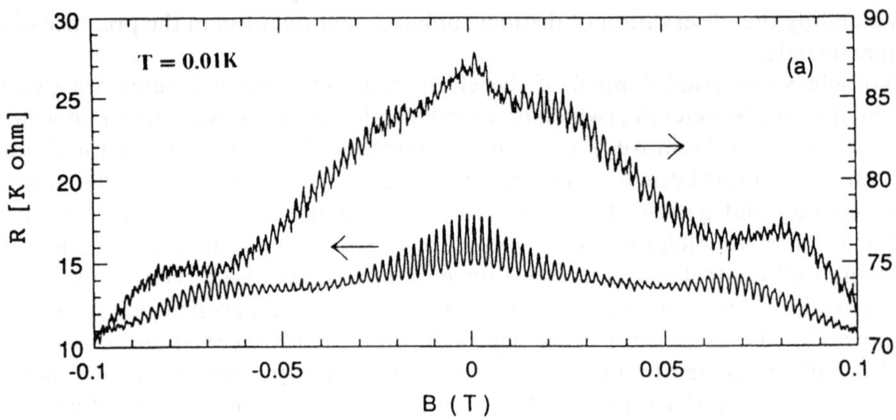

Fig. 2. Magneto-resistance sweep in two-coupled rugs containing AB oscillations in the low and high resistance regimes

could be observed in our samples under very high resistive conditions corresponding to up to 200 K ohm per ring (or 16 $h/^2e$) [5]. Under such conditions, in very high mobility samples, one expects the fluctuating potential induced by the randomness of the donor atoms to result in reduction of the electron mobility by more than two orders of magnitude. Correspondingly, localization effects would start dominating the transport behavior in the conducting channels. In our case, however, this does not deter the observation of clear AB oscillations.

Second, there are peaks in the envelope of the AB oscillations, symmetric around zero magnetic field. This background modulation can be explained in terms of electron focusing [5, 7]. Different from the point contact case [8], electron focusing appears periodically in a ring geometry, whenever the classical cyclotron orbit matches the loop diameter. It is interesting to note that this focusing effect also appears in the extremely high resistance samples. The fact that we could observe up to five resistance peaks satisfying the focusing condition indicates that there is no mode mixing in the loops.

The presence of distinct modes in the wires and rings has been also confirmed with conductance measurements at 4.2K. Although our structures were more than 2μm long, distinct conductance steps were observed quantized at multiples of $2^2e/h$ in a single wire, $2^2Ne/h$ in N parallel wires, 0.8 e/h in a single ring,[2] 1.2 e/h in two parallel rings, and $2^2e/h$ in four parallel rings. The results for the wires (Fig. 3 shows the conductance steps for one and ten wires) are consistent with what one would expect from quantum point contacts which are typically an order of magnitude shorter than our wires. It has been previously predicted that conductance steps would not be observable in such long wires. Invoking a randomness in the Si donor distribution, one would gradually see the conductance steps smear out and eventually disappear. Our experimental results contradict such predictions, which indicates that the assumption about the randomness of donors is not accurate (at least in the high mobility layers we used in these experiments).

The steps in conductance in the ring samples at values different from $2^2e/h$ are attributed to the connected geometry of the loops and can be explained in terms of

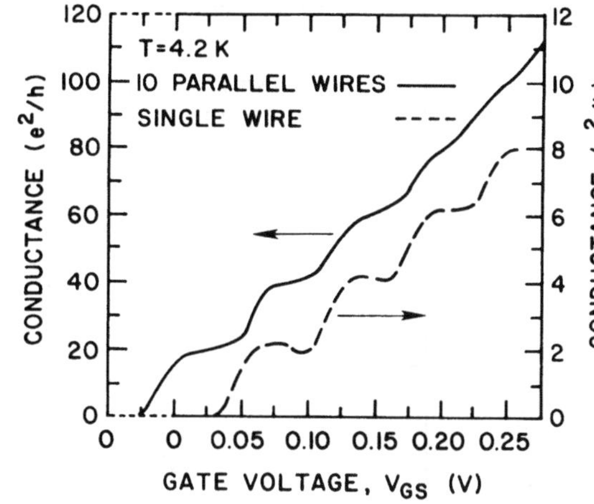

Fig. 3. Quantized conductance steps as a function of gate bias in a single wire and ten parallel wires. The wires are 2μm long and 0.4 μm wide. The estimated effective wire width is below 0.1μm

classical addition of quantized conductances (M. Buettiker, 1990, private communication). The focusing data, which are very clear even at 4.2 K, and indicate that the modes are distinct in the loops, provide further evidence that the quantized conductance steps are correlated with modal transport in the connected structure.

At about 20K, no more conductance steps were discernible in the wire structures. This is due to thermal smearing and incoherent scattering resulting in mode mixing. However, a new and interesting effect was observable in the temperature range of 50–250K. At a distinct gate bias, the conductance, rather than monotonically increasing, exhibits a drop (Fig. 4). The effect is observed in the dark and after light emitting diode (LED) illuminations. The only difference in the two cases is the threshold voltage shift, and the corresponding shift in the conductance curve. In a gated Hall bar (5 μm × 100 μm) made of the same material this effect was not discernible, indicative that the effect is caused by the quantum confinement. The interesting thing is that the effect appears at a relatively high temperature as opposed to various other quantum effects. Plotting the peak-to-valley ratio of the conductance as a function of temperature (Fig. 5) shows that the effect is strongest around 170K and dies out at above 250K. One possible explanation of this effect is that there is a phonon resonance between the quantum wire energy subbands. At a distinct gate bias, the subband separation (not necessarily between one and the next) matches the phonon absorption energy. In this case, an enhancement of phonon-assisted scattering between those two subbands occurs, resulting in a mobility reduction and a corresponding drop in conductance. Such an effect has been previously predicted by theoretical work [9, 10], and may be promising for future applications.

Conclusion

We have described a fabrication technique based on wet chemical etching of quantum wires and rings, which results in minimal damage and preserves the high purity of the starting material. In devices made by this technique, we have observed

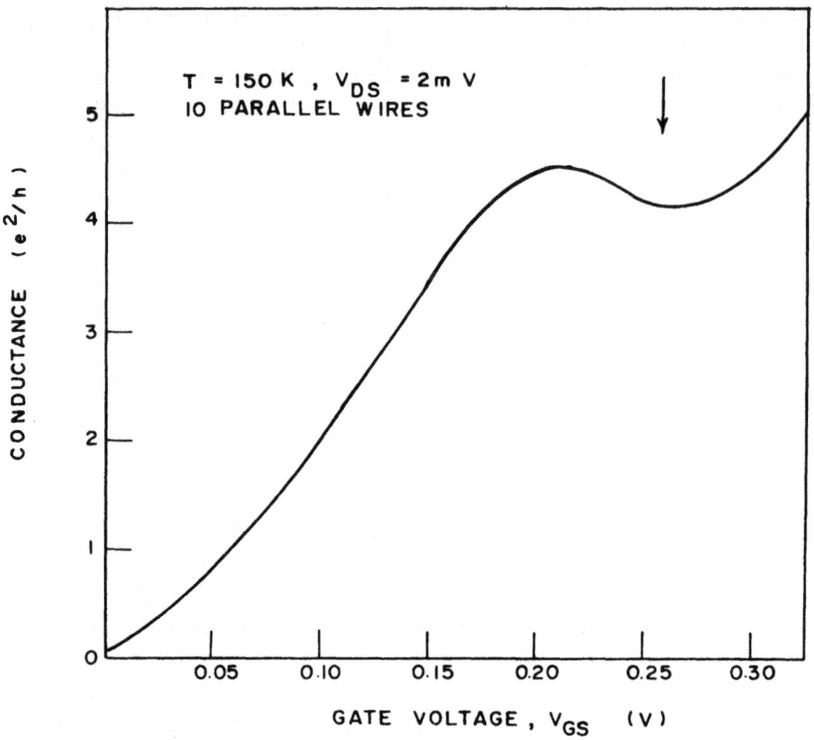

Fig. 4. Conductance as a function of gate bias in ten parallel wires at 150K. The *arrow* indicates the position of the new resonance effect

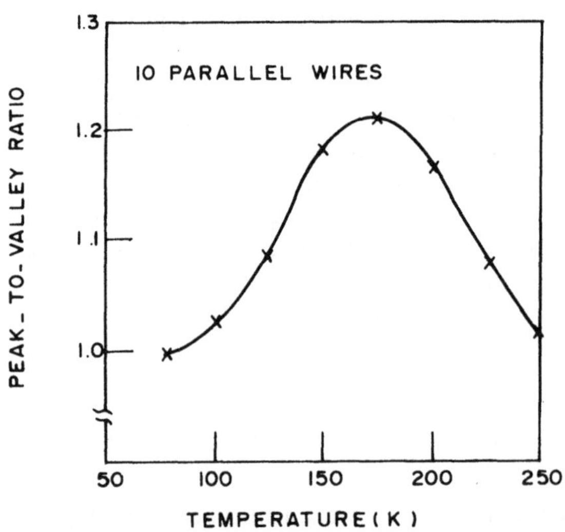

Fig. 5. Peak-to-valley ratio of the new resonance effect as a function of temperature

a variety of interesting effects, such as strong Aharonov-Bohm oscillations, even in the extremely high resistance limit, where localization effects were expected to be dominant. We have observed electron focusing around the diameter of the rings in the temperature range of 0.1–4.2 K. This, together with the observation of quantized conductance steps in the loops, provide the first clear evidence of preservation of modes in ring geometries. Clear modal transport is also observed in relatively long quantum wires, indicating that some assumptions about the randomness of the donor potential were incorrect. In addition, a new resonance effect is observed for the first time and is attributed to an intersubband phonon assisted resonance.

Acknowledgment. This work has been performed in collaboration with S. Washburn, currently at the Dept. of Physics and Astronomy, University of North Carolina, K.Y. Lee, currently at IBM East Fishkill, and J.M. Hong, at IBM T.J. Watson Research Center.

The author would like to thank D. Kern and S. Rishton for their help with the electron beam lithography, and M. Buettiker for interesting discussions and model calculations of the conductance in the loops.

References

1. Washburn S (1989) Am J Phys 57:1069
2. Wharam DA, Thornton TJ, Newbury R, Pepper M, Ahmed H, Frost J, Hasko D, Peacock D, Richtie DA, Jones G (1989) J Phys C21:L209
3. van Wees BJ, van Houten H, Beenakker CW, Williamson JG, Kouwenhoven LP, van der Marel D, Foxon CT (1988) Phys Rev Lett 60:848
4. Ismail K, Burkhardt M, Smith HI, Karam NH, Sekula-Moise PA (1991) Appl Phys Lett 58:1539
5. Ismail K, Washburn S, Lee KY (1991) Appl Phys Lett 59:1998
6. Sakaki H (1980) Jpn J Appl Phys 19:735
7. Washburn S, Ismail K, Lee KY (1992) International workshop on quantum-effect physics, electronics and applications, Luxor, Egypt, January 1992; to be published in Quantum-effect physics electronics and applications, (eds) Ismail K, Ikama T, Smith HI, Hilger A: 1992
8. van Houten H, Beenakker CWJ, Williamson JG, Broekhart MEI, van Loosdrecht PHM, van Wees BJ, Mooij JE, Foxon CT, Harris JJ (1989) Phys Rev B39:8556
9. Leburton JP (1991) Talk, Int conf on hot electrons, Nara, Japan, June 1991
10. Mori N, Momose H, Taniguchi K, Hamaguchi C (1992) International workshop on quantum-effect physics, electronics and applications, Luxor, Egypt, January 1992; to be published in Quantum-effect physics electronics and applications, (eds) Ismail K, Ikama T, Smith HI, Hilger A: 1992

Lateral Electronic Superlattices on Semiconductors

16

Jörg P. Kotthaus[1]

Summary. The effects of a voltage-tunable lateral superlattice potential on the electronic properties of a two-dimensional electron system are reviewed in the ballistic regime in which the electronic mean free path is longer than the superlattice period for the particular case of square superlattices. Field-effect devices with laterally periodic gates serve to tune the ratio of Fermi energy to superlattice potential and make it possible to investigate all regimes between weakly perturbing and strongly confining superlattices. A magnetic field applied perpendicular to the plane of the electron system is used to provide a tunable length scale, namely the cyclotron radius at the Fermi energy. Magnetoresistance features that are observed when the diameter of the cyclotron orbit is equal to or a multiple of the superlattice period are shown to be quantitatively explainable in a semiclassical trajectory model of deterministic diffusion of ballistic electrons. The intraband response of lateral superlattices at infrared frequencies is demonstrated to provide information about the unscreened confining potential. Here the trajectory model can also serve to give a better understanding of the effects of a lateral superlattice on the dynamic electronic properties.

Introduction

One of the outstanding properties of two-dimensional electron systems (2DES) that are created by one-dimensional confinement of electrons in semiconductor heterostructures is that they can have extremely high electron mobilities $\mu = e\tau/m^*$ at low temperatures. This and the recent developments in nanofabrication technologies have made it possible to experimentally investigate the effects of an artificial superlattice potential in the plane of the 2DES on the electronic motion in the ballistic regime in which the superlattice period a is smaller than the elastic mean free path $l = v_F\tau$ of electrons moving with Fermi velocity v_F. (For a recent review of this subject, see, for example, Ref [1].) The concept of such studies is illustrated in Fig. 1 for the special case of a square lateral superlattice that will be the focus of this review. For amplitudes of the superlattice potential that are smaller than the Fermi energy E_F in the 2DES, as displayed in Fig. 1a, new lateral superlattice

[1] Sektion Physik, Universität München, Geschwister-Scholl-Platz 1, D 8000 München 22, Germany

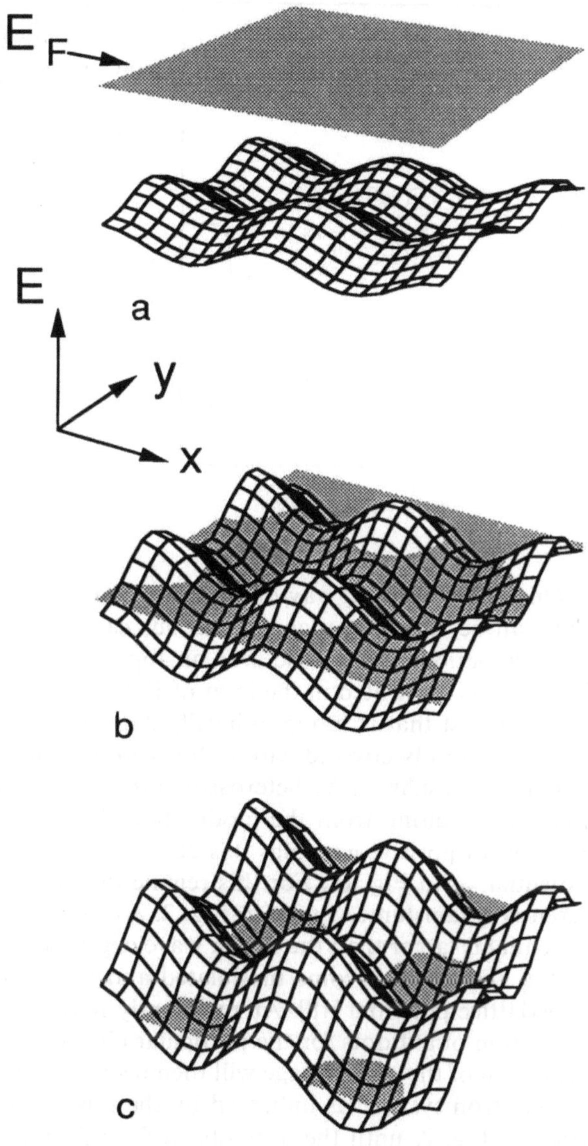

Fig. 1a–c. Sketch of a two-dimensional electron gas with Fermi energy E_F subjected to a two-dimensional sinusoidal superlattice potential for the characteristic cases of **a** weak modulation, **b** an array of antidots, and **c** an array of dots. (From [6] with permission)

transport phenomena have become observable [2–7]. For somewhat larger superlattice potential amplitudes, as sketched in Fig. 1b, the 2DES transforms in an array of so-called antidots, i.e., voids in the density distribution of the electron systems that can be considered as periodic scattering centers [7–11]. Finally, at sufficiently large superlattice potentials, as in Fig. 1c, one obtains arrays of electron disks in which the electronic motion can be quantized in all spatial directions [12–16]. Such quantum dot arrays cannot be easily accessed via transport experiments but can be studied in detail by intraband spectroscopy, as discussed below.

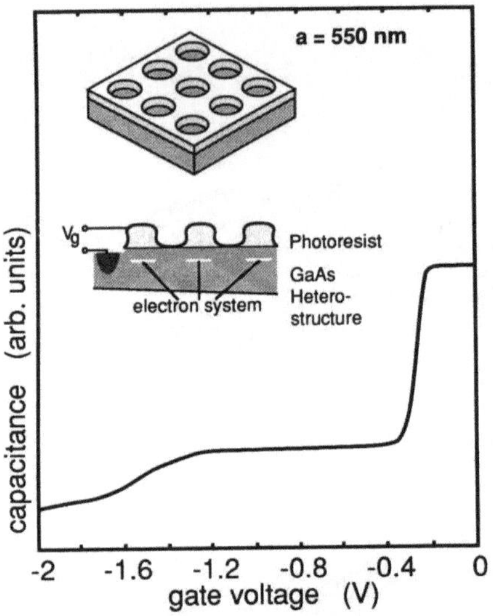

Fig. 2. Capacitance-voltage trace at T = 4.2 K for a field-effect device on GaAs designed to investigate square antidot arrays of period $a = 550$ nm, as sketched in the *inset*. (From [9] with permission)

One rather efficient way of imposing a lateral superlattice potential on a 2DES that can be tuned through the various cases sketched in Fig. 1 is by the use of suitable periodic gate structures on top of a modulation doped heterostructure [17]. This is illustrated in the inset of Fig. 2. Here a heterostructure is coated with a photoresist that is lithographically patterned to have arrays of holes and then rather uniformly covered with a thin semitransparent metallic gate. Without gate bias this AlGaAs/GaAs heterostructure contains a 2DES at the GaAs/AlGaAs interface resulting from the modulation doping of the AlGaAs. With moderate negative or positive gate bias this 2DES is subjected to a weak lateral superlattice potential, as in Fig. 1a. For this regime the differential gate capacitance remains essentially constant, as seen in Fig. 2. As long as the period a of the superlattice potential is larger than the Fermi wavelength in the 2DES the electron density N_S in the 2DES will become modulated and screen a large portion of the external superlattice potential [18]. At sufficiently negative gate bias this will result in the formation of antidots for the particular choice of resist pattern of Fig. 2. Further reduction of the gate voltage will then decrease the electron density in the remaining electron system, as indicated by the lower plateau region of the capacitance trace in Fig. 2, until the antidots sufficiently increase in size to finally yield an array of dots at the most negative gate voltages. Thus, with a device such as the one shown in Fig. 2 one can realize the various cases of Fig. 1 by just tuning a gate voltage. Employing a technology of two gates stacked on top of each other with an insulator in between one can create even more tunable superlattice structures, as has been demonstrated on Si Metal-Oxide-Semiconductor Field-Effect Transistor (MOSFET) [14, 19, 20]. In such stacked-gate devices one can control the superlattice potential rather independently of the density of the underlying 2DES.

In the devices discussed below the laterally periodic gate structures are patterned via holographic lithography [21]. Whereas with a single exposure one can pattern a line grating with period down to about 100 nm, two exposures in between which the sample is rotated are used to fabricate square or rectangular patterns. Alternatively, electron beam lithography [13] or focused ion beam implantation [8] may be employed to generate similar superlattice devices.

In the following sections two complementary methods of characterizing the intraband electronic properties of lateral superlattices will be discussed. One is the investigation of the magnetoresistance at low temperatures at which the electron system is degenerate with a well defined Fermi energy. The other is intraband magnetooptics in the far-infrared spectral regime [1]. In both cases a magnetic field B applied perpendicular to the plane of the 2DES is employed to introduce a tunable length scale, namely the cyclotron radius R_c at the Fermi energy E_F:

$$R_c = \frac{m^* v_F}{eB} = \sqrt{2\pi N_s l_B^2} = \sqrt{2N + 1}\, l_B \quad (1)$$

where N is the Landau level index at E_F and $l_B = \sqrt{\hbar/eB}$ the magnetic length. Of obvious interest is the magnetic field regime at which $R_c \approx a$, the superlattice period. For the devices considered in the following sections, the corresponding B-fields are relatively small such that the cyclotron energy $\hbar\omega_c$ is still small compared to E_F, and since N is relatively large Landau quantization may not be important.

Magnetoresistance in a Lateral Superlattice

In a moderately modulated one-dimensional superlattice, Weiss et al [3] first observed novel magnetoresistance oscillations that are directly related to commensurability between the cyclotron orbit and the superlattice period, a and occur at:

$$2R_c = (m + \phi)a, \qquad m = 1, 2, 3, \ldots \quad (2)$$

where ϕ is a characteristic phase factor. Though originally explained in a quantum mechanical model [4, 5] these oscillations are predominantly of classical origin as shown by Beenakker [22]. For a 1D superlattice they can be visualized as arising from a net drift of the center coordinate of the cyclotron orbit under the influence of the ExB force ,with the spatially oscillating electric field E generating the superlattice potential. For a relatively weak square superlattice potential generated in gated heterojunctions, as in Fig. 2, these commensurability oscillations also dominate the low field magnetoresistance, as shown in Fig. 3a. To determine whether the size and phase of these oscillations can also be described semiclassically Lorke et al [7] calculated the magnetoresistance in a simple but very flexible trajectory model based on the diffusion of electrons moving ballistically at the Fermi energy under the combined action of the superlattice potential and the Lorentz force. In such a model the equations of motion are used to calculate electron trajectories with randomly chosen intial spatial coordinates and velocity directions, but with the kinetic energy determined such that the sum of the kinetic and potential energy corresponds to the given Fermi energy. Sufficiently large sets (10^3–10^4) of such trajectories are calculated for an average time, corresponding to a typical momentum scattering time τ in the high mobility 2DES, to obtain a mean square distance

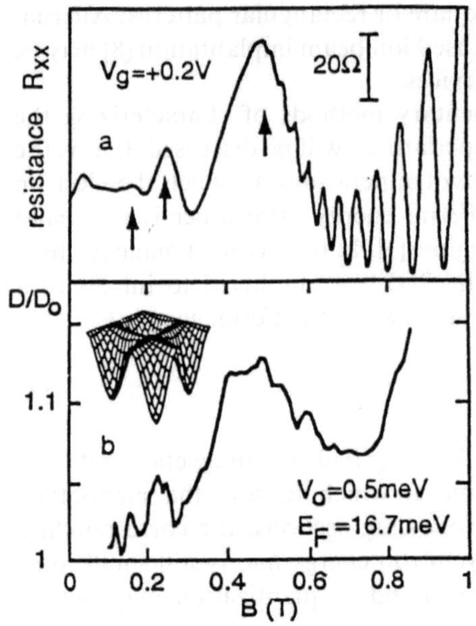

Fig. 3a, b. Measured (*arrows* in **a**) and calculated (**b**) commensurability oscillations in the low temperature magnetoresistance of a two-dimensional electron system on GaAs (density $N_s = 5.7 * 10^{11}$ cm^{-2}) subjected to a weak square superlattice (period $a = 460$ nm, peak to peak amplitude 4 V_0), as sketched in the *inset* in **b** (From [7] with permission)

$\overline{r^2}$, over which the electrons travel ballistically, and a two-dimensional deterministic diffusion coefficient $D = \overline{r^2}/(4\tau)$. For the case of a weak superlattice this then is added to the diagonal components D_o of the classical magneto-conductivity tensor of the unmodulated 2DES. For a sinusoidal superlattice of peak to peak amplitude $4V_o$ one obtains a calculated magnetoresistance, as displayed in Fig. 3b. The individual trajectories of the electronic motion in a square superlattice with a sinosoidal potential are calculated to be chaotic [23]. Though this is not visible in the present experiments it is the cause of the rather noisy calculated trace. This is in contrast to the case of a one-dimensional superlattice where the absence of chaotic trajectories is reflected in a much more rapid conversion of the similarly calculated magnetoresistance [6, 24].

When the superlattice modulation is increased such as to cause the formation of an antidot array [7–9], a quite different feature is observed in the magnetoresistance, as displayed in Fig. 4. With the formation of antidots one obtains a single magnetoresistance maximum at a magnetic field at which one has $2R_c \approx a$. Note that just before the formation of the antidots the magnetoresistance trace (dotted in Fig. 4c) is rather featureless with relatively strongly damped Shubnikov de Haas oscillations thought to arise from a rather inhomogeneous electron density distribution at such large superlattice amplitudes. Hence, the magnetoresistance maximum appears to be linked to the expulsion of electron trajectories from the area of the antidots, as visible in some representative calculated trajectories displayed in Fig. 5, where one can recognize both open and quasi-perodic trajectories. The magnetic field dependence of the resistivity, as calculated within the trajectory model for a sinusoidal superlattice potential, is shown in Fig. 6, and it exhibits a similar magnetoresistance maximum at $2R_c \approx a$ as the experimental traces. Essentially, this maximum reflects the localization of electron trajectories

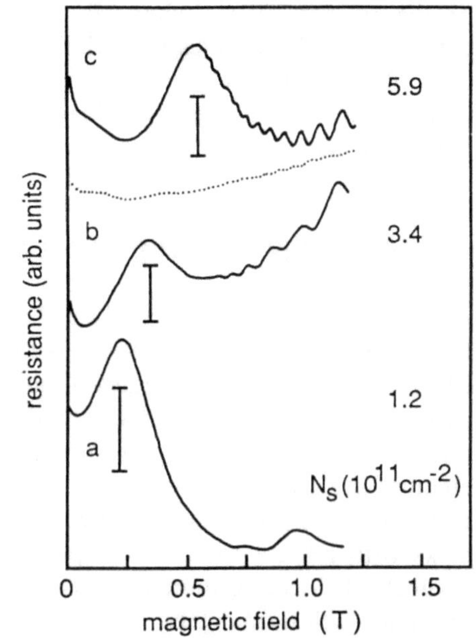

Fig. 4. Magnetoresistance of antidot arrays of different electron density N_s and period $a = 460$ nm. The strong maximum below the onset of Shubnikov-de-Haas oscillations occurs in all samples at a magnetic field where $a \approx 2R_c$. In (*c*) the *dotted trace* is measured just before (at gate voltage $-0.80V$) and the *solid trace* just beyond (at gate voltage $-0.90V$) the threshold where antidots are formed. The *vertical bars* mark 10% of the $B = O$ resistance. (From [7] with permission)

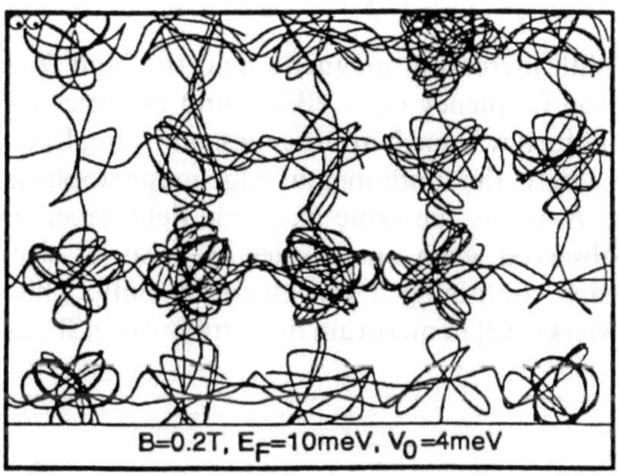

Fig. 5. Electron trajectories in a square antidot lattice of period $a = 460$ nm calculated with the parameters given in the *inset*. (From [6] with permission)

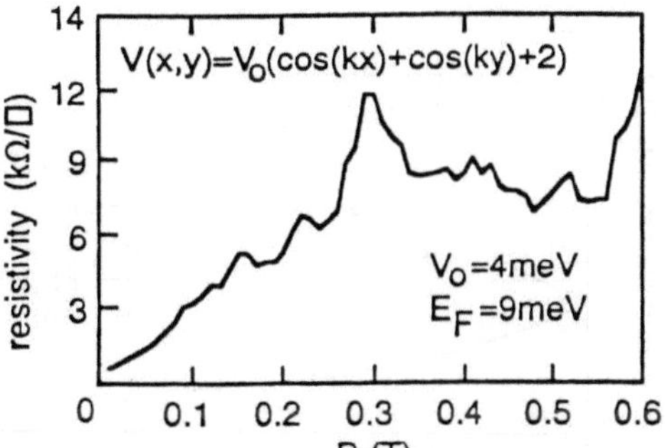

Fig. 6. Magnetoresistance of an antidot array of period $a = 2\pi/k = 460$ nm as calculated from the semiclassical model of ballistic diffusion for the parameters given (From [7] with permission)

on the antidot lattices when $2R_c \approx a$. In recent experiments on antidot lattices, defined by etching holes in a 2DES, additional magnetoresistance maxima have been observed that correspond to other localized orbits that are commensurate with the antidot lattice [11]. They are explained in a similar fashion, based on ballistic semiclassical electron trajectories.

Intraband Spectroscopy on Lateral Superlattices

A complementary way to analyze the electronic properties of electrons subjected to a lateral superlattice potential is by far-infrared transmission spectroscopy [1, 21]. Initial transmission studies of confined electrons arranged in a square array, as in Fig. 1c, were carried out with GaAs/AlGaAs heterostructures on which mesoscopic disks of quasi-two-dimensional electron systems were defined by mesa etching [25]. On such devices one observes a characteristic far-infrared resonance of frequency ω_0 that is explained by a plasma oscillation of all electrons in the disks with respect to the fixed positive background charge. In a perpendicular magnetic field these dimensional resonances split in a distinct manner into two branches with frequencies:

$$\omega_\pm = \sqrt{\omega_0{}^2 + \left(\frac{\omega_c}{2}\right)^2} \pm \frac{\omega_c}{2} \tag{3}$$

With increasing magnetic field the high frequency branch approaches the cyclotron frequency $\omega_c = eB/m^*$ and becomes increasingly bulk-like, whereas the low frequency branch approaches vanishing frequency and becomes a perimeter mode, namely, the fundamental edge magnetoplasmon mode [1].

Basically the same magnetic field dependence of infrared resonance energies is observed when one studies an array of electrostatically confined quantum dots, each containing a few electrons only. This was first observed by Sikorski and Merkt [12] in quantum dot arrays on InSb, as shown in Fig. 7. There the measured

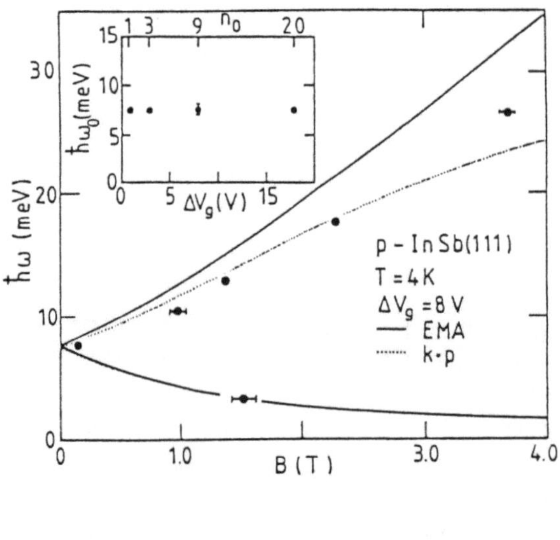

Fig. 7. Magnetic field dependence of the infrared resonance energies in a quantum dot array on InSb (period $a = 250$ nm, dot diameter about 100 nm). The experimental data (*dots*) are compared with a calculated dispersion assuming a parabolic conduction band (*solid line*, Eq. 3) and one including non-parabolicity (*dotted line*). The inset gives the dependence of the measured resonance energy at $B = O$ on gate voltage above threshold and the corresponding average electron number n_o per quantum dot. (From [27] with permission)

dispersion is compared with two calculated ones, one following eq. 3 (solid) and the other taking into account non-parabolicity (dotted). Here it is worth noting that the magnetic field dependence of Eq. 3 is also expected for dipole excitations of a single parabolically confined electron [26]. The most surprising result, however, which is displayed in the inset of Fig. 7, shows that the zero magnetic field frequency is found to be essentially independent of the number n_0 of electrons contained in each dot [27]. This result is now understood as reflecting the so-called generalized Kohn theorem [28]. This states that electrons that are confined in a parabolic bare confining potential respond to a spatially homogeneous infrared field only at the characteristic frequency that corresponds to the curvature of the bare confining potential. This resonance frequency reflects the center of mass motion of all electrons and is independent of the number of electrons that are confined, as well as of the dimensionality of the confined electron system [28, 29]. From the observation in Fig. 7 that ω_0 does not depend on electron number per dot one may now conclude that the bare confining potential in the InSb devices is essentially parabolic. This is not too surprising in devices in which the confining potential is generated electrostatically by gates that are spatially somewhat separated from the confined electron system so that only the lowest Fourier components of the gate potential provide the confinement [30, 31].

Essentially the same behavior is found in electrostatically confined quantum dots on Si [14], as well as on GaAs [16], as shown in Fig. 8. However, one consistently observes deviations from the magnetic field dispersion of eq. 3, particularly for the ω_+-mode at intermediate magnetic fields, as seen in Fig. 8b. To be able to understand the far-infrared response of electrons subjected to arbitrary superlattice potentials, Lorke used the semiclassical trajectory model discussed above to also calculate the spectrum of infrared excitations [24]. This is done by numerical Fourier analysis of the velocity components and an averaging procedure similar to the one used in computing the dc conductivity. In Fig. 9, the magnetic field dispersion of the far-infrared resonance position for the parabolic case is compared with one obtained from the trajectory model, and has a nonparabolic confining potential that softens towards the edges, as indicated in the inset. As in the experiment, the trajectory calculations yield a shift of the ω_+-mode above the parabolic model at intermediate fields. This makes it likely that the experimentally observed deviations from the parabolic case in fact reflect the nonparabolicity of the confining potential. Since the trajectory model is a semiclassical single particle model and does not include any collective effects or any lateral quantization, it is not surprising that it yields absolute frequencies that are much lower than those observed experimentally. In large part this is caused by using an approximative form of the screened confining potential for the model rather than the steeper bare confining potential as suggested by the generalized Kohn theorem. Expressed in a different manner, one may say that collective depolarization effects will shift the experimentally observed frequencies to values above those calculated in a single particle model based on the screened confining potential [1]. Thus, one can probably use the trajectory model only to understand dependencies at fixed electron numbers such as the magnetic field dispersion above, but one cannot expect it to correctly predict absolute frequencies or dependencies on electron density.

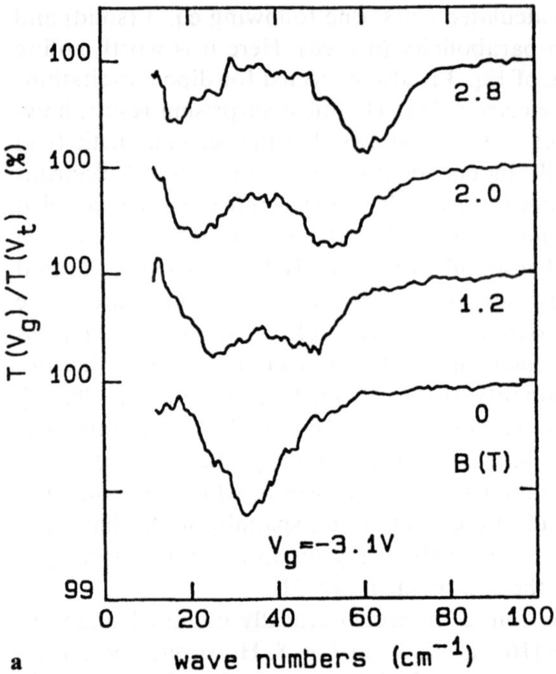

a

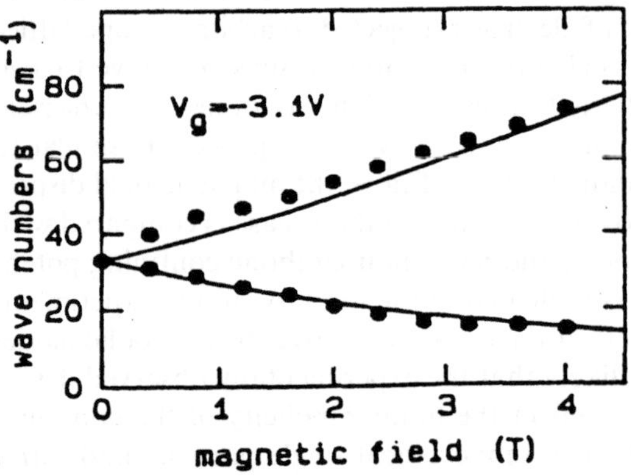

b

Fig. 8a, b. Magnetic field dependence of the far-infrared response of an array of quantum dots on GaAs (period $a = 450$ nm, electron number per dot $n_o \sim 50$): **a** normalized transmission spectra at T = 2K and **b** measured (*dots*) and calculated (*solid line*, Eq. 3) resonance positions. (From [16] with permission)

The far-infrared response of antidot lattices, as sketched in Fig. 1b, has recently been studied experimentally [9, 10] and yields a two-branch magnetic field dispersion of the resonance frequencies that is quite distinct from the one of parabolically confined quantum dots. Here the trajectory model is very useful, since it can be applied rather straightforwardly to a superlattice potential appropriate for an antidot lattice. The magnetic field dispersion of the infrared eigenfrequencies thus obtained is given in Fig. 10. It resembles the experimental findings quite well and has a high frequency mode that behaves in a manner similar to that in a

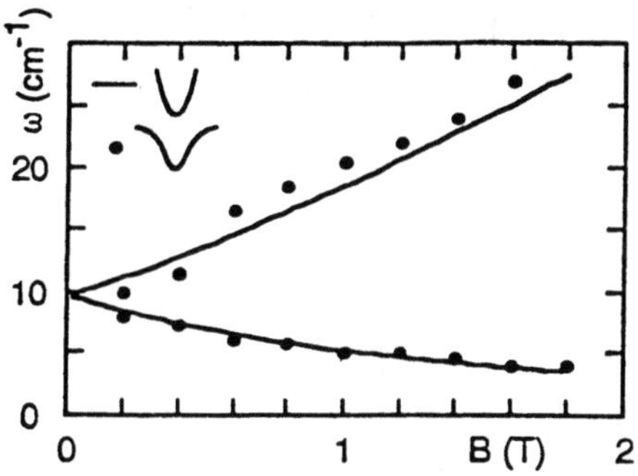

Fig. 9. Magnetic field dependence of the infrared resonance positions in an electron dot as calculated from a semiclassical trajectory model (*dots*) assuming nonparabolic confinement, as indicated in the *inset*, in comparison to the result expected for parabolic confinement (*solid line*, Eq. 3) (From [24] with permission)

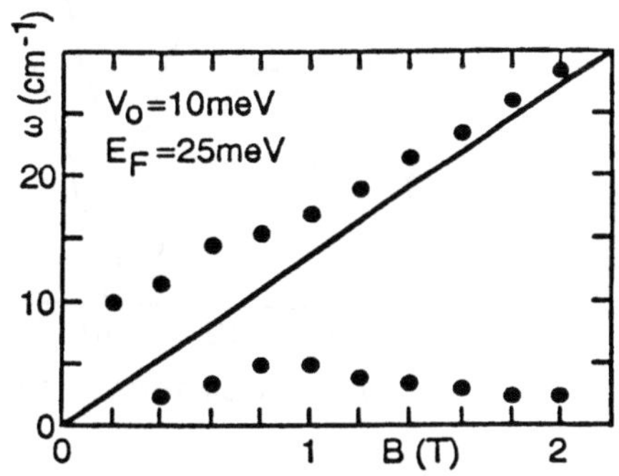

Fig. 10. Eigenfrequencies calculated in a semiclassical trajectory model for a square antidot array with sinusoidal potential on GaAs (*dots*) in comparison to the cyclotron frequency ω_c (*solid line*) (From [24] with permission)

quantum dot, but a low frequency branch that approaches zero frequency for both low and high magnetic fields. Again, predicted absolute frequencies are much lower than those found experimentally, the difference most likely reflecting collective effects.

Another characteristic superlattice can be formed if one starts with an array of electrostatically confined quantum dots and increases the Fermi energy somewhat in order to open narrow point-contact-like conducting constrictions between adjacent quantum dots [16]. Figure 11 shows the change in the infrared response as that happens, starting from isolated quantum dots at a gate bias of $V_g = -3.1$V. With the onset of conductive coupling, which can be deduced from the capacitance-voltage trace to take place at $V_g \geq -2.9$V, there occur characteristic changes in the infrared spectra. One is the appearance of another low-frequency mode below the edge-like mode of the isolated dot. Another is a high-frequency shoulder on the bulk-like mode of the isolated dot. The magnetic field dependence of these modes is illustrated in Fig. 12. The dispersion of the lowest frequency mode can be quantitatively described by the edge mode on a molecular-like perimeter marked as **B** in the inset of Fig. 12 that involves only two ballistic transmission events through a constriction between adjacent dots. Since all other similar perimeter modes involve at least four transmission events, they can be understood as happening with less probability and being more susceptible to inhomogeneous broadening. By increasing the gate voltage the highest frequency mode can be

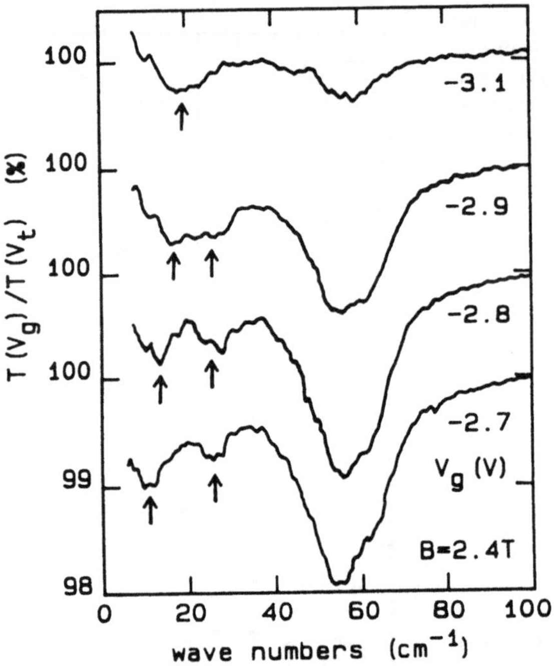

Fig. 11. Normalized transmission spectra of a square array (period $a = 450$ nm) of electron dots on GaAs at magnetic field B = 2.4 T and gate voltages where the dots change from being isolated (−3.1V) to being weakly coupled. Two edge-like modes (*arrows*) appear in the coupled case. (From [16] with permission)

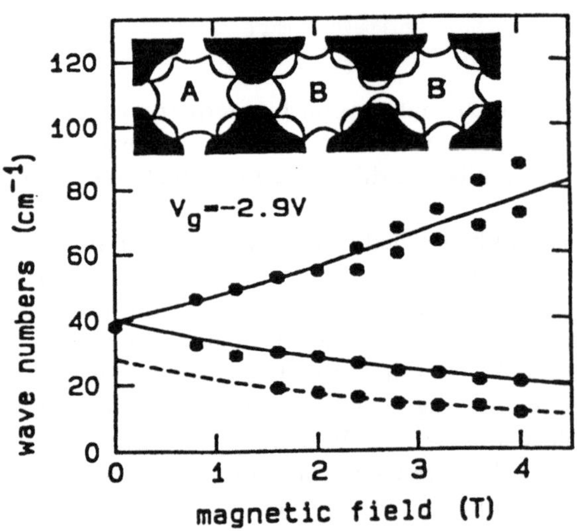

Fig. 12. Magnetic field dependence of the infrared resonance positions in a coupled quantum dot array on GaAs as measured from spectra, as in Fig. 11 (*dots*), and as expected for the A-type dot modes (*solid line*, Eq. 3) and the B-type edge-like mode (*dashed line*). (From [16] with permission)

followed to become the two-dimensional magnetoplasmon at wavevector $q = 2\pi/a$ for weak density modulation [6, 17]. It thus is a phonon-like mode of the coupled-quantum-dot solid. The observation that both modes that are characteristic for the coupled-dot array are observed only above certain threshold magnetic fields can be understood in a trajectory model as reflecting the fact that the transmission of electrons through a constriction only becomes likely if the cyclotron orbit shrinks to become comparable to or smaller than the width of the constriction [32].

Conclusions and Perspectives

The magnetotransport and magnetooptical properties of lateral electronic superlattices discussed above show that semiclassical trajectory models, originally proposed by Beenakker and van Houten [32] to explain anomalies of the Hall resistance in mesoscopic conductors, can explain many of the essential features observed here, although the confinement scales in the structures investigated are around 100nm and are thus comparable to those of laterally quantizing point contacts. It thus appears that the effects of lateral quantization in the static and dynamic lateral transport of electrostatically defined superlattices are still quite elusive, although they seem to be needed to explain magnetocapacitance data [13]. In addition, resonant magnetoconductance experiments have been reported that are not easily explained in a semiclassical fashion [33]. One thus can hope that by increasing feature definition, reducing feature sizes and fluctuations, and introducing more anharmonicity in the lateral superlattice potentials, quantum effects will become visible beyond doubt and allow us to experimentally study the fascinating miniband physics that has long been predicted for superlattice electrons in a magnetic field [34, 35].

Acknowledgments. This review is based on several collaborative studies and I want to particularly thank W. Hansen, A. Lorke, K. Ploog, and R. Winkler for their most valuable contributions. I also would like to thank J. Alsmeier, E. Batke, C.W.J. Beenakker, F. Brinkop, C. Dahl, K. Ensslin, D. Heitmann, U. Merkt, and D.A. Wharam for many stimulating discussions. This article was written while I was a visitor at the Center for Quantized Electronic Structures (QUEST) of the University of California in Santa Barbara, and I wish to thank all the members of the QUEST team for their outstanding hospitality and for providing a congenial research environment. The financial support of the ESPRIT Basic Research action and the Stiftung Volkswagenwerk is gratefully acknowledged.

References

1. Hansen W, Kotthaus JP, Merkt U (1991) Vol. ed. Reed M Semiconductors and semimetals, vol 35. Academic, New York: 279–380
2. Ismail K, Chu W, Antoniadis DA, Smith HI (1988) Appl Phys Lett 52:1071–1073
3. Weiss D, von Klitzing K, Ploog K, Weimann G (1989) Europhys Lett 8:179–184
4. Winkler RW, Kotthaus JP, Ploog K (1989) Phys Rev Lett 62:1177–1180
5. Gerhardts RR, Weiss D, von Klitzing K (1989) Phys Rev Lett 62:1173–1176
6. Lorke A (1991) Doctoral dissertation, University of Munich
7. Lorke A, Kotthaus JP, Ploog K (1991) Phys Rev B44:3447–3450
8. Ensslin K, Petroff PM (1990) Phys Rev B41:12307–12310
9. Lorke A, Kotthaus JP, Ploog K (1991) Superlattices and Microstructures 9:103–106
10. Kern K, Heitmann D, Grambow P, Zhang YH, Ploog K (1991) Phys Rev Lett 66:1618–1621
11. Weiss D, Roukes ML, Menschig A, Grambow P, von Klitzing K, Weimann G (1991) Phys Rev Lett 66:2790–2793

12. Sikorski C, Merkt U (1989) Phys Rev Lett 62:2164–2167
13. Hansen W, Smith TP, Lee KY, Brum JA, Knoedler CM, Hong JM, Kern DP (1989) Phys Rev Lett 62:2168–2171
14. Alsmeier J, Batke B, Kotthaus JP (1990) Phys Rev B41:1699–1702
15. Demel T, Heitmann, D, Grambow P, Ploog K (1990) Phys Rev Lett 64:788–791
16. Lorke A, Kotthaus JP, Ploog K (1990) Phys Rev Lett 64:2559–2562
17. Hansen W, Horst M, Kotthaus JP, Merkt U, Sikorski C, Ploog K (1987) Phys Rev Lett 58:2586–2589
18. Kotthaus JP, Heitmann D (1982) Surf Sci 113:481–484
19. Warren AC, Antoniadis DA, Smith HI (1986) Phys Rev Lett 56:1858–1861
20. Alsmeier J, Batke E, Kotthaus JP (1989) Phys Rev B40:12 574–12 576
21. Kotthaus JP (1991) In: Ferry DK, Barker JR, Jacobini C (eds) Granular nanoelectronics. Plenum, New York
22. Beenakker CWJ (1989) Phys Rev Lett 62:2020–2023
23. Geisel T, Wagenhuber J, Niebauer P, Obermair G (1991) Phys Rev Lett 64:1581–1584
24. Lorke A (1991) Proc EP2DS-9 Conf Surf Sci (1992) 263:307
25. Allen SJ Jr, Störmer HL, Hwang JCM (1983) Phys Rev B28:4875–4877
26. Fock V (1928) Z Phys 47:446
27. Sikorski C, Merkt U (1990) Surf Sci 229:282–286
28. Brey L, Johnson NF, Halperin BI (1989) Phys Rev B40:10647–10649
29. Maksym PA, Chakraborty T (1990) Phys Rev Lett 65:108–111
30. Laux SE, Stern F (1986) Appl Phys Lett 49:91–93
31. Kumar A, Laux SE, Stern F (1990) Phys Rev B42:5166–5175
32. Beenakker CWJ, van Houten H (1989) Phys Rev Lett 63:1857–1860
33. Toriumi A, Ismail K, Burkhardt M, Antoniadis DA, Smith HI (1990) Phys Rev B41:12 346–12 349
34. Langbein D (1969) Phys Rev 180:633–648
35. Hofstadter DR (1976) Phys Rev B14:2239–2249

Electron Scattering Experiments in Mesoscopic Conductors

17

Michael L. Roukes, Kenneth L. Shepard*, and
Bart P. Van der Gaag[1]

Summary. We have developed an experimental approach to direct evaluation of the entire transmission matrix of a mesoscopic conductor. This enables new investigations, previously not possible using standard transport measurements. Two initial applications are presented. In the first, we study perturbations to transport in small conductors caused by the measurement probes themselves. Comparing measurements in the complementary limits of *weak* and *strong* probe coupling, we elucidate mechanisms underlying transport anomalies observed in ballistic junctions. In the second, we employ a collinear pair of quantum point contacts to "image" the distribution function of electrons emerging from one constriction in the quantized regime. Modal structure observed in this outgoing electron "beam" provides new insights into details of the conductance quantization phenomenon and provides direct experimental evidence for *non*-adiabatic transport.

Introduction: Conduction and the T-Matrix

In the mesoscopic regime, we study transport on such short length scales and in samples so pure that electrical conduction involves only a small number of scatterers. Landauer has envisioned conduction in this limit as a *transmission problem*, i.e., a conventional quantum mechanical scattering problem [1]. Current incident upon a scatterer can be viewed as a superposition of partial waves. These evolve into outgoing waves according to a matrix of transmission probabilities determined by the local potential of the scatterer itself. Restriction of the scattering phase space to the simplest case of one dimension, representing transmission between idealized single mode leads, yields a simple relation between transmission probability and conductance:

$$G = \frac{e^2}{h} \frac{T}{1 - T} \tag{1}$$

This approach has been generalized and further developed in a large body of subsequent theoretical work (for useful reviews, see [2]).

[1] Bellcore, 331 Newman Springs Road, Red Bank, NJ 07701, USA
* Doctoral Candidate, Dept. of Electrical Engineering, Stanford University, Stanford, CA.

Transport experiments on very small structures are typically executed using a *multiprobe* configuration. For an experimentalist, a four-probe resistance measurement provides a means of isolating and studying a particular segment of the total voltage drop along the current path. Büttiker has generalized Landauer's picture to describe this configuration, partitioning the phase space for scattering into a multi-lead geometry in which each lead contains a complete set of quasi-1D states. Real connections from these leads to the outside world are approximated through the artifice of an *ideal electron reservoir* [3]. In this model the reservoirs are envisioned as separately feeding the eigenstates of each of the quasi one-dimensional (1D) leads. These leads converge and are coupled at a scattering region whose transmission matrix, **T**, subsumes the entire physics of the conductor. The transmission probabilities are fundamental to the theory, from them complicated expressions for the sample's resistance are derived. In experiments, however, it is *resistances* which are most directly obtained, but these provide little insight into the properties of **T**. Our study is motivated by the desire for a direct experimental probe of these scattering coefficients, and the detailed information they contain.

Phase Coherent Multiprobe Conductors

Büttiker approximates real conductors as being composed of three *distinct* entities: the *reservoirs*, the *leads*, and a *scattering region* (Fig. 1). Each reservoir, assumed to remain in internal equilibrium in the presence of current flow, is characterized by a single chemical potential, μ_i. According to these μ_i, each reservoir acts on its attached lead, populating the quasi-1D eigenstates which propagate towards the scatterer. This incident flux is redirected to scattered states, which carry flux back through the leads to the reservoirs. The reservoirs are assumed to constitute "perfect contacts", i.e., to absorb all returning flux without reflection. In steady state, assuming linear response, these approximations yield a linear relationship between the net currents flowing into the reservoirs, I_i, and the chemical potentials that ultimately develop at the reservoirs, μ_j:

$$\frac{h}{2e} I_i = -N_i \mu_i + \sum_j T_{ij} \mu_j \tag{2}$$

Here i and j are indices labelling the leads. In this model, voltage drops arise *only between reservoirs*, and all dissipation implicitly occurs within them. Despite its apparent simplicity, this model has proven to be a remarkably useful bridge between the transmission picture and the resistances measured in experiments (for useful reviews, see [2]).

The constants of proportionality in Eq. (2) are the *transmission coefficients*, T_{ij}. These coefficients represent the *total* probability of transmission from leads $j \rightarrow i$ for states at the Fermi energy, and are given by the double sum involving all occupied modes (indices k,l) in both leads: $T_{ij} = \sum_{kl} |t_{ij,kl}|^2$. N_i is the number of propagating modes; $N_i = \sum_j T_{ij}$. Unlike the transmission probabilities, $|t_{ij,kl}|^2$, the coefficients T_{ij} can exceed unity: $T_{ij} \in [0, N_i]$.

In experiments, external connections determine the role played by the *contacts*

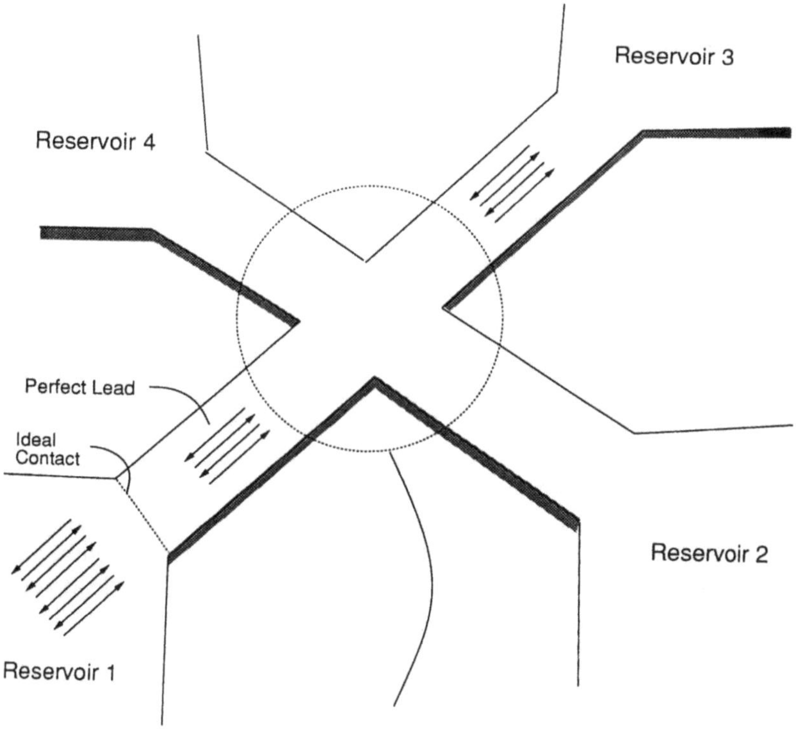

$$\mathbf{T} = \begin{bmatrix} T_{11} & T_{12} & T_{13} & T_{14} \\ T_{21} & T_{22} & T_{23} & T_{24} \\ T_{31} & T_{32} & T_{33} & T_{34} \\ T_{41} & T_{42} & T_{43} & T_{44} \end{bmatrix}$$

Fig. 1. Model for a four-probe conductor. Four multi-mode reservoirs equally populate the quasi-1D states of the leads. These carry flux to the "scatterer" (*encircled region*) whose transmission matrix determines how flux is apportioned to the outgoing states. This scattered flux returns to the reservoirs via the leads. The contact between the leads and the reservoir is assumed to be ideal; returning flux is absorbed without reflection

(modelled as reservoirs). By controlling the impedance of this external circuitry, and its ability to source or sink current, one can force the contacts to serve, ideally, in one of two complementary modes. A *voltage contact* is constrained so as to exchange negligible electron flux with the external circuitry. For a *current contact*, by contrast, this exchange is regulated to be of a constant, prescribed value. Assuming these ideal conditions, complicated four-probe resistances can be derived from Eq. (2), of general form

$$R_{kl,mn} = \frac{h}{2e^2} \frac{T_{mk}T_{nl} - T_{ml}T_{nk}}{D} \tag{3}$$

Here D is any cofactor of the matrix defined by Eq. (2) [3] and indices reflect current flow from reservoirs $k \to l$, yielding a potential drop from reservoir m to n. Although the most general four-probe geometry can involve up to 16 independent T_{ij}, there are only three unique four-probe resistances that can be measured on such an object. Clearly, a significant collapsing of information (contained in the individual T_{ij}) goes into the "formation" of a resistance.

Ballistic Multiprobe Conductors

Obviously, real samples are not so cleanly partitioned into separate components identifiable as reservoirs, leads, and the "scatterer". Instead, scattering generally occurs throughout the conductor. Additionally, in real samples, regions playing the role of electron reservoirs are, at best, geometrically rather ill-defined (and certainly not characterized by one uniform value of μ). A multiprobe ballistic conductor, such as that illustrated in Fig. 2, however, represents a special exception to these generalizations; it constitutes a simple, *almost literal realization* of Büttiker's model. Wide two-dimensional (2*D*) regions form reservoirs which connect to short, narrow leads. In these leads, current is carried essentially without scattering by electrons occupying only the lowest few transverse modes. The electrostatic potential in the locale of the junction region where the leads converge plays the role of the "scatterer". Its *geometric profile*, engineered by the microfabrication process, determines how incoming electrons are ultimately converted to outgoing flux in the leads. These samples provide the vehicle for our studies.

Direct Measurement of the Transmission Matrix

Conceptual Picture

Viewed from the perspective of Eq. (2) and the preceeding discussion, conventional resistance measurements implicitly involve the imposition of boundary conditions upon the *currents*, I_i. If, instead, we fix the *chemical potentials*, Eq. (2) becomes immediately simplified. By sourcing a current I_1 into reservoir 1 while *equalizing* μ_i at the other reservoirs, Eq. (2) is reduced to the form

$$\frac{h}{2e}(I_1 I_2 I_3 I_4) = -N_1(\mu_1\, 0\, 0\, 0) + (\mu_1\, 0\, 0\, 0)\mathbf{T} \tag{4}$$

Here, we have chosen to define chemical potentials such that those at current-sinking reservoirs vanish: $\mu_2 = \mu_3 = \mu_4 \equiv 0$. As depicted in Fig. 3, we can equalize these μ_i by ideal short-circuit connections linking the reservoirs. Under these conditions the off-diagonal T_{ij} become directly proportional to the currents,

$$T_{i1} = \frac{h}{2e}\frac{I_i}{\mu_1} \tag{5}$$

while the diagonal elements become

$$(T_{11} - N_1) = \frac{h}{2e}\frac{I_1}{\mu_1} \tag{6}$$

Fig. 2a–d. Electron micrographs of samples enabling direct measurement of **T** **a** Twelve-lead device approximately 1 *mm* long, showing 20 *μm Au* wirebonds attached to a "lead frame" fabricated by optical lithography. The leads converge into the active region of the device. **b** The active region of the device is a ~20 × 40 *μm* mesa etched from the *2DEG* heterojunction. Four of the twelve fingertips connect to separate *Au* gate electrodes. The *2DEG* beneath these is depleted upon application of sufficient negative bias isolating four separate rectangular *2D* "reservoirs". **c**, **d** *Open*-, *pinched*- cross junction regions. Gates separated by 400 *nm* define the leads. Weak coupling is obtained in *pinched*-cross junctions by additional 150-*nm*-wide constrictions formed between two 50-*nm*-wide protrusions at the ends of the gate fingers

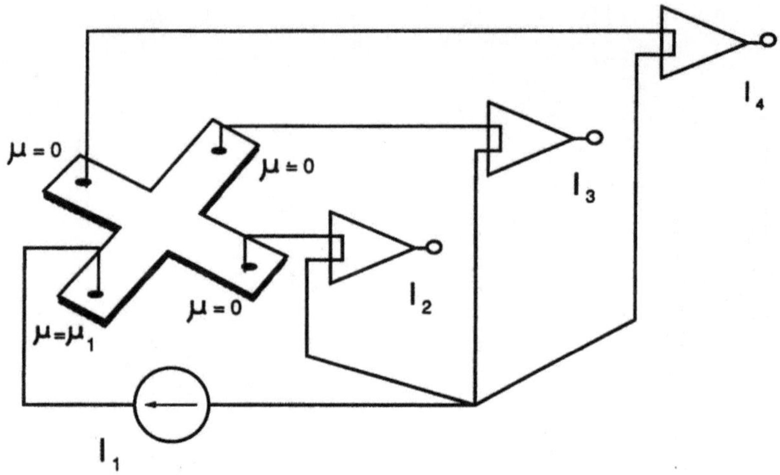

$$\frac{h}{2e}\begin{pmatrix} I_1 \\ -I_2 \\ -I_3 \\ -I_4 \end{pmatrix} = \begin{pmatrix} T_{11} & T_{12} & T_{13} & T_{14} \\ T_{21} & T_{22} & T_{23} & T_{24} \\ T_{31} & T_{32} & T_{33} & T_{34} \\ T_{41} & T_{42} & T_{43} & T_{44} \end{pmatrix} \begin{pmatrix} \mu_1 \\ 0 \\ 0 \\ 0 \end{pmatrix} - N_1 \begin{pmatrix} \mu_1 \\ 0 \\ 0 \\ 0 \end{pmatrix}$$

$$T_{i1} = \frac{h}{2e}\frac{I_i}{\mu_1} \quad (i \neq 1)$$

$$(T_{11} - N_1) = \frac{h}{2e}\frac{I_1}{\mu_1}$$

Fig. 3. Conceptual basis for direct experimental measurement of the transmission matrix. One specific set of boundary conditions leads to simplification of Büttiker's equations: current is sourced into reservoir 1 and simultaneously sunk via 2, 3, and 4. External, ideal, short-circuit connections which link the current-sinking reservoirs *equalize* their chemical potentials. In this configuration, the magnitude of the currents which flow are linearly related to one column of **T**. Other columns are obtained by disconnecting the cross, rotating it by multiples of 90°, and reconnecting to repeat the process

To obtain other columns of **T** ($i \neq 1$) one changes the source lead and follows the same procedure, equalizing the remaining three reservoirs.

Experimental Implementation, I

Practically, of course, real short-circuits between reservoirs will have finite resistance. Depending on the transmission probabilities (which change with external fields) the flux scattered into each of the reservoirs varies. Additionally, each path from a reservoir to the external point defined as "ground" will, in general, have a unique resistance. Both conspire to generate path-specific potential drops which render external short-circuits ineffective at equalizing the μ_i which are internal to the sample. We can circumvent this difficulty by including a separate I-contact and V-contact at *each* reservoir, and by introducing external, variable terminating impedances between each (sink) I-contact and "ground" (Fig. 4). This enables us to simultaneously monitor *all* the μ_i in the presence of current flow and null differences between them by iteratively adjusting the terminating impedances. This procedure is outlined in the flow chart of Fig. 4.

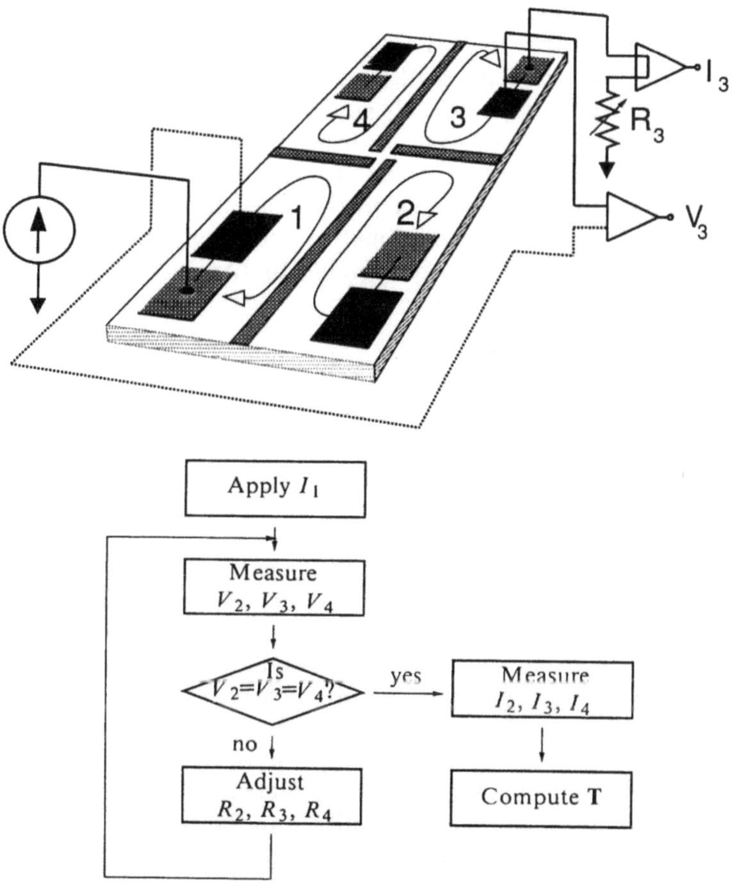

Fig. 4. Configuration for dynamic measurement of **T** while external fields are varied. For simplicity, only connections to leads 1 and 3 are shown. Each reservoir is separately contacted by a I-contact (*light shading*) and a V-contact (*dark shading*). Circuitry at 2 and 4 is identical to those at 3; these current-sinking reservoirs are terminated through a current amplifier (i.e., ammeter) to a variable impedance path to "ground". Voltage amplifiers, referenced to reservoir 1, allow measurement of the chemical potentials at 2, 3, and 4. These are monitored while iteratively adjusting the variable terminating impedances until their equalization is realized. This process, summarized in the flow diagram, must be carried out for each value (point) of the external field investigated

In the experiments it proves important to correctly assign the role of the contacts at each reservoir. This acts to minimize contributions to the measured potentials, V_j, from Hall potentials developed in the presence of a finite magnetic field (Fig. 4). Similar considerations have been discussed previously for the case of a two-probe geometry [4].

Non-Idealities, Errors and Associated Unpleasantness

Figure 5 schematically depicts the voltage drops which occur in our samples. Two undesirable internal voltage drops persist in our measurements, complicating evaluation of the T_{ij}. These arise from the geometrical "spreading" resistance between each reservoir and lead, and from scattering within the leads themselves. These can be minimized, but cannot be completely eliminated. In both cases their magnitude can be estimated from various two-probe measurements [5]. The special geometrical configuration of our samples reduces them to the level of several percent corrections to the quantized conductance of the leads themselves, in the few-mode limit of our investigations. One additional complication remains—at finite temperature, the T_{ij} actually obtained are *thermally averaged.* As a consequence, from our experiments we actually obtain

$$\langle T_{ij}(\mathrm{T})\rangle \sim \int T_{ij}(E)\left(-\frac{\partial f(E,\mathrm{T})}{\partial E}\right)dE \tag{7}$$

where $f(E)$ is the Fermi function and T is the temperature.

Experimental Implementation, II

In experiments, we are interested in studying the evolution of **T** in response to swept external "fields" (e.g., magnetic field, gate bias, temperature). In this situation the μ_i must be equalized *dynamically.* As mentioned previously, since the transmission to the reservoirs varies with external field, the potential drops along connections to an external "ground" will not remain constant. This precludes any static method of inferring T_{ij} based solely upon current measurements (a static approach was employed in [6]); to proceed one must dynamically adjust the μ_i to maintain the equalization condition as the external field varies. In practice, we find that self-consistent adjustment of terminating impedances is less convenient than an alternate procedure which allows us to employ static terminations.

System of Linear Equations

This alternate approach involves building a set of linear equations through successive measurement (sweeps) which we ultimately solve to obtain **T**. We accomplish this as follows: current is sourced into the *I*-contact at one reservoir, e.g., reservoir 1, while sinking it through various terminations of the *I*-contacts at the other reservoirs. *All* voltages and currents are simultaneously measured with swept field for seven different configurations of the *static* terminations. We denote these seven configurations by index α. In the case where current is *sourced* into lead 1 of an

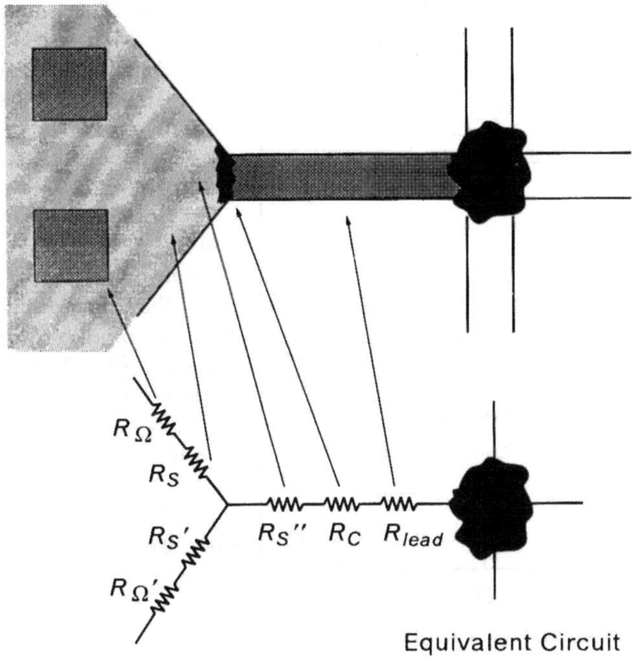

Fig. 5. Sketch and equivalent circuit diagram depicting internal voltage drops occuring in one arm of a real four-probe sample. *Top*: A wide reservoir, on the left, contains voltage- (*top square*) and current- (*bottom*) contacts; they are formed by metal alloyed into the semiconductor to form heavily doped 3*D* regions. These provide ohmic contact to the wide, planar 2*D* electron gas "reservoir" region (*shaded*) which narrows down to a few-mode "lead". The lead carries current to the "scatterer" (*dark blob*), to which all other leads also converge. *Bottom*: The equivalent circuit models all resistances as lumped elements: R_Ω and R_Ω' are the $3D \rightarrow 2D$ contact resistances; R_S, R_S', and R_S'' are "spreading" resistances associated with the necking-down of the ballistic 2*D* conductor; R_C is the $2D \rightarrow 1D$ contact resistance contained within the Büttiker equations; and R_{lead} represents residual scattering along the *quasi*-ballistic lead itself. Use of two ohmic contacts circumvents R_Ω, R_Ω', R_S, and R_S'. The flared geometry reduces R_S'' to the level of a few percent correction to the quantized lead resistance, R_C, for measurments in the few-mode limit. Finally, use of very short leads preserves nearly ballistic transport and minimizes R_{lead}. In the ideal sample, all resistances, except for R_C, would be zero

open-cross junction (discussed below), these represent the complete set of separate configurations (cfg.) by which current can be *sunk* from one (3 cfg.), two (3 cfg.), or all three (1 cfg.) of the remaining leads. For each lead *i*, this generates an overdetermined set of equations, labeled by the configuration index α, of form:

$$\frac{h}{2e^2} I_i^{(\alpha)} = [T_{i2} V_2^{(\alpha)} + T_{i3} V_3^{(\alpha)} + T_{i4} V_4^{(\alpha)}] \qquad (8)$$

The least-squares solution to these systems of equations provides the transmission coefficients T_{i2}, T_{i3}, and T_{i4} which are "best fit" to the complete set of experimen-

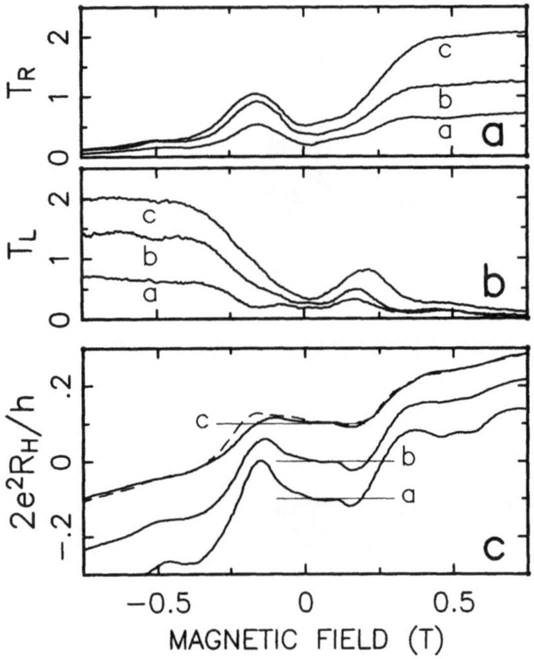

Fig. 6a–c. Data from a *pinched*-cross junction at three gate voltages labelled *a*, *b*, and *c*. At these biases; $V_G \sim -0.90$, -0.84 and -0.76 V, respectively. The corresponding quantized conductance through the constrictions, as shown in the *inset of Fig. 7b*, indicates transmission of ~0.5, 1.5, and 2.5, respectively. **a, b** "Turning coefficients", T_R and T_L, into the right and left probes. Because of the device symmetry, $T_R = T_{21}(B) \sim T_{43}(B)$ and $T_L = T_{41}(B) \sim T_{23}(B)$. **c** Hall resistance, $R_H = R_{13,24}(B)$, reconstituted from measured T_{ij} (solid curves). Also shown is R_H directly measured at bias *c* (*dashed*). Curves for *a* and *c* are offset by +0.1 and −0.1, respectively

tal data. Furthermore, since our system is overdetermined, we obtain information about the variances in the T_{ij}. As described below, these provide an additional window on the physics [5].

Experimental Details

The experiments are carried out using four-probe samples specifically designed for these measurements [5]; these require eight contacts. A *GaAs/AlGaAs* two-dimensional electron gas (2DEG) heterojunction ($\mu = 1.0 \times 10^6\ cm^2/V \cdot sec$, $n_S = 3.0 \times 10^{11}\ cm^{-2}$) is electrostatically gated to form reservoirs, leads, and a "scatterer" (Fig. 2). Four gate electrodes at the surface of a heterojunction can be independently biased to deplete the *2DEG* beneath. This allows us to electrostatically isolate four wide rectangular regions which serve as *2D* reservoirs. From these, narrower channels converge to the central junction region. The experiments reported here are carried out at ~2*K* using a 10 *nA* drive current.

Our approach provides a set of transmission coefficients satisfying the reciprocity relation [3], $T_{ij}(B) = T_{ji}(-B)$, to within a few percent. This is the level of the variances obtained from our least squares fit to obtain the T_{ij}. Furthermore, we verify that "reconstituted" resistances, formed using *measured* T_{ij} in expressions for $R_{kl,mn}$ from Eq. (2), are consistent with results from direct four-terminal measurements (Figs. 6b, 7e).

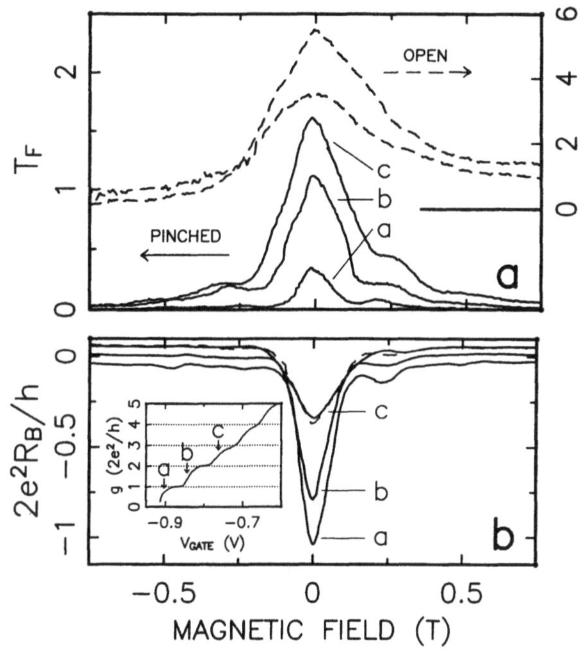

Fig. 7a, b. Forward transmission coefficient, $T_F = T_{42}(B)$. **a** The bottom curves (*solid*) show T_F for electrons propagating through both constrictions of a *pinched*-cross junction, following the trajectory depicted in Fig. 8a. Three bias conditions, labelled *a*, *b*, and *c* (*inset*, *panel b*) were employed; these yield transmissions ~0.5, 1.5, and 2.5 through the constrictions, respectively. The top (*dashed*) curves show T_F at two separate bias voltages for electrons propagating through an open-cross junction. For the two conditions chosen (which closely correspond to biases *a* and *c* used for the *pinched*-cross case), T_F is approximately 3 and 5 at zero field, respectively. **b** Bend resistance, $R_{B,} = R_{12,43}(B)$ for the *pinched*-cross junctions, reconstituted from measured values of T_{ij} (*solid*). Also shown is R_B directly measured at bias *c* (*dashed*). Curves *a* and *c* are offset by 0.05 and −0.05, respectively, for clarity of presentation

Application 1: Perturbation of Probes upon Measurements

Background

With macroscopic samples, one usually thinks of a non-perturbative measurement as involving an ideal potential probe, one that allows no net current to flow from the sample. In this limit, equipotential contours along the current path within the sample are not disturbed by the measurement. In the mesoscopic regime, the situation is much more subtle. From the standpoint of Büttiker's model, the distinction between current and voltage probes is made at the *reservoirs*. But the coupling of a "probe" to the "measurement region" occurs at the *scattering region* itself. Clearly, the manner in which the few-mode "leads" are attached to the "scatterer" is critical. Since transport is strongly non-local, special care is required in designing an experiment where differentiation between "probe" and "scatterer" is meaningful.

In the case of a ballistic four-probe junction, *strong coupling* is always obtained when, at the junction, little or no geometrical distinction exists between "voltage" and "current" probes. This is exemplified by the symmetric *open*-cross junctions of Fig. 2. Previous experimental and theoretical work have elucidated an entire family of transport anomalies which arise from what has become known as "junction scattering" in this configuration (for a recent review, see, e.g., [7]).

To realize the opposite extreme requires voltage probes almost totally decoupled from the main current path. Engquist and Anderson [8] define a weakly-coupled probe as accepting only *infinitesimal* transmission, t, compared to transmission along the main conductor itself, T. In this limit, $t/T \ll 1$, it is envisioned that the local chemical potential and, thereby, the intrinsic transport properites of quasi-1D wires might be measured non-perturbatively [8–10]. We achieve weak coupling in these experiments utilizing pinched-cross junctions (Fig. 2). These employ gated constrictions to separate probes from the main conductor. As described below, our findings illustrate that "weak coupling" is *not* synonomous with "non-perturbative coupling".

Experimental Results with Variable Coupling

We summarize our studies by a representative set of results from a *pinched*-cross junction (Fig. 6). In panels $a - c$, we display measurements obtained at three gate voltages (labelled a, b, and c). These biases permit only small transmission through the constriction, t $\sim$ 0.5, 1.5, and 2.5 (cf. Fig. 7b, inset), respectively, while allowing much larger transmission along the main conductor, $T \sim 7$, 8, and 10. Bias a provides the weakest coupling, $t/T \sim 0.07$. Our approach allows us to adjust t/T while closely maintaining the electrostatic symmetry of the junction. (We achieve this by monitoring the symmetry of **T** while adjusting the bias on each of the four separate gates.) This flexibility is essential to the success of these experiments.

We first discuss the "turning coefficients" for electrons injected into a *pinched*-cross junction and subsequently collected by weakly-coupled right and left leads, T_R and T_L. With sufficient magnetic field (applied normal to the plane of the junction), all flux is directed toward one of the side leads at the expense of the other, the choice dependent upon the sign of B. This behavior is evident in Fig. 6a, b. However, additional features appear at fields where the Lorentz force should deflect almost all injected flux into the opposite side probe. Most prominent are the peaks in T_R (and T_L) at $B \approx -0.2$ T ($+0.2$ T). *Open*-cross junctions exhibit similar behavior [5]. This anomalous transmission significantly impairs the development of the Hall effect across the junction at low-B. The (reconstituted) Hall resistance, R_H, is first seen to quench [11] then reverse sign [12] as transmission into the probes is reduced (biases $c \rightarrow a$, Fig. 6c).

We ascribe the unexpected features we have observed in the turning coefficients to electron flux specularly reflected from the junction boundary into the opposite probe (Fig. 6c, inset). Among numerous mechanisms proposed to explain these anomalies in R_H [7], only this "rebound" effect, first inferred from resistance measurements [12], appears to be consistent with the structure we observe in the T_{ij}.

It is worth noting that only for bias *c* can R_H be directly measured (Fig. 6c, *dotted curve*). For biases *a* and *b*, weak coupling precludes direct resistance measurements, which utilize voltage probes. In the laboratory, voltage measurements always involve some amount of current flow, because real voltmeters have finite impedance. Furthermore, parasitic leakage currents always accompany any external circuitry attached to a voltage reservoir (e.g., leads running down the cryostat, etc.). In general, the steady-state μ_i attained by any *voltage* reservoir fed by a weakly-coupled probe can easily be rendered meaningless by this non-ideal leakage of electron flux to the external circuitry.

Microfabrication-Imposed Potentials, I: Fact Vs Fancy

What is especially surprising is that this quenching effect *grows* as the probes become decoupled. At bias *a*, the anomalous peaks in T_L and T_R acquire similar magnitude to the total flux ultimately collected by the "correct" probe at higher B. Quenching of R_H is thereby *enhanced* by progressive probe decoupling. Presumably, if junction scattering is the mechanism underlying these phenomena, they should instead vanish in the weak coupling limit. Calculations have confirmed this expectation [9, 10], but closer comparison shows the contradiction arises solely because idealized potentials are used to model the weakly-coupled probes. In our data, signatures of junction scattering persist even in the limit of weak coupling. Even with almost completely constricted side probes, transmission through the main conductor manifests a related "magneto- size effect" [5]. These observations signify that the main conductor continues to be perturbed by the proximity of a probe, even in the weak coupling limit.

Our observations illustrate behavior generic to small conductors patterned from a *2DEG*. Because the electron density in these systems is low, screening lengths become appreciable. Furthermore, optimized heterojunction geometrics always involve "burying" the heterointerface far from deleterious surfaces; this enables attainment of high electron mobility. This constrains the electrostatics of confinement to be imposed remotely. Both of these factors conspire to yield "soft" (i.e., spatially smoothed) imposed potentials. Thus, invariably, it appears that some vestige of the potential intended to define a "probe" will always also affect the "channel".

Application 2: Momentum Spectroscopy on a Quantum Point Contact

Series Arrangement of Quantum Point Contacts

The *pinched* cross junction shown in Fig. 1 constitutes a collinear pair of point contacts embedded within a four-probe geometry equipped for T_{ij} measurement.

This gives us the ability to perform new studies to explore the details of the conductance quantization phenomenon. Experiments utilizing a simple series geometry [13] have provided an experimental verification that the electron distribution emerging from a constriction is collimated, favoring forward propagation, as had been argued previously [14]. Our approach permits us to undertake similar studies, but with the added ability here to carry them into the regime of nearly vanishing emitted flux.

These experiments can be viewed as involving ballistic constrictions which act as an *injector* and a *collector* of electron flux. We employ the collector to "image" the momentum distribution emerging from the injector, which is biased into the regime where conductance quantization is observed. Figure 7a shows T_F, the forward transmission coefficient for electrons propagating through both constrictions, i.e., from injector to collector. Upon application of finite magnetic field, B, the "beam" emerging from the injector is deflected from the collector and, thus, T_F falls. This allows us to evaluate the angular profile of the injected beam. For *open*-cross junctions we find a monotonic decay with increasing B, but in *pinched*-cross junctions we find a *non-monotonic* decay. New features emerge, which are offset from the maximum at $B = 0$ [5]. This neatly demonstrates the enhanced resolution of transport mechanisms provided by direct measurements of T_{ij}. One might expect to observe corresponding features in the bend resistance [15], $R_B = R_{13,42}$, which can be obtained by a straightforward four-probe measurement. (The earlier studies of electron beam collimation involved what was, in essence, a measurement of R_B [13].) In the limit of weak coupling, however, this becomes increasingly difficult. We display reconstituted traces of R_B (*solid curves*, Fig. 7b) and, for the case of bias *c*, a direct four-probe measurement of R_B (*dashed curve*, Fig. 7b). In our experiments, only for bias *c* is data from direct four-probe measurements of R_B meaningful. As seen, this trace is almost featureless except for the central dip at zero field.

Momentum Spectroscopy

The B dependence of the forward transmission, T_F, provides us with a sensitive momentum spectroscopy on the emerging beam. To deduce its angular profile we assume that, after injection, electrons propagate classically within the junction [13]. This approximation is valid at the temperatures of these studies, $T \sim 2K$, where significant phase averaging occurs. If we assume the opposite constriction acts as a *point* collector, electrons emitted at angle θ are collected when $r_c \sin\theta = d/2$. Here $r_c = (h/eB)\sqrt{n_s/2\pi}$ is the cyclotron radius and d is the distance between the constrictions. Figure 8a depicts this classical analysis. We obtain n_s from the quantum Hall effect, whereas d is extracted, assuming hard-wall potentials, from the quantized conductance along the main channel. Figure 9a shows the profile of the collected beam, $T_F(\theta)$, analyzed in this fashion. Our sample geometry permits two additional modes of spectroscopy via the turning coefficients [5]. A small segment of the rounded electrostatic boundary near the collector constriction specularly reflects flux into a side probe (via a "rebound" trajectory, as shown

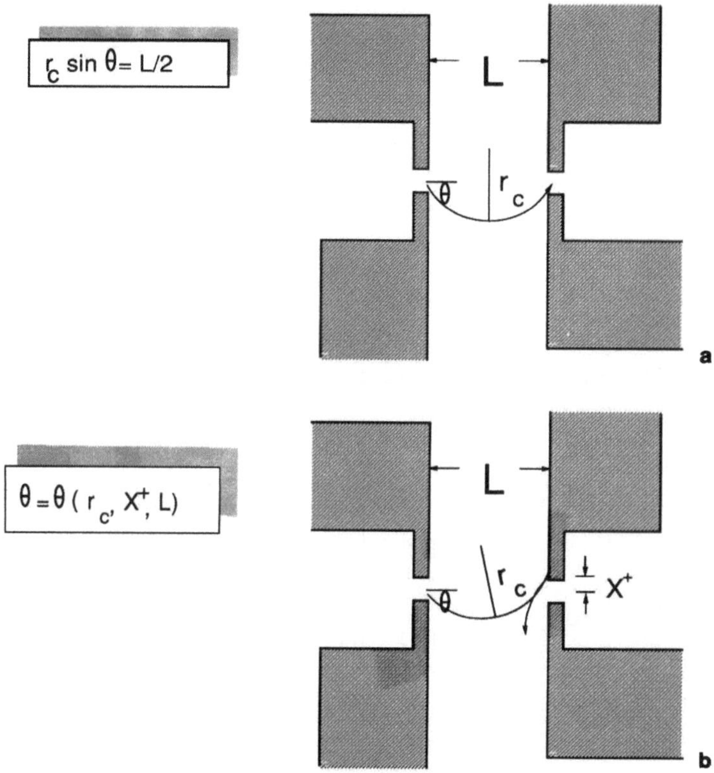

Fig. 8a, b. Classical analysis of the injected beam profile from the magnetic field dependence of transmission coefficients. **a** Angular dependence from the forward transmission. Schematic representation of the semiclassical trajectory determining T_F. **b** Angular dependence from the turning probabilities. An additional parameter X^+ locates the reflecting portion of the boundary with respect to the center of the constriction opposite the injector. Imaging via the other turning coefficient involves an analogous parameter, X^-

in Fig. 8b). This makes it clear that the rebound feature occurs at a magnetic field where the *central maximum* of the beam reflects into a side probe. As the beam continues to be swept past this reflecting spot, the angular profile of one side of the beam is imaged in the turning coefficient. Changing the sign of the magnetic field sweeps the beam directly *into* the side probe, which overwhelms the reflected image. (However, in this situation, the second half of the angular profile appears in the *other* turning coefficient, since reflection also occurs at an analogous spot on the other side of the collector, cf. Fig. 8b). As illustrated in Fig. 9b, c, the full beam profiles separately imaged by T_F and by the turning coefficients, T_L and T_R, are in qualitative agreement. Use of the turning coefficients to image the beam involves a slightly more complicated classical geometric analysis, with an extra parameter (Fig. 8b) [5]. In this latter case, we aim not so much for strict quantitative agreement as for a consistent overall picture.

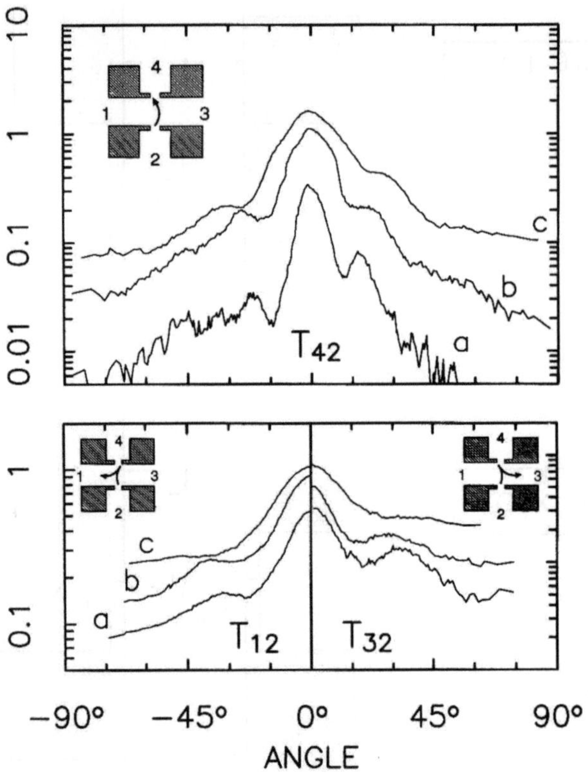

Fig. 9. Angular profile of the "beam" emerging from one constriction at biases *a*, *b*, and *c*. *Top*: T_F shown as a function of angle of emission. At these biases result, the separation between points contacts, *d*, are ~210, ~220 and ~270 *nm*, respectively. The corresponding densities, n_s, within the junction are ~1.8, ~2.0, and ~2.2 × 10^{11} cm^{-2}, respectively. *Left, Right*: Angular profile imaged utilizing the turning probabilites. At one sign of the magnetic field the electron beam is "steered" toward one side probe, at the expense of the other. In this situation, the probe with the smallest transmission can be used to image the beam profile. This is possible because only a small part of the junction boundary is correctly oriented to reflect flux to this probe via a "rebound" trajectory, shown in the *insets*

Observation of Structure in the Electron Distribution Function

Although propagation within the junction is essentially classical in these experiments, transport at the quantum point contacts is *not*. We constrict them to the point where, at most, only a few modes propagate. In this quantum mechanical regime, side lobes appear in the emerging beam (Fig. 10a). This structure cannot be attributed to diffraction. We have calculated the relative strength of secondary diffraction maxima for the most extreme case of electrons emerging from an abruptly terminated constriction into a wide 2*D* region. We find these maxima to be several orders of magnitude smaller in intensity than the central beam [5]. Smoother terminations reduce these maxima even further. On the other hand, features of the magnitude seen in these experiments can result from injection of a

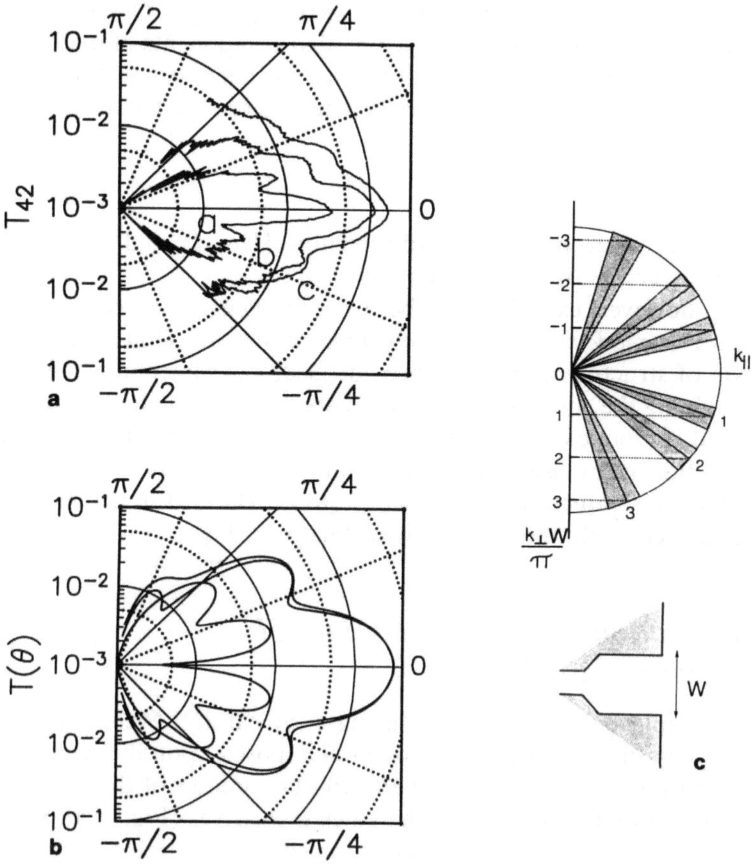

Fig. 10. a Polar representation of the emerging "beam" for biases *a*, *b*, *c*, employing a logarithmic radial coordinate. Only the angular profile obtained from T_F is shown. **b** Calculated angular profile of the momentum distribution emerging from a constriction allowing essentially only a single mode to propagate through its narrowest region. The model potential generating this data, used and described in connection with trace *a* of Fig. 12, induces intermode scattering because of its spatial non-adiabaticity. The outer curve representing the outgoing distribution is decomposed into its *odd* and *even* parity components which clearly display modal structure in the form of *three* and *four* peaks, respectively, on each side of zero angle. As described in the text, the odd peaks for $\theta > 0$ arise from the intermode scattering. (The much weaker even peaks arise from residual evanescent mode coupling through the short model constriction.) **c** Simple k-space diagram showing how a multimode beam, because of the transverse quantization, contains modal components which emerge at discrete angles when injected into a wide 2D region

multimode distribution into a semi-infinite (*2D*) region [5]. One expects from k-space arguments, such as shown in Fig. 10c, that components of multimode beam emerge at specific angles in this situation. This has been demonstrated by explicit calculations [16]. In this case the nth mode emerges centered at angles

$$\Theta^{(n)}{}_{\pm} \sim \pm \sin^{-1}\left(\frac{k_{\perp}^{(n)}}{k_F}\right) \tag{10}$$

symmetrically offset from the forward direction. Here, k_F and $k_{\perp}^{(n)}$ are the Fermi and (nth) transverse wave vectors, respectively.

Our data, however, exhibit side lobes even for bias a, i.e., *even when only the lowest transverse mode propagates through the constriction* (Fig. 10a). In this case, observation of structure appears to provide clear experimental evidence that the potential defining the constriction is non-adiabatic; i.e., it scatters outgoing flux to higher modal components of the emerging beam. Our calculations, described below, support this hypothesis. Furthermore, they confirm that intermode scattering need not destroy conductance quantization (as is exhibited in the experiments), a point recently demonstrated by explicit calculations [17].

Microfabrication-Imposed Potentials, II: Another Dose of Reality

Although the split-gate electrodes inducing a typical *2DEG* constriction may be morphologically smooth, the actual potential imposed upon the electrons will not be. The depletion region defining the *effective* boundaries of the constriction will not follow the smooth contours of the gates themselves (Fig. 11a). Calculations indicate that ionized donors, randomly situated nearby, induce strong potential fluctuations [18]. These will have there most profound effect in regions which are very nearly depleted, i.e., at regions defining the electrostatic boundaries of the conductor. We believe that, in a ballistic constriction close to pinch-off this, *in general*, results in transport controlled by a single, low density, narrow region (Fig. 11a): i.e., a *critical path*.

Model of a Critical Path

We illustrate this by recursive Green's function calculations (see, e.g., [7] or [19]) employing the model potential depicted in *Fig. 11b*. A narrow critical path (I), connects to a "transition" region (III) of width W_{III} and length L which then opens into the center of the junction itself. The narrowest (I) and transition (III) regions are connected by a short zone (II) where the channel widens abruptly. In this region (II) the potential is spatially *non*-adiabatic and intermode scattering can occur. Yacoby and Imry show that a channel of varying width becomes non-adiabatic if

$$\frac{dW}{dx} > \frac{1}{N(x)} \sim \frac{\pi}{k_F W(x)} \tag{11}$$

where $W(x)$ is the (transverse) width at (longitudinal) position x [20].

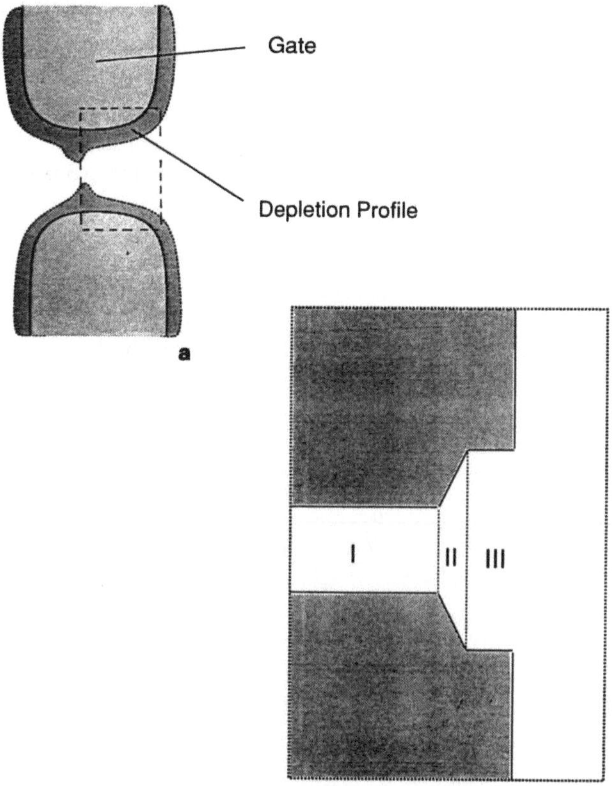

Fig. 11a, b. Critical path model of a quantum point contact. **a** The depletion region defining the *effective* boundaries of the constriction (*dark shading*) do not follow the smooth contours of the gates themselves (*light shading*) **b** Model potential for the half of the constriction contained within the *dashed box* of the sketch in **a**. The narrowest region (I) of width W_I and potential V_I empties into the transition region (III) of fixed width $W_{III} = 3.2\lambda_F$, length $L_{III} = 1.6\lambda_F$, and potential $V_{III} = 0$. These are connected by an non-adiabatic region (II), of fixed length, $0.8\lambda_F$, with width and potential *both* linearly graded to match at regions I and III. For curve *a*, *b*, and *c* of Figs. 10b and 12, $W_I = 0.95$, 1.3 and $1.6\lambda_F$, respectively; while $V_I = 0.38$, 0.04 and $0.00E_F$, respectively

Calculations of Injected Distributions in the Few-Mode Limit

Our calculations yield $T(\theta)$, the angular profile of the emerging beam. In Fig. 12a, three such curves are displayed, corresponding to one, two, and three modes propagating through the narrowest region (I). In each case, modal features are exhibited, at

$$\Theta^{(n)}_{\pm} \sim \pm \sin^{-1}\left(\frac{n\pi}{k_F W_{III}}\right) \tag{12}$$

similar to those observed experimentally. The transition region (III) not only determines the angular position of these features, but also their angular spread.

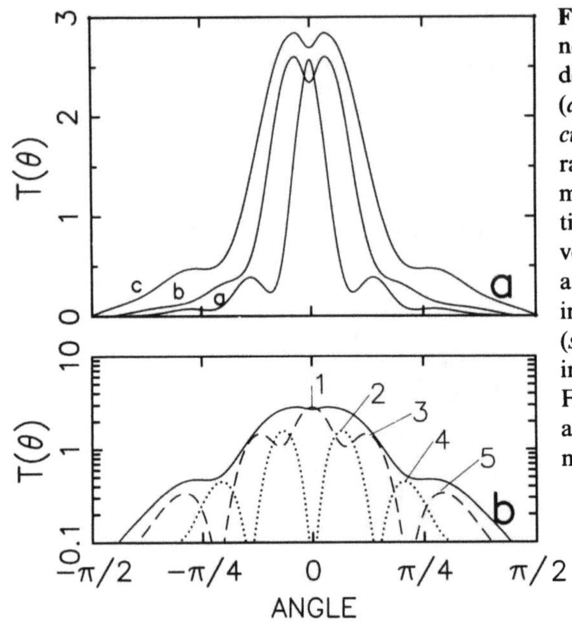

Fig. 12. **a** $T(\theta)$, flux per unit angle normalized by $\hbar k_F/m^*$. **b** Modal decomposition of curve *c* into even (*dotted curve*) and odd parity (*dashed curve*) components. These are separately conserved because of the mirror-plane symmetry of the potential. Peaks corresponding to transverse modes 1 through 5 of region III are labelled. For curve *c*, the predominant "shoulders" of the sum curve (*solid*) arise from intermode scattering into the fourth and fifth modes. For curve *a* analogous "shoulders" arise *solely* from scattering from mode 1 → 3 (*not shown*)

For these calculations our model consists of a narrow region (I) of width W_I and potential V_I which empties into the transition region (III) of fixed width $W_{III} = 3.2\lambda_F$, length $L = 1.6\lambda_F$, and potential $V_{III} = 0$. These are connected by an non-adiabatic region (II), of fixed length, $0.8\lambda_F$, with width and potential *both* linearly graded to match at regions I and III. For curve *a*, *b*, and *c*, $W_I = 0.95$, 1.3 and $1.6\lambda_F$, respectively; while $V_I = 0.38$, 0.04 and $0.00E_F$, respectively.

For ballistic propagation,

$$T_F(\theta) \propto \min[T_i(\theta), T_c(\theta)] \tag{9}$$

where subscript *i* (*c*) represents injector (collector). The proportionality factor involves the effective solid angle subtended by the collector. Here, we do not attempt to model the collection efficiency [5], thus the amplitudes of the calculated curves of Fig. 12 do not follow those of experiments. We present calculations solely for the *injected* distribution, $T(\theta)$, to demonstrate qualitative agreement for three relevant configurations of the constriction potential.

Modal structure is imaged in the calculated traces because those electrons which pass from I → III in a given mode, *n*, become collimated, despite the fact that intermode scattering occurs. For those electrons which remain in the same mode as they pass into III, the increased width of III lowers their transverse momentum, $k_\perp^{(n)}$, thereby increasing their forward momentum,

$$k^{(n)}{}_{\parallel} = \sqrt{k_F{}^2 - (k_\perp^{(n)})^2} \tag{13}$$

A related effect may play an important role in STM imaging [21]. We find that region III must be of length $L \geq \lambda_F$ for side lobes to emerge. This is clearly demonstrated in Fig. 13, where we display the case for curve *a* where only the length of

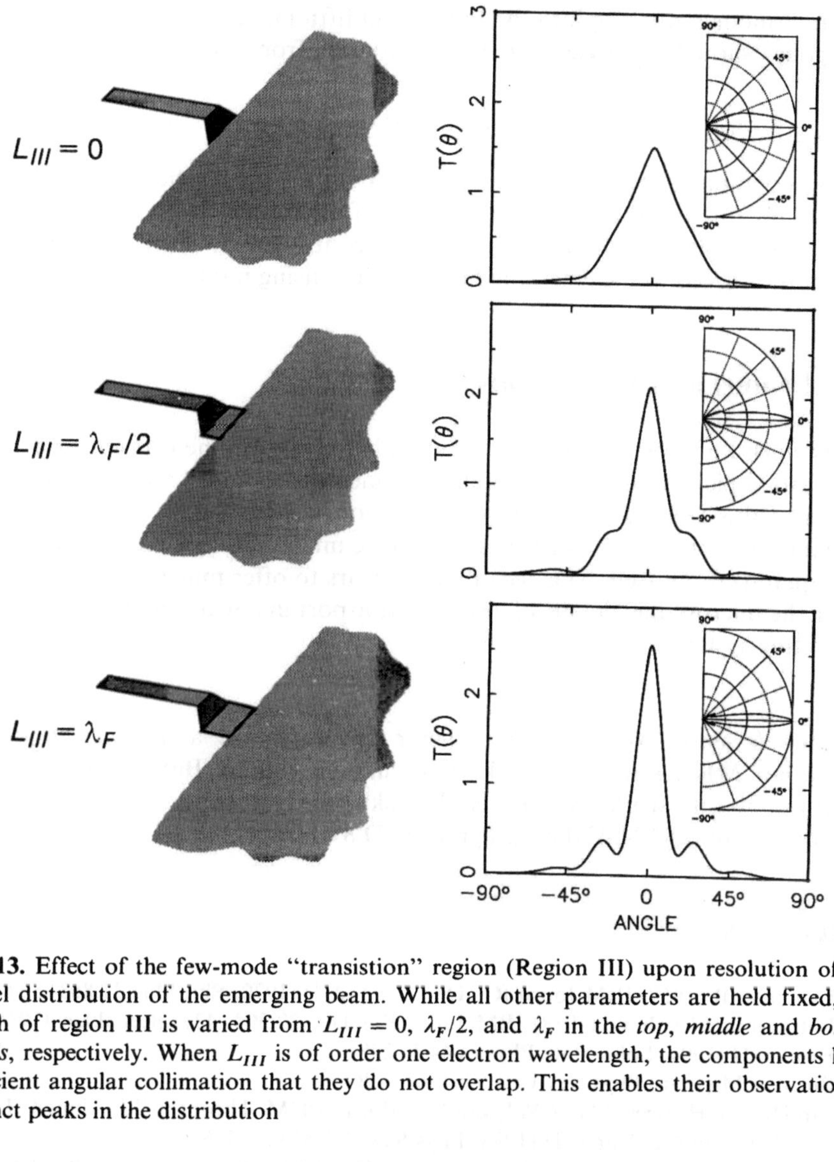

Fig. 13. Effect of the few-mode "transistion" region (Region III) upon resolution of the model distribution of the emerging beam. While all other parameters are held fixed, the length of region III is varied from $L_{III} = 0$, $\lambda_F/2$, and λ_F in the *top*, *middle* and *bottom panels*, respectively. When L_{III} is of order one electron wavelength, the components have sufficient angular collimation that they do not overlap. This enables their observation as distinct peaks in the distribution

III is varied. For $L < \lambda_F$, diffraction broadens the modal components into a single peak [5, 22].

Because of the mirror-plane symmetry of the potential (through both constrictions), even and odd parity momentum states do not mix. This allows us to perform a modal decomposition of the calculated traces to identify the origin of the shoulders imaged in the beam profile. Figure 12b shows this decomposition for curve *c*. Peaks corresponding to transverse modes I through 5 of region III are labelled. For curve *c*, the predominant "shoulders" of the sum curve (*solid*) arise from intermode scattering into the fourth and fifth modes. For curve *a*, however, analogous "shoulders" arise *solely* from scattering from mode 1 → 3.

Non-Adiabaticity and Intermode Scattering

Our observations show that conductance through a constriction can remain approximately quantized even in the presence of intermode scattering. This is consistent with a picture where conductance is controlled by a critical path which induces transitions in the emerging beam while causing minimal backscattering.

Conclusions and Projections

In this work we have shown that it is possible to measure the transmission matrix of a mesoscopic conductor in carefully designed experiments. Our first two applications of this technique have given us a more detailed view of the underlying mechanisms of transport anomalies in ballistic microjunctions and of the conductance quantization itself. This technique appears to offer much promise for elucidating the microscopic details of electron transport in a wide variety of ultrasmall conducting systems.

Acknowledgments. We thank A. Scherer for assistance and advice regarding fabrication. MR acknowledges helpful discussions with M. Büttiker in 1988 at the initiation of this work. KS gratefully acknowledges support from the Hertz Foundation and the NSF through grant EID 87-11693.

References

1. Landauer R (1957) IBM Journ Res Dev 1:223–231; (1970) Phil Mag 21:863–867
2. Entire journal volume (1988) IBM Journ Res Dev 32:303–437; Beenakker CWJ, van Houten H (1991) Solid State Physics 44:1–228
3. Büttiker M (1986) Phys Rev Lett 57: 1761–1764
4. van Houten H, Beenakker CWJ, van Loosdrecht PHM, Thornton TJ, Ahmed H, Pepper M, Foxon CT, Harris JJ (1988) Phys Rev B37:8534–8536
5. Shepard KL, Roukes ML, Van der Gaag BP (1992) Phys Rev Lett 68:2660–2663; Shepard KL, Roukes ML, Van der Gaag BP (1992) Phys Rev B, Oct. 15, 1992.
6. Greene SK, Pepper M, Wharam DA, Peacock DC, Ritchie DA, Frost JEF, Hasko DG, Ahmed H, Jones GAC (1991) J Phys: Condensed Matter 3:1961–1965.
7. Baranger HU, DiVincenzo DP, Jalabert RA, Stone AD (1991) Phys Rev B44:10637–10675
8. Engquist H-L, Anderson PW (1981) Phys Rev B24:1151–1154
9. Peeters FM (1988) Phys Rev Lett 61:589–592
10. Akera H, Ando T (1989) Phys Rev B9:5508–5511
11. Roukes ML, Scherer A, Allen SJ, Craighead HG, Ruthen RM, Beebe ED, Harbison JP (1987) Phys Rev Lett 59:3011–3014

12. Ford CJB, Washburn S, Büttiker M, Knoedler CM, Hong JM (1989) Phys Rev Lett 62:2723–2727
13. Molenkamp LW, Staring AAM, Beenakker CWJ, Eppenga R, Timmering CE, Williamson JG, Harmans CJPM, Foxon CT (1990) Phys Rev B41:1274–1277
14. Beenakker CWJ, van Houten H (1989) Phys Rev B39:10445–10448
15. Takagaki Y, Gamo K, Namba S, Ishida S, Ishibashi K, Aoyagi Y (1988) Solid State Comm 68:1051–1054; Timp G, Baranger HU, deVegvar P, Cunningham JE, Howard RE, Behringer R, Mankiewich PM (1988) Phys Rev Lett 60, 2081–2084
16. Nakamura A, Maki Y, Okiji A (1991) Phys Soc Jpn 60:749–752
17. Castano E, Kirczenow G (1992) Phys Rev B45:1514–1517
18. Nixon JA, Davies JH (1990) Phys Rev B41, 7929–7932
19. Shepard K (1991) Phys Rev B44: 9088–9091 (and references therein)
20. Yacoby A, Imry Y (1990) Phys Rev B41: 5341–5350
21. Lang ND, Yacoby A, Imry Y (1989) Phys Rev Lett 63:1499–1502
22. van der Marel D, Haanappel EG (1989) Phys Rev B39:7811–7820

Quantum Wires: From Ballistic Conduction to Fluctuations

18

TSUNEYA ANDO, HIROYUKI TAMURA, AND HIROSHI AKERA[1]

Summary. A review is given of recent numerical studies of quantum transport phenomena in semiconductor quantum structures in magnetic fields. The topics covers the ballistic conduction at quantum point contacts, conductance fluctuations in quantum wires, and the crossover of the quantum Hall effect from two dimension to one dimension.

Introduction

Recent developments in the technology of ultrafine lithography and crystal growth have enabled fabrication of quantum point contacts and ultra-narrow wires at semiconductor heterostructures. The size of such quantum structures is much smaller than the electron mean free path and is comparable to the Fermi wavelength, giving only a few occupied subbands. Numerical calculations of the conductance are required in order to fully understand quantum transport in such ballistic structures in the presence of a magnetic field. The purpose of the present paper is to give a brief review of some examples of such studies.

Below, the numerical method is reviewed briefly. Then, results of quantum-mechanical calculations of the ballistic transport at quantum point contacts are presented. It will be shown that the channel wave function can be directly observed through the magnetic-field dependence of the conductance at a series of two point contacts. Following this, conductance fluctuations in quantum wires are studied and a crossover between quantum and metallic wires and effects of boundary roughness scattering are also discussed. Finally, the transmission probability of edge states in quantum wires in high magnetic fields is discussed and its width dependence is shown to give a critical exponent for localization length of bulk 2D Landau levels.

Numerical Method

We consider a square lattice with lattice constant a and an isotropic nearest-neighbor transfer integral $-t$. We choose the gauge such that

$$(j,l|\mathscr{H}|j+1,l) = -t\exp(2\pi i\tilde{H}l), \qquad (j,l|\mathscr{H}|j,l+1) = -t, \tag{1}$$

[1] Institute for Solid State Physics, University of Tokyo, 7-22-1 Roppongi, Minato-ku, Tokyo, 106 Japan

where $\tilde{H} = \Phi/\Phi_0$ with $\Phi = Ha^2$ being the magnetic flux passing through a unit cell and $\Phi_0 = ch/e$ being the magnetic flux quantum. A confining potential $v(j, l)$ is introduced through the diagonal element of the Hamiltonian and effects of scattering are introduced through randomness of site energy $v(j, l)$ distributed uniformly with a certain width. For $a \ll \lambda_F$, the model simulates a continuum system containing a high concentration of scatterers with δ potential, where λ_F is the Fermi wavelength in two-dimensions.

In order to discuss a scattering problem, we first consider a wire infinitely long in the x direction and consisting of M lattice sites in the y direction (corresponding to a continuum system width $(M + 1)a$). The equation of motion can be written as

$$(E - \mathscr{H}_0)\mathbf{C}_j + tP\mathbf{C}_{j-1} + tP^*\mathbf{C}_{j+1} = 0 \tag{2}$$

where $\mathbf{C}_j$ is a vector describing the amplitudes of the jth cell consisting of M sites and P is an $M \times M$ matrix consisting of phase factors,

$$P_{ll'} = \exp(2\pi i\tilde{H}l)\delta_{l,l'}, \qquad (l, l' = 1, \ldots, M) \tag{3}$$

The Hamiltonian $\mathscr{H}_0$ is an $M \times M$ matrix within a cell.

To obtain independent solutions for Eq. (2) we first put $\mathbf{C}_j = \lambda^j\mathbf{C}_0$. Substituting this into Eq. (2), we have

$$\lambda\mathbf{C}_j = t^{-1}P(\mathscr{H}_0 - E)\mathbf{C}_j - P^2\mathbf{C}_{j-1} \tag{4}$$

which, being combined with $\lambda\mathbf{C}_{j-1} = \mathbf{C}_j$, leads to the following eigenvalue problem:

$$\lambda\begin{pmatrix}\mathbf{C}_j \\ \mathbf{C}_{j-1}\end{pmatrix} = \begin{pmatrix} t^{-1}P(\mathscr{H}_0 - E) & -P^2 \\ 1 & 0 \end{pmatrix}\begin{pmatrix}\mathbf{C}_j \\ \mathbf{C}_{j-1}\end{pmatrix} \tag{5}$$

The wave number k is related to the eigenvalue λ through $\exp[ika + i\pi\tilde{H}(M + 1) = \lambda$. This equation has $2M$ eigenvalues and $2M$ eigenvectors, which are classified into M right- and left-going waves. The M right-going solutions consist of traveling waves with velocity in the positive x direction and evanescent waves decaying exponentially in the positive x direction. Similarly, the M left-going solutions consist of traveling waves with velocity in the negative x direction and evanescent waves decaying exponentially in the negative x direction.

Let $\mathbf{u}_1(-), \ldots, \mathbf{u}_M(-)$ be $\mathbf{C}_0$ of the left-going solutions corresponding to $\lambda_1(-), \ldots, \lambda_M(-)$ and $\mathbf{u}_1(+), \ldots, \mathbf{u}_M(+)$ be $\mathbf{C}_0$ of the right-going solutions corresponding to $\lambda_1(+), \ldots, \lambda_M(+)$. Define

$$U(\pm) = (\mathbf{u}_1(\pm) \cdots \mathbf{u}_M(\pm)) \tag{6}$$

and

$$\Lambda(\pm) = \begin{bmatrix} \lambda_1(\pm) & & \\ & \ddots & \\ & & \lambda_M(\pm) \end{bmatrix} \tag{7}$$

Any left- and right-going waves are written, at $j = 0$ for example, as $\mathbf{C}_0(\pm) = U(\pm)\mathbf{C}(\pm)$, where $\mathbf{C}(\pm)$ is an appropriate vector consisting of expansion coefficients. For general j we have

$$\mathbf{C}_j(\pm) = U(\pm)\Lambda(\pm)^j\mathbf{C}(\pm) \tag{8}$$

which leads to the relation:

$$\mathbf{C}_j(\pm) = F(\pm)^{j-j'}\mathbf{C}_{j'}(\pm) \tag{9}$$

with

$$F(\pm) = U(\pm)\Lambda(\pm)U(\pm)^{-1} \tag{10}$$

Note that $U(\pm)$ is, in general, not a unitary matrix.

Now, let us consider the scattering problem for the wire with length N (from cell 1 to cell N) to both sides of which an ideal wire is attached. First, we separate the amplitude $\mathbf{C}_0$ at cell 0 into the right-going (incident) and left-going (reflected) solutions, $\mathbf{C}_0 = \mathbf{C}_0(+) + \mathbf{C}_0(-)$. Using Eq. (10), the amplitude at cell -1 is written as

$$\mathbf{C}_{-1} = F(+)^{-1}\mathbf{C}_0(+) + F(-)^{-1}\mathbf{C}_0(-) \tag{11}$$

This gives

$$\mathbf{C}_{-1} = F(-)^{-1}\mathbf{C}_0 + (F(+)^{-1} - F(-)^{-1})\mathbf{C}_0(+) \tag{12}$$

Therefore, the equation of motion at cell 0 becomes

$$(E - \tilde{\mathscr{H}}_0)\mathbf{C}_0 + tP^*\mathbf{C}_1 = -tP(F(+)^{-1} - F(-)^{-1})\mathbf{C}_0(+) \tag{13}$$

with

$$\tilde{\mathscr{H}}_0 = \mathscr{H}_0 - tPF(-)^{-1} \tag{14}$$

Note that $\tilde{\mathscr{H}}_0$ is not Hermitian.

On the other hand, only right-going (out-going) waves can exist at cell $N + 1$. Therefore, we have

$$\mathbf{C}_{N+2} = F(+)\mathbf{C}_{N+1} \tag{15}$$

which gives the equation of motion at cell $N + 1$

$$(E - \tilde{\mathscr{H}}_{N+1})\mathbf{C}_{N+1} + tP\mathbf{C}_N = 0 \tag{16}$$

with

$$\tilde{\mathscr{H}}_{N+1} = \mathscr{H}_{N+1} - tP^*F(+) \tag{17}$$

Note that $\mathscr{H}_{N+1} = \mathscr{H}_0$ and that $\tilde{\mathscr{H}}_{N+1}$ is not Hermitian.

Define the Green function $G = (E - \tilde{\mathscr{H}})^{-1}$ with

$$\tilde{\mathscr{H}} = \begin{bmatrix} \tilde{\mathscr{H}}_0 & -tP^* & 0 & \dots & 0 & 0 \\ -tP & \tilde{\mathscr{H}}_1 & V & \dots & 0 & 0 \\ 0 & -tP & \tilde{\mathscr{H}}_2 & \dots & 0 & 0 \\ \vdots & \vdots & \vdots & \ddots & \vdots & \vdots \\ 0 & 0 & 0 & \dots & \tilde{\mathscr{H}}_N & -tP^* \\ 0 & 0 & 0 & \dots & -tP & \tilde{\mathscr{H}}_{N+1} \end{bmatrix} \tag{18}$$

where $\tilde{\mathscr{H}}_j = \mathscr{H}_j$ for $j = 1, \dots, N$. Immediately, we can derive

$$\mathbf{C}_{N+1}(+) = \mathbf{C}_{N+1} = -t(N+1|G|0)P(F(+)^{-1} - F(-)^{-1})\mathbf{C}_0(+) \tag{19}$$

and

$$\mathbf{C}_0(-) = \mathbf{C}_0 - \mathbf{C}_0(+) = [-t(0|G|0)P(F(+)^{-1} - F(-)^{-1}) - 1]\mathbf{C}_0(+) \tag{20}$$

From these equations we can obtain the transmission coefficient $t_{\mu\nu}$ for incident channel ν with velocity v_ν and out-going channel μ with velocity v_μ as

$$t_{\mu\nu} = \sqrt{\frac{v_\mu}{v_\nu}}[-tU(+)^{-1}(N+1|G|0)P(F(+)^{-1} - F(-)^{-1})U(+)_{\mu\nu} \tag{21}$$

and the reflection coefficient $r_{\mu\nu}$ for incident channel ν and out-going channel μ as

$$r_{\mu\nu} = \sqrt{\frac{v_\mu}{v_\nu}}[U(-)^{-1}[-t(0|G|0)P(F(+)^{-1} - F(-)^{-1}) - 1]U(+)]_{\mu\nu} \tag{22}$$

In terms of the transmission coefficient $t_{\mu\nu}$, the conductance G is given by the multi-channel version of Landauer's formula [1]

$$G = \frac{e^2}{\pi\hbar}\sum_{\mu\nu}|t_{\mu\nu}|^2 \tag{23}$$

The Green function can be calculated with use of the well-known recursion method (see, for example [2], and references cited therein).

Quantum Point Contact in Magnetic Fields

The quantum point contact is an ideal system where quantum effects on the transport can be directly studied. The typical example is the quantization of the conductance into integer multiples of $e^2/\pi\hbar$ [3, 4]. The conductance quantization has been a subject of many theoretical investigations (both analytical and numerical) [2].

We consider a wire with length L_x and width L_y ($-L_x/2 < x < L_x/2$ and $-L_y/2 < y < L_y/2$) and introduce a parabolic potential which confines electron motion in the y direction in the vicinity of the center of the wire. At $x = 0$ the potential is parabolic, i.e., $V(y) = E_F(y/\Delta)^2 + V$, and the effective width becomes greater and the bottom energy becomes lower away from $x = 0$. The parameter Δ/λ_F characterizes the width of the channel of the point contact and V represents the maximum bottom energy, and the effective width becomes greater and the bottom energy becomes lower away from $x = 0$. The wire length L_x determines the length scale over which the potential varies. This model is expected to simulate the potential variation in the vicinity of actual quantum point contacts for appropriate values of the parameters. The potential profile for $\Delta/E_F = 1$ is shown in Fig. 1.

Figure 2 gives an example of the conductance as a function of V for different magnetic fields. The degree of the quantization depends sensitively on L_x/λ_F. In the absence of a magnetic field, the conductance varies almost linearly with V for $L_x/\lambda_F = 4$, while it becomes well quantized for $L_x/\lambda_F = 16$. The conductance quantization becomes better in higher magnetic fields, as has been observed in experiments and discussed theoretically. This is closely related to the appearance of edge

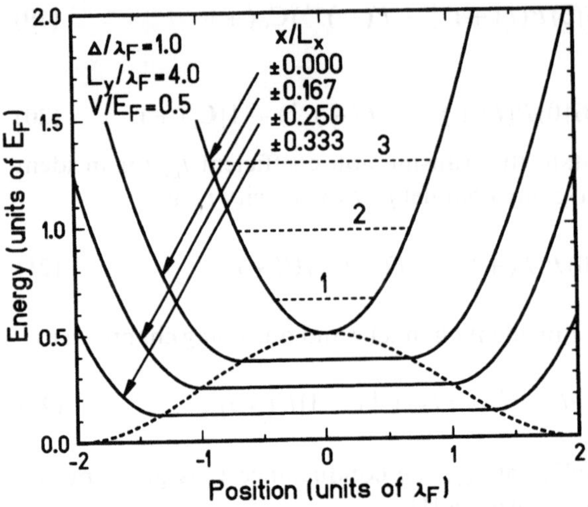

Fig. 1. The confining potential in the vicinity of a model quantum point contact. The *solid lines* represent the potential at $x/L_x = 0$, $\pm 1/6$, $\pm 1/4$, and $\pm 1/3$ as a function of y. The *horizontal dashed lines* represent the bottom of the one-dimensional subbands at the contact, i.e., for the parabolic potential at $x = 0$

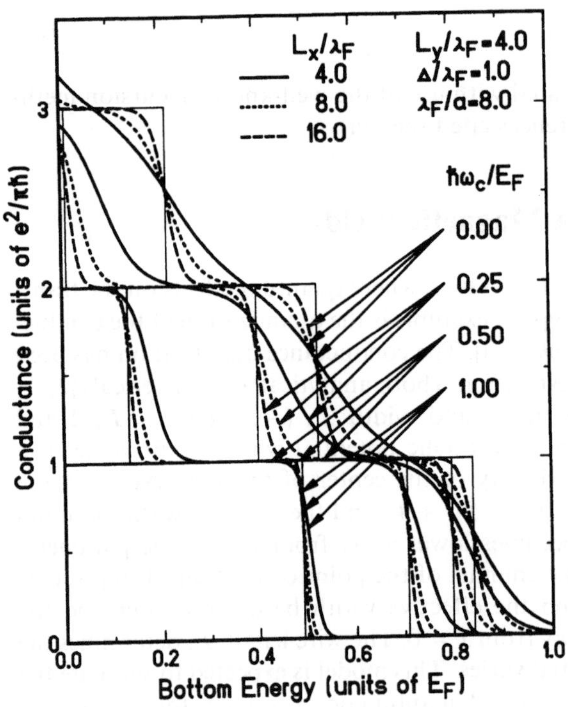

Fig. 2. An example of calculated conductance at a quantum point contact as a function of energy V of the point-contact bottom. The *dotted lines* represent the quantized conductance in the adiabatic limit, given by the number of the allowed channels

channels, which reduces backscattering probabilities at the point contact; the electron motion then becomes more adiabatic.

Figure 3 gives the current carried by each out-going channel for the case where the conductance is well quantized [$L_x/\lambda_F = 16$ at $\hbar\omega_c/E_F = 0$]. An appreciable part of the current is carried by channels different from those at the point contact, although the total conductance is well quantized. Inter- as well as intra-channel

Fig. 3. Contribution of different channels to the conductance across a point contact as a function of energy V of the point-contact bottom. The *dotted lines* represent the quantized coductance in the adiabatic limit, given by the number of the allowed channels

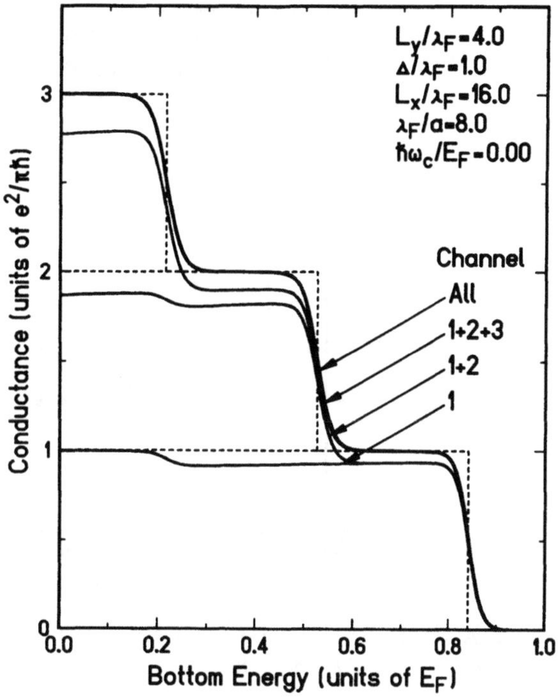

scattering due to nonadiabaticity of the confining potential has been shown to be exponentially small in the slowly-varying case. Therefore, the present result shows that the conductance is quantized even when the potential is far from being in the adiabatic limit.

Combinations of several point contacts are particularly interesting, because they provide means to measure the subband wave functions directly. Quite recently, Okada et al. measured the conductance across a series of two-point contacts as a function of magnetic field; they observed a single peak as a function of the magnetic field for the case of a single emitter channel and a splitting into two peaks when the channel number was two [5].

Figure 4 schematically shows a model for a series of two point contacts. The confining potential at both emitter (left, $-$) and collector (right, $+$) contacts is parabolic, i.e., given by $E_F(y/\Delta_\pm)^2 + V_\pm$. The boundary between the region of flat and parabolic potential changes according to $(x/L_\pm)^p + (2y/L_y)^p = 1$ with $p = p_\pm$ in the vicinity of the contacts. Infinitely long wires with width L_x are attached at both top and bottom sides, i.e., only out-going electron waves are allowed at $y = +L_y/2$. The distance between the point contacts is $D = L_x + L_+ + L_-$. For typical $\lambda_F \sim 500$ Å, we have $D \sim 1$ μm, which is close to the distance of the two point contacts used by Okada et al. [5].

Figure 5 gives the calculated conductance as a function of the distance of the two-point contacts divided by the classical cyclotron radius R_c. The channel number at the collector contact is fixed as three and that of the emitter is varied. The

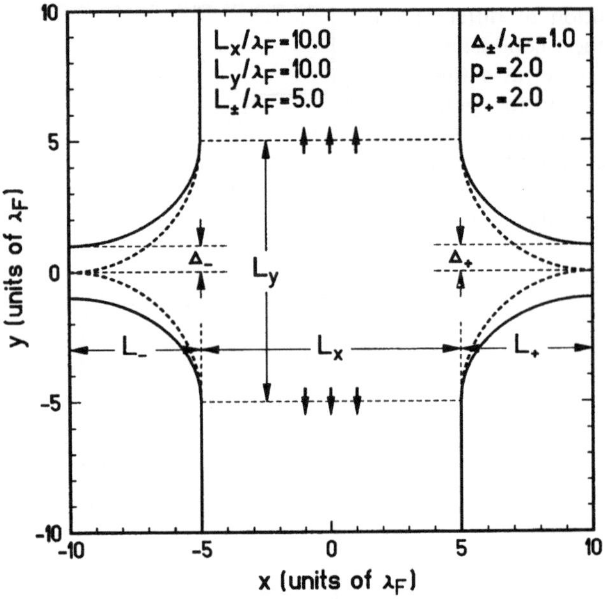

Fig. 4. A schematic illustration of a series of two-point contacts. Electrons are injected *from* the *left hand side* and collected *at* the *right hand side*. Only out-going electron waves are allowed at the *top* and *bottom*

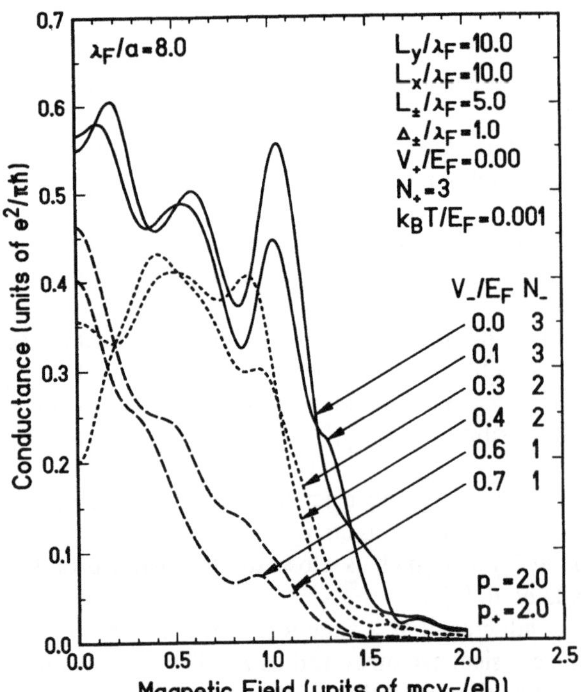

Fig. 5. Example of calculated conductance across the two point contacts as a function of a magnetic field for different channel numbers at the emitter point contact at low temperature $k_B T/E_F = 0.001$. The channel number at the collector contact is fixed as three

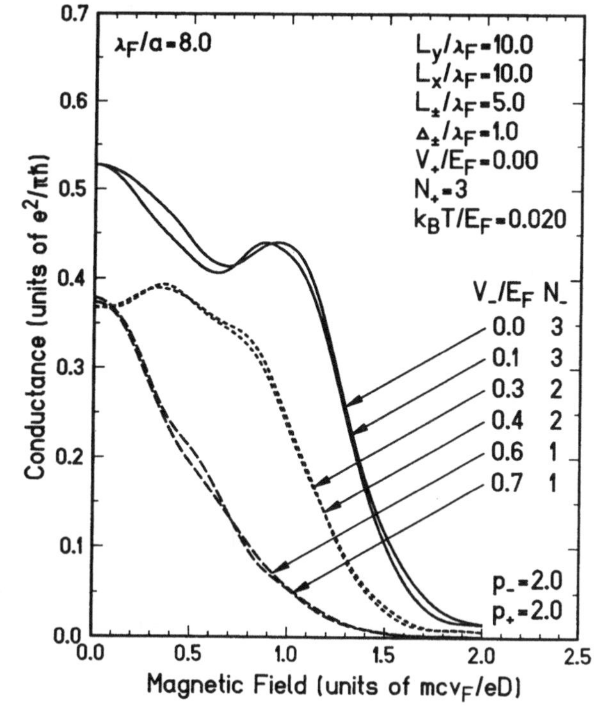

Fig. 6. Examples of calculated conductance across a series of two quantum point contacts as a function of magnetic field at a nonzero temperature $k_B T/E_F = 0.02$. The channel number at the collector contact is fixed as three

conductance is determined essentially by the channel number of the emitter, except for the presence of oscillations due to quantum interference.

At nonzero temperatures, averaging due to the broadening of the electron energy distribution must be taken into account. An example of the results of such energy averaging at $k_B T/E_F = 0.02$ is shown in Fig. 6, which demonstrates the disappearance of the fluctuations and shows clearly the features corresponding to the channel wave functions. The conductance depends also on the channel number of the collector point contact. An example is given in Fig. 7 for the channel number $N_+ = 2$. The difference of the conductance for $N_- = 1$ and 2 is still appreciable, but the conductance remains essentially the same for $N_- = 2$ and 3. The same feature can be seen when $N_+ = 1$, where the conductance for different N_+ is very close to each other. This interesting dependence on the collector channel has not been reported previously.

Conductance Fluctuations

Aperiodic oscillations of the conductance measured in metallic wires as a function of a magnetic field have been known to be a manifestation of universal conductance fluctuations. The universality has been shown theoretically in the diffusive regime where the system size is much larger than the mean free path [6, 7]. We can

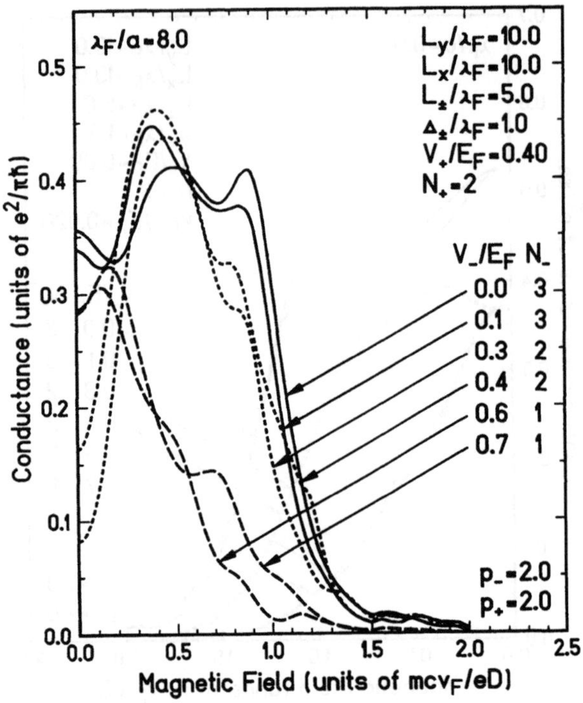

Fig. 7. Examples of calculated conductance across a series of two quantum point contacts as a function of magnetic field at zero temperature. The channel number at the collector contact is fixed as two

expect a different behavior in quantum wires where only a few well-defined 1D subbands are occupied.

Figure 8 shows the fluctuation as a function of wire length for several wires with different width W [8]. Note that the number of occupied 1D subbands is given by $2W/\lambda_F$, where λ_F is the Fermi wavelength. The fluctuation first increases with increasing length, takes a maximum, and then starts to decrease gradually. The initial increase corresponds to the simple fact that the conductance is quantized into the integer multiple of $e^2/\pi\hbar$ in the ballistic regime. The gradual decrease is due to the localization effect that is very important in narrow wires. The localization length calculated from the length dependence of the conductance is denoted by vertical arrows in the Fig. For sufficiently wide wires, the fluctuation exhibits a universal behavior, i.e., independent of the length, in the range much larger than the mean free path Λ in 2D and smaller than the localization length.

Figure 9 gives the results for the case where the broadening Γ is larger than the subband energy separation. In such 2D cases, the fluctuation becomes larger and shows the length dependence close to the curve obtained by perturbation calculations.

Boundary roughness is inherent in artificially confined wires. It is quite a peculiar scattering mechanism in the sense that the mean free path of the lowest 1D subbands is much larger than that of the higher subbands. This strong channel dependence disappears in the presence of weak magnetic fields, giving rise to a

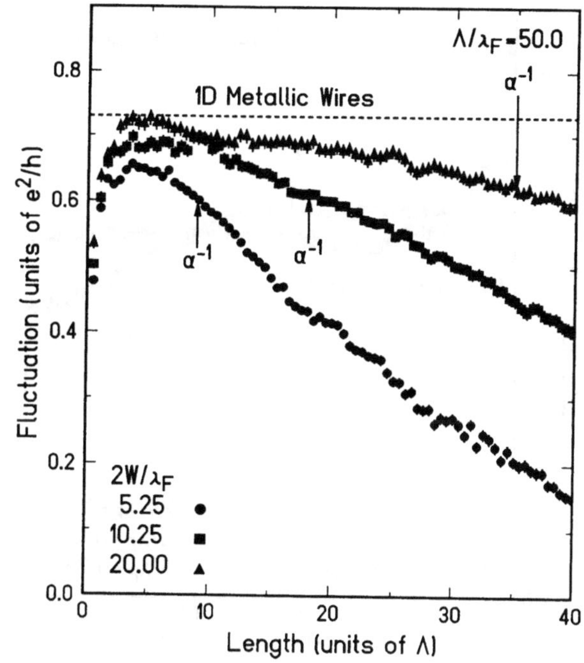

Fig. 8. Calculated conductance fluctuations as a function of wire length measured in units of Λ for different widths (the channel number is 5, 10, and 20). The *vertical arrows* indicate the localization length and the *horizontal arrow* indicates the universal value in 1D wires calculated perturbationally

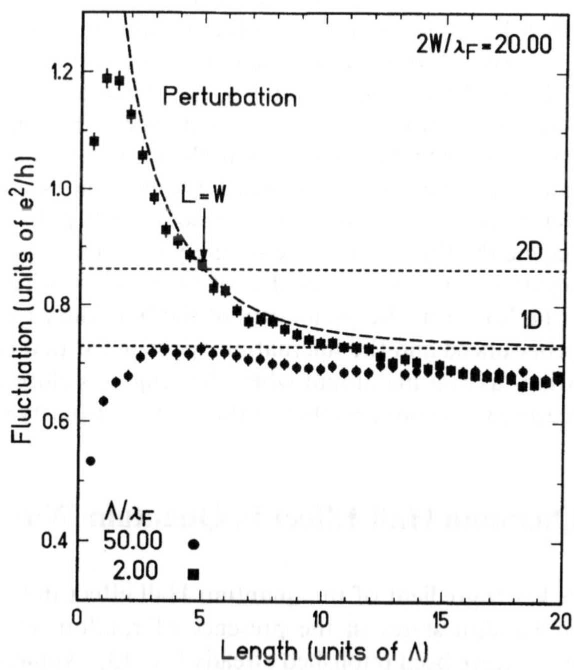

Fig. 9. Calculated conductance fluctuations for wide wires with channel number 20. The *squares* represent the results in the case that the broadening is larger than the subband energy separation and the *vertical arrow* shows the length corresponding to a 2D square. The *circles* are the same as the *triangles* given in Fig. 8

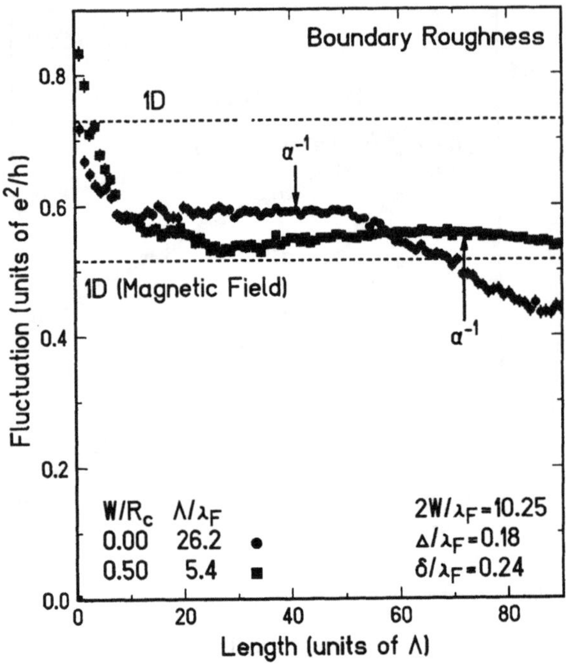

Fig. 10. Calculated conductance fluctuations as a function of length in quantum wires with boundary roughness (average displacement Δ and correlation length δ) in the absence and presence of a magnetic field

strong positive magnetoresistance [9] and a suppression of the weak-field Hall effect [10]. The conductance fluctuations in the presence of boundary roughness are expected to be quite different from those for bulk impurity scattering.

Figure 10 gives an example of calculated conductance fluctuations for boundary roughness scattering. There is a universal region much longer than the average mean free path, but shorter than the localization length, where the fluctuations are independent of the length, which is quite analogous to the case of impurity scattering. However, two features are note-worthy: In short wires in an early ballistic regime the fluctuations are strongly enhanced and the fluctuation in the universal region is not the same as that of 1D metallic wires. In the presence of a magnetic field, however, the peculiarity of the boundary roughness scattering becomes less pronounced and the fluctuation approached that of 1D metallic wires. The fluctuation takes a maximum when the length is close to the localization length. This feature is the same as that in the case of impurity scattering.

Quantum Hall Effect in Quantum Wires

A key ingredient of the quantum Hall effect in two dimension is the localization of Landau states in the presence of random potential fluctuations. Several reviews have been published already [11–13]. Numerical Thouless-number study has shown that states are all localized except those lying in the vicinity of the Landau-level center [14]. The critical exponent s defined by $\alpha \propto |E|^s$, with α being the inverse localization length, was obtained as $s \gtrsim 2$ for the lowest Landau level. This result was consistent with that obtained in a lattice model [15] and was also confirmed by later finite-size scaling calculations [16]. More recent finite-size calculations within a random-matrix model gave the exponent $s \sim 2.3$ [17].

There have been several experimental attempts to determine the critical exponent through the temperature dependence of the width of the Hall steps, i.e., the region where the Hall conductivity changes from one quantized value to another [18, 19]. However, the temperature dependence includes another unknown exponent describing the temperature dependence of the phase coherence length due to inelastic scattering. In quantum wires, on the other hand, the Hall effect is understood in terms of the transmission probability of edge channels. In particular, the Hall-step width is determined by the backscattering of edge states which are degenerate with broadened bulk Landau states.

The present numerical method can be used for the purpose of elucidating the connection of these two different descriptions, i.e., the localization of bulk Landau states and backscattering of edge channels. We consider a rectangular system with width W and length L containing scatterers with short-range potential. At $x = \pm L/2$, the system is connected to an infinitely long ideal wire whose bottom energy is slightly lowered (by ΔE). This is required because current carrying edge channels disappear in ideal wires when the energy is smaller than that of the lowest Landau level and the transmission vanishes identically. As will be shown in actual numerical calculations, this shift causes negligible effects on the transmission probability of edge states for wires with sufficiently large width.

Calculations are performed for $L/W = 1/2$, 1, and 2 with size $W/l = 5$, 10, 15, 20, 25, 30, 35, 40, and 50, where l is the magnetic length. The number of different samples is typically 10^4. The broadening of the Landau level is $\Gamma/\hbar\omega_c = 0.2$ and the energy shift is chosen as $\Delta E = \hbar\omega_c/2$. Figure 11 gives some examples of calculated energy dependence of the transmission probability T for the lowest Landau level. The dotted lines (close to a step function) represent the transmission probability for square systems with $L/W = 1$ in the absence of randomness. The transition probability in the presence of randomness exhibits much slower increase and the energy dependence becomes steeper with increase of the system size, except in the smallest system with $W/l = 5$.

In order to deduce the effective energy width over which the transmission probability increases from zero to unity, we fit the results to the function,

$$T(E) = \frac{1}{\exp[\beta(E - E_{\rm th})] + 1} \tag{24}$$

The parameter β or $\Gamma(dT/dE)_{\rm max}$ represents the inverse of the effective energy width and $E_{\rm th}$ the effective threshold energy. The fitted results are shown by solid lines in Fig. 11. Although the fitting is not perfect, particularly in the region where $T \sim 0$ and 1, this procedure is expected to be sufficient in determining the effective width and threshold. The resulting $\Gamma(dT/dE)_{\rm max}$s are listed in each figure.

Figure 12 gives the steepness parameter as a function of the width in logarithmic scales. It is clear that the results can be fitted to a straight line, except those for

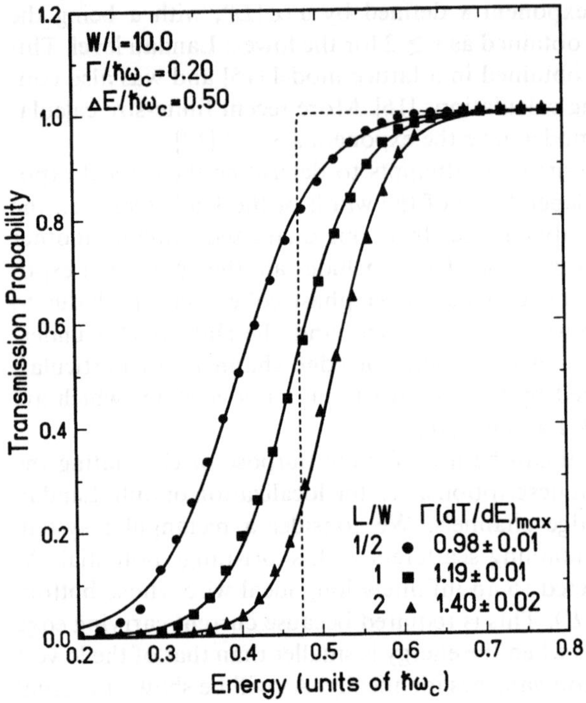

a

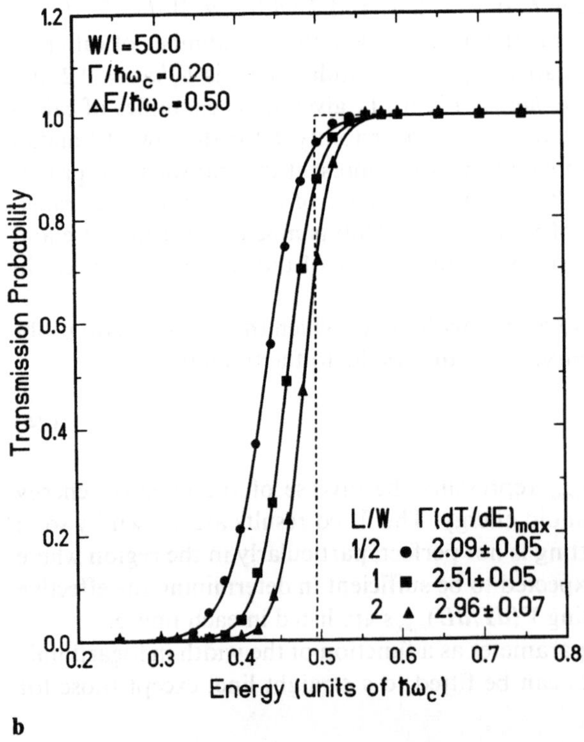

b

Fig. 11a, b. Calculated transmission probability as a function of energy for the lowest Landau level. **a** $W/l = 10$ and **b** 50

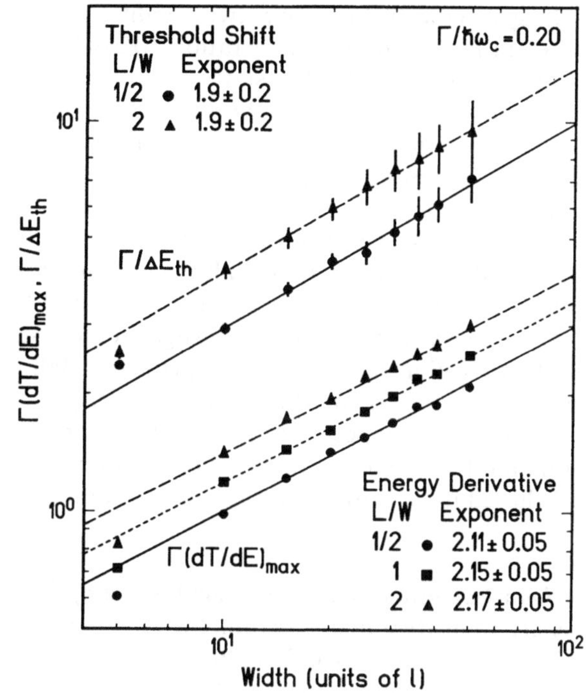

Fig. 12. Log-log plot of the maximum energy derivative of the transmission and the difference of the threshold energy for different L/Ws as a function of system width W. A fitting to a straight line gives the critical exponents s and s' in agreement with each other within statistical errors

$W/l = 5$, showing that $\Gamma(dT/dE)_{\max} \propto (W/l)^{1/s}$. The exponent is almost independent of the form of the system and is obtained as $s = 2.2 \pm 0.1$.

The threshold energy becomes larger with increasing L for a given W. In quantum wires with L much larger than W, this threshold approaches the spectral edge of the broadened Landau levels in two dimensions. For systems with comparable L and W, however, the threshold for systems with different L/Ws becomes closer with increasing system size. Figure 12 also contains the difference in the threshold energy $\Delta E_{\mathrm{th}}^2 = E_{\mathrm{th}}^{L/W=2} - E_{\mathrm{th}}^{L/W=1}$ and $E_{\mathrm{th}}^{1/2} = E_{\mathrm{th}}^{L/W=1} - E_{\mathrm{th}}^{L/W=1/2}$. Although statistical errors are much larger, ΔE_{th} is also fitted to a power law dependence, $\Delta E_{\mathrm{th}} \propto (W/l)^{-1/s'}$. The resulting $s' = 1.9 \pm 0.2$ is slightly smaller than s but is highly likely to be same as s.

For energies lying below the Landau level, the transmission occurs through bulk states because there are no edge states for $-L/2 < x < +L/2$. Because bulk states are localized exponentially, the transmission becomes appreciable for energies satisfying the condition $\alpha(E_-)L \lesssim 1$, where α is the inverse localization length. For energies higher than the Landau level, on the other hand, backscattering takes place through bulk Landau states. Therefore, the backscattering becomes appreciable for energies satisfying $\alpha(E_+)W \lesssim 1$. For square systems ($L = W$), therefore, the width of the energy region over which the transmission probability varies from zero to unity is given by $\alpha(E_\pm)W \lesssim 1$. If we use $\alpha(E) \propto |E|^s$, we have $\Gamma(dT/dE)_{\max} \propto (E_+ - E_-)^{-1} \propto W^{1/s}$. Therefore, the exponent obtained above is nothing but the exponent for the inverse localization length of bulk 2D Landau states.

In systems with finite size, two conditions, $\alpha(E_-)L \lesssim 1$ and $\alpha(E_+)W \lesssim 1$, give different $E_\pm$ depending on the ratio L/W. In infinitely large systems, they converge to the energy of extended states at the center of the Landau level. Thus, the exponent s', defined by the threshold shift for different system forms, is also expected to be the same as that of the inverse localization length of bulk Landau states. The present numerical results are consistent with the picture currently accepted that states are all localized exponentially, except those at the center of the Landau level. The critical exponent obtained in the present study is ~ 2.2. This is consistent with that of almost all the existing theory either numerical or analytic, since actual errors are likely to be larger than ± 0.1 estimated statistically.

Quite recently, Koch et al. investigated the system-size dependence of the width of the transition region between quantized Hall plateaus; they deduced a critical exponent ($s \sim 2.3$) for the localization length of bulk Landau states [20]. The present result is expected to justify this experimental procedure of determining s, although the lateral confining potential in actual GaAs/AlGaAs heterostructures is believed to be slowly-varying rather than the abrupt potential assumed above.

Acknowledgments. This work has been supported, in part, by the Industry-University Joint Research Program, "Mesoscopic Electronics" and by the Grants-in-Aid for Scientific Research on Priority Areas, "Electron Wave Interference Effects in Mesoscopic Structures" and "Computational Physics as a New Frontier in Condensed Matter Research", from the Ministry of Education, Science and Culture of Japan.

References

1. Landauer R (1957) IBM J Res Dev 1:233; (1970) Philos Mag 21:863
2. Ando T (1991) Phys Rev B44:8017
3. van Wees BJ, van Houten H, Beenakker CWJ, Williamson JG, Kouwenhoven LP, van der Marel D, Foxon CT (1988) Phys Rev Lett 60:848
4. Wharam DA, Thornton TJ, Newbury R, Pepper M, Ahmed H, Frost JEF, Hasko DG, Peacock DC, Ritchie DA, Jones GAC (1988) J Phys C21:L209
5. Okada M, Saito M, Takatsu M, Kosemura K, Yokoyama N (1992) Semicond Sci Technol 7:B223
6. Altshuler BL (1985) JETP Lett 41:648
7. Lee PA, Stone AD (1985) Phys Rev Lett 55:1622
8. Tamura H, Ando T (1991) Phys Rev B44:1792
9. Akera H, Ando T (1991) Phys Rev B43:11676
10. Akera H, Ando T (1990) Surf Sci 229:268; (1990) Phys Rev B41:11967
11. Ando T (1985) Prog Theor Phys Suppl 84:69
12. The quantum Hall effect (1987) Prange RE, Girvin SM (eds) Springer, New York
13. Aoki H (1987) Rep Prog Phys 50:655
14. Ando T (1983) J Phys Soc Jpn 52:1893; (1983) 53:3101, 3126
15. Ando T (1989) Phys Rev B40:5325; (1990) Phys Rev B42:5626

16. Aoki H, Ando T (1985) Phys Rev Lett 54:831; Ando T, Aoki H (1985) J Phys Soc Jpn 54:2238
17. Huckestein B, Kramer B (1990) Phys Rev Lett 64:1437
18. Wei HP, Tsui DC, Paalanen MA, Pruisken AMM (1988) Phys Rev Lett 61:1294
19. Wakabayashi J, Yamane M, Kawaji S (1989) J Phys Soc Jpn 58:1903
20. Koch S, Haug RJ, von Klitzing K, Ploog K (1991) Phys Rev Lett 67:883

Magnetophonon Resonances in Quantum Wires

19

Nobuya Mori, Hideki Momose, and Chihiro Hamaguchi[1]

Summary. Magnetophonon resonances are studied in quantum wires with square confinement potential. Magnetoconductivity calculated by using the Kubo formula is found to have two components; one related to the current carried by electron hopping motion between the localized cyclotron orbits through electron-phonon interaction, and the other caused by the current carried by electron motion affected by the confinement potential. The latter, σ_{po}, is found to be two orders of magnitude greater than the former, σ_{ep}, around the fundamental resonance condition in the case of a GaAs quantum wire whose width is 200Å and barrier height is 100meV.

Introduction

Magnetophonon resonance (MPR), which is the oscillatory variation of magnetoconductivity with a high magnetic field, is known to be a powerful tool for the study of electron-optical-phonon interaction in semiconductors [1–4]. Recent developments in microfabrication technology have made it possible to obtain quantum wire structures by introducing a confinement potential in a two-dimensional electron system. In such systems, various new phenomena associated with MPR are expected. Vasilopoulos et al. [5] have studied MPR in quantum wires with a weak parabolic confinement potential of frequency Ω; they predicted that the usual resonance condition $\omega_0 = P\omega_c$ would be modified by $\omega_0 = P\tilde{\omega}_c$, where P is an integer, ω_0 and ω_c are the longitudinal-optical-phonon (LO phonon) frequency and cyclotron frequency, respectively, and $\tilde{\omega}_c$ is given by $(\omega_c^2 + \Omega^2)^{1/2}$. We followed the model for quantum wires and calculated the conductivity σ_{xx}, using the Kubo formula [6]. Our calculation revealed that σ_{xx} can be expressed by the sum of two types of contributions; one related to the current carried by electron hopping motion between the localized cyclotron orbits through electron-phonon interaction, and the other caused by the current carried by electron motion affected by the confinement potential. The former, σ_{ep}, is directly proportional to the coupling constant α of the electron-optical-phonon interaction and the latter, σ_{po}, is found to be inversely proportional to α. At resonances σ_{ep} exhibits maxima and σ_{po}

[1] Department of Electronic Engineering, Osaka University, 2-1, Yamada-oka, Suita City, Osaka, 565 Japan

minima. For weak confinement potential, i.e., for wide quantum wires, σ_{ep} is dominant and total magnetoconductivity σ_{xx} shows maxima at resonances, as pointed out by Vasilopoulos et al. On the other hand, in the case of strong confinement potential, or in narrow quantum wires, σ_{po} becomes large and σ_{xx} is not always maximum at resonances, resulting in minima with an extremely strong confinement potential. In this paper we study MPR in quantum wires with square confinement potential.

Model for a Quantum Wire

We employ a simple model for a quantum wire in which ideal two-dimensional electron gases are confined by square confinement potential and electrons are only free along one direction. We assume that the heterointerface is normal to the z axis, and that the confinement of electrons is in the y direction. A magnetic field **B** is applied in the z axis. One-electron Hamiltonian H_e for the two-dimensional system is written as

$$H_e = \frac{1}{2m^*}(\mathbf{p} + e\mathbf{A})^2 + V(y) \tag{1}$$

where $V(y)$ is the confinement potential:

$$V(y) = \begin{cases} V_0, & |y| \geqslant \frac{1}{2}a \\ 0, & \text{otherwise} \end{cases} \tag{2}$$

a is the wire width and V_0 the barrier height. Using the Landau gauge $\mathbf{A} = (-By, 0)$, we evaluate the eigenvalues $E_N(Y)$ and eigenstates $e^{ikx}\varphi_N(y - Y;\, Y)$ numerically, N being the Landau level index, $Y = \ell^2 k$ the center coordinate of cyclotron motion, and $\ell = (eB)^{-1/2}$. In Fig. 1 we show $E_N(Y)$ as a function of Y for a quantum wire of GaAs in which we take $a = 200$Å, $V_0 = 100$meV, and $m^* = 0.07m_0$.

Magnetoconductivity

Next we calculate the magnetoconductivity σ_{xx} in the manner described in Ref. [6]. We find that σ_{po} is expressed as

$$\sigma_{po} = \frac{ne^2\tau_{po}}{\tilde{m}^*} \tag{3}$$

where n is the one-dimensional electron density, $\tilde{m}^* = (\omega_c/V_0)^2 m^*$, and the relaxation time caused by electron-optical-phonon interaction, τ_{po}, is given by

$$\tau_{po} = \frac{\beta\omega_c \displaystyle\int dY \frac{1}{W_0(Y)} \{\ell\varphi_0^2(-\tfrac{1}{2}a - Y;\, Y) + \ell\varphi_0^2(\tfrac{1}{2}a - Y;\, Y)\}^2 e^{-\beta E_0(Y)}}{\displaystyle\int dY e^{-\beta E_0(Y)}} \tag{4}$$

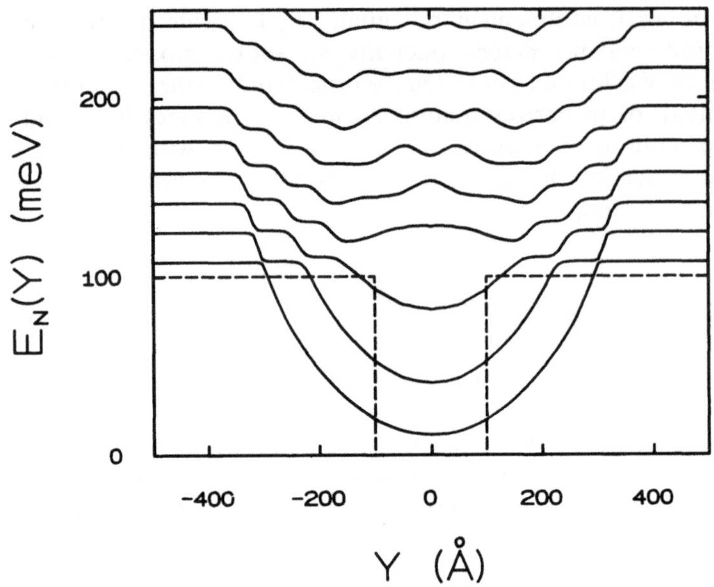

Fig. 1. Electron energy $E_N(Y)$ as a function of the center coordinate of cyclotron motion Y at $B = 10$T for $N = 0$ to 8 in a quantum wire of GaAs of $a = 200$Å and $V_0 = 100$meV. The *dashed line* shows the square confinement potential

Note that we consider the low density case and retain the term related to the absorption processes of lowest Landau level. In Eq. (4), $W_0(Y)$ is the scattering rate of electrons in the lowest Landau level and it is given by

$$W_0(Y) = \alpha\omega_0 N_0 \sum_{N'} \int dq_x dq_y \frac{\omega_0}{k_0(q_x^2 + q_y^2)^{1/2}} J_{N'Y}^2(q_x, q_y)\delta(E_0(Y) + \omega_0 - E_{N'}(Y + \ell^2 q_x)) \tag{5}$$

α being the Fröhlich coupling constant of electron-optical-phonon interaction, ω_0 the LO phonon frequency, N_0 the occupation number of LO phonons, $k_0 = (2m^*\omega_0)^{1/2}$, and

$$J_{N'Y}^2(q_x, q_y) = \left| \int \varphi_{N'}(y - \ell^2 q_x; Y + \ell^2 q_x) e^{iq_y y} \varphi_0(y; Y) dy \right|^2 \tag{6}$$

Next we show the expression for σ_{ep}. In the case of low density, σ_{ep} is reduced to

$$\sigma_{ep} = \frac{ne^2\tau_{ep}}{m^*} \tag{7}$$

with

$$\tau_{ep} = \int dY \tau(Y) e^{-\beta E_0(Y)} \Big/ \int dY e^{-\beta E_0(Y)} \tag{8}$$

and

$$\tau(Y) = \alpha N_0 \beta \left(\frac{\omega_0}{\omega_c}\right)^2 \sum_{N'} \int dq_x dq_y \frac{\omega_0 q_y^2}{k_0^3 (q_x^2 + q_y^2)^{1/2}} J_{N'Y}^2(q_x, q_y) \delta(E_0(Y) + \omega_0 - E_{N'}(Y + l^2 q_x)) \tag{9}$$

Results and Discussion

In Fig. 2 we show the calculated magnetoconductivity at $T = 100$K as a function of applied magnetic field around the fundamental resonance condition for a quantum wire of GaAs as treated in Fig. 1. We used the following values for the material parameters for GaAs: $\alpha = 0.07$, $\omega_0 = 36.2$meV. In our calculation, the delta functions in Eqs. (5) and (9) are replaced by the Lorenzian function with constant width $\Gamma = 2$meV to take into account the Landau level broadening effect phenomenologically. At resonances σ_{ep} exhibits maxima and σ_{po} minima. We have calculated the magnetoconductivity of quantum wires with slightly different V_0 or a and found that if V_0 increases or a decreases, the ratio of σ_{po} and σ_{ep} becomes larger and the resonance field moves to a lower value. This behavior is the same as in the case in a quantum wire with parabolic confinement potential. It is for this reason that electrons at any position in a quantum wire of $a \sim 200$Å feel a confinement potential barrier, resulting in the almost parabolic dependence of $E_0(Y)$ with

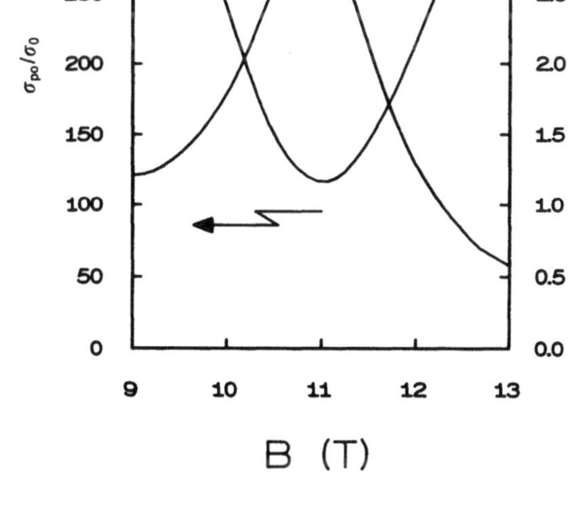

Fig. 2. Calculated magnetoconductivity at T = 100K as a function of an applied magnetic field B for a quantum wire of GaAs as treated in Fig. 1. The conductivity caused by the current carried by electron motion affected by the confinement potential is represented by σ_{po} and the conductivity related to the current carried by electron hopping motion between the localized cyclotron orbits through electron-phonon interaction is represented by σ_{ep}

respect to Y, as shown in Fig. 1, the same dependence found in the case of parabolic confinement potential. However, if a increases enough to make the region around the center of the quantum wire flat, or if V_0 decreases and electrons over the barrier potential become able to contribute to the conductance, the behavior of σ_{po} and σ_{ep} may be different from that in the parabolic confinement potential case.

Acknowledgments. This work has been supported, in part, by a Grant-in-Aid for Scientific Research on the Priority Area, "Electron Wave Interference Effects in Mesoscopic Structures", from the Ministry of Education, Science, and Culture of Japan.

References

1. Stradling RA, Wood RA (1968) J Phys C1:1711–1733
2. Peterson RL (1975) In: Willardson RK, Beer AC (eds) Semiconductors and semimetals, vol 10. Academic, New York, p 221
3. Nicholas RJ (1985) Prog Quantum Electronics 10:1–75
4. Hamaguchi C, Mori N (1990) Physica B164:85–96
5. Vasilopoulos P, Warmenbol P, Peeters FM, Devreese JT (1989) Phys Rev B40: 1810–1816
6. Mori N, Momose H, Hamaguchi C (1992) Phys Rev B45:4536–4539

Spin-Dependent Nonlocal Quantum Transport Influenced by Gate Voltage in GaAs/AlGaAs Wires

20

KAZUHITO TSUKAGOSHI, KENICHI OTO, SADAO TAKAOKA, KAZUO MURASE[1], YUKIHIKO TAKAGAKI*, KENJI GAMO, AND SUSUMU NAMBA[2]**

Summary. The magneto-transport properties of two-dimensional electron gas (2DEG) at high magnetic fields and low temperatures are related to the edge and bulk states. These states cause nonlocal effects, such as nonlocal Shubnikov-de Haas (SdH) oscillations, and the influence of extra probes on the four-terminal resistance. We observed the selective diminishing of the down spin peak and the up spin peak in nonlocal SdH oscillations and investigated the influence of the extra probes which ostensibly have nothing to do with the four-terminal resistance. The influence of the extra probes on normal and nonlocal SdH measurement is explained by a decrease in the chemical potentials of the edge states at the extra probes.

Introduction

Nonlocal SdH resistance is defined as four-terminal resistance, where the nominal current path is separated from the voltage probes. In the classical view, nonlocal resistance is very small. However, nonlocal SdH oscillations have been observed in high-mobility GaAs/AlGaAs wires at high magnetic fields and low temperatures [1–6]. The nonlocal SdH oscillations are considered to be due to the mixing of the edge states and the bulk state at the electrode [2–6]. In order to observe nonlocal SdH oscillations, it is necessary that there exist both edge and bulk states and that the chemical potentials of the edge states do not change, at least over the separation length ΔL, between the current and voltage path. According to recent experiments, nonlocal SdH oscillations have been measured up to $\Delta L = 1.5$ mm, and the energy relaxation length of the edge states will be on the order of mm. The mechanism of such a long-lived edge current has attracted much attention [4, 5].

[1] Department of Physics, Faculty of Science, Osaka University, 1-1 Machikaneyama, Toyonaka, 560 Japan
[2] Department of Electrical Engineering, Faculty of Engineering Science and Research Center for Extreme Materials, Osaka University, 1-1 Machikaneyama, Toyonaka, 560 Japan
* Present address: Center for Solid State Electronics Research, Arizona State University, Temple, AZ 85287-6206, USA
** Present address: Nagasaki Institute of Applied Science, 536 Amibacho Nagasaki-shi, Nagasaki, 851-01 Japan

In this brief report, we report the magneto-transport properties of high-mobility multi-terminal GaAs/AlGaAs wires with four Schottky gates. Nonlocal SdH oscillations were observed with various gate voltages, where the gate lay across the nonlocal path. Further, the nonlocality of normal and nonlocal SdH oscillations at high magnetic fields was studied by connecting and/or disconnecting the extra probes which ostensibly have nothing to do with SdH measurement.

Experimental Result

Samples

Our devices were fabricated by UV lithography and chemical etching from a wafer of modulation doped GaAs/AlGaAs heterostructure. The Schottky gates were formed by evaporating Au on the GaAs cap layer. Two gates lay across the current path to control the edge and bulk currents beneath the gate. There were two additional gates on the Hall probes to connect or disconnect these probes electrically, as shown in Fig. 1a [7]. The electrodes were made by alloying AuGe/Ni/Au layers and the contact resistance was typically 100 Ω. The carrier density (n_s) and mobility (μ) of this wafer were $n_s = 3.0 \times 10^{11}$ cm^{-2} and $\mu = 6.0 \times 10^5$ cm^2/Vs, respectively, at 4.2 K. Magneto-resistance up to 8 T was measured at 1.7 K and

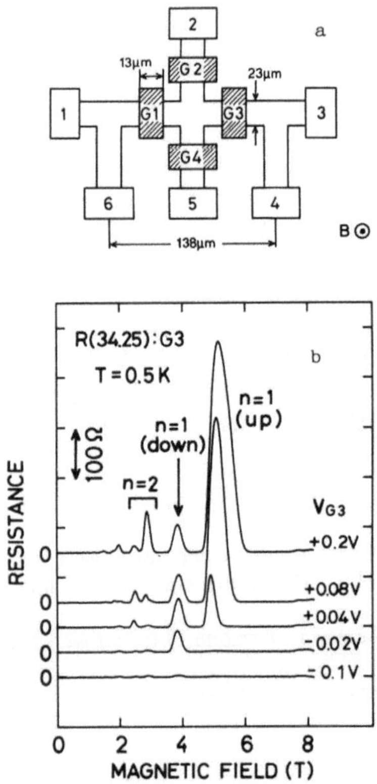

Fig. 1. a Schematic view of sample dimension and probes; gate No. Gi(i = 1 ~ 4) denotes Schottky gate. **b** Nonlocal SdH oscillations with various gate voltages (V_{G3})

0.5 K by applying AC current (300 nA $\sim$ 3 μA) at 15 Hz. The direction of the magnetic fields is denoted in Fig. 1, where the edge currents circle around clockwise. The measurement was made after illumination by light-emitting diode (LED) in order to increase the carrier density.

Nonlocal SdH Oscillations

The gate voltage (V_{G3}) dependence of the nonlocal SdH oscillations was measured by applying 500 nA at 0.5 K, as shown in Fig. 1b, in order to investigate the influence on the edge and bulk currents. With decreasing V_{G3}, that is, as the carrier density underneath the gate area decreased, the peaks due to the spin up state decreased at first, and then those due to the spin down diminished. This selective diminishing was observed in Landau levels number 1 and 2. At 1.7 K, similar behaviors were also observed. This selective diminishing may be related to the spin-splitting enhancement which is due to exchange interactions. Spin-dependent behavior may be explained in the following way: When the Fermi level intersects with the extended state of the down spin, the energy levels of the down spin state and the up spin state are almost the same, due to the small spin splitting energy. With decreasing gate voltage, the Fermi level comes between these spin states. The spin-splitting beneath the gate area is then enhanced due to exchange interactions. At the gate, the down spin state electrons cannot go through the gate, whereas the up spin state electrons can go through the gate as the bulk current. On the other hand, when the Fermi level intersects with the up spin state, with decreasing gate voltage, the up spin state electrons cannot go through the gate area and the bulk current does not flow. The peaks of the nonlocal SdH oscillations do not appear. The peak of the down spin state remains at smaller gate voltages than that of the up spin state.

Influence of Extra Probes

In the classical view, the influence of extra probes, which are regarded as not being concerned with four-terminal resistance, is little. However, in the magnetotransport properties of 2DEG at high magnetic fields and low temperature nonlocal effects appear [6,7]. The extra probes affect the normal SdH oscillation, as well as the nonlocal SdH oscillations. The influence of the extra probes was investigated by connecting and/or disconnecting them (Fig. 2). When the extra probes were disconnected electrically, the amplitude of the normal SdH oscillations became smaller (Fig. 2a). This phenomenon can be explained by the mixing of the edge states and the bulk state at the electrode (Fig. 3). When the extra probe is connected (Fig. 3a), the edge and bulk states go through the gate and enter the electrode. Their chemical potentials are mixed and equilibrated. As a result, the chemical potential of the edge states outgoing from the electrode is smaller than the incoming, since the chemical potential of the incoming bulk state has decreased in the conducting path. However, when the extra probe is disconnected, the edge states would not be mixed with the bulk state. The resistance due to the normal SdH oscillations becomes smaller when the extra probes are disconnected. The

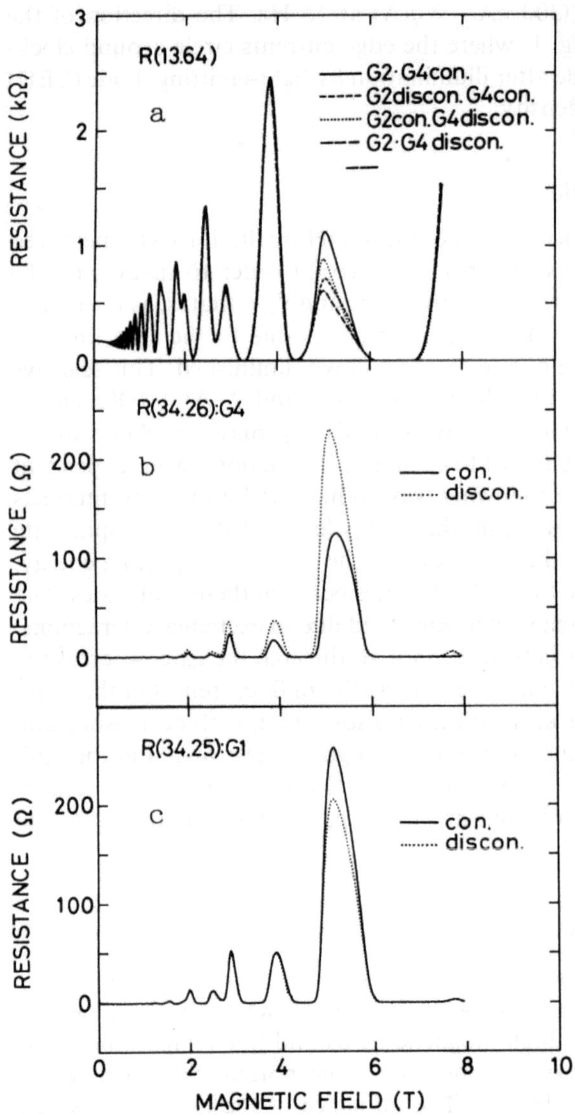

Fig. 2a–c. Normal and nonlocal SdH oscillations with connected (*con.*) and/or disconnected (*discon.*) extra probes. These extra probes ostensibly have nothing to do with four-terminal resistance: **a** Normal SdH oscillations; **b** and **c** Nonlocal SdH oscillations

nonlocal SdH oscillations are also greatly affected by the extra probes. Figure 2b, c shows the nonlocal resistance R(34.26) with connected or disconnected probe 5 obtained by changing the gate voltage (G4), and the nonlocal resistance R(34.25) with probe 1 and 6 by changing G1. The amplitude of the nonlocal SdH oscillations is larger when the extra probe 5 is disconnected. When the extra probe is connected, the chemical potentials of the edge states become smaller at the extra probe than in the disconnected case. The nonlocal amplitude in the connected case becomes smaller, since the total difference of the chemical potential is decided by

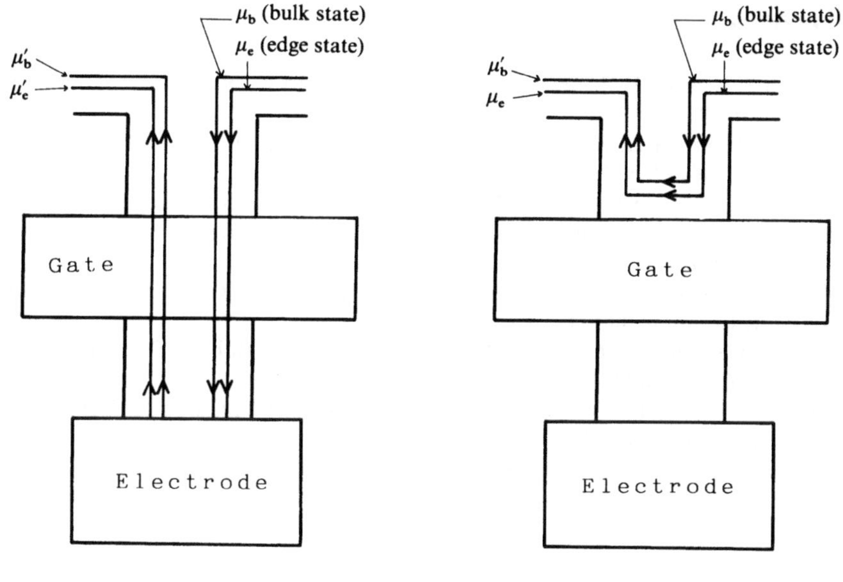

Fig. 3a, b. Influence of extra probe on chemical potentials of edge and bulk states: **a** Extra probe is connected; **b** Disconnected. When the extra probe is connected, the chemical potential of the edge state changes

probes 3 and 4. In the case of R(34.25), the amplitude of oscillation was larger when extra probes 1 and 6 were connected. In the connected case, the chemical potentials of the edge states decreased at these extra probes, and the voltage difference between probes 2 and 5 became larger.

Conclusion

Our experimental results confirm that nonlocal SdH oscillations are caused by mixing of the edge and bulk states. The selective scattering of the bulk states with different spins is induced by the gate. In the magneto-transport properties of 2DEG at high magnetic fields the effect of all electrodes linked to the sample have to be taken into consideration.

Acknowledgments. We thank S. Nishi of Oki Electric Industry Co. Ltd. for providing the molecular beam epitaxy (MBE) wafer. This work was supported, in part, by a Grant-in-Aid for Scientific Research on the Priority Area, "Electron Wave Interference Effects in Mesoscopic Structures", from the Ministry of Education, Science, and Culture of Japan. One of the authors (S.T) acknowledges the support of the Murata Science Foundation.

References

1. Takaoka S, Kubota H, Murase K, Takagaki Y, Gamo K, Namba S (1990) Solid State Commun 76:293
2. McEuen PL, Szafer A, Richter CA, Alphenaar BW, Jain JK, Stone AD, Wheeler RG, Sacks RN (1990) Phys Rev Lett 64:2062
3. Takaoka S, Tsukagoshi K, Oto K, Sawasaki T, Murase K, Takagaki Y, Gamo K, Namba S (1992) Proc 5th international conference modulated semiconductor structures, Nara. Surface Science 267:282
4. Takaoka S, Sawasaki T, Tsukagoshi K, Oto K, Murase K, Gamo K, Namba S (1991) Solid State Commun 80:571
5. Komiyama S, Nii H, Ohsawa M, Fukatsu S, Shiraki Y, Itoh R, Toyoshima H (1991) Solid State Commun 80:157
6. Tsukagoshi K, Oto K, Takaoka S, Murase K, Takagaki Y, Gamo K, Namba S (1991) Solid State Commun 80:797
7. Müller G, Weiss D, Koch S, von Klitzing K, Nickel H, Schlapp W, Lösch R (1990) Phys Rev B42:7633

Transition from Weak to Strong Spin-Orbit Scattering Effect on Conductance Fluctuation in InGaAs/InAlAs Quantum Wires

21

Yoshino K. Fukai[1], Hayato Nakano, Shunji Nakata, Seigo Tarucha[2], and Kunihiro Arai[1]

Summary. Conductance fluctuations (CFs) in GaAs/AlGaAs and InGaAs/InAlAs quantum wires were investigated at temperatures ranging from 2 down to 0.06 K. The temperature dependence of the amplitude of CF $\langle \delta g \rangle$ did not increase monotonically with decreasing temperature. The temperature dependence of the InGaAs/InAlAs quantum wire followed a one-half reduction of $\langle \delta g \rangle$, which was theoretically predicted to appear in the strong spin-orbit scattering limit. We found that the observed temperature dependence correlated well with the transition from weak to strong spin-orbit scattering regimes at very low temperatures. This transition was not as clearly observed in the GaAs/AlGaAs wire as in the InGaAs/InAlAs wire, probably due to the weaker spin-orbit scattering effect.

Introduction

Conductance fluctuations (CFs) have recently been observed in mesoscopic diffusive systems. This phenomenon, which arises from the quantum interference of randomly scattered coherent electrons in disordered conductors, gives an important insight into the scattering mechanisms in such systems. Random matrix theory (RMT) [1] and diagrammatical analyses [2] both predicted reductions in the amplitude of CF at various energy scales, with changes in the electron scattering mechanisms. These changes are summarized in Table 1. This change in the CF amplitude was recently confirmed experimentally to be a function of the external magnetic field [3]. Changes induced by spin-orbit scattering in the CF amplitude, namely, a simple one-half reduction in the strong limit have not yet been evidenced in mesoscopic systems. However, in two-dimensional films, evidence of spin-orbit scattering can be exemplified by observing positive magnetoresistance at very low temperature [4].

This paper reports the first time the transition from a weak to a strong spin-orbit scattering regime was observed in a mesoscopic diffusive wire with decreasing temperatures. A clear one-half reduction in $\langle \delta g \rangle$ was observed at low temperatures in a weak magnetic field, corresponding to the middle column in Table 1. The

[1] NTT LSI Laboratories, 3-1 Morinosato Wakamiya, Atsugi, Kanagawa, 243-01 Japan
[2] NTT Basic Research Laboratories, Midori-cho 3-9-11, Musashino, Tokyo, 180 Japan

Table 1. Variation of $\langle \delta g \rangle$

	$B = 0$	$(h/e)/L_{in}^2 < B < hD/(g\mu_B L_{in}^2)$	$2hD/(g\mu_B L_{in}^2) < B$
Weak s-o coupling ($\tau_{in} \ll \tau_{so}$)	δg_0	$(1/\sqrt{2}) \cdot \delta g_0$	$(1/2) \cdot \delta g_0$
Strong s-o coupling ($\tau_{in} \gg \tau_{so}$)	$(1/2) \cdot \delta g_0$	$(1/2\sqrt{2}) \cdot \delta g_0$	$(1/2\sqrt{2}) \cdot \delta g_0$

τ_{in}, inelastic scattering time; τ_{so}, spin-orbit scattering time; $\delta g_0 \sim (e^2/h)(L_{in}/L)^{(4-d)/2}$ $L_{in} < L$, L_{in}, inelastic scattering length (Ref. [2]); d, sample dimension

temperature dependence of $\langle \delta g \rangle$ was analyzed theoretically, and reasonable scattering times were obtained. Results reveal that spin-orbit scattering has a longer time scale than inelastic scattering appears only in very low temperature ranges.

Quantum Wires

The spin-orbit scattering time, τ_{so}, is estimated as $\tau_{so} \propto \tau_0/(g-2)^2 k_F^2$, where τ_0 is the elastic scattering time and k_F is the Fermi wave number [5]. Successful observation of the strong scattering is attributed to the use of InGaAs/InAlAs two-dimensional diffusive wires whose g-factor is 7.45 instead of GaAs/AlGaAs wires whose g-factor is 0.52. InGaAs/InAlAs wires have shorter spin-orbit scattering times than other time scales. In wire fabrication, we used selective crystal growth for GaAs/AlGaAs wires [6] and Ga ion implantation for InGaAs/InAlAs wires [7]. Large carrier concentration, n_s, and relatively low mobility μ also cause short spin-orbit scattering times: $n_s = 3.46 \times 10^{12}$ cm^{-2} and $\mu = 34000$ cm^2/Vs for GaAs/AlGaAs and $n_s = 2.18 \times 10^{12}$ cm^{-2} and $\mu = 2300$ cm^2/Vs for InGaAs/InAlAs. The measured wire lengths L and widths W were: $L = 1.7$ μm and $W = 0.3$ μm for GaAs/AlGaAs and $L = 1.0$ μm and $W = 0.4$ μm for InGaAs/InAlAs. These wires were treated as mesoscopic diffusive systems.

Low Temperature Measurement

Antilocalization properties within the measured temperature range were confirmed by positive magnetoresistance in GaAs/AlGaAs wires [6] and by the log T dependence of the two terminal zero field resistances in the InGaAs/InAlAs wire (shown in Fig. 1). These are related to the strong spin-orbit scattering [7].

The observed conductance fluctuations of the InGaAs/InAlAs wire are shown in the inset of Fig. 1. These CF have typical characteristics of reproducibility and sample dependence. The CF amplitudes $\langle \delta g \rangle$ at measured temperatures, denoted as $\langle \delta g \rangle_{exp.}$, are shown in Fig. 2 by circles. Similar plots for GaAs/AlGaAs are indicated by triangles. The value for $\langle \delta g \rangle_{exp.}$ was determined by the data from 0.4 to 1.2 T. This range in a magnetic field is low enough to satisfy the Zeeman energy of $\zeta = g\mu_B B/2$ smaller than $\hbar/\tau_0$, where μ_B is the Bohr magneton. In the InGaAs/InAlAs wire $\langle \delta g \rangle_{exp.}$ increases as temperature decreases from 2.0 to 1.2 K. It is nearly constant from 1.2 to 0.1 K and again increases below 0.1 K. However, in the GaAs/AlGaAs wire, $\langle \delta g \rangle_{exp.}$ seems to saturate at temperatures around 0.1 K.

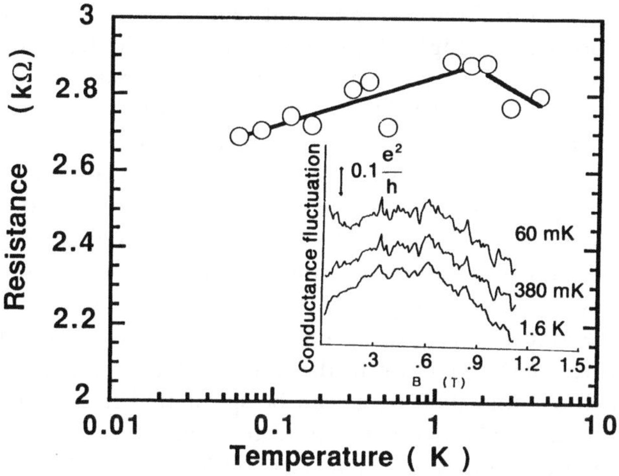

Fig. 1. Temperature dependence of resistance of the InGaAs/InAlAs quantum wire under zero magnetic field. *Inset*, conductance fluctuations

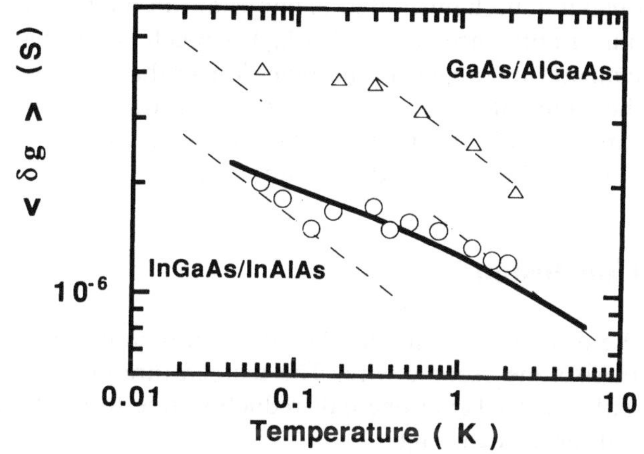

Fig. 2. Temperature dependence of conductance fluctuation amplitude of the GaAs/AlGaAs wire (*triangles*) and the InGaAs/InAlAs wire (*circles*). *Solid curve* shows the theoretical fit. *Broken lines* show $T^{-1/3}$ dependence of $\langle \delta g \rangle$ at strong ($T < 0.1$ K) and weak ($T > 1.2$ K) spin-orbit scattering regimes

This difference is explained well in terms of τ_{so} and inelastic scattering time τ_{in}. In the high temperature region, τ_{in} is much shorter than τ_{so}. Therefore, $\langle \delta g \rangle_{exp.}$ is dominated by inelastic scattering, and is given by $\langle \delta g \rangle_{exp.} \propto \alpha T^{-p}$, which is within the limit of weak spin-orbit scattering. When τ_{in} becomes much larger than τ_{so} at very low temperatures, $\langle \delta g \rangle_{exp.}$ is reduced by a factor of two due to spin-orbit scattering, and is given by $\langle \delta g \rangle_{exp.} \propto \alpha/2\ T^{-p}$, which is within the strong limit. The broken lines in Fig. 2 show the temperature dependency of $\langle \delta g \rangle_{exp.}$ within the limits of weak and strong spin-orbit scattering regimes, taking into consideration factors above one-half. The $\langle \delta g \rangle_{exp.}$ of the InGaAs/InAlAs wire showed a transition from weak to strong spin-orbit scattering regimes. This behavior was not clear in the GaAs/AlGaAs wire, probably due to insufficient spin-orbit scattering effect in this temperature range. The reasons will be discussed later. The temperature dependence of $\langle \delta g \rangle_{exp.}$ for the InGaAs/InAlAs wire was analyzed using Feng's [8] expression, applicable under various energy conditions, in two-dimensional cases. The inelastic scattering length $L_{in}(T)$ was first determined assuming

$\tau_{so} = \infty$ in the weak spin-orbit region above 1.2 K. In this region, $L_{in}(T)$ is proportional to $\langle \delta g \rangle_0(T)$, which represents the temperature dependence of $\langle \delta g \rangle_{exp.}$ above 1.2 K. In addition, analysis shows that $\langle \delta g \rangle_0 = 1.47 \times 10^{-6} T^{-1/3}$, and that $L_{in}(T) = 3.5 \times 10^{-8} T^{-1/3}$. This L_{in} satisfies the criteria for the diffusion channel: $2\hbar D/g\mu_B L_{in}^2 > B$ and $\phi_0/L_{in}^2 < B$, where $\phi_0 = h/e$. Therefore our measurement was swept to the level shown in the middle column of Table 1. Using this $L_{in}(T)$, $\langle \delta g \rangle_{exp.}$ fits the theoretical curve well when $\tau_{so} = 7.0 \times 10^{-14}$ s, as indicated by the solid line in Fig. 2. In the high temperature regime $\langle \delta g \rangle$ is governed only by the nature of inelastic scattering with electron-electron scattering, while in the very low temperature regime it is governed by spin-orbit scattering, as indicated by the theoretical curve fitting the broken line in two regimes. For the GaAs/AlGaAs wire, $\langle \delta g \rangle$ reduction in the limit of strong spin-orbit scattering can be observed below about 0.02 K, as can be speculated from Fig. 2. The condition $\tau_{so} \ll \tau_{in}$ is not satisfied and insufficient transition occurs in the measured temperature range. Figure 3 shows the characteristic lengths of the InGaAs/InAlAs wire obtained from the present analysis of $L_{so} = \sqrt{D \cdot \tau_{so}}$. Thermal diffusion is another phase breaking mechanism that is important in metals. However, it is not dominant in our sample since $L_T = \sqrt{\hbar D/k_B T}$ is much larger than L_{in} and L_{so} in the measured temperature ranges. L_{in} becomes longer than the spin-orbit scattering length L_{so} in the temperature range where antilocalization and CF reduction occurs. This also explains well the transition from the weak to the strong scattering regime below 1 K in our InGaAs/InAlAs quantum wire.

Conclusion

We observed CF amplitude reduction with decreasing temperature in a InGaAs/InAlAs mesoscopic wire. The temperature dependence of CF amplitude agreed with the predicted one-half reduction due to the transition from weak to strong spin-orbit scattering.

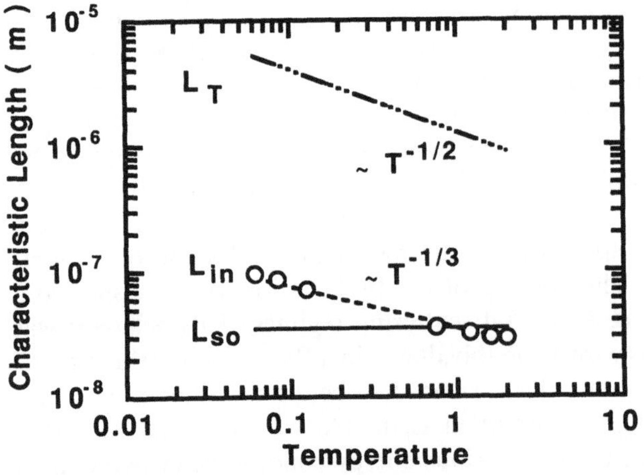

Fig. 3. Temperature dependence of inelastic scattering length, L_{in}, spin-orbit scattering length, L_{so}, and thermal diffusion length, L_T, of the InGaAs/InAlAs quantum wire

Acknowledgments. The authors are indebted to T. Kimura, Y. Horikoshi, T. Izawa, K. Hirata, S. Yamada, A. Matuda, S. Kurihara, and Y. Ishii for their continuous encouragement throughout this work. The authors are also indebted to H. Asai for the metal organic chemical vapor deposition (MOCVD) growth of facet quantum wires and to H. Fukuyama and H. Yoshioka for their valuable discussions.

References

1. Stone AD (1989) Phys Rev B39:10 736–10 743
2. Lee PA, Stone AD, Fukuyama H (1987) Phys Rev B35:1039–1070
3. Debray P, Pichard JL, Vicente J, Tung PN (1989) Phys Rev Lett 63:2264–2267
4. Kawaji S, Kuboki K, Shigeno H, Nambu T, Wakabayashi J, Yoshino J, Sasaki H (1985) In: Chad: JD, Harrison (eds) The WA 17th international conference on physics of semiconductors. pp 413
5. Elliott RJ (1954) Phys Rev 96:266
6. Fukai YK, Yamada S, Nakano H (1990) Appl Phys Lett 56:2123–2125
7. Fukai YK, Nakano H, Nakata S, Tarucha S, Arai K to be submitted
8. Feng S (1989) Phys Rev B39:8722–8724

Tunneling

Trajectory Representations, Fluctuations, and Stability of Granular Electronic Devices

22

JOHN R. BARKER, SCOTT ROY, AND SHARIF BABIKER[1]

Summary. Granular nanoelectronics or single-electronics is concerned with the development of device technologies which approach the information theoretic limit of one bit on one electron. Only recently, with the demonstration of a single-electron shift register device based on correlated single electron tunneling in coupled metal-insulator-metal capacitors, has a serious candidate emerged which has the appropriate device and system properties. These system properties are shown here to include: pronounced fault tolerance to variations in capacitor parameters but sensitivity to cross-talk. The regimes of logic stability are described. Simple global clocking schemes are shown to lead to incoherence and synchronization failure unless precautions are taken. The analysis is based on the Fokker-Planck projection of the density matrix in the charge diagonal representation and by Monte Carlo and linear programming solutions to the coupled circuit equations. Transfer of these circuit concepts to semiconductor configurations based on strongly confined squeezed two-dimensional electron gas (2DEG) channels is discussed. Finally, the Bohm-de Broglie pilot field model of quantum mechanics is examined as a useful interpretive and computational aid to understanding the granular electronic regime. Apparent contradictions with conventional quantum transport formalisms are resolved and the connection with conventional Wigner distributions is established.

Introduction

Exciting developments in both experiment and theory in the last few years [1] have made it possible to seriously consider a future electronics technology based on systems of devices functioning at the limit of **one bit to one carrier**: the information theoretic limit. Granular electronic devices include devices which contain a very small number of carriers; devices which are subject to fluctuations due to discrete charges perturbing the confinement potential; and devices for which charging effects such as the Coulomb blockade become important. Although there are well-known examples of potentially useful single electron switching processes, for example, electron waveguide interferometer switches, the controllable phenomena

[1] Nanoelectronics Research Centre, Department of Electronics and Electrical Engineering, University of Glasgow, Glasgow G12 8QQ, Scotland, UK

are limited by coherence and fluctuation problems to very low temperatures and the configurations have little integratability. Only recently, with the demonstration of a single-electron shift register device based on correlated single electron tunneling in coupled metal-insulator-metal capacitors [2], has a serious candidate emerged which has the appropriate device and system properties for a single-electronics technology.

In the present paper we report briefly on the theoretical system properties of such devices, paying attention to fault tolerance, stability, logic stability, and clock coherence.

Even at this system level of description it is necessary to use quantum mechanical considerations in modeling the circuit properties. At a deeper level, and particularly for analagous developments in semiconductor environments, it is advantageous to use a formalism which gives some idea of the phase-space dynamics of individual electrons. Traditionally, the Wigner distributions have been used for this purpose, but they contain serious interpretive and computational problems. Here, we report briefly on using a non-standard interpretive framework for quantum transport, based on the pilot field picture developed originally by Bohm and de Broglie.

Single-Electron Capacitative Devices

Let us first briefly view the salient developments in single-electronic switching based on tunneling capacitors. Modern lithography provides the opportunity to fabricate metal-insulator-metal (MIM) tunnel junctions with an area below (100 nm)^2, and consequently a capacitance of less than 10^{-15} F (1 fF). When one electron is transferred across the junction by tunneling, the charging energy $E_c = e^2/2C$ is about 1 K, in temperature units, for $C = 1$ fF and leads to non-trivial *discrete* charging effects at low temperatures (we define the charging transition temperature by $T_c = E_c/k_B$). A range of new phenomena has been predicted theoretically, including the *Coulomb blockade, single-electron-tunneling oscillations and correlated single-electron-tunneling via soliton states*. These phenomena have been observed experimentally in the last 3 years (recent reviews are given by Likhaerev [3] and Averin and Likhaerev [4]). Most experimental effort has been directed at relatively large capacitances ($T_c < 1$ K) and involves single junctions, circuits with two or three junctions (shift register devices [2]), and longer linear arrays. Studies of 2D arrays have been reported [5, 6], including evidence for a Kosterlitz-Thouless-Berezinskii phase transition [7] (below the transition temperature there exist charge-anti-charge bound pairs, above the transition there are just free charges $+e$ and $-e$).

The transition temperature becomes higher as the capacitance C is diminished. This may be achieved by reducing the effective area of the capacitor. To date, relatively large area capacitances have been reported with transition temperatures in the mK–3K range. High temperature demonstrations have been reported using scanning tunneling microscope probes as one electrode to ultra-small capacitors fabricated with inorganic and organic components (operating temperatures between mK and 300K have been reported). Significantly, single electron charging phenomena are not confined to metal insulator systems [8] and have been observed in squeezed geometry 2DEG and other semiconductor systems [9–14].

Single-Electronics Systems Technology

The principal new development in single electronics *systems* technology has been the demonstration of the "clocking" of single electrons in a shift-register system or turnstile [2] based on an array of four small tunneling capacitors in a MIM system driven through a non-tunneling capacitor by an applied voltage applied to the center of the array. Such a system functions by the correlated tunneling of single electrons, which occurs when the capacitor charging energy $E_c = e^2/2C$ exceeds the thermal supply energy $k_B T$.

In recent theoretical studies of single electron shift registers (see following section) we find that correlated tunneling of individual electrons through a cascade of tunneling capacitors is very stable against fluctuations, including variations in the design parameters of the capacitors. This remarkable stability appears to be associated with the existence of a long-range order parameter which derives from the induced macropolarization of the electrode array; indeed, soliton charge states are predicted theoretically [15]. Interestingly these have some analogies with the well-known soliton states in electroactive polymers [9, 16, 17].

The demonstration of the turnstile device [2] allows the control of single electrons moving through the device by an external gate. Such devices will find practical applications in *quantum metrology* [3], and possibly as *electrometers* of extreme sensitivity ($i_n < 2 \times 10^{-4}$ $eHz^{-1/2}$) [18]. But perhaps of most importance, single-electron-tunnel devices (SET) are attractive as a technology for *digital electronics*. The basic reasons are:

(i) *Device performance*—noise immunity, speed, and operating temperature *all improve with reduced device size*, leading to the possibility of very high packing densities in the 1 bit per carrier limit at reasonable operating temperatures.
(ii) *Quantization of charge*—the fact that discrete charge is held within the SET structure offers a natural 'regeneration' of logic levels during operation.
(iii) *Speed-power product*—this is substantially better than any known electronic logic family. The SET power-delay product is of the order of $k_B T_0$, where T_0 is the 'maximum operating temperature' of the device. For $(30\ \text{nm})^2$ junctions this would give $k_B \times 30\text{K}$ or 2.5 meV. This may be compared with, for example, complementary metal-oxide silicon (CMOS) $\sim 10^{10}$ eV; single flux quantum logic $\sim 10^4$ eV. SETs thus easily break the critical 1 fJ barrier (6×10^3 ev) by several orders of magnitude.
(iv) *Good structural tolerances*—the device performance is relatively insensitive to poorly-controlled aspects of the fabrication, such as tunnel barrier thickness.

It appears, therefore, that we are poised on the threshold of a genuine limiting-scale nanoelectronics technology in which the following criteria are satisfied or potentially satisfied:

(a) High temperature operation
(b) Stability against intrinsic and fabrication-induced fluctuations
(c) System integrability (planar technology)
(d) Very high bit density
(e) Good power dissipation properties
(f) Demonstrated system capability

None of these criteria are presently satisfied by quantum interference devices and other proposed low-dimensional semiconductor devices. However, these system advantages are offset by the severe difficulties of handling the tunnel junction technology. There are also limits to system complexity inherent in the hanging resist technology. MIM systems are therefore useful now as a development tool for exploring single-electronic systems; but in the long term it is unlikely that MIM is the best materials configuration. Provided it is possible to solve the problem of controlling the fluctuation potential [19, 20] and of obtaining strong low dimensional electron confinement, the long term prospects favor semiconductor heterostructures, as we shall discuss later.

The opportunity for single-electronics technology is also tempered by the fact that our understanding of these devices is such that we cannot answer some of the most basic engineering questions concerning their *in situ* operation. Among these questions are the following:

(i) Can we construct a system in which the operation of one single electron switch modifies the operation of another? In other words, can we build a two-component single electron digital integrated circuit?
(ii) How far apart do single electron devices have to be to ensure no interaction?
(iii) How well-shielded do the circuits have to be? (Modification of circuit properties has been demonstrated for induced charges as small as e/30, which is the basis for the electrometry aspirations).
(iv) What is the best material system for the realisation of these devices? (Metal/oxide, 2DEG, quantum dots, molecular ...)
(v) How can we join these devices together?
(vi) How do you perform input/output (I/O) to a SET integrated circuit?

Modelling Techniques for Single-Electronic Systems

In the following sections we report briefly on detailed analytical and computer models of existing and planned experimental single-electronic devices based on the Coulomb blockade, correlated tunneling, and related effects. Most of the modeling has been applied to the hanging-resist MIM Al-Al_2O_3-Al devices, but the more general models also have applicability to 2D3G geometries and self-aligned vertical geometry devices (developed by R. Cheung and J. Weaver at Glasgow, to be published). Studies have been made of the *system properties* of arrays of coupled single-electronic capacitor devices. Particular emphasis has been placed on understanding the effects of *cross-talk and noise immunity*, which are particularly problematic in densely packed, capacitively coupled systems.

At a microscopic level, the Coulomb blockade, single-electron-tunneling (SET) oscillations, correlated tunneling, and soliton formation may be understood using the thermodynamic density matrix approach in the charge diagonal approximation, as established by Averin and Likhaerev [21]. This approach leads to a description of the time-dependence of the charge distribution f(Q) on a capacitor junction via a Fokker-Planck-like equation augmented by difference equation terms which represent tunneling. We have extended the theory to include tunnel-time delays, using the phase-delay model described by Barker [22], and to include the description of several coupled junctions. The advantage of the density matrix approach is that we can study microscopic effects and the influence of fluctuations very closely.

At a more macroscopic level it is possible to describe extended circuits of SET devices in a similar formalism to that of Bakhvalov et al [23] for a linear array of junctions, but using a matrix representation of the various voltages, currents, and circuit elements. The circuit matrix equations may be solved by Monte Carlo methods. In our Monte Carlo simulation a general network solver is used to discover values of critical charge $q_{critical}$ for each of the tunneling junctions of the system. At these charge values tunneling through a junction becomes possible. The simulation then repeatedly characterizes the circuit for a given set of input voltages and clocked charge positions; discovers which tunneling event will occur next; updates the voltage and charge values due to this event and any external changes in device bias; and re-characterizes the device.

In the case of modeling shift-register or turnstiling devices, the Monte Carlo simulation runs the device for a specified number of cycles of gate voltage and counts the number of transferred electrons. This is done for each required value of bias voltage and modulation amplitude. To discover the frequency response of the device, the simulation repeats the counting process over different sets of gate voltage cycle times, then translates the gate cycle times into frequency.

The Monte Carlo simulations, particularly for locating the stable operating regimes, are very compute-intensive and a much faster scheme based on linear programming has been developed. The linear programming approach at first follows the Monte Carlo method in obtaining the critical charges. Thereafter, the method is the reverse of the Monte Carlo approach. Instead, we decide first which transition events we wish to allow or disallow. This defines clocked electron positions. For each event the allow/disallow barrier ocurs at a value of critical charge which is higher than the critical charge for allowed events or lower otherwise, thus defining charge values. Knowing these, it is possible to solve for gate and bias voltages to obtain a line which demarcates allowed events from forbidden events in the parameter space V_A–V_G formed by the potential V_A at a tunneling electrode and the gate voltage V_G. The scheme functions by dividing the parameter space into a large number of discrete cells to obtain the legal range of device operation.

Full technical details of these computational schemes will be reported elsewhere.

Fault Tolerance and Operational Regimes of Single-Electronic Systems

Linear Arrays of Junctions

Detailed numerical studies based on the application of linear programming and Monte Carlo techniques to the circuit analysis predict the existence of significant *domains of stability* for coupled tunnel junctions. Figure 1a illustrates a linear tunnel junction circuit in terms of the junction capacitance C and capacitance to ground C_0. Arrays with between 7 and 21 junctions have been studied. It is easiest to characterize a tunnel junction array by the threshold voltage Vthresh, defined as the point at which the current voltage curve of the junction characteristic turns

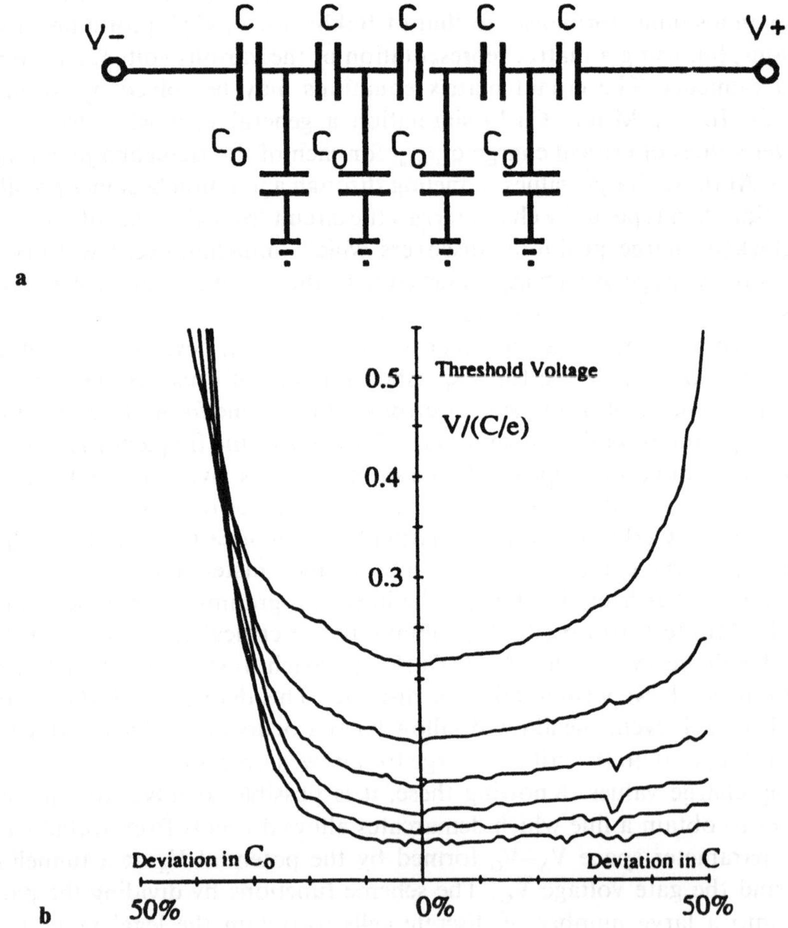

Fig. 1. a A linear tunnel junction array. **b, c** Coulomb blockade threshold voltage as a function of the standard deviation of random capacitance values

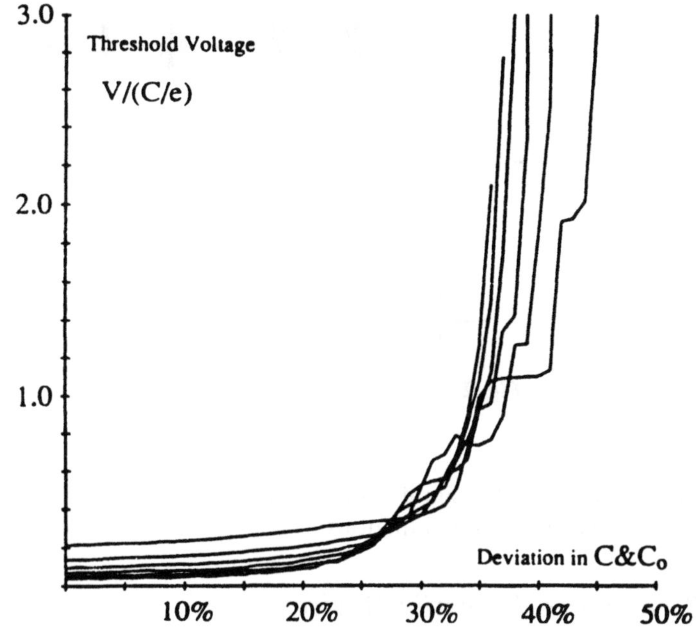

c

Fig. 1 (cont.)

on half the blockade voltage. Figure 1b, 1c shows the Coulomb blockade threshold voltage as a function of the standard deviation for an ensemble of arrays configured from a random (Gaussian) distribution of capactances for the mean junction capacitances C and mean array capacitance C_0. Tolerances of up to 30% standard deviation in the capacitor parameters in an array appear feasible without loss of system performance, as measured by the number of clock-shifted electrons. Beyond this deviation there is a dramatic rise in threshold voltage indicating the presence of very small (compared to the mean) capacitor values in the array, which leads to instability and breakdown of the correlations. A study of the temperature dependence of linear arrays shows that optimum results are obtained for correlated single electron tunneling when $T < 30\ T_c$. In the regime of large capacitor devations, the onset and scale of the instabilities are practically unaffected by lowering the temperature below the optimum value.

Where $C_0 > C$, the threshold curves become more noisy. Small changes in the number of electrons moving through the device then have a more marked effect on the position of V_{thresh}.

The effect of introducing correlations into the random values for the capacitances has little effect on the fault tolerance. For example, alternating high and low values of C, to simulate misalignment during fabrication, has no noticeable influence on the general results.

The effect of raising or lowering the value of just one capacitance in a long array was shown to have negligible effect until an order of magnitude change was effected. This result holds wherever the fault was placed, for example, at an edge or

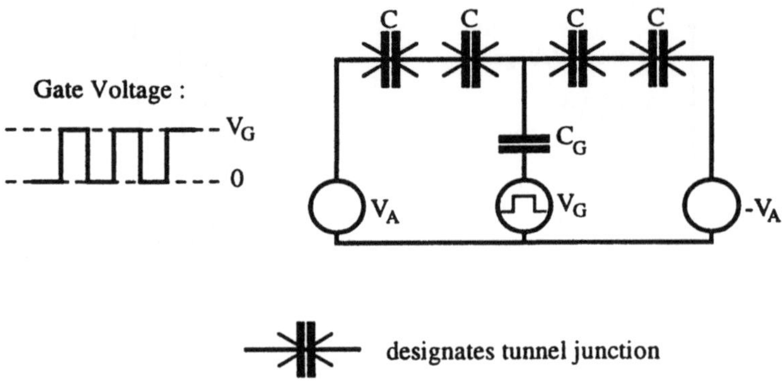

Fig. 2. A two-phase turnstile device

in the center of an array. No significant differences in fault tolerance were found in arrays with between 7 and 21 junctions.

Single Two-Phase Turnstile Devices

Figure 2 sketches the effective circuit for a two-phase turnstile device (Geerligs version [2].) The device functions by clocking electrons from the electrode at potential V_A through the tunnel junctions (marked with a cross) to electrode at potential $-V_A$ by the application of a gate signal V_G comprising an ac component superposed on a dc bias. The allowed regimes of operation for the turnstile were found by both Monte Carlo and linear programming to lie within iris-like areas in V_G–V_A space. For a large gate capacitance $C_G \gg C$ the area of turnstiling operation becomes extremely small. Our results confirm that C_G must be much less than C for turnstiling to occur (Geerligs original experiments [2] had $C_G = 0.5$ C).

Single Three-Phase Turnstile Devices

Figure 3 shows a three-phase device of a type under experimental study at Glasgow. To simplify the analysis we have chosen symmetric devices with identical outer junction capacitances, identical outer gate capacitances, and so on. Figure 4 shows a typical plot (from a set of plots taken over a large range of capacitor values) of the operating regime for turnstiling computed by Monte Carlo (circled points) and linear programming (lines). The allowed operating regions are much more complex than for two-phase devices. It was found that the turnstiling area was affected much more by the variation of the outer gate/shunt capacitances than by the inner ones: too large an outer gate capacitance will seriously degrade performance.

A detailed comparison of three-phase devices with their two-phase counterparts has been made where each device was compared with equivalent component values, each operating in the middle of its respective allowed range. The three-

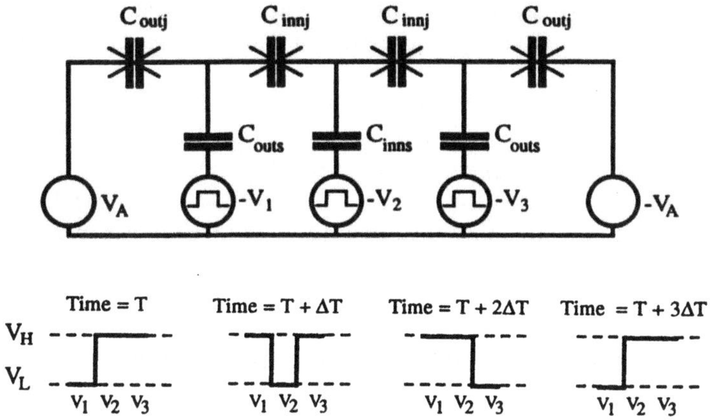

Fig. 3. A three-phase turnstile device

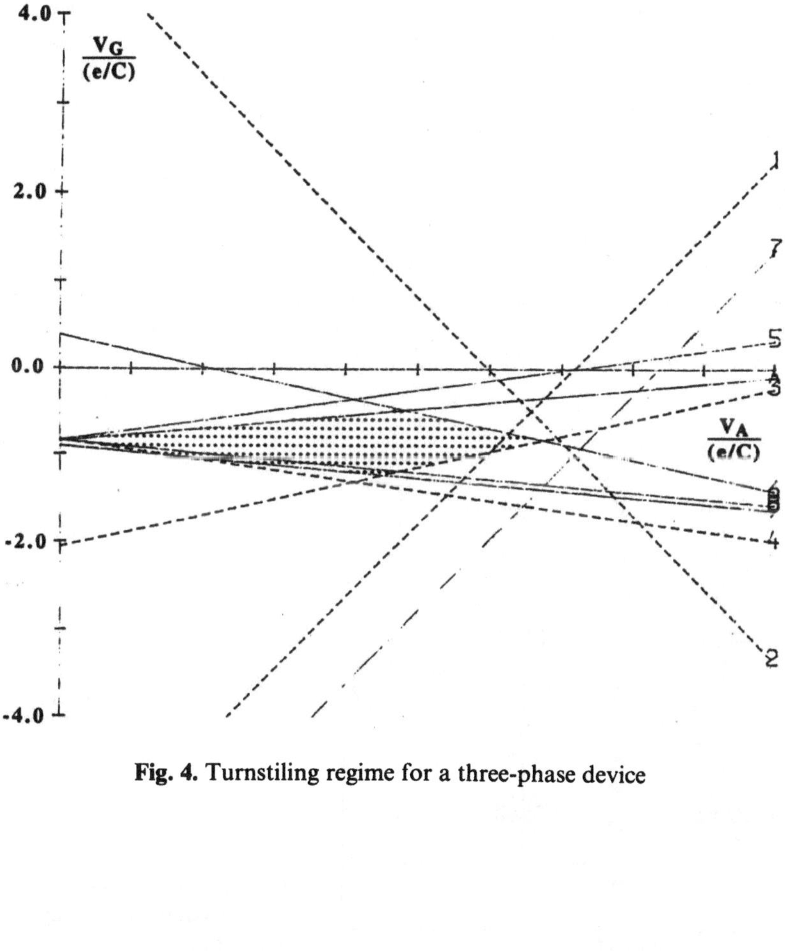

Fig. 4. Turnstiling regime for a three-phase device

phase device has a high frequency advantage compared to the two-phase device. However, the three-phase device has generally a smaller area of legal operation than the two-phase device.

The three-phase device has been mooted as potentially useful for allowing a larger gate voltage swing to haul electrons through the device and therefore increase reliability and operating speed. However, it was found that large voltage swings produce an avalanching effect on the electron flow, which breaks down the turnstiling action. We conclude that the advantages of three-phase over two-phase devices appears small.

Cross-Talk and Coupled Single Electronic Systems

Our studies are being extended to evaluate *cross-talk* between signals propagating on adjacent capacitor arrays. Cross-talk is a serious potential obstacle for the development of practical capactive circuits and it merits considerable study. Figure 5 illustrates two two-phase turnstiles coupled by stray capacitances C_S. Detailed modeling studies were made for this circuit to find the magnitude of the stray capacitance which would produce a significant change in the ideal circuit parameters. In particular, we have focused on the changes induced in the critical charges: the charge required on a junction to bring the onset of tunneling is not only dependent on the junction capacitance, but also on the effective impedance of the rest of the circuit, including strays which result in circuit coupling.

Figure 6 is a graph of the junction critical charge (both inner and outer junctions of a symmetric circuit) as the strays C_S are varied from 10^{-3}C to 10^3 C. At low C_S the critical charge is approximately C/3, the same as for an isolated two-phase turnstile. This value falls at about $C_S = 20$ C and continues until $C_S \sim C/6$, at which point tunneling becomes much easier and avalanching occurs.

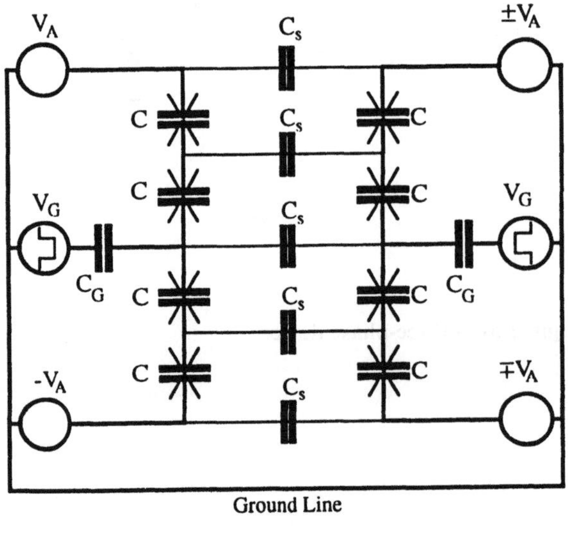

Fig. 5. Two two-phase turnstile devices coupled by stray capacitances C_S

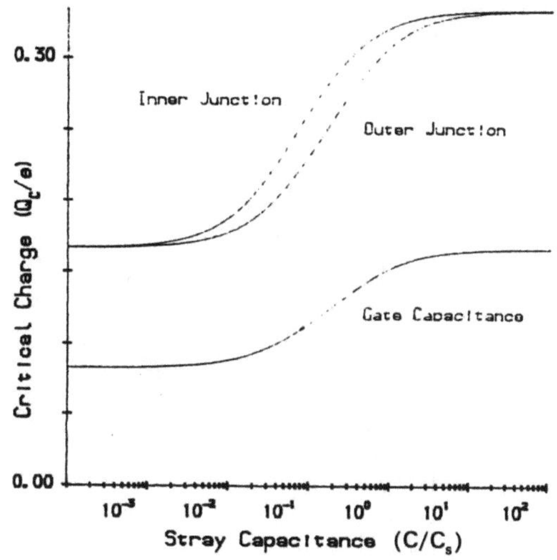

Fig. 6. Graph of junction critical charge as function of stray capacitance for two two-phase coupled turnstile devices linked by stray capacitances

Preliminary studies of the legal areas of turnstile operation when cross-talk is present indicate that it introduces an asymmetry into the operation of parallel turnstiles in which an electron may be locked into one turnstile while tunneling occurs in the adjacent turnstile. Thus, each turnstile clocks through a different number of electrons in a given time, depending on initial conditions. The frequency response shows a similar breakdown of symmetry. Further work is clearly required here and we are in the process of examining the cross-talk between information carrying bit streams. It seems likely that coding schemes may be required to overcome cross-talk interference.

Coherence and Clocking Problems

For the case of a single junction, we find analytically and numerically that the distribution f(Q) undergoes a damped oscillation with an asymptotic approach to a non-uniform distribution in charge space. Predicted SET charge oscillations for a single junction appear to damp out exponentially in time under the density matrix theory (see Fig. 7). This is a consequence of the ensemble statistics implied by that theory. An individual junction will admit undamped but fluctuating SET charge oscillations in time, but an ensemble of identical junctions becomes incoherent and the net effect appears to be a damping of the charge oscillations. This picture has been confirmed by detailed Monte Carlo calculations.

The consequences of incoherence are very serious for the synchronized switching of charge in single-electronic circuits; the effect is similar to the diffusion-delay problem in long VLSI paths. Thus a parallel arrangement of identical correlated electron sources driven by a single control voltage would become incoherent over a sufficient period of time. We have found, in a preliminary study, that this effect can be overcome by using capacitative cross-coupling between adjacent sources to

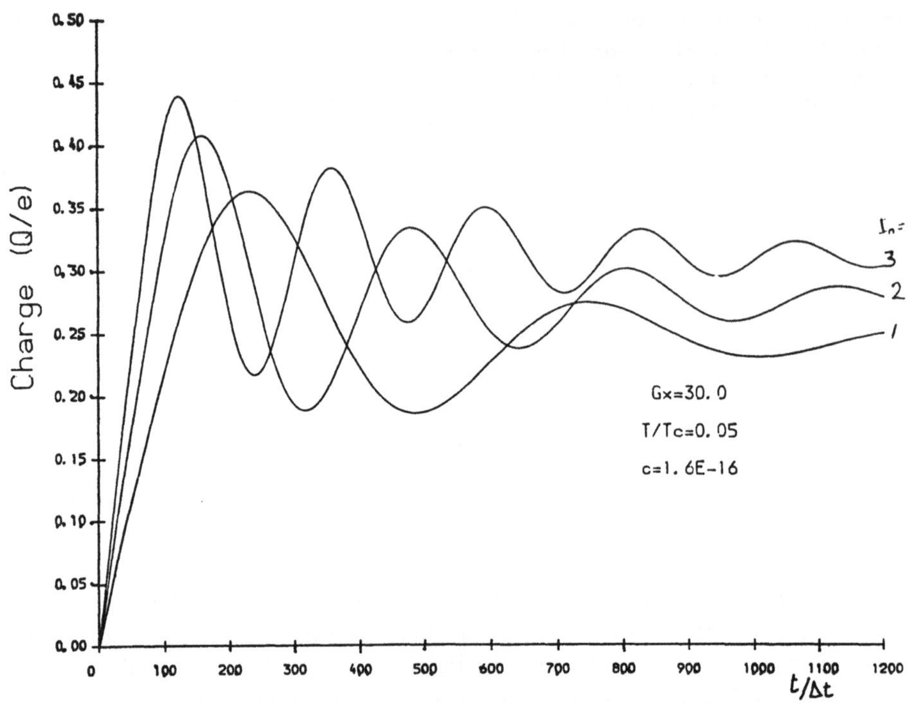

Fig. 7. Time evolution of charge on a single junction in the SET oscillation regime for different current bias on a single tunneling junction in the single electron tunneling oscillation regime, obtained by solving the density matrix equations in the charge diagonal limit

set up a lateral collective state analagous to the longitudinal correlated state in long capacitor chains. Such an arrangement, of course, would lead to cross-talk between modulated parallel electron sources, and the effects of this remain to be evaluated.

SET Devices in Semiconductor Configurations

The long-term prospects for single electronics favor semiconductor heterostructures using quantum networks. Here we define a quantum network as a connected set of quantum wires produced by the lateral patterning of a 2DEG. A very simple example is afforded by two coupled quantum point contacts which form an elementary capacitative device in which charging effects and resonant tunneling coexist [14]. These systems are potentially gateable to form analogues of the MIM turnstiles. There are considerable physical differences, however, which we will not discuss here. The major problem, and it affects most 2DEG-based devices, is how to reliably interconnect such devices and hence achieve integrability when one cannot, at present, control either the fluctuation potential or obtain strong low-dimensional electron confinement. But if these problems can be solved very much smaller capacitances and hence, higher operating temperatures would become available; moreover, there would much be much more scope for producing configurations which ameliorate problems with shielding and cross-talk.

There are also problems of understanding which impinge on basic concepts in quantum mechanics. These are to do with the apparent delocalization of charge in double-well systems and in rings and with whether coherence can be destroyed by charge trapping-detrapping in the vicinity of a one-electron system, i.e., can we define whether quantum measurement effects take place or not?

Trajectory Representations of Single Electrons

The Pilot Field Picture

Quantum mechanics normally deals with predicting the statistical outcome of a series of measurements on identical systems or the outcome of a measurement on an ensemble of quantum systems. An example of the former might be the prediction of the statistical pattern of electron counts in a low intensity two-slit experiment (which has a solid state analogue in Aharonov-Bohm semiconductor rings). An example of the latter might be a conventional current transport measurement on a device containg very many carriers. It has even been argued by some that quantum mechanics cannot describe single quantum systems [24]. Although quantum mechanics provides a perfectly satisfactory framework for modeling the statistics of single-electron systems there is little in the orthodox *interpretation* to provide much intuition or to suggest tests for the acknowledged problems with quantum mechanics, which become particularly apparent for single systems. Indeed, simple questions of the coherence of coupled Coulomb blockade oscillator devices and related problems on the synchronization of single electron circuits by clock signals require an understanding of individual events as well as ensemble averages. At a harder level there are difficult interplays between single systems and many-body dissipative systems when one considers phenomena ranging from the quantum image force arising from a single electron tunneling to the problem of the quantum measurement process itself.

In seeking a fruitful way forward we have found considerable interpretive power and some very useful new computational advantages in adopting the *pilot-wave* or pilot field alternative to orthodox Copenhagen theory put forward originally by de Broglie and later revived by Bohm [25, 26] and Vigier (reviewed in Jammer [27]). This approach predicts identical outcomes to conventional quantum mechanics but has the potential for significant deviation from the orthodox theory if "wave-packet collapse upon measurement" is rejected: still an open question in quantum measurement theory. Its practical significance, however, is that it (a) retains a considerable element of the physical realism implicit in classical physics, notably the existence of well-defined *particle trajectories* in phase space; and (b) through the device of the *quantum potential*, which describes the coupling of the pilot self-field to the electron, it provides a simple removal of very many of the paradoxes of quantum theory, such as the two-slit experiment dichotomy (see Fig. 9), the delayed-choice problem and a smple picture of how interference and tunneling phenomena occur.

The pilot-field picture provides a detailed picture of individual carrier histories which suggests a very useful framework for developing ideas in single-electronics. However, there have been a number of severe objections to the approach. In a preliminary study, we have been able to overcome most of these difficulties and to reconcile the approach with conventional many-body quantum transport theory. The details will be given elsewhere and just a few points will be presented in this paper.

Let us first outline some aspects of the standard pilot wave picture of quantum mechanics as a device for thinking about granular electronic structures from a non-conventional viewpoint (excellent reviews have been given in the collections edited by Schommers [28]). There are three reasons for doing this. The first is that it is philosophically interesting; the second is that it offers an alternative computational and visualization tool for some quantum transport problems in single electronics; and the third is that it might provide an excellent stimulus for devising tests for weaknesses in quantum mechanics and its interpretive framework. The pilot wave model exploits the original view of both de Broglie and Einstein and, later, Bohm that objectively real quantum waves exist that propagate in space-time according to the relevant wave equation. Moreover, this picture is a dual picture, in that an objectively real spatially localized particle (or particles for a many-body wave field) exists in association with the extended quantum wave $\phi = c\,\psi$ where c is a constant and ψ is the normalized wavefunction satisfying Schrödinger's equation. The apparently stochastic behavior of electrons undergoing phenomena such as diffraction or interference is then identified with deterministic particle trajectories resulting from the classical action of the classical forces on the electron, plus the influence of a quantum force derived from a potential function V_Q determined by the wave field $\psi = R\exp(iS/\hbar)$ according to:

$$V_Q = (-\hbar^2/2m)R^{-1}\nabla^2 R$$

This result (which is shown here in the restricted form for the non-relativistic Schrödinger equation) is in accordance with the existence of a Hamilton-Jacobi-like equation for the action S of a particle which can be derived by separating the real and imaginary parts of the Schrödinger equation in terms of R and S:

$$-\partial S/\partial t = (\nabla S)^2/2m + V + V_Q$$

$$\partial R^2/\partial t + \nabla\cdot\{R^2[\nabla S]/m\} = 0$$

The second equation is just the usual continuity equation. The possible real average dynamics of an electron may be described by classical trajectories on which the momentum is given by

$$\mathbf{p} = \nabla S$$

Philippidis et al. [29, 30] and Vigier et al. [31] have performed exact calculations for the particle trajectories present in two-slit interference and in two-slit interference

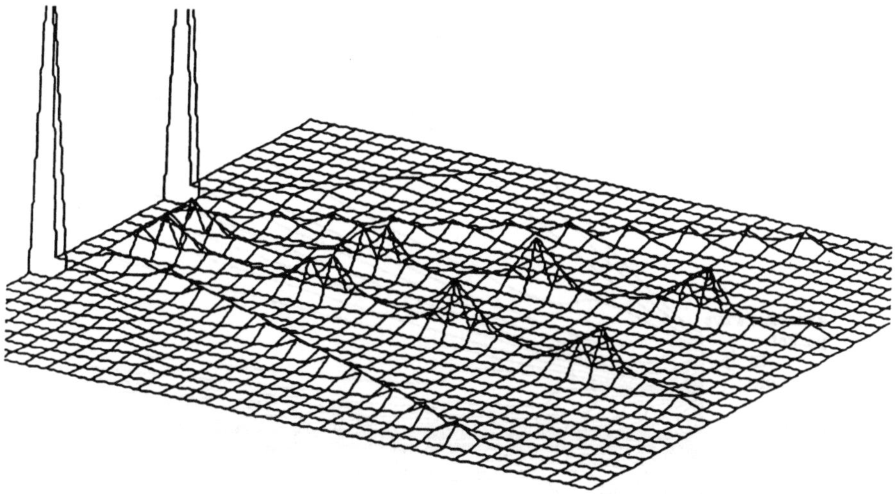

Fig. 8. Quantum potential for a two-slit experiment

in the presence of the Aharonov-Bohm effect. Figures 8 and 9 illustrate a similar calculation by us for the quantum potential and a sample of electron trajectories for a twin source problem set up in a hypothetical squeezed 2DEG geometry. The intensity distribution depends on the occupation probability of each classical trajectory in the complicated quantum potential and, by construction, of course, agrees precisely with conventional quantum theory.

Relation to the Wigner Phase Space Distributions

There is an obvious conflict between conventional density matrix theory as represented by the non-positive definite Wigner phase space distributions $f_W(\mathbf{R}, \mathbf{P})$ of many body quantum transport theory (or equivalent thermodynamic Green functions) and the positive definite phase space probability distributions $f_B(\mathbf{R}, \mathbf{P})$ implied by the existence of well-defined trajectories in the pilot field picture. Indeed, explicit expressions for the two in the case of a single electron pure state give:

$$f_W(\mathbf{R}, \mathbf{P}) = (2\pi\hbar)^{-N} \sum_{\mathbf{r}} R(\mathbf{R} - \mathbf{r}/2) R(\mathbf{R} + \mathbf{r}/2) \exp\{i/\hbar[S(\mathbf{R} + \mathbf{r}/2) - S(\mathbf{R} - \mathbf{r}/2)]\} \exp(-i\mathbf{P}.\mathbf{r}/\hbar)$$

where N is the number of dimensions and

$$\begin{aligned} f_B(\mathbf{R}, \mathbf{P}) &= R^2(\mathbf{R})\delta[\mathbf{P} - \nabla S(\mathbf{R})] \\ &= (2\pi\hbar)^{-N} \sum_{\mathbf{r}} R(\mathbf{R})R(\mathbf{R})\exp\{i/\hbar[\nabla S(\mathbf{R})]\}\exp(-i\mathbf{P}.\mathbf{r}/\hbar) \end{aligned}$$

However, the resolution of this problem is obtained by taking account of the coupling of the electron to its pilot field in deriving the statistical dynamics. In-

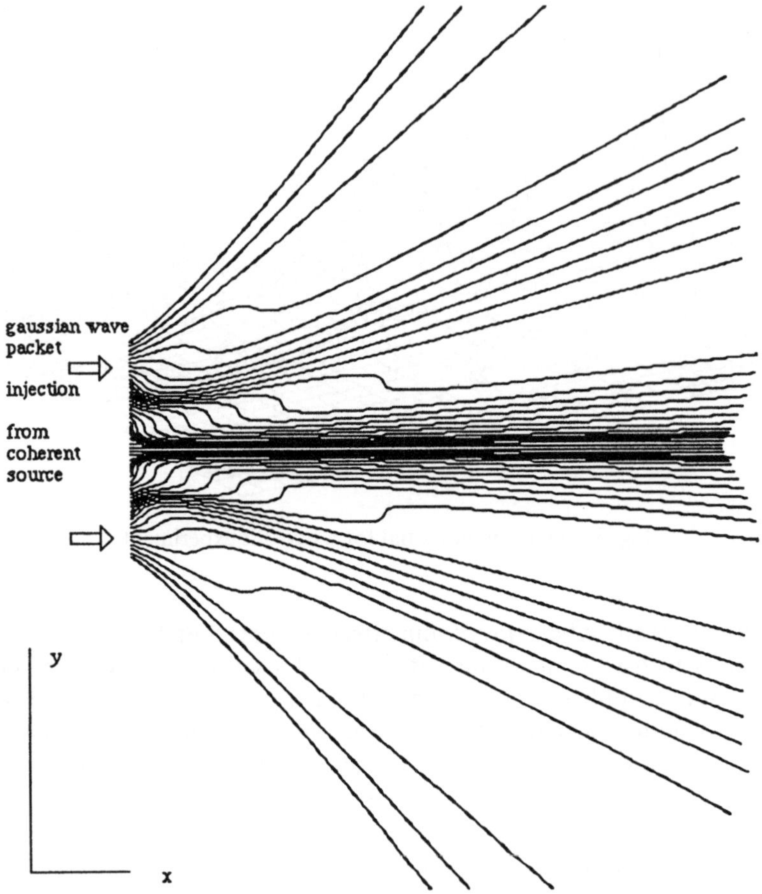

Fig. 9. Pilot field trajectories for a two-slit experiment based on 100 test electrons emitted from a coherent source

deed, one of the great advantages of the pilot field approach is that it does not possess the severe difficuties of non-compactness and numerical instability found recently to be inherent with Wigner function descriptions of quantum transport.

To illustrate, let us use $f_W(x,p)$ and $f_B(x,p)$ to compute the Heisenberg uncertainty relations or statistical dispersion relations in the two approaches:

Orthodox approach — *Pilot field approach*

$$\langle\Lambda\rangle_W = \int \Lambda f_W(x,p)\,dx\,dp$$

$$\Delta_W x = \{\langle(x - \langle x\rangle_W)^2\rangle_W\}^{1/2}. \qquad \langle x\rangle_B = \int x f_B(x,p)\,dx\,dp = \langle x\rangle_W$$

$$\langle x^2\rangle_B = \int x^2 f_B(x,p)\,dx\,dp = \langle x^2\rangle_W$$

$$\Delta_B x = \{\langle(x - \langle x\rangle_B)^2\rangle_B\}^{1/2} = \Delta_W x$$

$$\Delta_W p = \{\langle(p - \langle p\rangle_W)^2\rangle_W\}^{1/2}. \qquad \langle p\rangle_B = \int p f_B(x,p)\,dx\,dp = \langle p\rangle_W$$

$\Delta_W x \Delta_W p > \hbar/2.$

$$\langle p^2/2m\rangle_B + \langle V_q\rangle_B$$
$$= \int (p^2/2m) f_B(x, p) dxdp + \int V_q(x) f_B(x, p) dxdp$$
$$= \langle p^2/2m\rangle_W$$
$$\text{let } \Delta_B p = \{\langle p^2 + 2mV_q\rangle_B - \langle p^2\rangle_B\} = \Delta_W p$$
$$\therefore\ \Delta_B x \Delta_B p > \hbar/2,$$

where the averages $\langle\cdots\rangle$W and $\langle\cdots\rangle$B refer to the orthodox quantum mechanical expectation value and the average value computed from the pilot field picture statistical distribution f_B, respectively. Here we obtain identical results by remembering to compute the interaction energy with the pilot field, i.e., the quantum potential in determining the conventional kinetic energy. In fact an exact nonlinear relationship exists between the Wigner function and the pilot field probability distributions in the case of pure states. We shall not develop that here, but we note that

$$R^2(\mathbf{x}) = \int f_W(\mathbf{x}, \mathbf{p}) d^3p \quad \text{and} \quad \nabla S(\mathbf{x}) = \int \mathbf{p} f_W(\mathbf{x}, \mathbf{p}) d^3p$$

so that the wavefunction and the pilot field distributions are obtained from the first few moments of the Wigner function. Indeed, we believe this provides a route to a more general use of phase-space trajectories when the Wigner function is derived from a mixed state density matrix. R and S then appear as parameters in the quantum Hamilton Jacobi equation and the continuity equation.

The pair of equations for R and S give a perfectly viable computational approach to computing quantum transport problems by solving the problem as a coupled particle problem (to obtain $p = \nabla S(x)$) and a field problem to determine R. Details of such a scheme, including the handling of gauge problems and quantization of action, will be outlined elsewhere.

Among the applications, we have found this picture useful in modeling the Aharonov-Bohm effect in ring geometries (1D multiply connected model), resonant tunneling (with a simple picture of the decay of the resonant state), electron-photon/phonon scattering, 2D two-slit experiments, and correlations in a two-electron gas in a confinement potential.

Non-locality remains stubbornly in the theory, as might be expected from the current understanding of quantum mechanics.

The pilot-field picture has valuable interpretive power, possibly fundamental significance, and practical computational advantages for single-electronics. However, as we will discuss in detail elsewhere, there are other, deeper and more semantically consistent ways of formulating such a theory, in which, for example, a stochastic element re-appears (the pilot field trajectories then represent, for example, the mean of an underlying random process). But these lead to identical predictions for measurable quantities if the orthodox view of measurement is preserved, and they are thus not testable.

There is evidently some advantage in applying these different interpretive approaches to basic problems in single-electronics. The exploration of what differences occur (if any) with orthodox theory (especially in the quantum measurement problem) will help studies in the foundations of quantum mechanics. In the latter case, it is interesting to consider the possible utilization of controlled coherent single electron sources and the fractional charge sensitivity of single-electron electrometry to devise direct tests for empty pilot fields along similar lines to recent proposals by Selleri [32]. This possibility is discussed further by one of us elsewhere [33].

Conclusions

We have shown by explicit calculations that, at least in MIM structures, there are genuine possibilities for developing systems of single electron devices. The consideration of single electron phenomena raises questions of interpretation of quantum transport theory, and we conclude that the pilot wave picture has useful properties over the conventional Wigner distribution approach.

Acknowledgments. We would like to acknowledge studentship support from the Science and Engineering Research Council (SR) and the Sudanese Government (SB).

References

1. Ferry DK, Barker JR, Jacoboni C (eds) (1991) Plenum, New York
2. Geerligs LJ, Anderegg VF, Holweg PA, Mooij JE, Pothier H, Esteve D, Urbina C, Devoret MH (1990) Phys Rev Lett 64:2691
3. Likhaerev K (1988) IBM J R and D 32:144
4. Averin D. Likhaerev K (1990) In Altshuler BL, Lee P, Webb P (eds) Quantum effects in small disordered systems. Elsevier, Amsterdam
5. Geerligs LJ, Mooij JE (1990) In: Ferry DK, Barker JR, Jacoboni C (eds) Granular Nanoelectronics Plenum, New York, pp 393–412
6. Mooij JE, van Wees BJ, Geerligs LJ, Peters M, Fazio R, Schon G (1990) Phys Rev Lett 65:645
7. Kosterlitz JM, Thouless DJ (1973) J Phys C6:1181
8. Barker JR (1990) In: Shaw DW, Bean JC, Keramidis VG, Peercy PS (eds) Materials Research Society Proceedings, vol 198, epitaxial heterostructures, pp 3–16
9. Kuzmin L, Delsing P, Claeson T, Likhaerev K (1989) Phys Rev Lett 60:309
10. Williamson JG, Staring AAM, Kouwenhoven LP, van Houten H, Beenakker CWJ, Timmering CE, Mabesoone, M Foxon CT (1991) In: Kirk WP, Reed MA (eds) Proc int symp nanostructures and mesoscopic systems, Santa Fe. Academic Santa Fe
11. Williamson JG, van Houten H, Beenakker CWJ, Broekaart MEI, van Wees BJ, Foxon CT (1990) Phys Rev B41:1207 Melrav U, Kastner M, Wind S (1990) Phys Rev Lett 65:771
12. Flemsberg K, Jonson M (1991) Phys Rev B43:7586
13. McEuen PL, Foxman EB (1991) Phys Rev Lett 66:1926
14. Staring AAM, Williamson JG, van Houten H, Beenakker CWJ, Kouwenhoven LP,

Foxon CT (1991) In: van Haeringen W, Lenstra D (eds) Proc int symp analogies in optics and microelectronics, North Holland. (1992) Amsterdam
15. Likhaerev K, Bakhvalkov NS, Kazacha GS, Serdyukova SI, (1989) IEEE Trans Magn 25:1436
16. Yu Lu (1988) Solitons and polarons in conducting polymers. World Scientific, Singapore
17. Roth S, Bleier H (1989) In: A Aviram (ed) Molecular electronics science and technology. Engineering Foundation, New York
18. Kuzmin LS, Delsing P, Claeson T, Likhaerev K (1989) Phys Rev Lett 62:2539
19. Nixon JA, Davies JH, Baranger HU (1991) Phys Rev B43:12638
20. Laughton MJ, Barker JR, Nixon JA, Davies JH (1991) Phys Rev B44:1150–1153
21. Averin A, Likhaerev K (1986) J Low Temp Phys 62:345
22. Barker JR (1986) In: Kely MJ and Weisbuch C (eds) Physics of microstructures. Springer Proc in Physics 13:210–230
23. Bakhvalov NS, Kazacha GS Likhaerev K, Serdyukova SI (1989) Sov Phys JETP 68:581
24. Ballentine L (1970) Rev Mod Phys 42:358–381
25. Bohm D (1952) Phys Rev 85:166–179
26. Bohm D (1952) Part II 85:180–193
27. Jammer M (1974) The philosophy of quantum mechanics. Wiley, New York
28. Schommers W (ed) (1989) Quantum theory and pictures of reality. Springer, London, pp 279–331
29. Philippidis C, Bohm D, Kaye, RD (1982) Nuovo Cimento B71:75
30. Philippidis C, Dewdney C, Hiley, BJ (1979) Nuovo Cimento B52:15
31. Vigier JP, Dewdney C, Holland PR, Kyprianidis A (1987) In: Hiley BJ, Peat FD (eds) Quantum implications. Routledge and Kegan Paul, London
32. Selleri F (1989) In: Schommers W (ed) Quantum theory and picture of reality. Springer, London, pp 279–331
33. Barker JR (1991) In: Ferry DK, Barker JR, Jacoboni C (eds) Granular nanoelectronics. Plenum, New York, pp 327–342

23

Superconducting Quantum Point Contacts and Other Mesoscopic Josephson Junctions

CARLO BEENAKKER AND HENK VAN HOUTEN[1]

Extended Abstract

Superconductor-two-dimensional electron gas-superconductor (S-2DEG-S) junctions derive much of their recent interest from potential applications in a three-terminal transistor based on the Josephson effect. The ability of an electric field to penetrate into a 2DEG would allow one to modulate the critical supercurrent of the Josephson junction by means of the voltage on a gate electrode, in much the same way as one can modulate the conductance of a field-effect transistor in the normal state. Of particular interest are junctions which show a quantum-size effect on the conductance, since one would then expect a relatively large field-effect on the critical current [1]. Quantum-size effects on the conductance have been studied extensively in nanostructures such as quantum point contacts and quantum dots. The corresponding effects on the Josephson current have received less attention.

A quantum point contact (QPC) is a narrow and short, impurity-free constriction in a 2DEG, of dimensions comparable to the Fermi wavelength. In the normal state, the conductance G of a QPC is quantized in units of $2e^2/h$, due to the quantization of transverse momentum in the point contact region [2, 3]:

$$G = N\frac{2e^2}{h}, \quad N = 1, 2, \ldots \tag{1}$$

Each transverse mode contributes e^2/h (per spin direction) to G. We have recently shown that this quantization has an analogue for the DC Josephson effect [4]:

$$I_c = N\frac{e\Delta_0}{\hbar} \tag{2}$$

The critical current I_c of a superconducting QPC increases *stepwise* as a function of the contact width, with step height $e\Delta_0/\hbar$ *independent* of the parameters of the junction—but only dependent on the energy gap Δ_0 in the bulk superconductor. To obtain such universal behavior it is essential that the point contact is short compared to the superconducting coherence length $\xi_0 \equiv \hbar v_F/\pi\Delta_0$ (with v_F the Fermi velocity). The case of a longer junction is of great practical interest, but then one does not obtain a junction-independent behavior [5].

[1] Philips Research Laboratories, 5600 JA Eindhoven, The Netherlands

Whereas transport through a QPC is *ballistic*, transport through a quantum dot occurs by *resonant tunneling*. Consider a quantum dot in a 2DEG, which is weakly coupled via tunnel barriers to two electron reservoirs. At low temperatures and small applied voltages the conductance of the quantum dot is given by the Breit-Wigner formula, and has a lorentzian lineshape as a function of the energy ε_R of the resonant level relative to the Fermi level. We have investigated the question: What does the Breit-Wigner lineshape for the conductance imply for the lineshape of the critical current? That problem has not been considered in earlier related work on Josephson tunnel-junctions containing resonant impurity levels in the tunnel barrier [6].

We have found [7] that the lineshape is strongly non-lorentzian: It has a cusp at $\varepsilon_R = 0$ in the case of equal tunnel barriers, and can decay either linearly or quadratically in $1/\varepsilon_R$, depending on the relative magnitude of the resonance width and the superconducting energy gap. The critical current on resonance (for equal tunnel rates $\Gamma/\hbar$ through the two barriers) equals

$$I_c = \frac{e}{\hbar} \frac{\Delta_0 \Gamma}{\Delta_0 + \Gamma} \tag{3}$$

For $\Gamma \gg \Delta_0$ the critical current reaches its maximum value of $e\Delta_0/\hbar$. This is the same value as the discretization unit of the critical current of a QPC. A similar relation between ballistic transport and resonant tunneling exists in the normal state (where the conductance quantum e^2/h appears instead of the supercurrent unit $e\Delta_0/\hbar$).

A third transport regime which we have considered is the regime of *diffusive* transport through a point contact which is long compared to the mean free path l for scattering with impurities. The quantum $e\Delta_0/\hbar$ now appears as the root-mean-square (rms) amplitude of sample-to-sample fluctuations in the critical current [8]:

$$\text{rms}\, I_c \simeq \frac{e\Delta_0}{\hbar} \tag{4}$$

This phenomenon is the analogue of the "Universal Conductance Fluctuations" (UCF) in normal metals [9, 10] (where rms $G \simeq e^2/h$). Both the supercurrent and the conductance fluctuations are "universal" in the sense that the magnitude of the fluctuations is independent (up to a numerical coefficient of order unity) on the size of the junction or on the degree of disorder.

A necessary condition for the universality of the supercurrent fluctuations is that the length L of the junction, while being much greater than the mean free path l, is still short compared to the superconducting coherence length $\xi \equiv (\xi_0 l)^{1/2}$ in the dirty limit. The opposite regime that both l and ξ are $\ll L$ was studied previously by Al'tshuler and Spivak [11]. The amplitude of the critical current fluctuations,

$$\text{rms}\, I_c \simeq \frac{e v_F l}{L^2} \tag{5}$$

then depends on both l and L and is therefore *not* universal.

References

1. van Houten H (1991) Appl Phys Lett 58:1326
2. van Wees BJ, van Houten H, Beenakker CWJ, Williamson JG, Kouwenhoven LP, van der Marel D, Foxon CT (1988) Phys Rev Lett 60:848
3. Wharam DA, Thornton TJ, Newbury R, Pepper M, Ahmed H, Frost JEF, Hasko DG, Peacock DC, Ritchie DA, Jones GAC (1988) J Phys C21:L209
4. Beenakker CWJ, van Houten H (1991) Phys Rev Lett 66:3056
5. Furusaki A, Takayanagi H, Tsukada, M (1991) Phys Rev Lett 67:132
6. Glazman LI, Matveev KA (1989) JETP Lett 49:659
7. Beenakker CWJ, van Houten H (1992) In: Koch H, Lübbig H (eds) Proc 4th int conf on superconducting and quantum effect devices and their applications (SQUID '91). Springer, Berlin
8. Beenakker CWJ (1991) Phys Rev Lett 67:3836
9. Al'tshuler BL (1985) JETP Lett 41:648
10. Lee PA, Stone AD (1985) Phys Rev Lett 55:1622
11. Al'tshuler BL, Spivak BZ (1987) Sov Phys JETP 65:343

Mesoscopic Josephson Junctions

24

SHUN-ICHI KOBAYASHI[1]

Summary. Two-dimensional random networks of mesoscopic Josephson junctions were studied from the viewpoint of superconductor-insulator transition. The universality of the critical sheet resistance for global coherence is discussed.

Introduction

We have studied the transport properties of a system in which superconducting particles of mesoscopic size are packed two-dimensionally to form Josephson junctions [1]. This system involves many interesting problems. They may be classified into four major ones:

The first problem is the superconductivity in a very small particle in which the spacing of the one-electron energy levels δ can exceed the superconducting energy gap Δ. The second is that the Josephson effect in a very small junction, of which the charging energy associated with the transfer of a Cooper pair $(2e)^2/2C$, where C is the capacitance of the junction, can be comparable to or larger than the Josephson coupling energy. The third is the role of the normal resistance or the dissipation of the junction. The fourth is the cooperative effects of the neighboring junctions in the network.

Regarding the first problem, one may expect that the superconductivity in an isolated particle will be considerably modified when $\delta > \Delta$. This happens when the mean particle diameter is smaller than 100A. However, nuclear magnetic resonance experiments on the Knight shift and the spin-lattice relaxation time show that the superconductivity survives even when $\delta > \Delta$. In the limit of $E_J \rightarrow 0$, the network of small junctions can be used to study how far the superconductivity survives with the reduction of the size [2].

With regard to second problem, the total Hamiltonian of the system, including the Josephson energy $-E_J\cos\theta$ and the charging energy $4e^2n^2/2C$, is equivalent to that of a fictitious particle in θ-space with the mass $h^2C/2e^2$ moving in the one-dimensional potential $-E_J\cos\theta$, provided that the c-number n is replaced with $-id/d\theta$. The eigenstates for this Hamiltonian are then the Bloch states [3]. As can be seen in the tight binding approximation of the band theory, the group velocity of the fictitious particle is strongly dependent on its mass. When the mass is greater

[1] Faculty of Science, University of Tokyo, 7-3-1 Hongo, Bunkyo-ku, Tokyo, 113 Japan

than a certain value, the dwell time of the particle at each site becomes much longer than the experimental time scale and, practically, the particle can be regarded as at rest. Thus, $d\theta/dt = 0$ means that the junction can carry zero voltage supercurrent, otherwise the voltage appears across the junction. The critical value of the capacitance, or equivalently the mass, for the localization is of the order of 10^{-15}F [4].

The third problem may be the most delicate, as it involves the question of how to include the dissipation into the quantum mechanics. Presently, there are three different theories. The first is the shunt resistor model, in which the resistor is considered to be an ensemble of harmonic oscillators coupled to the Josephson junction [5]. Next is the quasiparticle tunneling model, which treats Josephson tunneling together with quasiparticle tunneling [6]. The last is the Boson localization model [7]. While the first two apply to a single junction, the last applies to a microscopically disordered film. Although the applicability of the two-dimensional Josephson network is not evident, this is closely related to the last problem, i.e., the cooperative effects.

It is particularly interesting that all these three theories claim that the critical resistance for determining the ground state to be superconductive or nonconductive is about $R_Q = h/4e^2 = 6.45$kohm. To be more precise, for the first model, R_Q applies to the resistance of the shunting resistor and for the second, to the tunneling resistance in a normal state. In both cases, the capacitance of the junctions should be smaller than 10^{-15} F to observe universal critical resistance. For the last, R_Q is the sheet resistance at the phase boundary between superconductor and insulator in the two-dimensional phase space of the magnetic field and the disorder at $T = 0$.

Orr et al., 1985, were the first to mention universal critical resistance, in a report on a quench-condensed tin film experiment [8]. Thc authors expected that the film would not be continuous but that it would have many microcracks to form a network of Josephson junctions, although the quench-condensation method is usually adopted to obtain continuous films.

In the present study, we used samples which we call film of oxidized metallic particles FOMPs. The fabrication method is as follows [9]: In the early stage of the vacuum deposition of metal from the evaporating source onto a relatively high temperature substrate, island-shaped small particles are formed instead of a continuous film. As the deposition is continued, smaller particles aggregate to form bigger ones. To cut the growth, light oxidation of the particle surface is effective. With oxidation, the thermal diffusion of metal atoms on the substrate is suppressed, so that further deposition makes new particles with a size proper to the additional mass thickness. The oxidation plays another important role. When the oxide layer is as thin as sub-monolayer, it acts as a scattering center for electrons moving from particle to particle. When it is more than 10Å, it is actually a tunneling barrier. Thus, by changing the degree of oxidation, the electron transmission probability can be varied over many orders of magnitude, without modifying other properties.

We looked at the old data for Sn FOMPs of which the particle diameter and the mean thickness are both 300Å [10] (Fig. 1). The sheet resistance 6kohm can be the

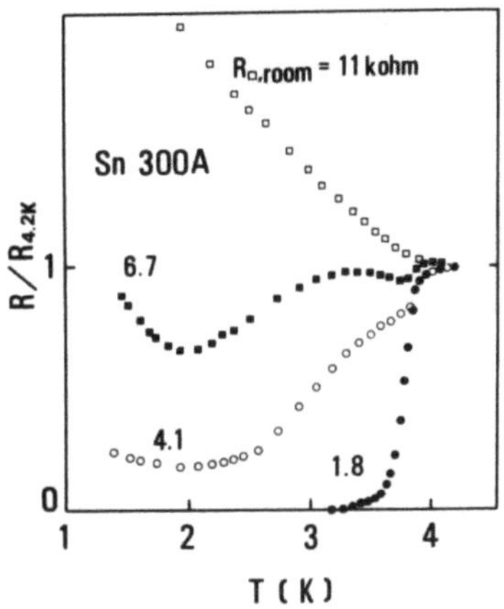

Fig. 1. Temperature dependence of the sheet resistance of 300Å Sn FOMP. Values are normalized at 4.2K (from [10] with permission)

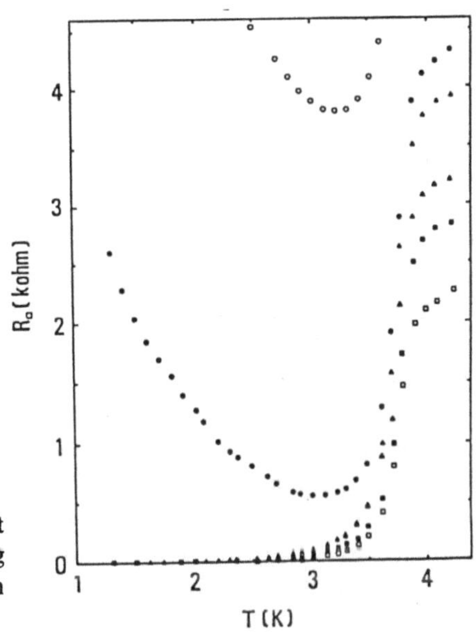

Fig. 2. Temperature dependence of the sheet resistance of 300Å Sn FOMP prepared using temperature gradient oxidation (from [11] with permission)

boundary to divide the samples into two groups, i.e., superconductors and semiconductors. In order to obtain FOMPs with much finer variation of the sheet resistance, the method of oxidation with temperature gradient was adopted [11]. The particle size and the mean thickness are the same as in Fig. 1. The results are shown in Fig. 2. In this case, the critical sheet resistance is smaller than 6.45kohm, being about 4kohm. As is clear in the figure, the normal sheet resistance depends considerably on the temperature. Therefore, there exist serious ambiguities in determining the critical resistance. The same applies for quench condensed films.

Figure 3 shows the result for an Al FOMP of which the mean particle diameter is as small as 15Å [12]. We note that, if an Al particle of this size is isolated, it is hardly superconductive, because the mean spacing of the one-electron energy

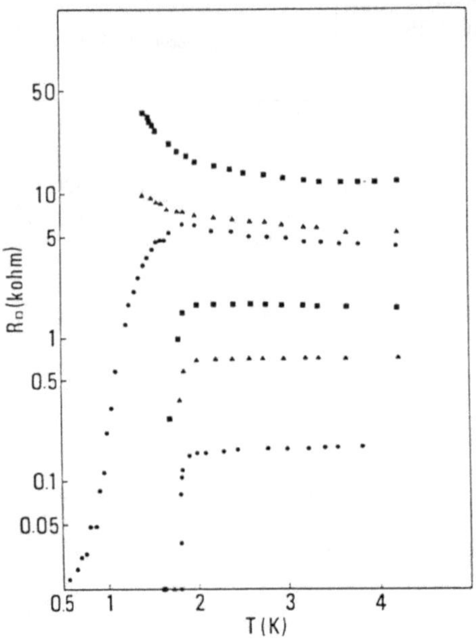

Fig. 3. Temperature dependence of the sheet resistance of 15Å Al FOMP (from [12] with permission)

levels will reach as high as 300K, comparable with the Debye temperature. Therefore, for the Al particles to be superconductive coupling of the particles to the neighbor is indispensable. The critical normal resistance just above the transition point is, again, about 6kohm.

Critical resistances all close to 6.45kohm have also been reported in several quench condensed films, such as Pb [13], Ga [14], and Nb [15]. Therefore we may conclude that the critical sheet resistance is universal, or at least nearly universal, for quench condensed films and FOMPs. In all of these, the decrease in transition temperature is gradual with increasing disorder until the sheet resistance reaches the critical value around 6kohm, and beyond that it rapidly goes to zero.

This behavior is clearly different from that of the continuous films. Continuous films with high resistance have been extensively studied from the viewpoint of the Anderson localization [16]. In these films, the transition temperature decreases smoothly with the increase of the sheet resistance, in agreement with the localization theory. The theory is perturbative so that we cannot know the sheet resistance at which T_c reaches zero. For the experiments reported so far, the sheet resistances are smaller than 6.45kohm.

Among FOMP experiments, those with Al [12] and Zn [17] show an appreciable decrease of T_c with increases of the sheet resistance. The common feature in these two is that the particle size is very much smaller than for the others (10Å for Zn and 15Å for Al). With regard to the limit of small size, the FOMPs are nothing but extremely dirty films. Therefore, these results mean that a length scale of this order is the boundary which determines the applicability of pictures of Josephson networks or microscopically dirty films. However, we do not know why it is 10Å, nor

Fig. 4. Temperature dependence of the sheet resistance of 15Å Al FOMP in magnetic fields (from [12] with permission)

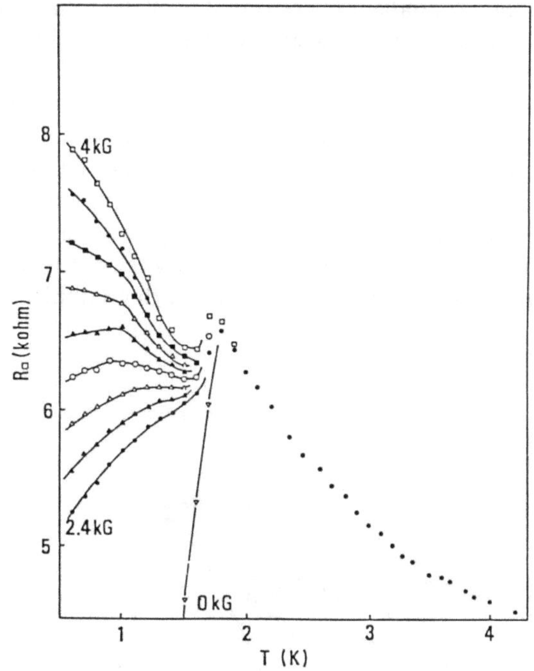

whether the relevant parameter is the length or the energy. Detailed studies to investigate systematically how the dependence of the transition temperature on the resistance changes with the particle size are desirable.

Whereas many experiments suggest the existence of universal resistance, it is difficult to specify which model is to be adopted. This is partly because there are no experiments with a single junction, and because the theories address only the phase diagrams but not the temperature dependence of the resistance. However, we can get some idea from studies in magnetic fields.

Figure 4 shows the temperature dependence of the sheet resistance of the Al FOMP measured in magnetic fields applied perpendicularly to the film [12]. The sample is identical with the one shown in Fig. 3 which has the largest normal resistance among those which undergo the superconducting transition. All curves show maxima with decreasing temperature. When the maximum resistance R_m exceeds about 6.45kohm with increase of the field, the low temperature behavior switches from superconductive to insulating. If we define the critical field H_c as the field at which the slope of the resistance changes sign, we have $H_c = 3.3$kG. By extrapolating the resistance that is horizontal in the figure to absolute zero, we have $R_{c,0} = 6.7$kohm.

This value of the critical field H_c at which the system switches from superconductive to insulating suggests that the transition is not of a single junction nature. The magnetic flux in the area surrounded by neighboring particles is smaller than the flux quantum by one order of magnitude. The flux in the individual junctions is much smaller. Therefore, it is evident that the junctions are strongly interacting with each other, and the coherence length of the network is of the order of 0.1μm, which is two orders of magnitude larger than the particle diameter.

The field-induced transition is also seen in Sn FOMPs with much larger particles [18]. In Fig. 5, results for a Sn FOMP with 280Å diameter are shown. Here again the extrapolated $R_{c,0}$ is about 6kohm, being not far from R_Q. Judging from this gradual decrease of the transition temperature, we apply the Josephson network picture to the Sn FOMPs of this particle size. Therefore, from the theoretical point of view, R_n of the Josephson network nature and $R_{c,0}$ of dirty film nature, both of which are close to R_Q, coexist in this sample.

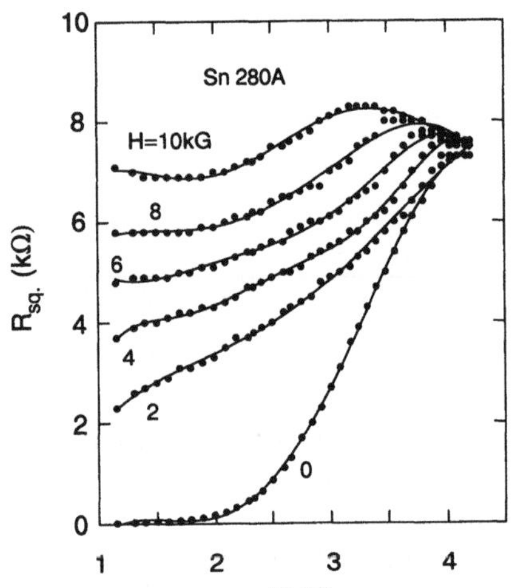

a

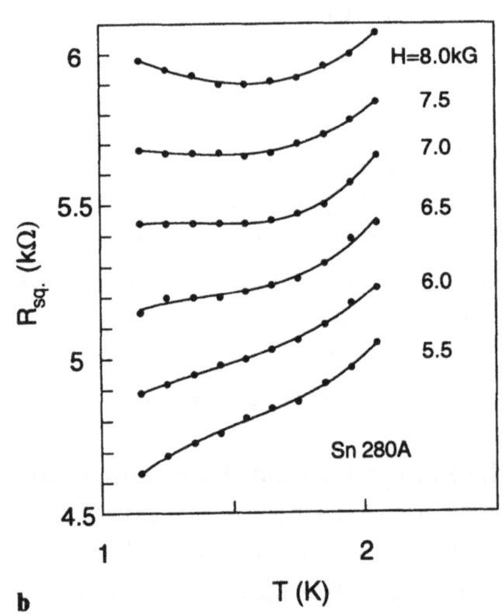

b

Fig. 5. a Temperature dependence of the sheet resistance of 300Å Sn FOMP in magnetic fields. **b** Enlarged low-temperature part with finer variation of field (from [18] with permission)

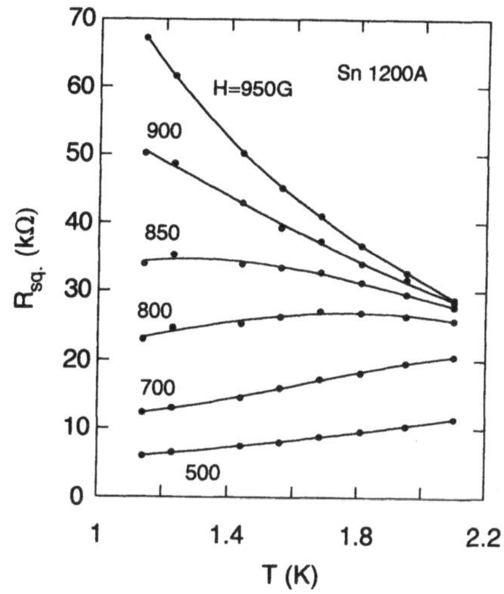

Fig. 6. Temperature dependence of the sheet resistance of 1200Å Sn FOMP in magnetic field (from [18] with permission)

It is not easy to estimate the capacitance of the junctions in FOMPs, and this is an important parameter for the application of theories of the universality of critical sheet resistance. Assuming that the thickness of the oxide barrier is 10Å and that the junction area is (300Å)^2, the capacitance is calculated to be of the order of 10^{-16}F. This is smaller than 10^{-15}F, the theoretical upper-limit to observe the universal critical resistance. Actually, preliminary results with particles as large as 1200Å indicate that the critical sheet resistance is considerably higher than 6.45kohm. The value of $R_{c,0}$ is also larger than R_Q, as shown in Fig. 6 [18]. As for the Boson localization model, no discussion of the capacitance is given [7].

Field-induced transition has also been reported for InO_x amorphous films [19], in which the disorder is claimed to be microscopic.

Conclusion

We have measured superconductor-insulator transition in the mesoscopic networks of Josephson junctions, which we call the FOMPs. By changing the disorder and the magnetic field, we found that, at the phase boundaries, the sheet resistance is approximately 6kohm, which is close to $R_Q(=h/4e^2)$. The result with field-tuned transition supports the Bose condensation model theory; however, we cannot identify the relevant model for disorder-tuned transition. Experiments with single junctions will be the key to elucidating this problem.

Acknowledgment. The author thanks R. Yamada, A. Nakamura, S. Katsumoto, F. Komori, T. Karasawa, S. Okuma, S. Tada, Y. Oostuka and W. Sasaki, with whom the studies reviewed here were carried out.

References

1. Kobayashi S (1988) Physica B152:223 Kobayashi S (1991) In: Kramer B (ed) "Quantum coherence in mesoscopic systems. NATO ASI Series B: Physics 254. Plenum, New York, p 485
2. Kobayashi S (1990) Phase Transit 24–26:463
3. Likharev KK, Zorin AB (1985) J Low Temp Phys 59:347
4. Schon G, Zaikin AD (1990) Phys Reports 198:237
5. Chakravarty S (1982) Phys Rev Lett 49:681
6. Eckern U, Schon G, Ambegaokar V (1984) Phys Rev B30:6419
7. Fisher MPA (1990) Phys Rev Lett 65:923
8. Orr BG, Jeager HM, Goldman AM (1985) Phys Rev B32:4920
9. Kobayashi S, Komori F (1985) Prog Theor Phys (Kyoto) Supplement 84:224
10. Kobayashi S, Tada Y, Sasaki W (1980) J Phys Soc Jpn 49:2075
11. Kobayashi S, Komori F (1988) Phys Soc Jpn 57:1884
12. Kobayashi S, Nakamura A, Komori F (1990) J Phys Soc Jpn 59:4219
13. Jeager HM, Haviland DB, Orr BG, Goldman AM (1989) Phys Rev B40:182
14. Jeager HM, Haviland DB, Goldman AM, Orr BG (1986) Phys Rev B32:4920
15. Nishida N, Okuma S, Asamitsu A (1991) Physica B169:487
16. Graybeal JM, Beaseley MR (1984) Phys Rev B29:4167
17. Okuma S, Komori F, Kobayashi S (1983) J Phys Soc Jpn 52:2639
18. Yamada R, Katsumoto S, Kobayashi S (1992) J Phys Soc Jpn 61:2656
19. Hebard AF, Paalanen MA (1990) Phys Rev Lett 65:927

One- and Zero-Dimensional Tunneling Diodes Fabricated by Focused Ion Beam Implantation

25

SEIGO TARUCHA, YOSHIRO HIRAYAMA*, TADASHI SAKU, AND YASUHIRO TOKURA[1]

Summary. One-dimensional and zero-dimensional tunneling diodes were fabricated using focused Ga ion beam implantation. The current vs voltage (I-V) curves of one-dimensional diodes showed broad steps near the onset of the tunneling current, followed by a series of small current peaks at higher voltages. The I-V curves of the zero-dimensional diodes, however, showed a series of peaks at the onset of the tunneling current. These differences are well reproduced by calculating the tunneling of a three-dimensional electron through the one-dimensional or zero-dimensional well levels. In the zero-dimensional diode, the amplitude difference between neighboring current peaks was constant because of the degeneracy of the zero-dimensional well levels confined by a harmonic lateral potential. One-dimensional diodes with strong lateral confinement in the contact region were fabricated by implanting a low dose of ions. This tunneling current is generated by the mixing of one-dimensional (emitter) and two-dimensional (well) subbands whose eigenfunctions have the same parity.

Introduction

Etching [1–4] and ion implantation [5–7] techniques have been used to construct an electron system within small lateral dimensions, with the confining potential imposed by sidewall depletion. When the sidewall depletion is large enough to pinch off the system, the laterally confined subbands are well defined by a harmonic potential. A series of current peaks due to resonant tunneling (RT) through these subband levels has been observed in double barrier (DB) diodes with one and two restricted lateral dimensions [1, 2, 5–7], which are, respectively, 1D and 0D RT diodes (1D RTDs and 0D RTDs). These diodes have had good lateral confinement in the contact region, as well as in the well region. Consequently, the current peaks have been assigned to tunneling from the laterally confined emitter levels through the laterally confined well levels, i.e., to 2D-1D tunneling for 1D RTDs [5–7] and to 1D-0D tunneling for 0D RTDs [1, 2, 8, 9]. When the lateral confinement becomes insignificant in the contact region, 3D-1D and 3D-0D tunneling take the place of 2D-1D and 1D-0D tunneling. A theoretical analysis of the tun-

[1] NTT Basic Research Laboratories, Midori-cho 3-9-11, Musashino, Tokyo 180 Japan
* Present address: Max-Planck-Institut für Festkörperforschung, 7000 Stuttgart 80, Germany

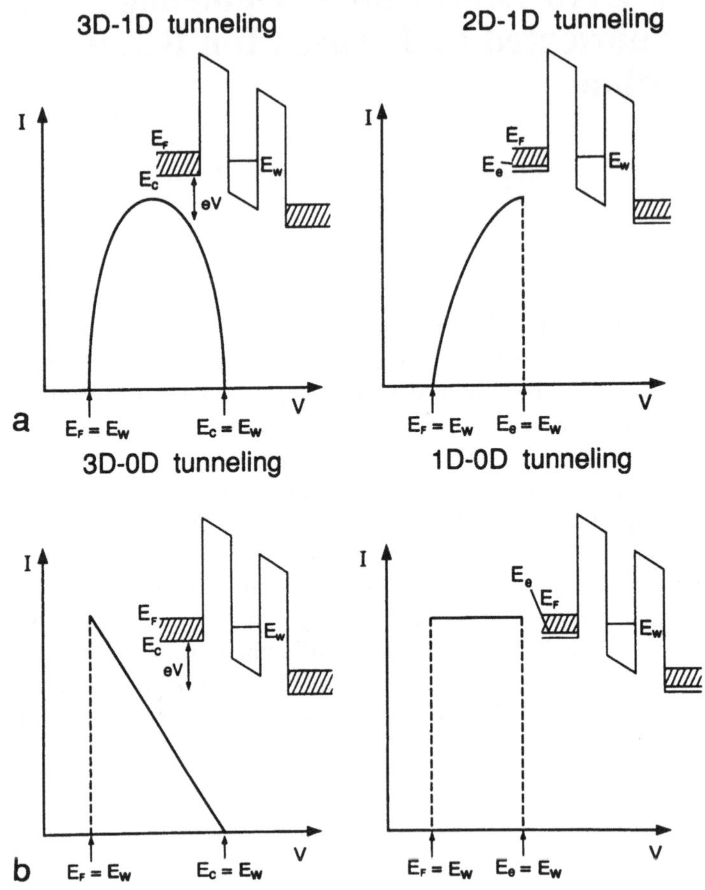

Fig. 1a, b. Qualitative I-V curves for **a** 3D-1D and 2D-1D tunneling and **b** 3D-0D and 1D-0D tunneling

neling supply function [10] predicts that the 3D-1D tunneling current will show a broad peak, which seems similar to the peak for 2D-1D tunneling (see Fig. 1). For the 2D-1D case, however, the current characteristics are generated by the mixing of 2D (emitter) and 1D (well) subbands [5–7]. This subband-mixing-induced tunneling is allowed for the 2D and 1D subbands whose transverse wavefunctions have the same parity. For 0D RTDs, it is postulated that the 3D-0D characteristics will show a current peak at the Fermi energy $E_F = E_w$, the 0D well level [10], whereas the 1D-0D tunneling characteristics will show a current step at $E_F = E_w$ [9] (see Fig. 1). Interesting for 0D RTDs is that the peak or step current is related to the degeneracy of the 0D level. In this paper we discuss the tunneling characteristics of 1D and 0D RTDs with and without lateral confinement effects in the contact region. We use focused Ga-ion-beam (Ga-FIB) implantation into DB diodes to fabricate 1D and 0D RTDs. The implantation-induced depletion depth increases in the well region with increasing ion dose, whereas it is nearly constant in the contact region due to the high doping density. We use a high ion dose to make 1D and 0D RTDs with least lateral confinement in the contact region. The 1D RTD exhibits a series of broad steps near the onset of the tunneling current, and these are followed by a series of small current peaks in the high-voltage range. The 0D RTD, on the other hand, exhibits a series of current peaks near the tunneling current onset. These features are well reproduced by calculation of 3D-1D and 3D-0D tunneling current. The amplitude of the current peak observed in the 0D RTD is related to the degeneracy of the 0D well level confined by a harmonic lateral potential. The lateral quantization effect in the contact region is observed in a 1D RTD prepared in such a way that the contact region, as well as the well region, is nearly pinched off. We show that the observed tunneling current characteristics are well assigned to 2D-1D subband mixing with parity conservation.

Sample Preparation

We used two kinds of DB structures to fabricate 1D and 0D RTDs. The first, DBS1, had a 6-nm GaAs well and two 10.8-nm $Al_{.3}Ga_{.7}As$ barriers; the second, DBS2, had a 6-nm GaAs and two 7-nm $Al_{.5}Ga_{.5}As$ barriers. The GaAs contact regions (emitter and collector) were doped with Si to 1×10^{18} cm^{-3}. Electrons were injected from the lower contact (emitter) into the well region. The large-area ($>2 \times 2$ μm^2) diodes exhibited a tunneling current peak of 7 ± 2 A/cm^2 at 0.28 ± 0.05 V for DBS1 and 18 ± 2A/cm^2 at 0.3 ± 0.05 V for DBS2. Primarily due to the band bending effect [11], the peak voltages were larger than the values ($2E_0/e$) given by the calculated ground-state subband energies E_0: 62 meV for DBS1 and 71 meV for DBS2. The lateral confinement in the **x-y** direction was imposed by implanting a 0.1μm diameter Ga-FIB, leaving a single $L_x \times L_y$ region unimplanted, as shown schematically in Fig. 2. Details of the fabrication have

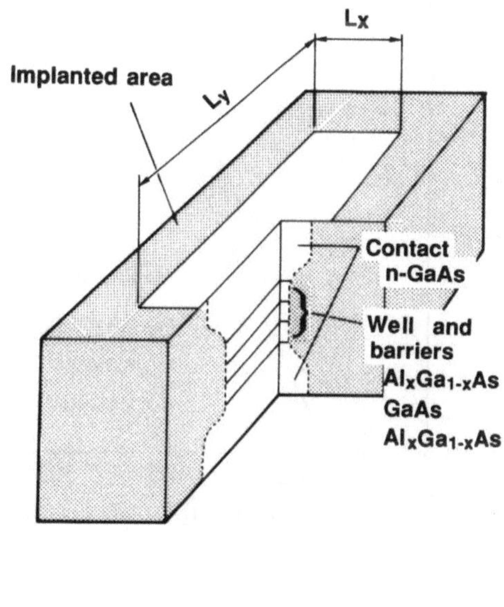

Fig. 2. Schematic diagram of the prepared device

been reported previously [5]. The value of L_x was reduced to fabricate 1D RTDs, whereas L_x and L_y were reduced equally to fabricate 0D RTDs. Ga ion implantation converts n-type material into highly resistive π-type or compensated n^--type material. The confining potential was formed by the implantation-induced depletion and could be approximated by a parabolic $n^+ - \pi$ junction profile with a flat region in the center. The depletion depth W_w in the well region varied from 35 ± 5 nm to more than 120 nm as a function of implantation dose ϕ ranging from 2.4×10^{12} to 5.0×10^{12} cm^{-2}. The depletion depth W_e in the contact region, however, was almost constant at 30 ± 10 nm over this range of ϕ. The DB region is assumed to be n^- due to the high doping in the contact regions. The large W_w at a high ϕ is probably caused by a significant number of electrons in the unimplanted DB region being captured by the implantation-induced damage, even though this damage is effectively repaired by thermal annealing [5]. W_w was evaluated from measurements of resonant current peaks I_p for diodes with a relatively large area $L \times L$ by assuming that the areal density of I_p is independent of diode area—i.e., $I_p^{1/2} \propto L - W_w$. Figure 3 shows plots of $I_p^{1/2}$ vs L for diodes prepared using different values of ϕ. Extrapolation to $I_p^{1/2} = 0$ gives the total depletion depth of $2W_w$. W_w for diodes prepared from DBS1 was larger than that for diodes prepared from DBS2, probably due to the thicker undoped region of DBS1. We used a high ϕ of 4 to 5×10^{12} cm^{-2} to fabricate nearly pinched-off well regions of the diodes having $L_x(L_y) \approx 2W_w$ with $W_w \geqslant 80 \pm 5$ nm. The high ion dose is important for leaving an emitter electron 3D-like. We used a low ϕ of 2 to 2.5×10^{12} cm^{-2} to fabricate nearly pinched-off well and contact regions of diodes having $L_x(L_y) \approx 2W_w \approx 2W_e$ with $W_w \geqslant 35 \pm 5$ nm. The low ion dose is important for achieving good lateral confinement in the emitter region. The lateral potential U in the x (or y) direction is approximated as

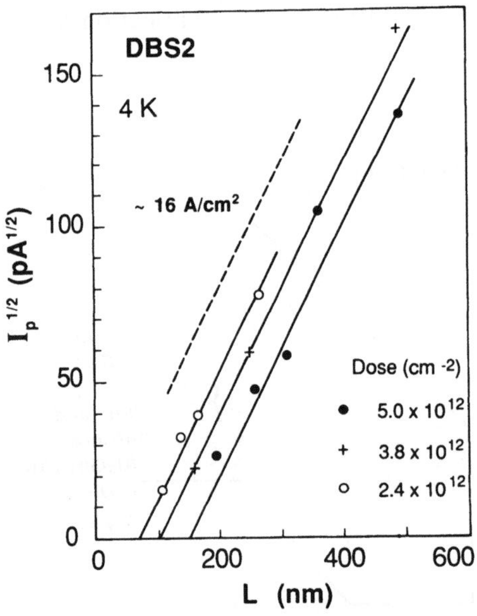

Fig. 3. $I_p^{1/2}$ vs L for diodes with a relatively large area L × L. (*Dashed line* shows the slope of the corresponding curve for a large-area diode)

$$U_{w,e}(x) = \begin{cases} \frac{1}{2}m^*\omega^2{}_{w,e}(|x| - W_{w,e})^2, & W_{w,e} \leqslant |x| \leqslant L \\ 0, & 0 \leqslant |x| \leqslant W_{w,e} \end{cases} \tag{1}$$

where the subscripts w and e represent the well and emitter regions, m^* is the effective electron mass, and $\omega_{w,e}$ is the frequency defining the lateral potential barrier height $\Delta U_{w,e} = \frac{1}{2}m^*\omega^2{}_{w,e}(L - W_{w,e})^2$. (The maximum lateral barrier height of 1.2 eV was assumed for an electron in the well region by taking into

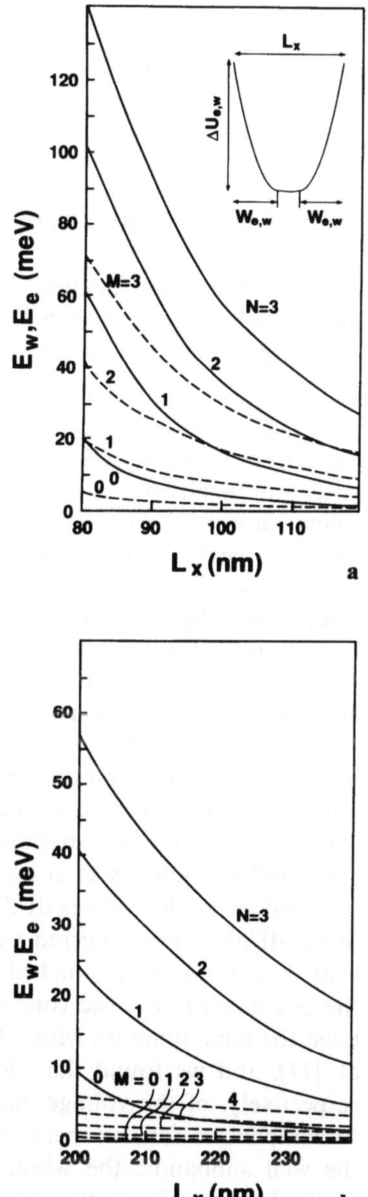

Fig. 4a, b. Calculated subband energies for a lateral potential shown in the *inset*. Parameters for the confining potentials: $\Delta U_w = 1.4$ eV, $\Delta U_e = 1.2$ eV, $W_e = 30$ nm, and **a** $W_w = 40$ nm and **b** $W_w = 100$ nm

account the band bending and the heterojunction confined subband energy [6, 7].) The calculated subband energies for the lateral potential $U_{w,e}(x)$ are shown as a function of L_x in Fig. 4 for a low ϕ (a) and for a high ϕ (b). The subband energy of E_w in the well region is defined by the quantum number N, and the subband energy of E_e in the emitter region is defined by the quantum number M. Under the nearly pinched-off condition of $L_x \approx 2W_w$, the laterally confined well levels are evenly spaced by $\hbar\omega_w = (2\Delta U_w/m^*)^{1/2}\ \hbar/L_x$. With a low ϕ, E_w and E_e are comparable because $L_x \approx 2W_w \approx 2W_e$, so that the lateral quantization effect in the emitter region becomes important. With a high ϕ case, on the other hand, the laterally confined emitter levels are much more closely spaced below E_F in the emitter region because there is a large flat region in the center of $U_e(x)$. The energy spacings are too small to provide well-defined subbands in the highly doped contact region.

One-Dimensional Tunneling Diodes

A tunneling supply function, which is proportional to the number of electronic states available for RT in the emitter region, describes the general nature of the tunneling characteristics. For 3D-1D tunneling this function is proportional to $\sqrt{(E_w - E_c)(E_F - E_w)}$ [10]; whereas for 2D-1D tunneling it is proportional to $\sqrt{E_F - E_w}$ with $E_F \geqslant E_w \geqslant E_e$, where E_e is the 2D emitter level [7]. The I-V curve for 3D-1D tunneling (Fig. 1) shows a gradual current increase near the current onset and then shows a peak at $E_w = (E_F + E_c)/2$. The I-V curve for 2D-1D tunneling, however, shows a peak at $E_W = E_e$, which is sharper than the 3D-1D tunneling peak. With 2D-1D tunneling, subband mixing allows an electron of a 2D emitter subband to tunnel through various 1D well levels when the lateral confinement differs between the well and emitter regions. Tunneling is allowed for the 2D and 1D subbands of the same parity in the x direction.

Figure 5 shows the I-V characteristics of the diodes fabricated from DBS1, prepared using a high ϕ. The well regions were nearly pinched off in the **x** direction by the depletion of $W_w = 80 \pm 5$ nm. The L_y values were large enough for us to assume free electron motion in the **y** direction at 4 K. As L_x was decreased, the initial main peak became small and shifted toward higher voltages. For the diodes with $L_x \leqslant 225$ nm, we observed a weak step-like feature near the current onset and small multiple peaks at higher voltages. These characteristics were more pronounced in the derivatives dI/dV. With the smaller L_x, the voltages at which dips in the dI/dV curve occurred were more widely spaced. The voltage spacing was nearly constant in the pinched-off diode with $L_x = 180$ nm. The arrows indicate the calculated resonance voltages corresponding to the 1D well levels (We calculated the band diagram without lateral restriction in the way reported by Reed et al. [11], and we found that, for DBS1 and DBS2, 42%–35% and 37%–32%, respectively, of the voltage increment applied in the resonance region was used to compensate for the energy separation between the emitter Fermi energy and the well subbands; the whole set of arrows was shifted to give the position of the lowest-voltage dip correctly). For this assignment, we calculated a one-

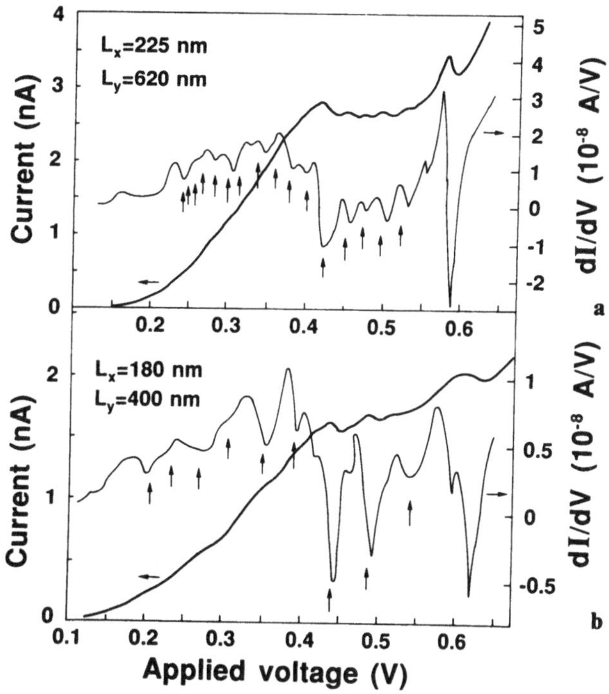

Fig. 5a, b. I-V characteristics of 1D RTDs prepared from DBS1. [$L_x \times L_y = 225 \times 620$ nm in (**a**) upper and $L_x \times L_y = 180 \times 400$ nm in (**b**) The evaluated W_w is 80 ± 5 nm. The *fine lines* are the derivatives dI/dV]

dimensional self-consistent band diagram to correct the band bending in the z direction (see immediately above for method of calculation) under the assumption that the band bending caused by the lateral restriction simply gives a voltage offset. The good agreement of the arrow locations with the dI/dV dips allows us to assign the observed current characteristics to 3D-1D tunneling. For the diode with $L_x = 180$ nm, the even-numbered dI/dV peaks (lowest-voltage peak = peak 0) are larger than the odd-numbered peaks. This is probably due to faint lateral quantization in the emitter region, as we have discussed previously [5].

Figure 6 shows the I-V curves of 1D RTDs prepared using a low ϕ in DBS2. The diode with $L_x \times L_y = 70 \times 400$ nm exhibited three pronounced current peaks about 200 mV apart. This spacing is nearly twice the value that corresponds to the calculated subband splitting in the well region, which suggests RT associated with alternate 1D subbands in the well region. The calculation of the emitter subband levels gives two 2D subbands below E_F (See Fig. 4), where E_F of 45 meV is assumed. (due to the lateral constriction imposed by Ga-FIB implantation, effective electron density is reduced by about 20% from the initial doping level). The three current peaks are well reproduced by the M–N = 0–0, 0–2, and 0–4 resonances generated by the subband mixing effect. The M = 1 emitter subband contributes least to the I-V curve because it is very close to E_F and therefore contributes little to the tunneling supply function. When L_x was slightly increased (as in the diode with $L_x \times L_y = 90 \times 400$ nm), small peaks accompanied the main peaks observed in the curve of the diode with $L_x \times L_y = 70 \times 400$ nm. Four 2D emitter subbands

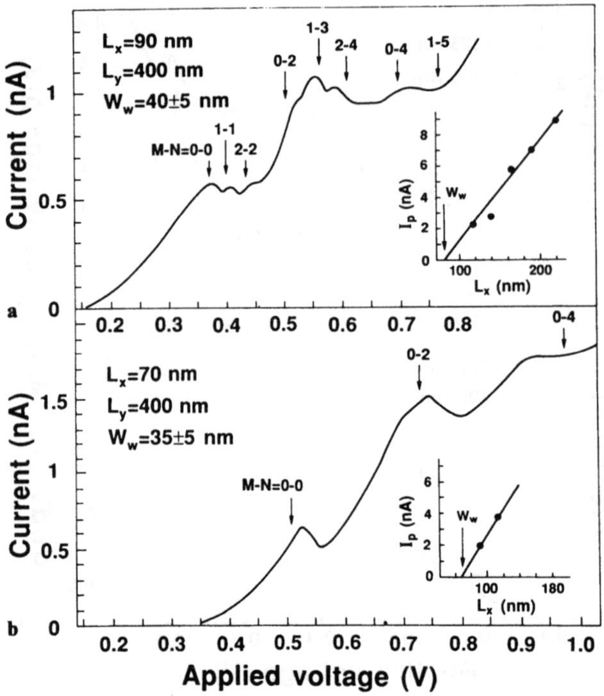

Fig. 6a, b. I-V curves for 1D RTDs prepared from DBS2. [$L_x \times L_y = 90 \times 400$ nm and $W_w = 40 \pm 5$ nm in: **a** and $L_x \times L_y = 70 \times 400$ nm and $W_w = 35 \pm 5$ nm in; **b** *Arrows* labeled "M-N" indicate the calculated resonance positions for tunneling of an electron from the Mth emitter subband through the Nth well subband]. *Insets* show the $I_p - L_x$ plots used to determine the values of W_w

are expected to be below E_F (see Fig. 4). As indicated by the arrows, the small peaks are well assigned to mixing of 2D and 1D subbands.

A voltage shift of the initial main resonance peak, like that observed in the restricted-area diode, has previously been attributed to the additional band bending imposed by the lateral restriction [2, 5]. We assume in interpreting our results that the additional band bending gives an offset to the resonance voltage. The main resonance peak is associated with the 2D ground state confined by the heterojunctions. In the DBS2 the second resonance associated with the 2D first excited state appeared in the high voltage range. We have examined the validity of the above assumption by measuring the voltage shifts of the first and second tunneling current peaks for various L × L area diodes with no significant lateral quantization effects. The results are shown in Fig. 7: when L was less than 550 nm, the current peaks for the first and second resonances shifted by the same voltage. This supports our assumption used for calculating the resonance voltages.

Zero-Dimensional Tunneling Diodes

The 3D-0D tunneling supply function of bias voltage jumps when $E_F = E_W$, the 0D well level, and decreases in proportion to $E_w - E_c$ [10], where E_c is the bottom of the emitter conduction band. The 1D-0D tunneling supply function, on the other hand, is nonzero and constant when $E_e \leqslant E_w \leqslant E_F$, where E_e is the 1D

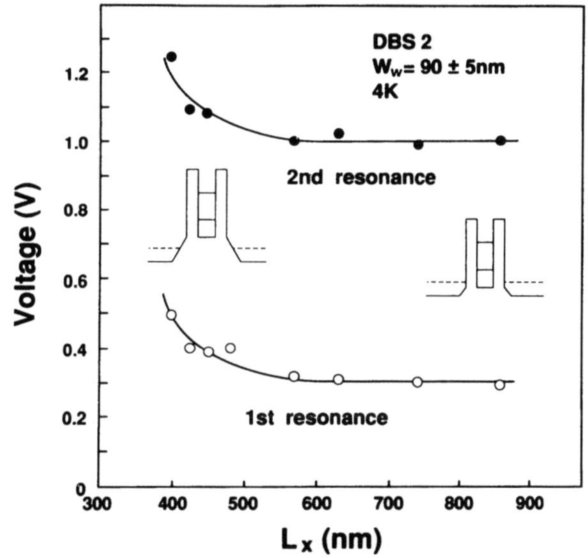

Fig. 7. First and second tunneling current peak voltages for the L × L diodes prepared from DBS2

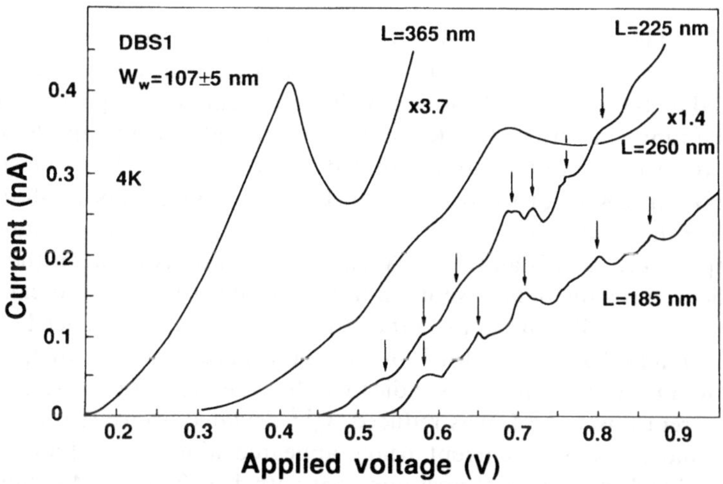

Fig. 8. I-V characteristics of 0D RTDs prepared from DBS 1. (*Arrows* indicate the series of small peaks near the current onset)

emitter level [9]. The I-V curve for 3D-0D tunneling has a peak at the current onset, whereas the 1D-0D tunneling curve has a step at the current onset (Fig. 1).

Figure 8 shows the I-V characteristics of diodes that have square unimplanted areas of various sizes $L(=L_x = L_y)$. The evaluated W_w was 107 ± 5 nm, as shown in Fig. 9. These diodes were fabricated using a high ϕ in DBS1. When L was reduced, the initial main RT current peak became small and shifted to a higher voltage, and a small undulation appeared on the low-voltage side of the peak. This

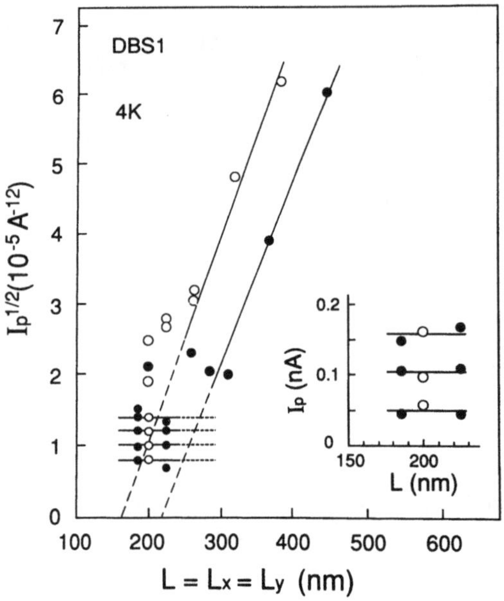

Fig. 9. $I_p^{1/2}$ vs. L for the L × L diodes. (*Open circles*, $\phi = 4.0 \times 10^{12}$ cm^{-2}. *Filled circles*, $\phi = 5.0 \times 10^{12}$ cm^{-2}. *Inset* shows I_p vs L for the series of current peaks at the current onset)

voltage shift is larger than that in 1D RTDs because of the stronger lateral restriction. With an even still smaller L, the undulation became a series of current peaks (as indicated by the arrows). The lowest-voltage current peak appeared around the onset of the tunneling current. Interestingly, we found that the amplitude of the Nth peak from the lowest is approximated by NΔI, where ΔI is constant. The current peaks observed for the diodes with L = 225 nm were about 38 mV apart, and those for the diode with L = 185 nm were 56 mV apart. These voltages correspond, respectively, to 15- and 22-meV energy separations between the resonant levels in the well region. When we use a harmonic approximation to calculate the lateral potential, the 0D subbands in the well region are equally spaced, at 14 meV for L = 225 nm and at 18 meV for L = 185 nm. Figure 9 shows how $I_p^{1/2}$ changes with L in RTDs. The small circles indicate either the peaks or the steps observed near the current onset. As L was reduced, $I_p^{1/2}$ deviated above the line along the $I_p^{1/2}$ plots and a series of current peaks appeared almost simultaneously. The amplitude of each peak was almost constant with $L \leqslant 225$ nm. As shown in the inset, the I_p value is approximately defined by ΔI = 50 pA. A similar I-V curve was observed for diodes prepared from DBS2 (Fig. 10). The 4K curve has a series of current peaks nearly identical to that of the L = 185-nm diode shown in Fig. 8, but the ΔI value of the diode prepared from DBS2 larger by a factor of 2.4. This factor compares well with the increase of I_p for large area diodes, suggesting that the ΔI value is determined from the one-dimensional tunneling probability in the z direction.

The observed I-V characteristics are well reproduced by the 3D-0D tunneling supply function shown in Fig. 1. The specific peak current NΔI is then related to the degeneracy for the 0D well levels, as discussed later. For a 0D RTD, Reed et al. [1] observed a series of current peaks superimposed on an initial main

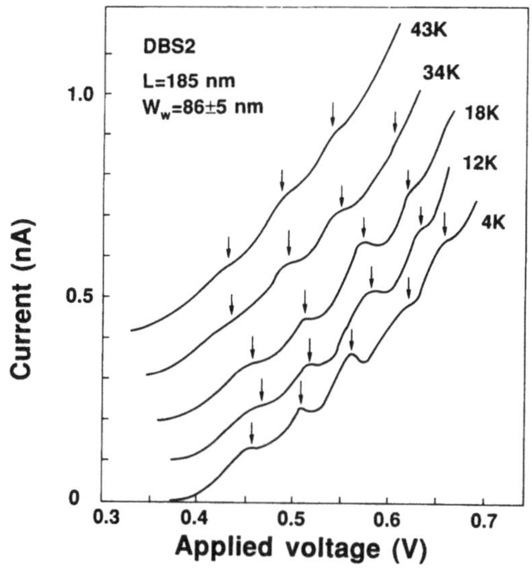

Fig. 10. I-V curves at various temperatures for diodes prepared from DBS2

resonance peak. This peak appears at $E_c = E_w$ in 3D-2D tunneling whose bias condition is similar to that in the 1D(ground)-0D(ground) case. These peaks were assigned to 1D-0D tunneling, since the lateral quantization is strong enough to generate 1D emitter subbands with a spacing greater than 20 meV [2]. At higher temperatures the small current peaks became obscure, leaving the main resonance peak unchanged. This suggests that the small peaks are dominated by tunneling of an emitter electron near the bottom of the 1D level. The current peaks with the lowest and second lowest voltages seem step-like, probably due to the lateral quantization effect in the emitter region. The detailed analysis for the I-V curve has been given by Bryant [9]. For our 0D RTDs, in contrast, the main resonance feature disappeared as the diode size was reduced, and all of the small current peaks, which were observed near the current onset, became obscure above 30 K. These behaviors are explained by the current peak being dominated by the tunneling of a 3D emitter electron at E_F.

3D-0D Tunneling Model

When the well region is nearly pinched off or $L \gg L - W_w$, a lateral potential $U_w(x, y)$ is approximated by a cylindrical dot with a parabolic potential $U_w(r)$ in the radial direction:

$$U_w(r) = \begin{cases} 0, & r_0 \geqslant r \geqslant 0, \\ \frac{1}{2}m^*\omega^2{}_w(r - r_0)^2, & R \geqslant r \geqslant r_0, \\ \frac{1}{2}m^*\omega^2{}_w(R - r_0)^2, & r \geqslant R, \end{cases} \tag{2}$$

where r_0 and R are, respectively, equivalent to $L/2 - W_w$ and L/2 (See Fig. 11).

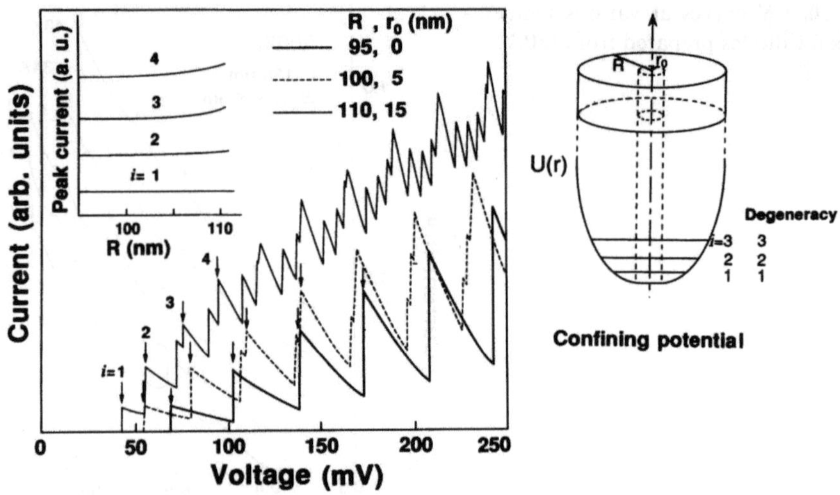

Fig. 11. Calculated tunneling currents for various 0D RTDs. *Inset* shows the amplitude of the main peak "i" with respect to R. The confining potential assumed for the calculation is shown at the *right*

The laterally confined energies are deduced from the Schrödinger equation of

$$\left[-\frac{\hbar^2}{2m^*}\Delta + U(r)\right]\Phi(r,\theta) = E\Phi(r,\theta) \tag{3}$$

in cylindrical coordinates (r, θ, z). This equation can be solved by separating variables as $\Phi = e^{im\theta}\,\varphi_n(r)$, where n and m are the principal and orbital quantum numbers, respectively, with $n = 0, 1, 2, \cdots$ and $m = 0, \pm 1, \pm 2, \cdots$. The eigenenergy is degenerate for $E_{n,m}$ with $k = 2n + |m|$ = nonnegative integer, so that the $(k + 1)(= 1, 2, 3, \cdots)$th level from the lowest has a degeneracy of $k + 1 (= 1, 2, 3, \cdots)$-, except for spin degeneracy.

We calculated the sequential tunneling current J through the 0D well levels, which are assumed to have separable wavefunctions with r, θ, and z. J is given to the first order by a simplified form of

$$J = \sum_{i=n,m} \eta_i S_i T_i \tag{4}$$

where η_i is the coupling intensity of wavefunctions between the emitter and well regions, S_i is the tunneling supply function, and T_i is the transmission coefficient in the **z** direction. We assumed here that J is mostly restricted by the electron tunneling through the emitter barrier. We also assumed that, at the limit of a strong confinement of the heterojunctions, η_i is constant with i, that each T_i has the same dependence on the emitter electron energy, and that this dependence is given by the WKB approximation for a 2D state confined by the heterojunctions without lateral constriction. Figure 11 shows the calculated tunneling currents for $R = 95$, 100, and 110 nm with $R - r_0 = 95$ nm. We assumed a DB structure equivalent to DBS1. The current has a sharp peak with a jump at $E_{n,m} = E_F$. For

$R \gg r_0$, the amplitude of the main peak indicated by the arrow "i" is proportional to the number of nearly degenerate 0D eigenstates at E_F when the overlap between the neighboring peaks is not very large. The inset shows that the amplitude of the main peak i = 1 to 4 is almost constant for R = 95 to 110 nm, and that the amplitude difference is nearly the same between adjacent peaks. The existence of an undepleted region lifts the degeneracy and makes the resonance peaks split; nevertheless, as shown for the R = 110 nm curve, periodic peaks are revealed by the nearly degenerate levels of $E_{n,m}$ with n > 2. These results correspond to the I-V curves for the 0D diodes shown in Fig. 8.

3D-1D Tunneling Model

A similar calculation was performed for the 3D-1D tunneling. Because there is no degeneracy of 1D levels, the subband structure in the well region is rather simple, even when $L_x > 2W_w$. The results calculated for the diodes with L_x = 225 and 200 nm and with W_w = 80 nm are shown in Fig. 12. As expected from Fig. 1, a current due to RT associated with a single 1D well subband has a rather broad peak. The overlap of such current peaks results in broad steps near the tunneling current onset, followed by a series of small peaks at higher voltages. These features are similar to the observed I-V characteristics.

Charging Effect

The transfer of one electron into a quantum dot through a small tunnel junction raises the electrostatic potential difference between the well and the contact reservoir, and it lifts the degeneracy of the 0D levels. The electrostatic energy E_s arises

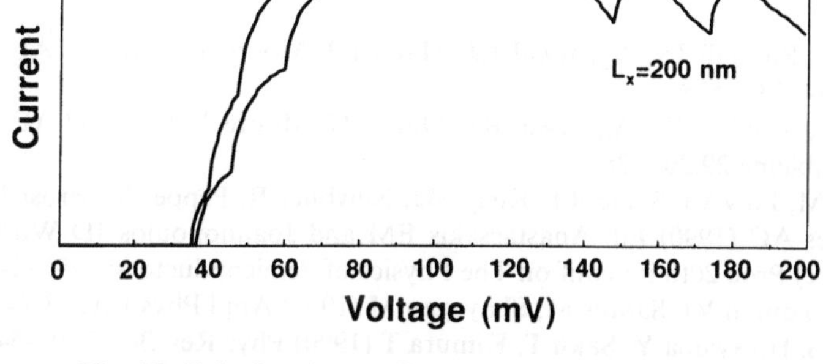

Fig. 12. Calculated tunneling currents for 1D RTDs

from a charging energy e^2/C in a junction capacitance (C) model as well as from a repulsive Coulomb interaction in the quantum box. We used a two-dielectric model to calculate E_s for the addition of one electron into a quantum dot via tunneling through a constricted single barrier. The calculated E_s is 8–10 meV for the tunneling into the ground 0D state, and this value is small compared to the present 0D level separation. The E_s for the excited state is even smaller, due to the larger spatial extension of wavefunction. The charging effect may therefore cause a shift or split of the resonance peak. This is not evident in our data, however, probably because in the relevant bias condition the transparency of the collector barrier is much larger than that of the emitter barrier, so that the charge is not accumulated in the well.

Conclusions

1D and 0D RTDs were fabricated using Ga-FIB implantation. We used a high ion dose for implantation to prepare 1D and 0D RTDs without a lateral quantization effect in the emitter region. The I-V curves of 1D RTDs show broad steps near the onset of the tunneling current, followed by a series of small peaks at higher voltages. The I-V curves of the 0D RTDs, however, show a series of current peaks at the onset of the tunneling current. These differences are well reproduced by calculation of 3D-1D and 3D-0D tunneling. In the 0D RTDs, we find that the peak amplitude is defined by the degeneracy of the 0D well level confined by a harmonic potential. We used a low ion dose for implantation to prepare 1D RTDs with good lateral quantization in the emitter region. These diodes also showed a series of small current peaks, but the peak positions did not correspond to the 1D well levels. These peaks arise from mixing of the 2D (emitter) and 1D (well) subbands whose eigenfunctions have the same parity.

Acknowledgment. We are grateful to Yoshiji Horikoshi and Tatsuya Kimura for their stimulating suggestions throughout this work.

References

1. Reed MA, Randall JN, Aggarwal RJ, Matyi RJ, Moore TM, Wetsel AE (1989) Phys Rev Lett 60:535–538
2. Reed MA, Randall JN, Aggarwal RJ, Matyi RJ, Moore TM, Wetsel AE (1989) Festkörperprobleme 29:267–283
3. Tewordt M, Law VJ, Syme RT, Kelly MJ, Newbury R, Pepper M, Frost JEF, Ritchie DA, Jones AC (1990) Ed. Anastassakis EM and Joannopoulos JD World Scientific (Singapore) Proc 20th int conf on The Physics of Semiconductors: 2455–2458
4. Su Bo, Goldman VJ, Santos M, Shayegan M (1991) Appl Phys Lett 58:747–749
5. Tarucha S, Hirayama Y, Saku T, Kimura T (1990) Phys Rev B41:5459–5462
6. Tarucha S, Hirayama Y (1991) Phys Rev B43:9397–9400

7. Tarucha S, Tokura Y, Hirayama Y (1991) Appl Phys Lett 58:1623–1625
8. Bryant GW (1989) Phys Rev B39:3145–3152
9. Bryant GW (1991) Phys Rev B44:3782–3786
10. Liu HC, Aers GC (1989) J Appl Phys 65:4908–4914
11. Reed MA, Fresley WR, Ducan WM, Matyi RJ, Seabaugh AC, Tsai H-L (1989) Appl Phys Lett 54:1256–1258

Logic Circuits Using Resonant-Tunneling Hot Electron Transistors (RHETs)

26

NAOKI YOKOYAMA, MOTOMU TAKATSU, KENICHI IMAMURA, HIROAKI OHNISHI, SHUNICHI MUTO, TOSHIHIKO MORI, TAKAMI ADACHIHARA, AND AKIHIRO SHIBATOMI[1]

Summary. This review paper has focused on the research and development of resonant-tunneling hot electron transistors. It is shown that an InGaAs-based RHET with a thin collector barrier structure improves current gain and peak-to-valley current ratio. RHET basic logic gates, including Exclusive-NOR, and three-input majority gates, and a state-holding circuit are proposed. We used these gates to produce working frequency dividers, a latch circuit, and a full adder. These circuits use fewer transistors than bipolar circuits and dissipate less power because of their lower supply voltage and total current.

In future integrated circuits, delay time due to wires connecting transistors is expected to be more significant than transistor speed. New functional devices, such as the RHET, promise a higher scale of integration giving breakthroughs in the speed and power limitations of conventional integrated circuits.

Introduction

Hot electron transistor using a Metal-Insulator-Metal-Insulator-Metal (MIMIM) structure was proposed by C.A. Mead [1] in 1960. With a positive bias on the base, electrons are injected from the emitter to the base through the thin insulator layer by a tunneling effect. Hot electrons, which survive scattering in the metal base, can reach the collector. Since the hot electrons keep their high speed, the device is expected to be extremely fast. Several variations [2] of this device were proposed. However, these efforts have had limited success because it was difficult to obtain a high-quality, very thin metal base.

Thanks to recent progress in MBE and MOCVD growth technology, we can now replace the metal base with high-quality n-type GaAs, and the thin insulator layers with non-doped AlGaAs layers with well-controlled heterointerfaces. Hot electron transistors (THETA) using GaAs/AlGaAs heterostructures were proposed by M. Heiblum [3] in 1981.

In 1984, we were the first to demonstrate the hot electron transistor, using MBE-grown GaAs/AlGaAs heterostructures [4], and, in 1985, we observed evidence of near-ballistic motion of electrons in this entity [5].

[1] Fujitsu Laboratories Ltd., Fujitsu Limited, 10-1 Morinosato-Wakamiya, Atsugi, 243-01 Japan

Having demonstrated that semiconductor heterojunctions can be used in hot electron transistors, our goal is to develop new functional hot electron transistors. This is because the use of new functional devices will allow for a large reduction in the number of components in integrated circuits, promising breakthroughs in their speed and power limitations.

Operating Principle of Resonant-Tunneling Hot Electron Transistor (RHET)

Figure 1 shows the operating principle of a RHET. The RHET that we proposed [6] in 1985 is essentially the same as a hot electron transistor, but has the important difference that the RHET uses a resonant tunneling barrier as a hot electron injector. This difference changes the device properties and the RHET exhibits a negative differential transconductance region due to this resonant-tunneling effect.

Figure 2 shows the collector current-voltage characteristics as a function of base voltage measured at 77K. Note that there was a peak in collector current, followed by a valley, indicating the presence of a negative transconductance region.

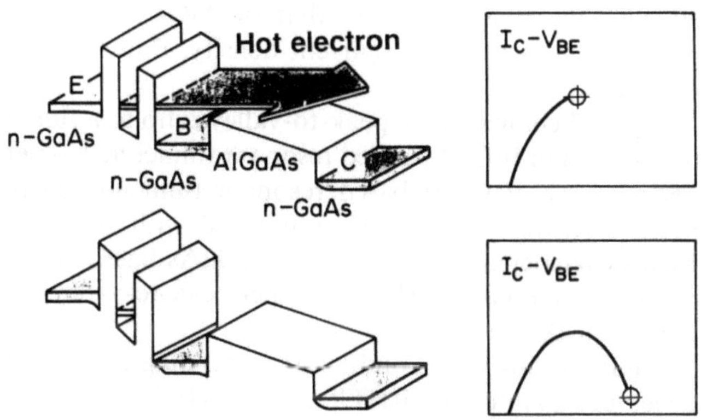

Fig. 1. Operating principle of RHET

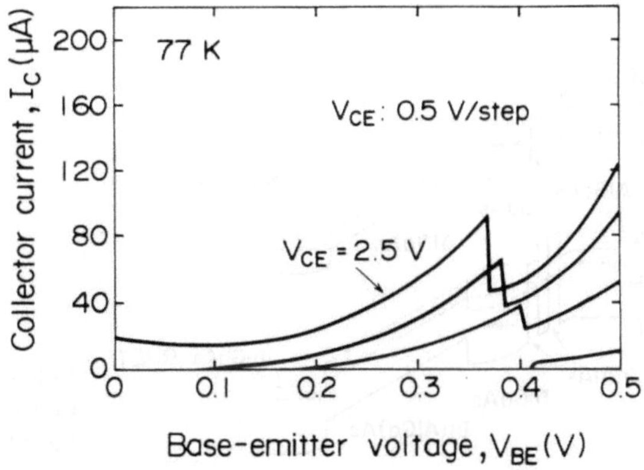

Fig. 2. Collector current measured as a function of base voltage at 77K for RHET fabricated using GaAs/AlGaAs heterojunctions

However, the peak-to-valley current ratio was very small, and the current gain was only approximately 1. By optimizing the resonant-tunneling barrier structure, and the collector barrier height, we were able to increase the peak-to-valley ratio to 2.6 and the current gain to about 5 [7]. However, these are still poor for device applications. To significantly improve these characteristics, we changed the compound semiconductor from GaAs to InGaAs-based materials.

RHETs Using InGaAs-Based Materials

Comparison of the conduction bands of the InGaAs-based RHET and the GaAs-based RHET (Fig. 3) shows why we changed from GaAs to InGaAs. The key to the InGaAs-based RHET is that the InGaAs base has a wider separation energy, 0.55 eV between points Γ and L. It also has a wider separation energy at the heterointerface with the collector barrier.

Because of its wider energy window, where hot electron transport is near-ballistic and electrons can reach the collector, this RHET has higher current gain than the GaAs based RHET. Another important factor is that the effects of indirect valleys in tunneling transport in the barrier are reduced due to their higher energies, decreasing the valley current. The InAlAs barrier also has a lighter effective mass, sixty percent that of AlGaAs barriers. This decreased effective mass increases tunneling-current density, thus increasing the peak-to-valley current ratio.

Figure 4 compares the peak-to-valley ratio as a function of peak current density for GaAs and InGaAs-based resonant tunneling barriers [8]. These plots show the performance of GaAs-based resonant tunneling barriers, InGaAs/InAlAs resonant tunneling barriers lattice-matched with InP, and pseudomorphic InGaAs/InAlAs barriers. It is clear that InGaAs-based materials, particularly pseudomorphic materials [9], show a higher peak-to-valley ratio with increased collector current peak density.

Figure 5 shows the conduction band diagram of the InGaAs-based RHET we developed to integrate RHETs. This RHET uses a pseudomorphic and asymmetric resonant-tunneling barrier to increase peak-to-valley ratio with a peak current

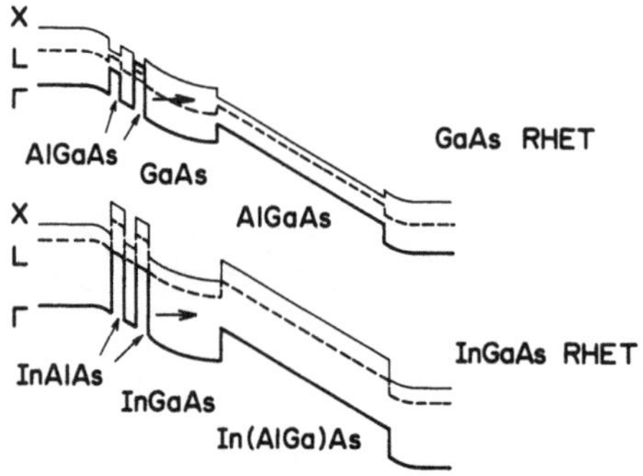

Fig. 3. Comparison of the conduction bands of the InGaAs-based RHET and the GaAs-based RHET

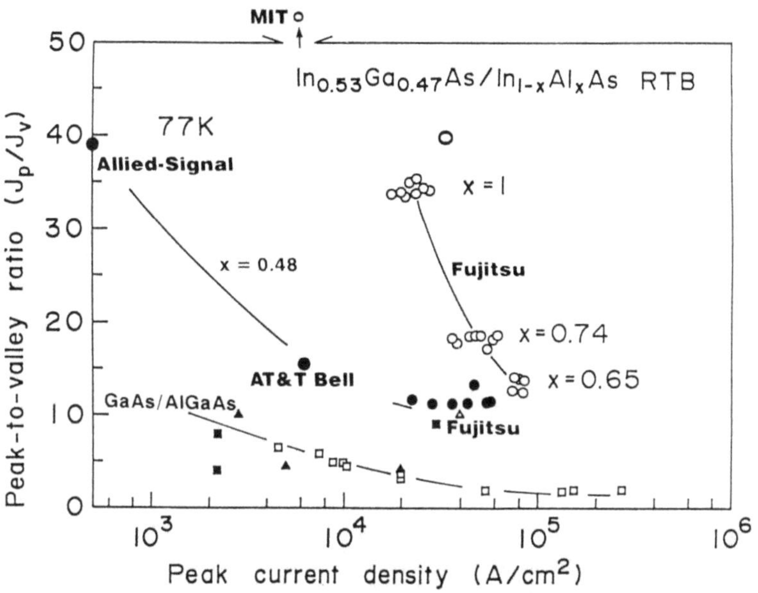

Fig. 4. Peak-to-valley ratio as a function of peak current density for GaAs and InGaAs-based resonant-tunneling barriers

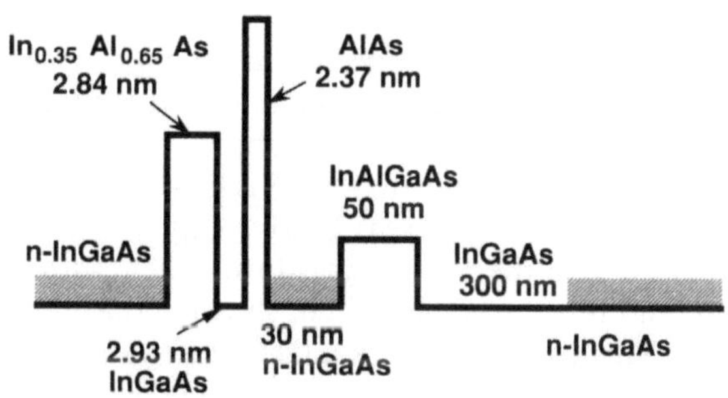

Fig. 5. Conduction band diagram of the InGaAs-based RHET which we developed to make integrated circuits

density on the order of 10^4 A/cm^2. The RHET uses very thin InAlGaAs collector barriers to reduce intervalley scattering and lessen the effect of collector band bending [10].

Notice that non-doped InGaAs regions are inserted between the collector barrier and the collector in both RHETs to increase collector breakdown voltage. This region is thicker in the lower structure to increase the operating margin of RHET integrated circuits.

Figure 6a shows the emitter current as a function of base-emitter voltage for the RHET intended for integration. The peak current density is 1.7×10^4 A/cm^2 and the peak-to-valley ratio is about 20. Figure 6b shows the collector current-voltage characteristics measured at 77K in a common base configuration. The common-

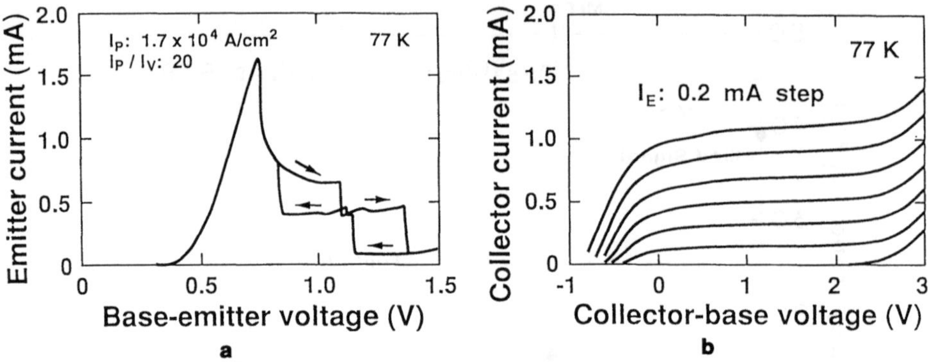

Fig. 6. a Emitter current as a function of base-emitter voltage of the RHET. **b** Collector current-voltage characteristics measured at 77K in a common base configuration

base current gain is over 0.95 at collector base voltages above 0.5. Please note that the current gain is high close to a collector voltage of zero.

Having demonstrated improved RHET performance, we turned to developing integrated circuits.

Basic Logic Gates Using RHET [10]

As the first step, we looked for a variety of basic logic gates. The circuit shown in Fig. 7 uses two input resistors for A and B. One more resistor is connected between the base and emitter as a level shifter.

When both inputs are low, RHET-1 is off, and the output is high. An emitter-follower RHET-2 increases drive capability. When either input is high, RHET-1 goes on, and the output goes low. When both inputs are high, the RHET-1 goes off, and the output goes high, showing that this circuit acts as an Exclusive-NOR gate. This Exclusive-NOR gate uses about one-fifth the number of transistors used by a bipolar transistor Exclusive-NOR gate.

To demonstrate other advantages of the RHET, we developed three input logic gates, shown in Fig. 8. Three input resistors are connected to the collector of RHET-1. When two or all inputs are low, the RHET-1 collector voltage goes

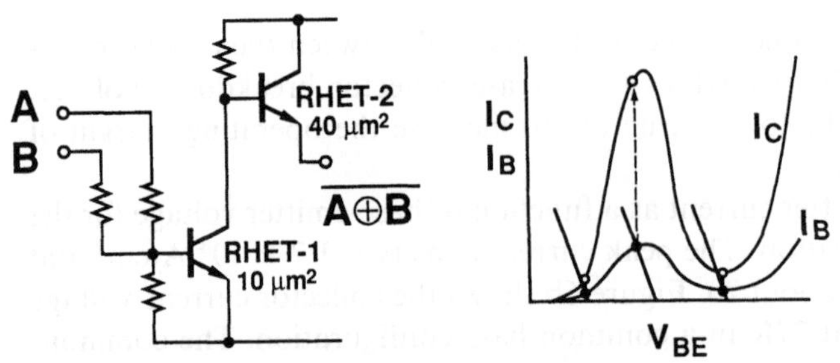

Fig. 7. Exclusive-NOR gate using two RHETs

Fig. 8. Three input majority logic gate using three RHETs

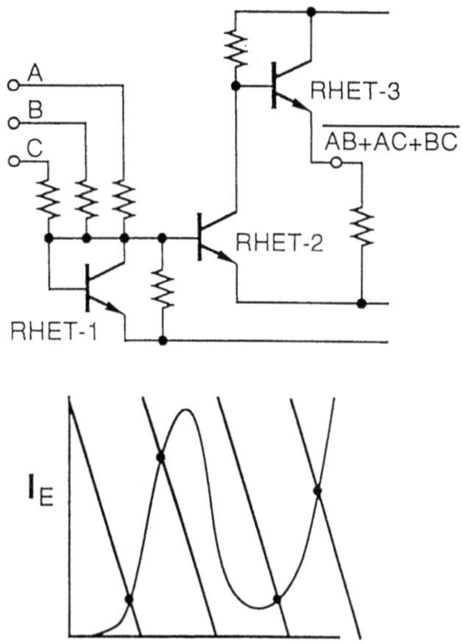

A	B	C	Out
0	0	0	1
0	0	1	1
0	1	0	1
0	1	1	0
1	0	0	1
1	0	1	0
1	1	0	0
1	1	1	0

low, RHET 2 goes off, and the output goes high. When two or all inputs are high, the RHET-1 collector voltage goes high, RHET-2 goes on, and the output goes low. RHET-3 operates as an emitter follower to increase the current drive. The truth table shows that this circuit acts as a three input majority logic gate. This three input majority gate uses about one-seventh the number of transistors used by a bipolar transistor gate.

Figure 9 shows a state-holding circuit which uses the hysteresis characteristics of a series circuit of a RHET and two input resistors. RHET-2 acts as a voltage amplifier and RHET-3 acts as an emitter follower to increase current drive. Please note that the output of the state-holding circuit changes only when both inputs are the same as the current output. Shaded areas indicate conditions when the output changes. The output after the change is the complement of the input.

RHET Integrated Circuits [11]

A chain of state-holding circuits is similar to a shift register, as shown in Fig. 10. If the input is the same as the clock signal, the complement of the input is fed to the next stage. This is indicated by the output levels in bold type. At the next stage the input will differ from the clock signal, and the output will not change. Data in the chain is passed to the next stage at each clock edge.

A frequency divider is produced by adding a feedback loop to this quasi-shift register. The three-stage state-holding circuit with feedback, shown in Fig. 11, acts as a 1/3 frequency divider. By putting an Exclusive-NOR gate in the clock line, we can produce a 1/2 frequency divider. We fabricated these circuits and confirmed that they work at 77K. Our 1/2 and 1/3 frequency dividers use fewer transistors than equivalent dividers using bipolar ECL gate circuits.

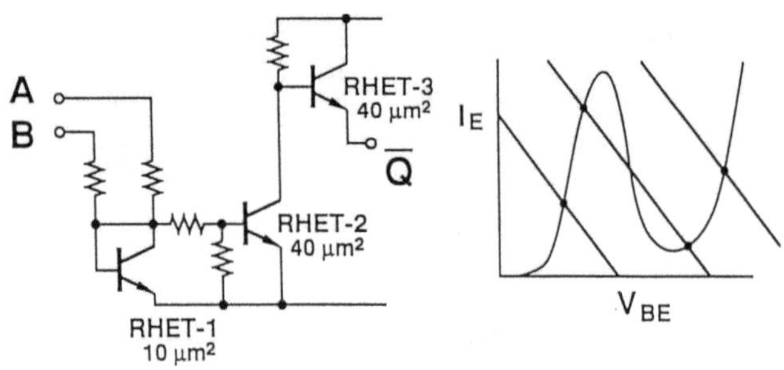

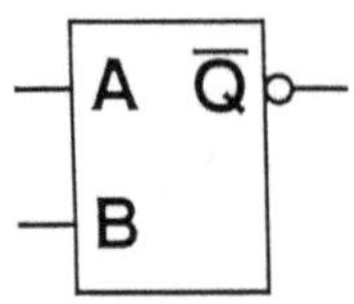

A	B	$\overline{Q}_n$	$\overline{Q}_{n+1}$
0	0	0	1
0	0	1	1
1	1	0	0
1	1	1	0
0	1	0	0
0	1	1	1
1	0	0	0
1	0	1	1

Fig. 9. A state-holding circuit using the hysteresis characteristics of a series circuit of a RHET and two input resistors

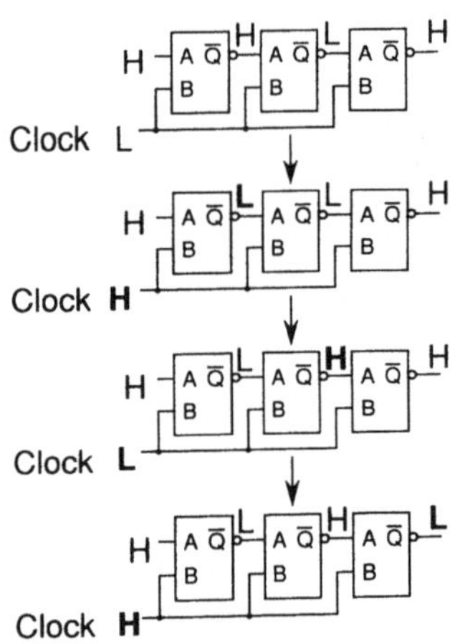

Fig. 10. A quasi shift register with a chain of state-holding circuits

In Fig. 12, a state-holding circuit holds data and an Exclusive-NOR gate controls the holding operation. When the clock is low, the Exclusive-NOR gate operates as an inverter in the feedback loop. The inverted output is fed to input A of the state-holding circuit. Both inputs cannot be the same as the current output, so the output remains unchanged. When the clock is high, the Exclusive-NOR gate feeds the output to input A. If the input is the same as the current output, the output changes to the complement of input D.

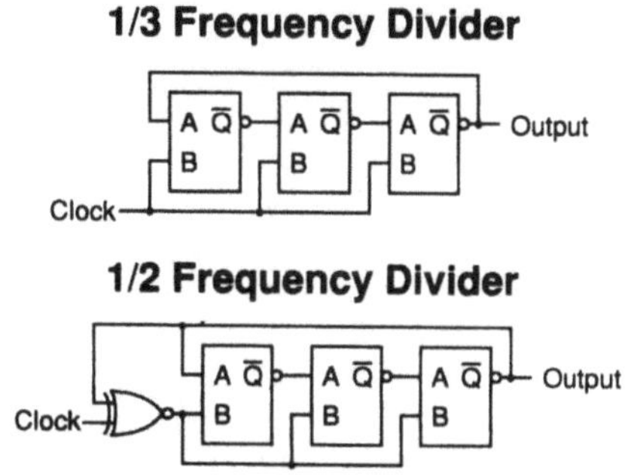

Fig. 11. Frequency dividers (1/2 and 1/3) using RHET state-holding circuits

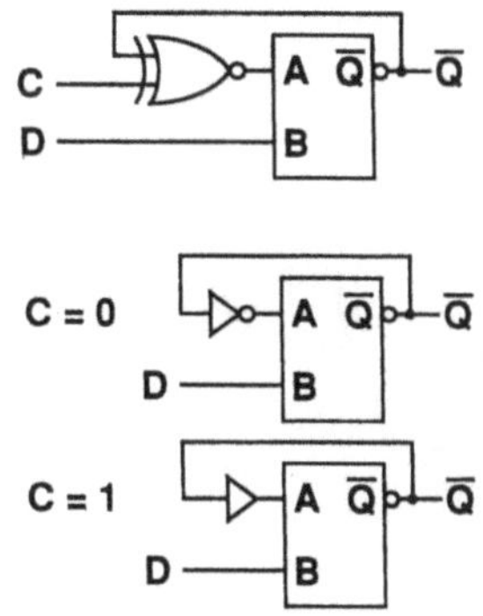

C	D	$\overline{Q}_n$	$\overline{Q}_{n+1}$
0	0	0	0
0	0	1	1
0	1	0	0
0	1	1	1
1	0	0	1
1	0	1	1
1	1	0	0
1	1	1	0

Fig. 12. A latch circuit using the state-holding circuit and Exclusive-NOR gate

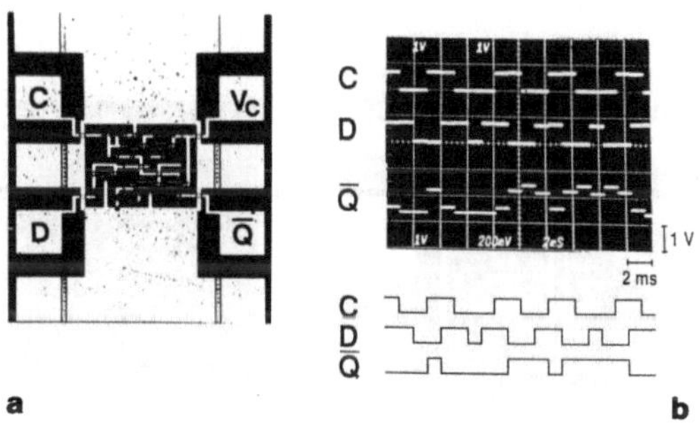

Fig. 13a, b. Completed latch circuit, and its correct operation measured at 77K.

The chip for the completed latch circuit shown in Fig. 13 measures 600 by 700 μm. The circuit's resistors are made of WSiN film deposited by reactive sputtering.

We tested the operation of the RHET circuit on the right (b). When the clock signal is low, the output does not change. When the clock signal is high, output changes depending on the input. This input sequence includes all state transitions of the latch circuit. This shows the RHET latch circuit works correctly. Figure 14 shows that the RHET latch circuit uses 5 transistors, compared to 11 used by the bipolar latch, demonstrating that using RHETs enables us to reduce the number of transistors in an integrated circuit.

Combining the three-input majority logic and two Exclusive-NOR gates, we can build the full adder circuit shown in Fig. 15. A, B, and Ci are the inputs. S is the

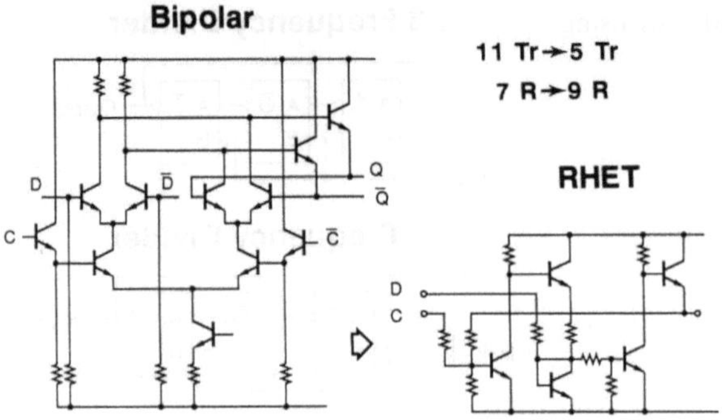

Fig. 14. Comparison of latch circuits

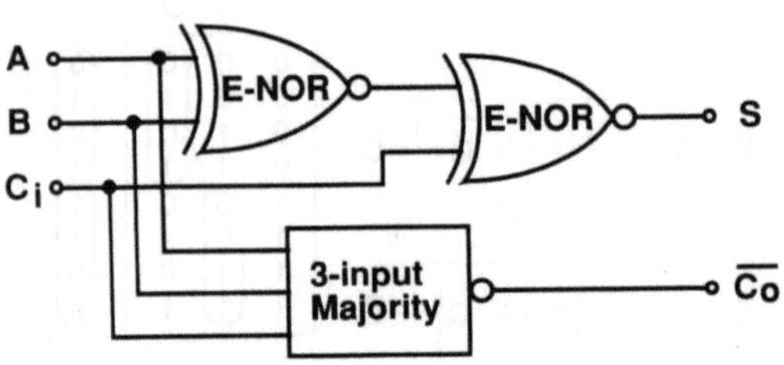

Fig. 15. A full adder using three-input majority circuit and two Exclusive-NOR gates

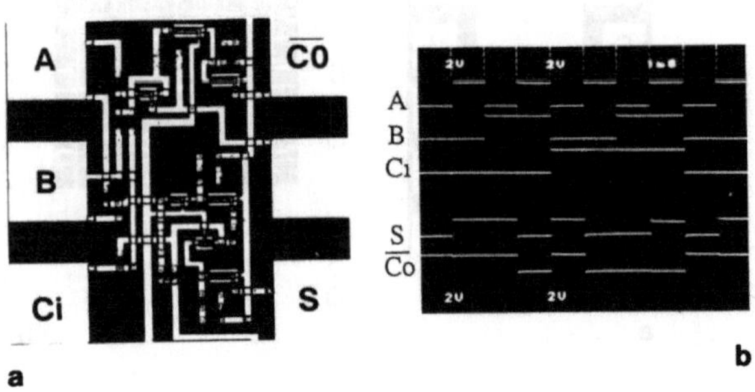

Fig. 16a, b. Completed RHET full adder, and its correct operation measured at 77K

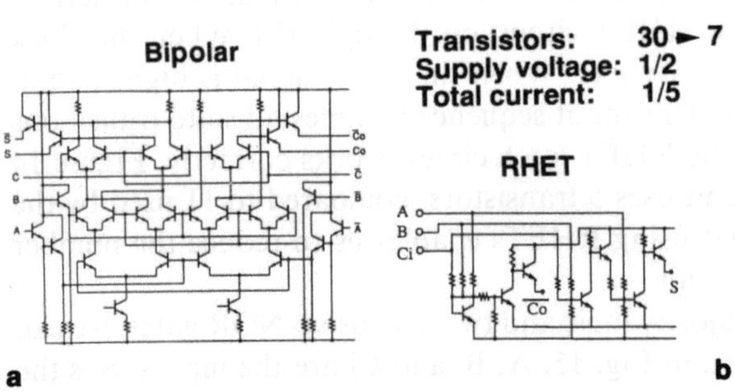

Fig. 17a, b. Comparison of full adders

sum signal, and Co is the complement of the carry signal. The microphotograph in Fig. 16 is the completed full adder circuit using RHETs. The image on the right (b) shows operation of the full adder observed at 77K with a supply voltage of 3 V. These traces show correct function of the full adder. The circuit in Fig. 17b is the RHET full adder. The circuit in Fig. 17a is the full adder using conventional bipolar transistors. Note that the number of transistors is reduced from 30 to 7. Compared to a bipolar transistor device, the supply voltage is about half, and the total current is about one-fifth, suggesting that power dissipation is ten times lower. We have not measured the RHET full adder's operating speed yet, but we expect it to be comparable to or better than that of the bipolar full adder. By using high-speed RHETs that we have developed [12], the operating speed will be about five times faster.

Acknowledgments. The research and development of RHET was carried out under the auspices of the R and D Association for Future Electron Devices as part of the R and D of Basic Technology for Future Industries sponsored by the New Energy and Industrial Technology Development Organization.

References

1. Mead CA (1960) Proc IRE 48:359
2. Sze SM Physics of semiconductor devices 2nd edn John Wiley, New York, p 559
3. Heiblum M (1981) Solid-State Elec 24:343
4. Yokoyama N, Imamura K, Ohnishi T, Nishi H, Muto S, Kondo K, Hiyamizu S (1984) IEDM Technical Digest, p 532
5. Muto S, Imamura K, Yokoyama N, Hiyamizu S, Nishi H (1985) Electron Lett, p 555
6. Yokoyama N, Imamura K, Ohnishi T, Nishi H, Muto S, Kondo K, Hiyamizu S (1985) Japan, J Appl Phys 23:L311
7. Mori T, Ohnishi H, Imamura K, Muto S, Yokoyama N (1986) Appl Phys Lett 49:1770
8. Inata Tsuguo, Muto S, Sasa S, Fujii T, Hiyamizu S (1987) Extended abstracts of the 19th conf on SSDM, Tokyo, 1987. Japan Society of Applied Physics, Tokyo p 359
9. Inata T, Muto S, Nakata Y, Sasa S, Fujii T, Hiyamizu S (1987) Japan. J Appl Phys 26:L1332
10. Takatsu M, Imamura K, Ohnishi H, Mori T, Adachihara T, Muto S, Yokoyama N (1990) Device research conference University of California, Santa Barbara, June 27, 1990
11. Imamura K, Takatsu M, Mori T, Adachihara T, Ohnishi H, Muto S, Yokoyama N (1991) Extended abstracts of the 1991 SSDM, Tokyo, 1991. Japan Society of Applied Physics, Tokyo, p 80; Takatsu M, Imamura K, Ohnishi H, Mori T, Adachihara T, Muto S, Yokoyama N (1991) Technical digest of 1991 GaAs IC symposium, Monterey. I000, New York p 95
12. Mori T, Imamura K, Ohnishi H, Takatsu M, Adachihara T, Muto S, Yokoyama N (1990) Device Research Conference, University of California, Santa Barbara, June 27, 1990

Electron Charge Distribution and Transport Properties of Narrow Channels

27

ARISATO KAWABATA[1]

Summary. We show that in one-dimensional systems the conductance at zero temperature is determined only by the electronic charge distribution, using an extended form of Friedel sum rule. We apply the theory to quantum dots and double point contacts. It is predicted that the transmission probability of the electrons through a quantum dot is unity when the dot has a magnetic moment of magnitude 1/2 due to an unpaired electron. This transparency of the dot is the manifestation of the Kondo effect and is the resonant transmission via a Kondo resonant state at the Fermi level. The case of a larger magnetic moment will also be discussed.

Introduction

We can calculate the transport coefficients of an electronic system, in principle, using a fundamental formula for the transport coefficients, if we know the scattering mechanism of the electrons in detail. In most cases, however, the calculations are no simple matter. It is particularly difficult to get reliable results when electron-electron interaction plays an important role. In fact, the periodic conductance oscillations in a narrow channel as a function of the electron density, observed recently by Scott-Thomas et al. [1], and by Meirav et al. [2] were explained in terms of a quantum dot model by van Houten and Beenakker [3], taking account of the effects of the Coulomb interaction. Meir et al. [4] pointed out the importance of the Coulomb interaction in the temperature dependence of the conductance peaks of quantum dots observed by Meirav et al. [5], and by McEuen et al. [6].

Many years ago, Friedel [7] pointed out that the phase shift of the electrons at the Fermi level is related to the charge distribution of the electrons. With the use of the Friedel sum rule combined with a symmetry argument, in some cases the conductance of the system is determined only by the charged distribution.

In this paper, we apply the Friedel sum rule to the calculation of the conductance of a mesoscopic system. For the sake of simplicity we will work on a one-

[1] Department of Physics, Gakushuin University, 1-5-1 Mejiro, Toshima-ku, Tokyo, 171 Japan

dimensional system, but the theory is applicable to cases of higher dimensions, if the number of the channel outside the dot which has a significant coupling with the states inside the dot is one.

We will extend the Friedel sum rule, making use of the symmetry of the system. Then, we will find that the conductance at zero temperature is determined only by the numbers of the electrons in the virtual bound states in the dot. The results given in the following are almost independent of the details of the model.

One-Dimensional Model of a Quantum Dot

The model Hamiltonian is composed of three parts, kinetic energy K, Coulomb interaction U between electrons, and one-electron potential energy V, which is assumed to form a dot;

$$\mathscr{H} = K + U + V \tag{1}$$

$$K = \sum_{k,\sigma} \varepsilon_k a_{k,\sigma}^{\dagger} a_{k,\sigma} \tag{2}$$

$$U = \frac{1}{2} \sum_{\sigma,\sigma'} \iint u(|x - x'|) \psi_\sigma^{\dagger}(x) \psi_{\sigma'}^{\dagger}(x') \psi_{\sigma'}(x') \psi_\sigma(x) \mathrm{d}x \mathrm{d}x' \tag{3}$$

$$V = \sum_{\sigma} \int_{-a}^{a} v(x) \psi_\sigma^{\dagger}(x) \psi_\sigma(x) \mathrm{d}x \tag{4}$$

where $\psi_\sigma(x)$ is the field operator of electrons with σ the spin index and $a_{k,\sigma}$ is its Fourier transform. Here we assumed that the one electron potential $v(x)$ vanishes in the regions $|x| \geqslant a$, and that $v(x)$ is symmetric around $x = 0$, i.e., $v(x) = v(-x)$. The symmetry of $v(x)$ plays a very important role in the following arguments.

First, we consider the case of noninteracting electrons, i.e., $U = 0$. Because of the symmetry of $v(x)$, the wave functions of the one-electron eigenstates are symmetric or antisymmetric about $x = 0$, and, in the regions $|x| \geqslant a$, they can be written in the forms

$$\phi_{k,s}(x) = \cos\{k|x| + \eta_s(k)\} \tag{5}$$

$$\phi_{k,a}(x) = \mathrm{sign}x \sin\{k|x| + \eta_a(k)\} \tag{6}$$

respectively, where the normalization factors are abbreviated. The Friedel sum rule [7, 8] relates the phase shifts $\eta_s(k)$ and $\eta_a(k)$ to the electronic charge distribution. We can easily show that in the present case it is of the form,

$$\eta_s(k_F) + \eta_a(k_F) = \frac{\pi}{2} \Delta N \tag{7}$$

where k_F is the Fermi momentum and ΔN is the number of electrons displaced into the region around $x = 0$ by the potential $v(x)$.

On the other hand, the conductance is generally determined by the phase shifts. In the present case, investigating the wave function

$$\phi_{k,t}(x) \equiv e^{i\eta_s(k)}\phi_{k,s}(x) + ie^{i\eta_a(k)}\phi_{k,a}(x) \tag{8}$$

we easily find that the probability of transmission through the region $|x| \leqslant a$ for an electron at the Fermi level is given by

$$T = \cos^2\{\eta_s(k_F) - \eta_a(k_F)\} \tag{9}$$

Then the conductance is given by Landauer formula [9]

$$G = \frac{2e^2}{h} T \tag{10}$$

where e is the electronic charge and h is the Planck constant.

We cannot determine $\eta_s(k)$ and $\eta_a(k)$ separately from Eq. (7) even if we know the value of ΔN, but here we can generalize the Friedel sum rule, making use of the symmetry of the system, as will be shown in the following section.

Generalized Friedel Sum Rule

The Friedel sum rule becomes very powerful if it holds independently for separate channels [e.g., for each spin component, or for each component of angular momentum (in three dimensions)]. The generalization of the Friedel sum rule in this context was discussed by Shiba [10] and by Yoshimori and Zawadowski [11]. In the present case, we can derive the generalized Friedel sum rule because of the simplicity of the system. In fact, in the case of noninteracting electrons, we can derive the relation Eq. (7) separately for symmetric and antisymmetric states, i.e.,

$$\eta_s(k_F) = \frac{\pi}{2}\Delta N_s \tag{11}$$

$$\eta_a(k_F) = \frac{\pi}{2}\Delta N_a \tag{12}$$

where ΔN_s and ΔN_a are the contribution to ΔN of the electrons in the symmetric and antisymmetric states.

The general proof of the Friedel sum rule for interacting electron systems was given by Langer and Ambegaokar [12] using of thermal Green function formalism. The Green function is defined by [13]

$$G_\sigma(x, x'; \tau, \tau') \equiv -\langle T_\tau(\psi_\sigma(x, \tau)\psi_\sigma^\dagger(x', \tau'))\rangle \tag{13}$$

where $\psi_\sigma(x, \tau)$ is the Matsubara representation of field operator $\psi_\sigma(x)$, T_τ is the time ordering operator, and $\langle\cdots\rangle$ represent the thermal average. Instead of the Fourier transform, we can expand the field operator $\psi_\sigma(x)$ in terms of the annihilation operators $b_{k,\sigma}$ and $c_{k,\sigma}$ associated with symmetric and antisymmetric free electron states, of which the wave functions are

$$\phi_{k,s}^0(x) = \cos kx \tag{14}$$

$$\phi_{k,a}^0(x) = \sin kx \tag{15}$$

respectively, where the normalization constants are abbreviated. Then, the Green function can be written in the form

$$G_\sigma(x, x'; \tau, \tau') = \sum_{k,k'} \{\phi^0_{k,s}(x)\phi^0_{k',s}(x')G^s_\sigma(k, k'; \tau, \tau') + \phi^0_{k,a}(x)\phi^0_{k',a}(x')G^a_\sigma(k, k'; \tau, \tau')\} \tag{16}$$

where $G^s_\sigma(k, k'; \tau, \tau')$ and $G^a_\sigma(k, k'; \tau, \tau')$ are the symmetric and the antisymmetric parts of the Green function, respectively, and are given by

$$G^s_\sigma(k, k'; \tau, \tau') \equiv -\langle T_\tau(b_{k,\sigma}(\tau)b^\dagger_{k',\sigma}(\tau'))\rangle \tag{17}$$

$$G^a_\sigma(k, k'; \tau, \tau') \equiv -\langle T_\tau(c_{k,\sigma}(\tau)c^\dagger_{k',\sigma}(\tau'))\rangle \tag{18}$$

Because of the symmetry of the system, we can apply the arguments in Ref. [12] separately for G^s_σ and G^a_σ, and we obtain Eq. (11) and Eq. (12), with

$$\Delta N_s = \sum_{k,\sigma} (\langle b^\dagger_{k,\sigma}b_{k,\sigma}\rangle - \langle b^\dagger_{k,\sigma}b_{k,\sigma}\rangle_0) \tag{19}$$

$$\Delta N_a = \sum_{k,\sigma} (\langle c^\dagger_{k,\sigma}c_{k,\sigma}\rangle - \langle c^\dagger_{k,\sigma}c_{k,\sigma}\rangle_0) \tag{20}$$

where $\langle \cdots \rangle_0$ denote the average in the absence of the potential $v(x)$. Here it is to be noted that the generalized Friedel sum rule is valid only at 0K and when the Ground state is not degenerate (i.e., there is no inelastic scattering of quasi-particles) and the phase shift is independent of the spin directions.

Conductance Through Quantum Dot-Resonant Tunneling

Hereafter, to be specific, we consider a case where the region $|x| \leqslant a$ is separated from the other regions by two potential walls of $v(x)$ (see Fig. 1). This region will be called a dot. We consider the case where the potential walls are high enough and thick enough and the transmission probability of an electron at the Fermi energy through each wall is much smaller than unity if interference effects are neglected.

Suppose that v_0, the potential energy of the bottom of the dot, is decreased. Then the virtual bound states will appear one by one in the dot at energies higher than the Fermi level, and they will come down into the Fermi sea. The transmission probability T is very small, except in the case of resonant transmission and, hence, from Eq. (9) we find that

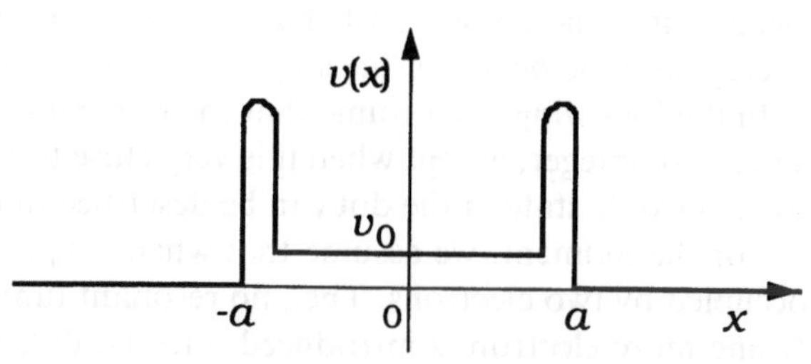

Fig. 1. The one-electron potential $v(x)$ which forms the quantum dot

$$\eta_s(k_F) - \eta_a(k_F) \cong \frac{\pi}{2} + n\pi \tag{21}$$

where n is an integer. It is to be noted that ΔN_s and ΔN_a are not the numbers of the electrons in the dot. When the dot is vacant, they are the numbers of the electrons displaced from the region of the dot, and hence $\eta_s(k_F)$ and $\eta_a(k_F)$ are not zero. We define ΔN_{0s} and ΔN_{0a} by

$$\Delta N_{0s} = \Delta N_s - \Delta N_{ds} \tag{22}$$

$$\Delta N_{0a} = \Delta N_a - \Delta N_{da} \tag{23}$$

where ΔN_{ds} and ΔN_{da} are the total numbers of the electrons occupying the symmetric and antisymmetric virtual bound states, respectively. Since $T \ll 1$ when there is no electron in the dot, from Eqs. (11), (12) and (21) it follows that

$$\Delta N_{0s} - \Delta N_{0a} = 1 + 2n \tag{24}$$

The resonant tunneling occurs when a virtual bound state crosses the Fermi level. The virtual bound states are symmetric or anti-symmetric because of the symmetry of the system. If a symmetric virtual bound state crosses the Fermi level, for instance, $\eta_a(k_F)$ changes little, and *vice versa*. Therefore, in the case of weak electron correlation, ΔN_{ds}, and hence ΔN_s in Eq. (11), changes by 2 when a virtual bound state crosses the Fermi level, and the argument of cosine in the rhs of Eq. (9) will change by π. Then the conductance will exhibit a peak at a value of v_0 at which

$$\eta_s(k_F) - \eta_a(k_F) \cong n\pi \tag{25}$$

From Eq. (10) we find that the peak values are

$$G_{\text{peak}} \cong \frac{2e^2}{h} \tag{26}$$

The width of the peak is roughly equal to that of the virtual bound state. Since the situation is the same when an antisymmetric virtual bound state crosses the Fermi level, the conductance G and the total number ΔN_d of the electrons in the dot (i.e. $N_{ds} + N_{da}$) behave as in Fig. 2 as functions of v_0.

Effects of Coulomb Interaction in the Dot

When the electron correlation in the dot is very strong, in order to introduce one electron into the dot we must change v_0 by an amount nearly equal to the charging energy, and the height of the steps of ΔN_d as a function of v_0 is unity, as in Fig. 3.

In the following, we assume that the number of the electrons in a virtual bound state is an integer, except when it is very close to the Fermi level. This is the case if the electronic state in the dot can be described in terms of Hartree-Fock theory.

For the moment, we assume that when ΔN_d is even each virtual bound state is occupied by two electrons. Then no resonant tunneling occurs and Eq. (21) holds. If one more electron is introduced into the dot, because of Eq. (11) or Eq. (12),

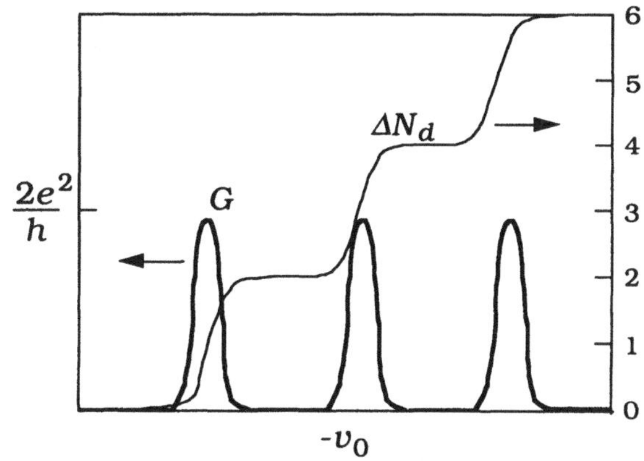

Fig. 2. Electron number ΔN_d in the dot and the conductance G through the dot as functions of $-v_0$, v_0 being the potential energy of the bottom of the dot, in the case when the effects of the Coulomb interaction in the dot are negligible (schematic)

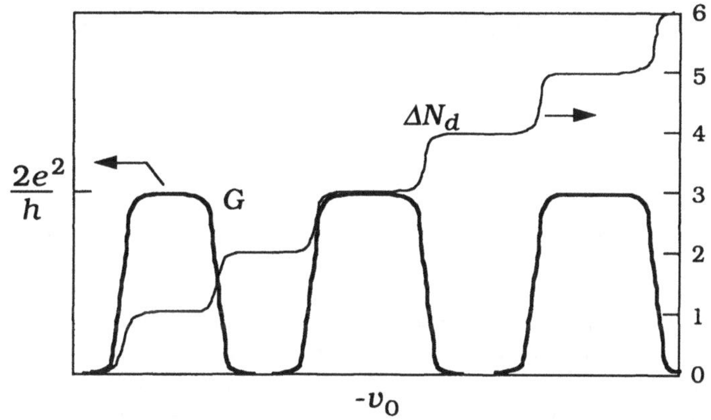

Fig. 3. Electron number ΔN_d in the dot and the conductance G through the dot as functions of $-v_0$, in the case when the strong Coulomb interaction in the dot gives rise to a localized moment for odd ΔN_d (schematic)

$\eta_s(k_F)$ [$\eta_a(k_F)$] increases by $\pi/2$ if that electron is put in a symmetric (antisymmetric) virtual bound state. Then, Eqs. (24) and (25) hold as long as ΔN_d is odd, and, instead of a sharp peak, we should observe a plateau or a broad peak of the conductance as a function of v_0, as in Fig. 3. One of the conditions for the Friedel sum rule to hold is that the ground state is not degenerate. This is the case if the localized spin is coupled antiferromagnetically with the electrons outside the dot (hereafter these electrons will be called "the conduction electrons"). Namely, the localized spin is compensated by the polarization of the conduction electrons to form a singlet ground state (Kondo effect [14]). It is easy to show that the coupling is antiferromagnetic within perturbation theory.

This remarkable result, that the dot is almost transparent when ΔN_d is odd, is due to resonant tunneling; when a localized moment of spin 1/2 exists in the dot, there is a sharp virtual bound state fixed at the Fermi level [15] and the resonant tunneling through this state gives rise to a large transmission probability. A similar phenomenon has been observed as the zero-bias anomaly in the case of tunneling in the metal-oxide-metal system (for a review, see [16]), in the presence of paramagnetic impurities.

Next we consider the case when the total spin S of the localized moment in the dot is larger than 1/2. In fact, the possibility of the large localized moment in a quantum dot has been pointed out recently by Häusler, Kramer and Mašek [17]. The behavior of Kondo systems is very much dependent on the ratio of $2S$ to f, f being the number of the channels of the conduction electrons coupled with the localized spin [18, 19]. Therefore there are several situations:

a) *One of two unpaired electrons is in a symmetric virtual bound state and the other is in an antisymmetric one.*

In this case, we have $S = 1$ and $f = 2$, for the localized spin couples with the symmetric and the antisymmetric channels. When $2S/f = 1$, the ground state is non-degenerate, and we can apply the arguments in the above. Since ΔN_{ds} and ΔN_{as} are odd, we find that

$$G \cong 0 \tag{27}$$

Thus, the Kondo resonance does not always lead to the transparency of the quantum dot.

b) *Two unpaired electrons are in two different symmetric virtual bound states.*

In this case, the localized spin couples with only one channel, and hence $f = 1$ and $S = 1$. When $2S/f > 1$, the ground states are degenerate: The localized spin is incompletely compensated by the conduction electrons and the free spin of magnitude $S - f/2(=1/2)$ remains. The coupling between this spin and the conduction electrons is renormalized to zero in the zero temperature limit and the spin can be oriented by an infinitesimally small magnetic field. The ground state is then not degenerate.

Here we assume that the generalized Friedel sum rule is valid separately for each spin direction[1], namely,

$$\eta_s^\sigma(k_F) = \pi \Delta N_s^\sigma \tag{28}$$

$$\eta_a^\sigma(k_F) = \pi \Delta N_a^\sigma \tag{29}$$

where $\eta_s^\sigma(k_F)$ and $\eta_a^\sigma(k_F)$ are the phase shifts of an electron with spin σ, and ΔN_s^σ and ΔN_a^σ are the contributions to ΔN_s and ΔN_a of the electrons with spin σ.

Suppose that the uncompensated spin is oriented to z-direction. Then, if we let S_z be the z-component of the localized spin, the state of the incompletely compen-

[1] Shiba [10] derived this form of generalized Friedel sum rule on the Anderson model, but, to the knowledge of the author, it has not yet been proven on more general models.

sated spin is the linear combination of the state with $S_z = 1$ and that with $S_z = 0$. When $S_z = 1$, the conduction electrons around the dot is polarized to partly compensate it. The z-component of the total polarization must be $-1/2$, and hence, if the charge neutrality around the dot is not violated, the total numbers of the excess up spin and down spin conduction electrons are $-1/2$ and $1/2$, respectively. When $S_z = 0$, the z-component of the total spin of the polarized conduction electron is $1/2$, and hence the signs of the total number of the excess electrons for each spin direction are reversed.

Then, from the above consideration, we find that

$$\Delta N_s^{\uparrow} = \Delta N_{0s} + \tfrac{3}{2} + n_s \tag{30}$$

$$\Delta N_s^{\downarrow} = \Delta N_{0s} + \tfrac{1}{2} + n_s \tag{31}$$

and

$$\Delta N_a^{\sigma} = \Delta N_{0a} + n_a \tag{32}$$

where n_s and n_a are the numbers of the symmetric and antisymmetric doubly occupied virtual bound states in the dot, respectively (note that when $S_z = 1$ the total numbers of the up and down spin electrons in the symmetric virtual bound states are $2 + n_s$ and n_s, respectively, and that when $S_z = 0$ they are both $1 + n_s$). Then, from Eqs. (24), (28) and (29), it follows that

$$\eta_s^{\sigma}(k_F) - \eta_a^{\sigma}(k_F) = m\pi \tag{33}$$

m being an integer, and hence from eqs. (2.9) and (2.10) we find that

$$G \cong \frac{2e^2}{h} \tag{34}$$

c) *Two unpaired electrons are in two different antisymmetric virtual bound states.*

We can apply the same arguments as in b) to this case and we obtain the same result.

d) *The magnitude of the localized spin is larger than unity.*

Since $f \leqslant 2$, in this case $2S$ is larger than f, and the localized spin is partly compensated. If the unpaired electrons occupy only the symmetric or antisymmetric virtual states, we can apply the arguments in b), and we obtain Eqs. 30–32 with 3/2 in the rhs of Eq. (30) replaced with $2S - 1/2$. Then, as is easily seen, we obtain the same results as in b).

Next we consider the case when the unpaired spin occupies m_s symmetric states and m_a antisymmetric states. From the same consideration as above, it follows that

$$\Delta N_s^{\uparrow} = \Delta N_{0s} + m_s - \tfrac{1}{2} + n_s \tag{35}$$

$$\Delta N_s^{\downarrow} = \Delta N_{0s} + \tfrac{1}{2} + n_s \tag{36}$$

$$\Delta N_a^{\uparrow} = \Delta N_{0a} + m_a - \tfrac{1}{2} + n_a \tag{37}$$

and

$$\Delta N_a^{\downarrow} = \Delta N_{0a} + \tfrac{1}{2} + n_a \tag{38}$$

and from Eqs. (24), (28), and (29), we obtain

$$\eta_s^{\sigma}(k_F) - \eta_a^{\sigma}(k_F) = \frac{\pi}{2} + m\pi \tag{39}$$

m being an integer. Hence, in this case the conductance is very small:

$$G \cong 0 \tag{40}$$

Thus, from the investigations in this section, we can derive a simple rule: The dot is transparent when the localized spin is coupled with one channel of the conduction electrons, and is opaque when the spin is coupled with two channels.

Conclusion

We have investigated the effects of the Kondo resonance on electronic conduction through a quantum dot. The transparency of a one-dimensional system due to Kondo resonance has already been investigated by Ng and Lee [20], and by Grazman and Raikh [21] in a specific model; however, in this paper we dealt with a more general model and with cases of various spin states in the dot.

In order to observe the Kondo resonance, several conditions, which have not been stated explicitly, have to be fulfilled. Firstly, to have a well-defined localized moment, the width of the virtual bound state without many-body effects must be smaller than the Coulomb repulsion energy. Secondly, the results given in the preceding sections are valid only at zero temperature, and the temperature must be low enough. Practically, the temperature must be lower than the Kondo temperature. The Kondo temperature is given by [14, 18]

$$T_K \cong \frac{\varepsilon_F}{k_B} e^{-1/\tilde{J}} \tag{41}$$

where ε_F is the Fermi energy of the conduction electron, k_B is the Boltzmann constant, and $\tilde{J}$ is the coupling parameter given by

$$\tilde{J} \cong \frac{\Delta}{U} \tag{42}$$

Δ and U being the width of the virtual bound states and the charging energy in the dot, respectively. These expressions are correct only when $\Delta \ll U$, and hence T_K is much smaller than ε_F. The possibility of observing the Kondo effect depends on the control of Δ, which represent the coupling between the electrons inside and outside the dot.

So far, we have considered one-dimensional systems. In the following, as an example of a two-dimensional system, we consider a quantum dot formed by two point contacts. Generally, in two-dimensional systems we have to introduce many phase shifts associated with each channel. If the point contacts are narrow, so that only one channel has a considerable coupling with the dot, the problem reduces to

Fig. 4. Model of a double point contact

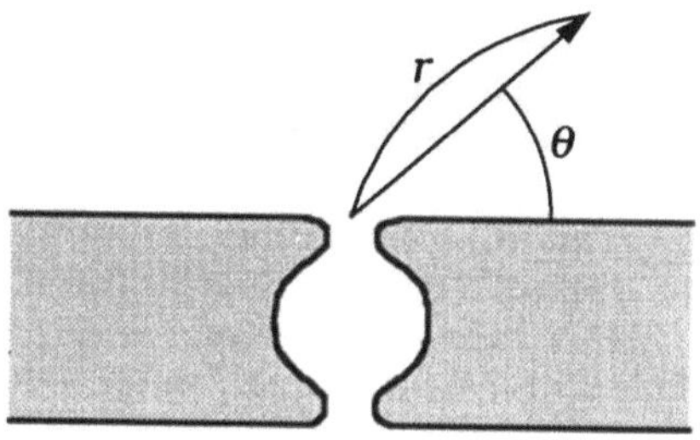

a one-dimensional problem, and the arguments given above are still valid. For instance, consider the system shown in Fig. 4. Except in the regions close to the point contacts, the eigenfunctions of each channel at the Fermi level are of the form

$$\phi(r,\theta) = AJ_n(k_F r)\sin n\theta \tag{43}$$

where n is a positive integer and J_n is the Bessel function of order n. Therefore, when r is very small, we have

$$\phi(r,\theta) \propto r^n \tag{44}$$

and we find that the amplitude of the wave function near the point contact, and hence the coupling with the dot, is smaller for the channels with larger n.

Acknowledgments. The author is grateful to Prof. H. Shiba, Prof. J. Kondo, and Prof. Y. Ono for their valuable discussions. This work has been partly supported by a Grant-in-Aid for Scientific Research in the Priority Area, "Electron Wave Interference Effects in Mesoscopic Structures", from the Ministry of Education, Science and Culture of Japan.

References

1. Scott-Thomas JHF, Field SB, Kastner MA, Smith HI, Antoniadis DA (1989) Phys Rev Lett 62:583–586
2. Meirav U, Kastner MA, Heiblum M, Wind SJ (1989) Phys Rev B40:5871–5874
3. van Houten H, Beenakker CWJ (1989) Phys Rev Lett 63:1893–1893
4. Meir Y, Wingreen NS, Lee PA (1991) Phys Rev Lett 66:3048–3051
5. Meirav U, Kastner MA, Wind SJ (1990) Phys Rev Lett 65:771–774
6. McEuen PL, Foxman EB, Meirav U, Kastner MA, Meir Y, Wingreen NS, Wind SJ (1991) Phys Res Lett 66:1926–1929
7. Friedel J (1952) Phil Mag 43:153–189
8. Friedel J (1954) Adv Phys 3:446–507
9. Landauer R (1987) Z Phys B68:217–228 (and references therein)
10. Shiba H (1975) Prog Theor Phys 54:967–981
11. Yoshimori A, Zawadowski A (1982) J Phys C (Solid State Physics) 15:5241–5253
12. Langer JS, Ambegaokar V (1961) Phys Rev 121:1090–1092

13. Abrikosov AA, Gor'kov LP, Dzyaloshinski IE (1975) Methods of quantum field theory in statistical physics. Dover, New York, pp 97–106
14. Kondo J (1969) In: Seitz F, Turnbull D, Ehrenreich H (eds) Solid state physics, vol 23. Academic, New York, pp 183–281
15. Yamada K (1975) Progr Theor Phys 53:970–986
16. Duke CB (1969) In: Seitz F, Turnbull D, Ehrenreich H (eds) Solid State Physics, suppl 10. Academic, New York, pp 1–353
17. Häusler W, Kramer B, Mašek J (1991) Z Phys B83:435–442
18. Cragg DM, Lloyd PL (1979) J Phys C (Solid State Physics) 12:L215–219
19. Nozieres Ph, Blandin A (1980) J Physique 41:193–211
20. Ng TK, Lee PA (1988) Phys Rev Lett 61:1768–1771
21. Grazman LI, Raikh MÉ: (1988) JETP Lett 47 452

Observation of the Charging Effect in I-V Characteristics Through a Quantum Dot Fabricated by a Focused Ion Beam at 4.2K

28

SHUNJI NAKATA[1]*

Summary. Small tunnel barriers were formed on AlGaAs/GaAs fabricated by Focused-Ion-Beam (FIB) implantation. These samples were measured using DC current at 4.2K. The current-voltage (I-V) curve had two distinct regions: a tunneling regime and a region which resulted from a thermal current over the barrier height. In the source-drain voltage (Vsd) range from −5 mV to 5 mV, a dip in the dI/dV curve was observed in the tunneling region. In a different sample, periodic and reproducible staircase steps with periodicity Vsd = 35 mV were observed in the tunneling region at 4.2K. This Coulomb staircase phenomenon occurs when the number of electrons in a quantum dot is changed.

Introduction

Microfabrication technology has recently advanced and submicrometer-order semiconductor structures can now be fabricated. Recent studies in the quantum transport field have enabled us to observe universal conductance fluctuations, quantized conductance, and 0D-tunnel diodes in mesoscopic systems [1–5].

Recently the charging effect has come to be thought of as very important when electrons tunnel through a small capacitance. This has been studied using a metal-insulator-metal junction [6, 7]. Since the capacitance in the small junction is very small, the charging energy e^2/C becomes very large. Suppression of single electron tunneling can then be observed (Coulomb blockade). Moreover staircase steps in the tunneling current in the double capacitance system were observed in an experiment using a scanning tunneling microscope [8].

Single electron tunneling is interesting not only in metals but also in semiconductors. Scott-Thomas et al. [9] and Staring et al. [10] observed periodic current oscillation changing upper gate voltage in a low mobility wire. Their explanation of this phenomenon was that a quantum dot structure was formed in a part of the quantum wire and that this structure created the periodic oscillations. Meirav et al. [11] fabricated a double constriction structure using the split gate method. They observed the same periodic oscillations changing gate voltage. However, the

[1] NTT Basic Research Laboratories, Musashino-shi, Tokyo, 180 Japan
* Present address: NTT LSI Laboratories, Atsugi-shi, Kanagawa, 243-01 Japan

Coulomb staircase is not yet well understood experimentally in semiconductors when Vsd is changing.

In this paper, we discuss tunneling phenomena about small constriction structures fabricated by a Focused-Ion-Beam (FIB). The confining potential profiles are very gentle in the split gate formed by a Schottky electrode evaporated on a wafer, but for the reason that Ga ion implantation regions are p-type like, having many acceptor levels, steep potential profiles are possible. This could lead to a small capacitance system, in which the charging energy would be very large and periodic characteristics would be observed at higher temperatures.

Sample Fabrication and Measurements

The starting wafer was a modulation-doped structure grown by molecular beam epitaxy (MBE). Ga ions were implanted into this wafer using FIB scanning; the implanted region became highly resistive. The sample structure is shown in Fig. 1; it has two constrictions, the distance between the two constrictions is 0.6 μm and the width of the constrictions is 0.4 μm. The depletion length extends nearly 0.2 μm from the implanted region.

Both the I-V curve and the derivative curve are shown in Fig. 2. The measurement temperature was 4.2K and DC current was used. The constant external voltage method was used for driving the current. Vsd was changed from -120 mV to 120 mV. The inset shows a magnification of the I-V curve in the Vsd range from -20 mV to 20 mV. Resistance in the tunneling region is almost 3 MΩ. A threshold voltage (Vth) is observed which changes the I-V characteristics drastically. Vth is equal to 40 mV in Fig. 2. The dI/dV above Vth is 30 times larger than that below Vth. From this drastic change it can be concluded that the current above Vth is mainly thermal current over the barrier height. Below Vth = 40 mV, dI/dV first increases but then immediately saturates and becomes constant (0.4 μA/V) until the voltage reaches about 40 mV. These characteristics differ from the usual tunneling situation in which dI/dV typically increases monotonically. The dI/dV re-

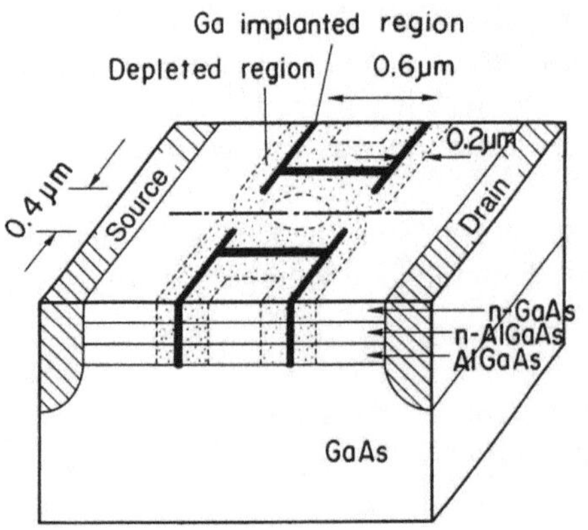

Fig. 1. Sample fabricated by Ga-Focused-Ion-Beam implantation. *Dotted areas* show the depleted regions

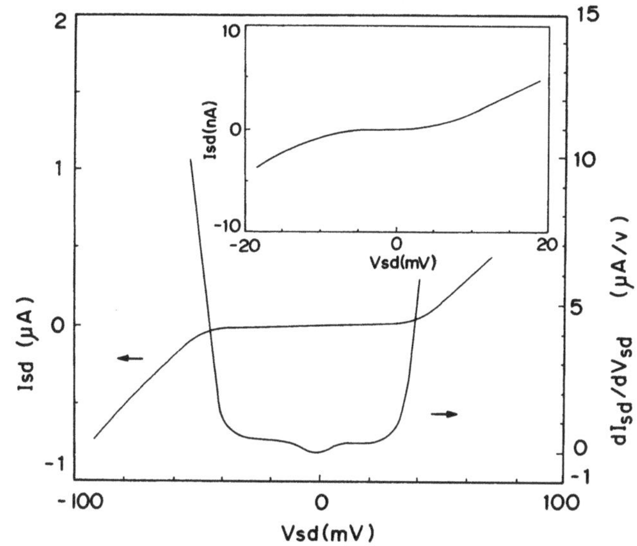

Fig. 2. I-V curve and dI/dV curve at 4.2K. The *inset* shows a magnification of the I-V curve

duces from the saturation value in the range -5 mV < Vsd < 5 mV although the Fermi energy is 15 meV. This dip in dI/dV curve may be caused by tunneling suppression due to Coulomb blockade. In this system, the sample has two constrictions (i.e., two capacitances) so that Coulomb blockade might be observed using the constant voltage one. According to Likharev [12] and Mullen et al. [13], the tunneling current is suppressed when the voltage Vsd is less than e/2C. From the experimental value of e/2C = 5 mV, capacitance is estimated to be 1.6×10^{-17} F.

The other I-V curve, shown in Fig. 3, is for a different sample. The measurement temperature was 4.2K. Vsd was changed from 0 to 140 mV, side gate voltage Vg was changed from 0 to 1.2 V. I-V curves for different Vg were plotted shifting each curve by 10 nA. The depletion length in the central region decreased to 0.1 μm as Vg increased. In this case, periodic structures were observed differently from those in Fig. 2 and the characteristics were reproducible. The periodicity ΔVsd was 20 mV. The periodic structures may be related to the charging effect through the small quantum dot. The current characteristics had a steplike shape at Vsd = 20 mV and 40 mV.

Next, to confirm the charging effect more clearly, another measurement was made after the sample was further illuminated. Then the potential profile of the constriction was able to be changed. After illumination once, the LED was not used again during measurements. The I-V curve is shown in Fig.4. Vsd was changed from -220 mV to 220 mV, and side gate voltage was changed from 0 to 1.2 V in steps of 0.3 V. Average resistance in the tunneling region was about 3 MΩ. The periodic staircase was clearly observed and reproducible with ΔVsd = 35 mV and ΔIsd = 10 nA. The observed Coulomb staircase was not largely affected by changes in Vg, although |Vth| did decrease as Vg got larger since the carrier

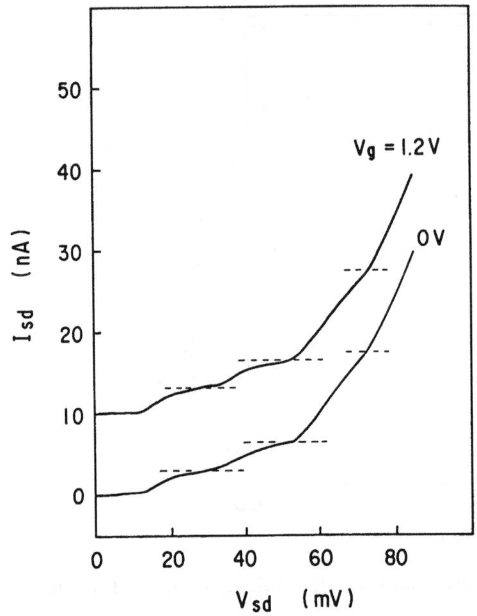

Fig. 3. I-V curve in a different sample at 4.2K. The curve for Vg = 1.2V is shifted by 10nA

density was increasing (Vth is about 180 meV in Fig. 4). The appearance of the Coulomb staircase results from the quantum dot structure, whose capacitance is estimated to be 5×10^{-18}(F) from ΔVsd. This small capacitance and its independence from the side gate voltage suggest that the quantum dot may be formed in one constriction in this case.

Theory

Here semiclassical simulation [13] was performed to understand the charging effect. Although the external voltage Vex is fixed, the voltage across each junction Vi is a function of its capacitance and the charge accumulated between them.

$$V1 = C2/(C1 + C2)Vex - dQ/(C1 + C2)$$

$$V2 = C1/(C1 + C2)Vex + dQ/(C1 + C2) \qquad dQ = (n_1 - n_2)e$$

where n_i is the number of electrons that have tunneled across the i-th junction. We assumed a stochastic process, so n_i can be written as:

$$n_i(t + \Delta t) = \begin{cases} n_i(t) + 1 & \text{with probability} \quad r_i\Delta t \\ n_i(t) - 1 & l_i\Delta t \\ n_i(t) & 1 - (l_i\Delta t + r_i\Delta t) \end{cases}$$

where r_i and l_i are the tunneling rates through the i-th tunnel barrier. The calculation was performed by iterating the above stochastic process until stable states

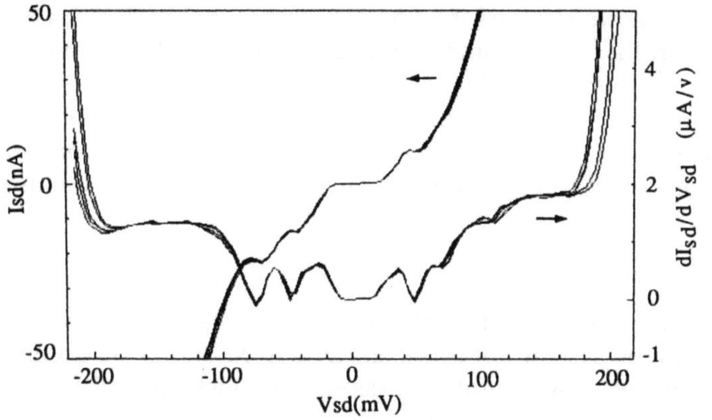

Fig. 4. I-V curve and dI/dV curve after changing the potential profile at 4.2K. Many steps are clearly observed and reproducible

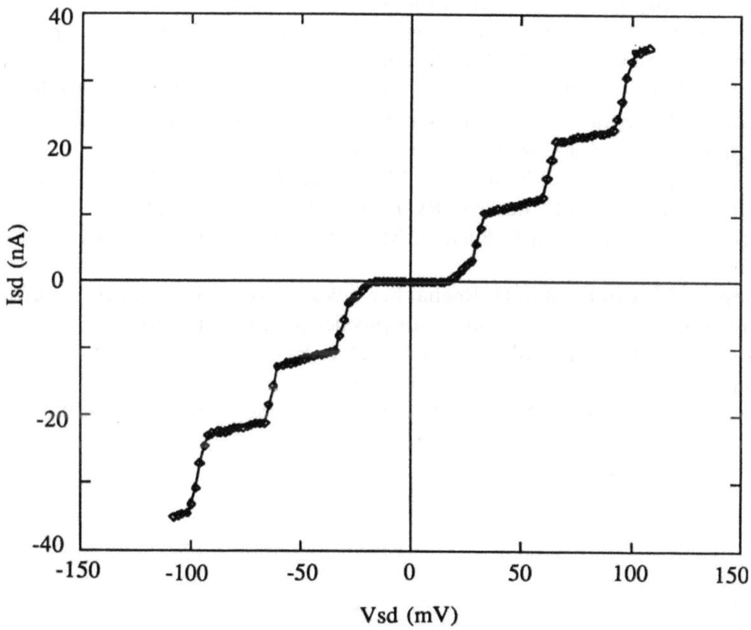

Fig. 5. The semiclassical I-V calculation. Parameters are R1 = 500kΩ, R2 = 2.5MΩ, C1 = 1aF, and C2 = 5aF

were reached. Here we assumed that the dot voltage in this process was more likely to be lower than the source contact voltage, as compared with the reverse condition. The results are shown in Fig. 5. The voltage in the central region can float to the source and drain voltage. Then the I-V characteristics are not linear but there is a staircase due to the charging effect of accumulated electrons in the center region. The change of electron number in the central region is quantized in steps to $\Delta n = 0, 1, 2, \ldots$

Acknowledgment. The author is grateful to S. Tarucha for his encouragement and valuable discussion. The author is also grateful to M. Stopa, Y. Tokura, M. Ueda, and J.R. Phillips for their helpful discussions and indebted to T. Saku for the high quality wafer.

References

1. Heinrich H, Bauer G, Kuchar F (eds) (1988) Physics and technology of submicron structures, vol 83. Springer, Berlin (Series in Solid State Sciences)
2. Nakata S, Yamada S, Hirayama Y, Saku T, Horikoshi Y (1990) Jpn J Appl Phys 29:32
3. Hirayama Y, Saku T (1990) Phys Rev B41:2927
4. Nakata S, Hirayama Y, Tarucha S, Horikoshi Y (1991) J Appl Phys 69:3633
5. Tarucha S, Hirayama Y, Saku T, Kimura T (1990) Phys Rev B41:5459
6. Fulton TA, Dolan GJ (1987) Phys Rev Lett 59:109
7. Kuzmin LS, Delsing P, Claeson T, Likharev KK (1989) Phys Rev Lett 62:2539
8. Wilkins R, Ben-Jacob E, Jaklevic RC (1989) Phys Rev Lett 63:801
9. Scott-Thomas JHF, Field B, Kastner MA, Smith HI, Antoniadis DA (1989) Phys Rev Lett 62:583
10. Staring AAM, van Houten H, Beenakker CWJ, Foxon CT (1990) In: Landwehr G (ed) High magnetic fields in semiconductor physics 3. Springer, Berlin
11. Meirav U, Kastner MA, Wind SJ (1990) Phys Rev Lett 65:771
12. Likharev KK (1988) IBM J Res Develop 32:144
13. Mullen K, Ben-Jacob E, Jaclevic RC, Schuss Z (1988) Phys Rev B37:98

Dimension Dependence of Elastic Scattering Effects on Resonant Tunneling

29

YASUHITO ZOHTA[1], KOICHI NAKAMURA[2], AND HIROSHI EZAWA[3]

Summary. We have shown that elastic scattering influences resonant tunneling in quite a different manner in the 1D and 3D cases. In 1D resonant tunneling, elastic scattering simply shifts the resonance energy and has little influence on the transmission peak and the full width. In contrast, in 3D resonant tunneling, elastic scattering reduces the transmission peak and broadens the full width, just as in the case of inelastic scattering.

Introduction

Recent advances in crystal growth technology have made it possible to fabricate thin layers with electron wave dimension in the growth direction. This has stimulated studies of resonant tunneling [1], known as a quantum-mechanical phenomenon since the birth of quantum mechanics. Further, recent progress in lithography has enabled nano-structures in the lateral direction. Consequently, the dimension-dependent properties of resonant tunneling are now one of the most stimulating problems in semiconductor physics from both the theoretical and experimental points of view [1, 2].

In this paper, we will discuss the dimension dependence of elastic scattering effects on resonant tunneling (RT) using the scattering (S) matrix method [3, 4] combined with the Feynman path integral [5–8]. We will deal with three cases; (1) a one-dimensional (1D) structure, (2) a set of 1D structures where each 1D structure has a scattering center at a different position, and (3) a three-dimensional (3D) structure. Case (2) corresponds to a device which is frequently realized in experiments.

In 1D RT, the scattered waves interfere constructively with each other, which leads only to a shift in resonance energy. On the other hand, in 3D RT, the scattered waves interfere destructively and, as a result, elastic scattering reduces the transmission peak and broadens the resonance width just as inelastic scattering does. Case (2) shows the average behavior of the 1D case.

[1] Toshiba R and D Center, 1 Komukai, Saiwai-ku, Kawasaki, 210 Japan
[2] Division of Natural Science, Meiji University, Izumi, Eifuku 1-9-1, Suginami, Tokyo, 168 Japan
[3] Department of Physics, Gakushuin University, Mejiro 1-5-1, Toshima, Tokyo, 171 Japan

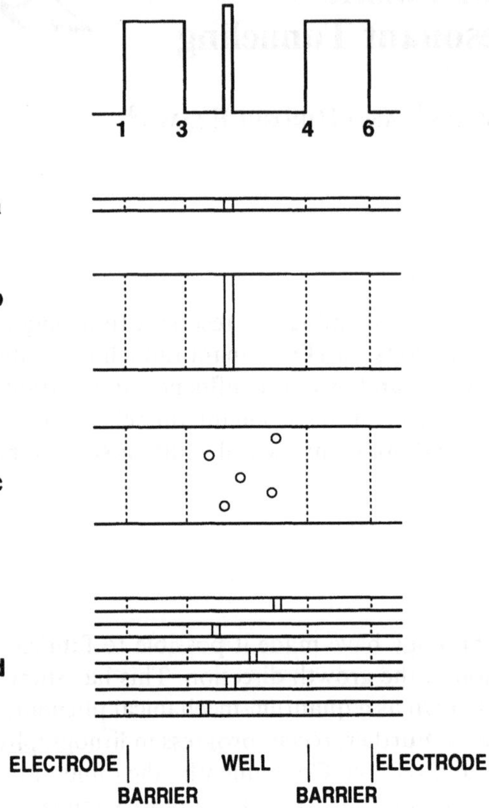

Fig. 1a–d. Models for scatterers. **a** 1D structure, **b** 1D structure (free particle motion in the y and z directions is allowed). **c** 3D structure. **d** set of 1D structures connected in parallel

Model

The models we will deal with are shown in Fig. 1, where a and b represent the 1D model, c is the 3D model, and d is a set of the 1D models connected in parallel, which is frequently used in experiments. The scatterers in the 1D model and in the 3D model are represented as solid lines and solid circles, respectively, and the scatterers are assumed to be distributed randomly in Fig. 1 c and d.

1D Structure

We start with a 1D double-barrier structure, as shown in Fig. 2. A quantum well III + IV ($x_3 \leqslant x \leqslant x_4$) of width L is sandwiched between two equal barriers (II + V) with potential V_b and thickness l. Outside them there exist two electrodes I and VI. We set a very thin rectangular barrier potential, of height V_s and thickness δ ($\ll$L), in the well representing an elastic scattering potential. Its position is L_1 from the left barrier.

Using the S matrix method, we can obtain the transmission coefficient T

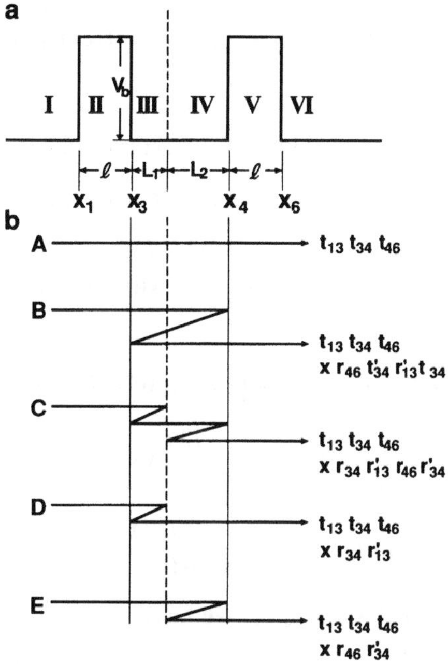

Fig. 2. a Energy band diagram for a double-barrier structure. **b** The Feynman path trajectories

through the double-barrier structure as

$$T = |t_{16}|^2$$

$$t_{16} = \frac{t_{13}t_{34}t_{46}}{1 - r_{46}t'_{34}r'_{13}t_{34} + r_{34}r'_{34}r_{46}r'_{34} - r_{34}r'_{13} - r_{46}r'_{34}} \tag{1}$$

Here, t_{ij}, t'_{ij}, r_{ij}, and r'_{ij} are elements of the S matrix which connects the wave amplitude at x_i and x_j.

Expanding the denominator of Eq. (1), we have

$$t_{16} = \sum_{i=0}^{\infty} t_{13}t_{34}t_{46}(r_{46}t'_{34}r'_{13}t_{34} - r_{34}r'_{13}r_{46}r'_{34} + r_{34}r'_{13} + r_{46}r'_{34})^i \tag{2}$$

The scattering matrix method can be viewed from the path integral point of view through this equation, which represents how the classical paths shown in Fig. 2 (b) contribute to transmission from 1 to 6. Considering the meanings of t, t′, r, and r′, we can relate each term appearing in Eqs. (1) and (2) to the Feynman diagram of Fig. 2 (b) as follows: (A) $t_{13}t_{34}t_{46}$ represents direct motion from left (1) to right (6), (B) $t_{13}t_{34}t_{46} \times r_{46}t'_{34}r'_{13}t_{34}$ represents a path including two reflections from the barriers (the Fabry-Perot Process), (C) $t_{13}t_{34}t_{46} \times r_{34}r'_{13}r_{46}r'_{34}$ represents a path including two reflections from the scatterer and two reflections from the barriers, and (D) $t_{13}t_{34}t_{46} \times r_{34}r'_{13}$ and (E) $t_{13}t_{34}t_{46} \times r_{46}r'_{34}$ represent paths including one reflection from the scatterer and one reflection from the barrier, respectively.

For rectangular barriers, we have

$$t_{13} = t'_{13} = t_{46} = t'_{46} \equiv t = |t| \exp\left[i\left(\frac{\pi}{2} - \chi\right)\right]$$

$$= \frac{i \sin 2\theta}{\cos 2\theta \sinh(\kappa l) + i \sin 2\theta \cosh(\kappa l)} \tag{3}$$

$$r_{13} = r'_{13} = r_{46} = r'_{46} \equiv r = |r| \exp(-i\chi)$$

$$= \frac{\sinh(\kappa l)}{\cos 2\theta \sinh(\kappa l) + i \sin 2\theta \cosh(\kappa l)} \tag{4}$$

where

$$|t|^2 = 1 - |r|^2 = [1 + \sinh^2(\kappa l)/\sin^2 2\theta]^{-1} \tag{5}$$

$$k = \sqrt{\frac{2m}{\hbar^2} E}, \quad \kappa = \sqrt{\frac{2m}{\hbar^2}(V_b - E)}, \quad \tan\theta = \frac{\kappa}{k} = \sqrt{\frac{V_b - E}{E}} \tag{6}$$

$$\tan\chi = \tan 2\theta \frac{\cosh(\kappa l)}{\sinh(\kappa l)} \tag{7}$$

For the scatterer, we have

$$t_{34} = t'_{34} = t_s \exp(ikL), \quad r_{34} = r_s \exp(i2kL_1), \quad r'_{34} = r_s \exp(i2kL_2) \tag{8}$$

where t_s and r_s can be obtained in a similar manner to Eqs. (3) and (4). Note that $|t| \ll 1$ and $|r| \simeq 1$ for the two barriers, while $|r_s| \ll 1$ and $|t_s| \simeq 1$ for the scatterer.

In a case with no scattering, Eq. (1) reduces to

$$t_{16} = \frac{|t|^2 \exp[i(kL - 2\chi + \pi)]}{1 - |r|^2 \exp[i2(kL - \chi)]} \tag{9}$$

Substituting Eqs. (3), (4), and (8) into Eq. (1) yields

$$t_{16} = \frac{t^2 t_{34}}{1 - |r|^2 \exp(i2\Delta) + i(2|r||r_s|)\exp(i\Delta)\cos[k(L_1 - L_2)]} \tag{10}$$

where

$$1 - r_{46} t'_{34} r'_{13} t_{34} + r_{34} r'_{13} r_{46} r'_{34} = 1 - |r|^2 \exp(i2\Delta) \tag{11}$$

$$r_{34} r'_{34} + r_{46} r'_{34} = -i2|r||r_s| \exp(i\Delta)\cos[k(L_1 - L_2)] \tag{12}$$

$$\Delta = kL - \chi + \pi/2 - \chi_s \tag{13}$$

For $\cos[k(L_1 - L_2)] = 0$, resonant tunneling occurs under

$$\Delta_r = kL - \chi + \pi/2 - \chi_s = n\pi (n = 0, 1, 2 \ldots)$$

It should be noted here that Eq. (11) has the same form as the denominator of Eq. (9). This means that the amplitude of the Fabry-Perot process (B) is attenuated by the scatterer, but this attenuation is supplemented by an equal amount by process (C).

In order to see the behavior around resonance for the case of $\cos[k(L_1 - L_2)] \neq 0$, we put $\Delta_d = \Delta - \Delta_r$ and expand Eq. (10) with respect to Δ_d. Then, we have

$$t_{16} = \frac{t^2 t_{34}}{|t|^2 + 2i|r|\{|r_s|(-1)^n \cos[k(L_1 - L_2)] - |r|\Delta_d\}} \tag{14}$$

From this equation, we can easily see that resonance transmission occurs under

$$\Delta_d = (-1)^n \frac{|r_s|}{|r|} \cos[k(L_1 - L_2)] \tag{15}$$

This indicates that the resonance energy is shifted by $\Delta E = (\hbar/L)(2E/m)^{1/2}\Delta_d$. It should be noted that this energy shift comes from processes (D) and (E) and depends upon the position of the scatterer in the well. We can also see that the full width remains unchanged in spite of the energy shift. Thus, we can say that, in the 1D structure, elastic scattering in the well shifts the resonance energy depending on the position of the scatter, but has no influence on the transmission peak and the full width at half maximum.

3D Structure

We now turn to the 3D structure. In order to proceed with the discussion in comparison with the 1D case, we add free particle motion in the y and z directions so that the 1D model can be depicted as in Fig. 1b, where the scatterer forms a plane. In the 3D model the scatterer is considered to be a ball, as shown in Fig. 1c. It is rare for an electron scattered by a certain scatterer to be subjected to the same process during its traverse [process (C) in Fig. 2b]. Interference between processes (D) and (E) seldom arises because there are many electron path trajectories in the 3D case. Accordingly, we can neglect terms corresponding to processes (C), (D), and (E). Then, only the term of $1 - r_{46}t'_{34}r'_{13}t_{34}$ corresponding to Fabry-Perot process remains in the denominator of Eq. (1). Using Eqs. (4) and (8), we have

$$t_{16} = \frac{t^2 t_{34}}{1 - r_{46}t'_{34}r'_{13}t_{34}} = \frac{t^2 t_{34}}{1 - |r|^2|t_s|^2\exp(i2\Delta)} \tag{16}$$

Expanding Eq. (16) with respect to Δ_d, we obtain

$$t_{16} = \frac{t^2 t_{34}}{|t|^2 + |r_s|^2 + i2\Delta_d} \tag{17}$$

We confine ourselves to what is called the strong scattering case [5, 6] where $|t|^2 \ll |r_s|^2 \ll 1$, since scattering has little influence on the coherent case where $|r_s|^2 \ll |t|^2 \ll 1$. Then we have

$$T_{Nr} = \frac{|t|^4}{|r_s|^4}, \quad \frac{\Gamma_N}{\Gamma_\infty} = \frac{|r_s|^2}{|t|^2}, \quad \frac{I_N}{I_\infty} = \frac{|t|^2}{|r_s|^2} \tag{18}$$

where suffix N denotes the strong scattering case. Noting the relationships

$$|t_s|^2 = 1 - |r_s|^2 = \exp(-L/l_F) \simeq 1 - L/l_F$$
$$l_F = v\tau = 2NL \tag{19}$$

between the transmission coefficient t_s through the scatterer, the mean free path l_F, the scattering time τ, the electron velocity $v = \sqrt{2E/m}$, and the number N of periodic motions in the well, we have

$$|r_s|^2 = \frac{L}{l_F} = \frac{1}{2N}$$

Substituting this into Eq. (18) and using $\tau_0 = h/\Gamma_\infty$ and Eqs. (17) and (19), we obtain

$$T_{Nr} = \frac{4N^2}{\exp(4\kappa l)} = \left(\frac{\tau}{\tau_0}\right)^2$$

$$\frac{\Gamma_N}{\Gamma_\infty} = \frac{\exp(2\kappa l)}{2N} = \frac{\tau_0}{\tau}$$

$$\frac{I_N}{I_\infty} = \frac{2N}{\exp(2\kappa l)} = \frac{\tau}{\tau_0}$$

It can be seen that, as scattering increases (that is, as τ decreases), the coherent transmission peak decreases by the square of τ, the full width is broadened in proportion to the inverse of τ, and, as a result, the integrated coherent transmission decreases in proportion to τ [9]. These results are the same, except for the prefactors, as the previous ones obtained by path integrals [5, 6]. It should be noted here that the reduction of the transmission peak is caused by the fact that process (C), shown in Fig. 2, rarely takes place in 3D.

A Set of 1D Structures Connected in Parallel

We finally discuss the case of a set of 1D structures connected in parallel, as shown in Fig. 1c. This structure is frequently fabricated in experiments. We assume that there are many segments and that the scatterers are distributed randomly so that the effect of the random distribution can be expressed by

$$T = \frac{1}{L}\int_0^L |t_{16}|^2 dL_1 \qquad (20)$$

This integration can be carried out analytically and gives

$$T_p \propto \left(\frac{\tau}{\tau_0}\right)^{1/2} (N)^{-1/4}$$

$$\frac{\Gamma_N}{\Gamma_\infty} \propto \text{const}$$

$$\frac{I_N}{I_\infty} \propto \left(\frac{\tau}{\tau_0}\right)^{1/2} N^{-1/4}$$

The peak position is shifted by $\Delta E = (\hbar/L)(2E/m)^{1/2}(|r_s|/|r|)$. This behavior can be explained as follows: As described earlier, each scatterer shifts the resonance energy, but does not change the shape of the resonance curve. In integrating the square of Eq. (1) [in Eq. (20)] the integral around $\cos[k(L_1 - L_2)] = 1$ for the ground state and the integral around $\cos[k(L_1 - L_2)] = \pm 1$ for the excited states contribute most effectively to T. This gives an energy shift and one peak for the ground state and two peaks for the excited states. The reduction in T_p comes from the effects of averaging.

References

1. Chang LL, Mendez EE, Tejedor C (eds) (1991) Proceedings of the NATO workshop on resonant tunneling: Physics and applications. Plenum, New York
2. Reed MA, Randall JN, Aggarwal RJ, Matyi RJ, Moore TM, Wetsel AE (1988) Phys Rev Lett 60:535
3. Zohta Y (1989) Jpn J Appl Phys Pt 2 28:L2144
4. Zohta Y, Nakamura K, Ezawa H (1991) Solid State Commun 8:885
5. Zohta Y (1990) Phys Rev B41:7879
6. Zohta Y (1989) Solid State Commun 72:931
7. Zohta Y (1990) Solid State Commun 73:847
8. Freed K (1972) J Chem Phys 56:692
9. Zohta Y and Ezawa H (1992) J Appl Phys to be published on Oct. 15

Observation of Rotating Angle Dependence of Tunneling Spectra in Heavily Doped *n*-Type Silicon

30

KAZUMI SHIIKUMA, TAKAYUKI EZAKI,* HIROYUKI ENOMOTO, AND HAJIME OZAKI[1]

Summary. We observed the rotating angle dependence of tunneling spectra in *n*-type Si, which behavior reflects the anisotropic nature of the Fermi surface of Si. This behavior confirms the conservation of the transverse component of the electron wave vector in tunneling.

Introduction

Tunneling phenomena in solids are now widely applied to the spectroscopic investigation of electronic states. In tunneling spectroscopy, tunneling characteristics have been used mostly in relation to the energy conservation of tunneling electrons. In the tunneling of Bloch electrons, however, the transverse component of the electron wave vector is also conserved. However, there have been few experiments which clearly show such an effect [1, 2]. One reason is that the crystal axes are fixed in a common direction in the entire tunneling system, which is epitaxially grown on a substrate with a fixed crystal axis by, for example, molecular-beam epitaxy (MBE). To observe such an effect, we proposed angle resolved tunneling spectroscopy (ARTS) [1, 3]. In ARTS, the tunneling current between the sample and the probe crystals is measured by rotating the probe crystal over the sample crystal in real space. By rotating the probe crystal, its k-space is also rotated and angle-dependent selective tunneling occurs by virtue of the conservation of the transverse component of the electron wave vector ($k\perp$). Therefore, the tunneling $I-V$ and its derivative characteristics have a strong dependence on the rotating angle, which comes from the change in the overlapping of the projections of the Fermi surface and a constant energy surface of the same energy as the sample and the probe crystals, respectively, on the junction plane.

To observe the above-mentioned $k\perp$ conservation, we perform angle resolved tunneling using single crystals of n-type Si both for the sample and for the probe crystals. The projection of the six constant energy ellipsoids in the conduction

[1] Department of Electrical Engineering, Waseda University, 3-4-1 Ohkubo, Shinjuku-ku, Tokyo, 169 Japan
* Present address: Sony Corporation, 4-14-1 Asahi-cho, Atsugi, 243 Japan

band of Si on the plane perpendicular to the [111] axis has six-fold symmetry. The Fermi surface of n-type Si is confined in a narrow region in the k-space, and is thus convenient for the observation of k-selective tunneling.

Experimental

The sample and the probe crystals were cut from the same heavily As doped Si (111) wafer with an electrical resistivity of 2.2 m$\Omega \cdot$cm at room temperature. After removing native SiO_2 from the surface with HF, a SiO_2 layer about 3 nm thick was thermally grown on the surface at 850°C in a dry Ar/O_2 atmosphere. Ohmic contact was made on the opposite face of the wafer by vacuum evaporation of Au:Sb (Sb:0.5%) alloy, followed by annealing at 240°C for 15 min. Contact resistance was less than 10^{-3} Ω at both room and liq. N_2 temperatures, which was negligible compared with a tunneling resistance of several hundred ohms.

Figure 1 shows the ARTS unit used in the present study. Firstly, the axes within the surfaces of the sample and the probe crystals were set to be parallel by adjusting the orientation flats. We defined this rotating angle as zero degrees. The probe Si was rotated over the sample Si after the former was lifted up so as not to break the thin SiO_2 layer. Then the two crystals were attached gently to fabricate the tunnel junction. Three springs were arranged symmetrically under the sample holder in order to get a stable junction.

Tunneling measurements were carried out by using the conventional modulation technique with fundamental modulation frequency of 1 kHz and modulation bias of 1 mV.

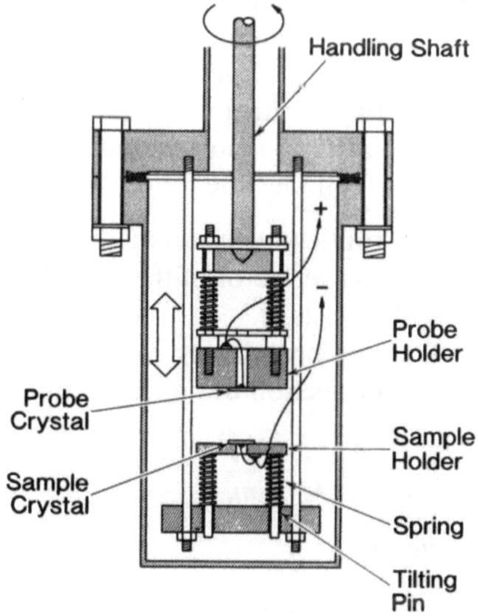

Fig. 1. Tunneling unit for ARTS measurements placed in a cryostat. The probe holder is rotated and lifted up and down through the handling shaft. Tilting of the sample is adjustable so that the two crystals are kept parallel with the tilting pins under the sample holder. The springs under the sample holder make the tunnel junction stable

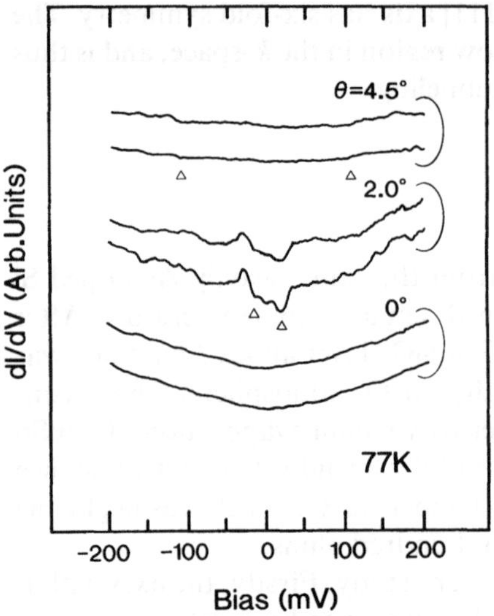

Fig. 2. $dI/dV - V$ characteristics in heavily doped *n*-type Si measured twice at rotating angles of 0°, 2.0°, and 4.5° at 77 K. *Open triangles* represent the critical bias voltages $\pm V_c$

Results and Discussion

Figure 2 shows the $dI/dV - V$ characteristics measured twice at each rotating angle at 77K. Reflecting the change of the overlapping of the projections of the two Si Fermi surfaces on the (111) plane, characteristic kink structures are observed in the tunnel conductance (dI/dV) at the critical bias voltage, $\pm V_C$, which is dependent on the rotating angle θ. The conductance curve at 0° is similar to that of a metal-insulator-metal tunnel junction [4], and no additional structures can be seen. On the other hand, a rapid increase of conductance was observed at the bias voltage of ± 18 mV in curves measured at $\theta = 2.0°$. At $\theta = 4.5°$, the conductance was almost constant up to biases of ± 107 mV, and increased gradually with increasing bias. Experimental plots of V_C as a function of θ are shown in Fig. 3.

To estimate the anisotropic nature of the Fermi surface of Si from this figure, theoretical $V_C - \theta$ curves and tunneling curves were calculated using the Bardeen formula [5] under the following assumptions:

a) The tunneling process is direct (elastic). Thus the total energy and the transverse wave vector of the tunneling electrons are conserved. Tunneling probability is constant.
b) Thermal broadening of the Fermi surface is neglected.
c) The projection out of six Fermi surfaces on (111) plane is expressed as

$$E_F = \frac{\hbar}{2}\left(\frac{k_x^2}{M} + \frac{k_y^2}{m_t}\right),$$

where

$$M = \frac{m_t + 2m_l}{3}$$

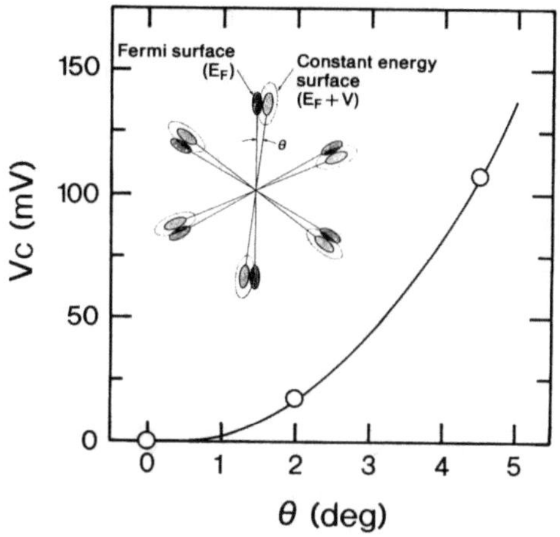

Fig. 3. Theoretical curve of V_c as a function of θ with the fitting parameters of $E_F = 0.80$ mV and $m_t/m_0 = 0.12$. *Open circles* represent the experimental data derived from the $dI/dV - V$ characteristics in Fig. 2. The *inset* is a schematic of the overlapping of the Fermi surface and the constant energy surface with the same energy on the (111) plane. The *dotted line* represents the constant energy surface, with energy equal to E_F of the sample, of the probe crystal. It is expanded by the applied bias voltage and overlaps with the Fermi surface of the sample crystal

Here E_F is the Fermi energy measured from the conduction band edge, $\hbar$ is the reduced Planck's constant, and m_t and m_l are the transverse and the longitudinal effective masses, respectively.

d) The tunneling current that flows between two crystals is proportional to the overlapping of the projections of a constant energy surface, occupied by the electrons, of the sample crystal and the constant energy surface of the same energy of the probe crystal, schematically shown in the inset of Fig. 3. Thus, the electrons, shown in the doubly hatched area in the Fig. 3 inset, can tunnel through the barrier satisfying energy and the $k\perp$ conservation.

As is well known, the effect of heavy doping is to introduce modifications in the band structure of the host crystal [6,7]. However, we neglected this effect, except for the changes in effective mass and Fermi level. Thus, fitting parameters are E_F and m_t in the calculation of the theoretical $V_C - \theta$ curve; m_l is kept constant, as it has little effect on the $V_C - \theta$ curve. It is to be noted that the information regarding the Fermi surface along the $\Gamma - X$ axes is difficult to obtain by using this arrangement, since the probe of ARTS measurements can be rotated only around the $\Gamma - L$ axes for the present. The theoretical $V_C - \theta$ curve best fitted to the experimental plots, which is shown in Fig. 3, is obtained with parameters E_F of 0.80 mV and m_t/m_0 of 0.12, where m_0 is the free-electron mass. The Fermi energy obtained in this study is almost equal to that obtained from the Hall effect measurements [8].

The theoretical $dI/dV - V$ curves using the values obtained from Fig. 3, are shown in Fig. 4. Sudden increases are reproduced in the dI/dV at bias voltages of $\pm V_C$. These behaviors are consistent with the experimental results shown in Fig. 2. Thus, we conclude that the tunnel conductance obtained in this study shows that the angle-dependence originated from the anisotropic nature of the Fermi surface in n-type Si and by virtue of the $k\perp$ conservation of the tunneling electrons.

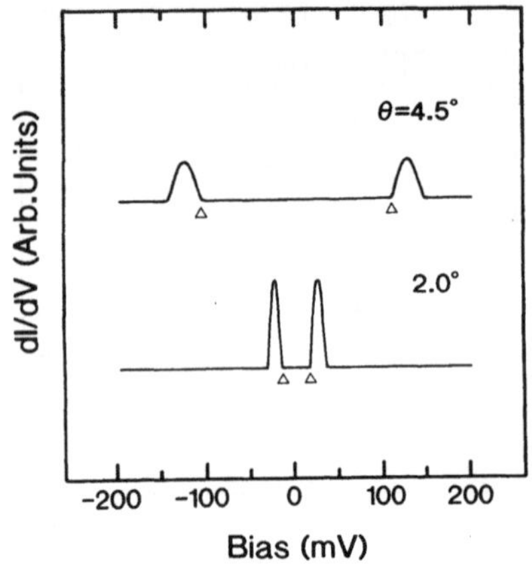

Fig. 4. Calculated $dI/dV - V$ characteristics at $\theta = 2.0°$ and $4.5°$, using the experimental values of $E_F = 0.8$ mV and $m_t/m_0 = 0.12$ derived from Fig. 3. The *open triangles* represent the critical bias voltages $\pm V_c$

Conclusions

We observed the rotating angle dependence of tunneling spectra in *n*-type Si, which behavior reflects the anisotropic nature of the Fermi surface of Si. A characteristic structure was observed in the tunneling curve at the bias voltage which is dependent on the rotating angle. This behavior is explained by conservation of the transverse component of the wave vector in tunneling.

Acknowledgments. We gratefully thank Y. Kato, N. Watanabe, J. Kudoh, and S. Kitazawa for their kind cooperation with regard to the experiments. We also wish to acknowledge K. Shirai of Fujitsu Limited for his useful discussion regarding sample preparation. This work has been supported by a Grant-in-Aid for Scientific Research on Priority Areas from the Ministry of Education, Science and Culture of Japan.

References

1. Enomoto H, Ozaki H, Suzuki M, Fujii T, Yamaguchi M (1991) J Vac Sci Technol B9:1022–1026
2. Demmerle W, Smoliner J, Berthold G, Gornik E, Weimann G, Schlapp W (1991) Phys Rev B44:3090–3104
3. Ezaki T, Enomoto H, Baba M, Ikeda Y, Oguro S, Kuramochi E, Ozaki H (1990) J Vac Sci Technol A8:182–185
4. Giaever I (1969) In: Burstein E, Lundqvist S (eds) Tunneling Phenomena in Solids. Plenum, New York, pp 19–30
5. Bardeen J (1961) Phys Rev Lett 6:57–59
6. Lowney JR (1988) J Appl Phys 64:4544–4578
7. Borghesi A, Stella A, Bottazzi P, Guizzetti G, Reggiani L (1990) J Appl Phys 67:3102–3106
8. Roy DK, Thakur RY (1979) Indian J Phys 53A:588–604

Capacitance in Single Electron Tunneling 31

Michael P. Stopa* and Yasuhiro Tokura[1]

Summary. The charging energy for single electron tunneling (SET) in mesoscopic semiconductor heterostructures is typically parameterized in terms of a single number C, the system "capacitance", as $e^2/2C$. This number, which is sensitively measured by spacings in the current versus gate-voltage oscillations, has not been independently computed, but is rather estimated from the simple geometric capacitances of the charge elements of the structure. We present a systematic study for the computation of the charging energy and capacitances in SET devices at various levels of approximation. First, assuming the capacitive elements of the structure to be perfect conductors and assuming constant contact voltages, we compute the charging energy and activation energy for SET. We show that the former energy is determined by the dot to gate capacity and the latter by the dot self-capacity. We calculate the screening effect of the contacts by Thomas-Fermi approximation. We generalize the notion of capacitance in order to treat the case of semiconductors and we use a self-consistent electronic structure calculation to investigate the "local capacities" of the quantum dot. Finally, we use the self-consistent energies from this calculation along with computed energy-dependent tunneling rates to compute the current-gate voltage characteristics of a realistic device. We compare the resonance spacings and activation energy with that determined from the simpler theory, as well as with that determined by experiment.

Introduction

Capacitance in an electrical system is strictly defined only for conductors, whose electrostatic potential can be specified by a single number, the voltage. Recent studies of Coulomb charging effect in single electron tunneling (SET) in mesoscopic semiconductor systems [1–3] have described the change in the electrostatic energy when a single electron is added to a quantum dot by introducing a phenomenological capacitance C, written, $\Delta E = \frac{e^2}{2C}$. In the canonical experiment a gate is employed to vary the dot electrostatic potential. Conservation of energy for

[1] NTT Basic Research Laboratories, 3-9-11, Midori-Cho, Musashino, Tokyo, 180 Japan
* Current address: The Institute of Physical and Chemical Research, Hirosawa 2-1, Wako, Saitama, 351-01 Japan

tunneling into and out of the dot in the low source-drain bias limit is then satisfied only at discrete "resonant" gate voltages, V_g^n. The spacing of these voltages is observed to be periodic. It has recently been realized [4] that the description of this periodicity requires at least one additional capacitance C_{gate}, between gate and dot. To date, these and other capacitances in the SET system have been either derived from the experimental data, specifically taking the resonance periodicity as $\Delta V_i = e/C_i$ as the various gate voltages are varied, or they have been estimated from elementary geometrical arguments.

It is the purpose of this paper (1) to develop the classical theory of Coulomb charging energy in a system with several capacitive elements, (2) to generalize the notion of capacitance to a semiconductor quantum dot system, (3) employing self-consistent (SC) electronic structure calculations, to compute the dot capacities in the static limit, (4) to modify a formula (Beenakker [5]) for SET conductance in the linear source-drain bias limit to the case of a self-consistently derived energy spectrum, (5) to employ a transfer matrix formalism to compute the energy dependence (although not the absolute values) of the the tunneling coefficients, and (6) by computing the current-gate voltage characteristics for a realistic device, to compare the resonance spacing and activation energy with that determined from the simpler theory as well as with that determined by experiment.

Classical Theory

We consider an idealized system (inset in Fig. 1) consisting of a perfect metal source (1), a drain (2), two gates (3 and 4), and a dot (5). The linearity of Poisson's equation in the charges Q_i and the voltages V_j of the elements implies:

$$Q_i = \sum_j C_{ij} V_j \tag{1}$$

The diagonal elements C_{ii} (which are positive) are self-capaci*ties* and the off-diagonal elements (which are all negative) are mutual inductances (see, for example, Ref. [6]) (collectively we will refer to them as capacities). Mutual capacit*ances* $C(\text{ij})$ are defined as the ratio of the change in the ith charge to the change in the ijth voltage difference when all other voltages are held constant. The symmetry of the C_{ij} and the arbitrariness of the voltage zero reduce the N^2 capacity components to $N(N-1)/2$ independent elements, (as with capaci*tances*). Also, $\sum_i C_{ij} = 0, \forall j$.

Spacing of the SET resonances with gate voltage is deduced as follows. Ignoring the contact/gate recharging time (screening time), V_i for $i = 1$ to 4 is the same before and after tunneling. One way to compute the charging energy due to tunneling is to evaluate the total electrostatic energy change, $\frac{1}{2}\sum_{i,j=1}^{5} C_{ij}(V_i(N+1)V_j(N+1) - V_i(N)V_j(N))$, where N is the number of electrons in the dot, and subtract the energy supplied by the power supplies due to change in the charges on the contacts, $\sum_{i=1}^{4} \Delta Q_i V_i$. We then use the resulting expression in the energy balance equation, including the bare level energies to determine the resonance condition. A

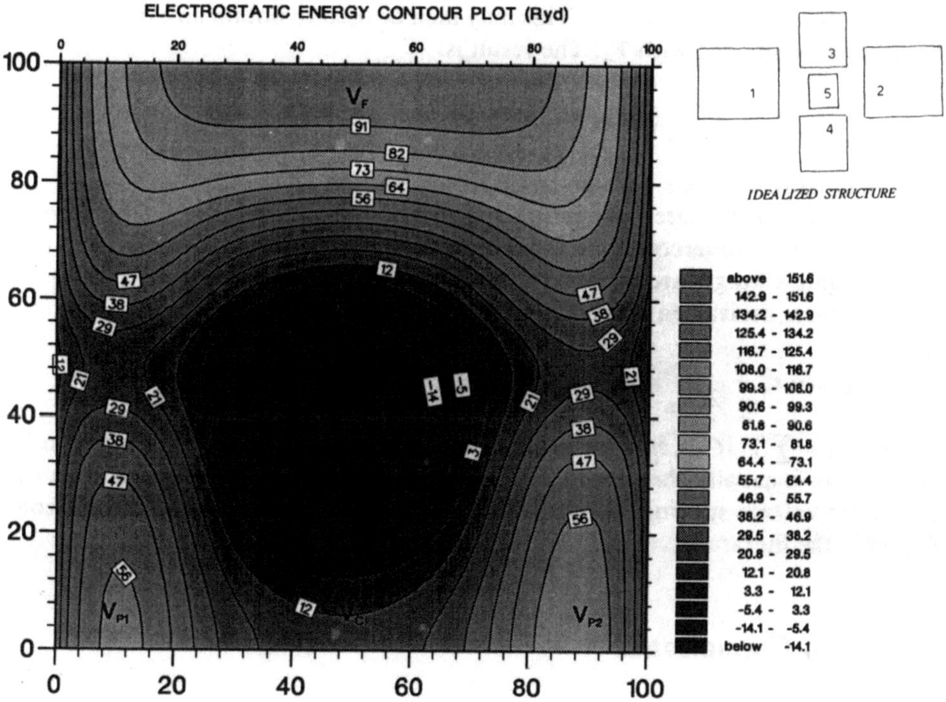

Fig. 1. Electrostatic potential contour plot for the SET quantum dot under study. Cross section at the 2 dimensional electron gas ($2DEG$) level. Contours are labelled in (effective) Ryd. Axis units are in (effective) Bohrs ($1Bohr \approx 100\ \mathring{A}$). The approximate positions of the four gates ($10Bohr$ above the 2DEG) are labelled. Positive dopant density is $2 \times 10^{11} cm^{-2}$ in a delta layer 200 $\mathring{A}$ above the 2DEG. Upper gate and point contact gates are at 1 eV. "Center" gate, V_C is at 0.45 eV. *Inset* is an idealized resonant tunneling structure composed of five elements: source and drain (1) and (2), two gates (3) and (4), and dot (5)

clearer approach is to equate the chemical potentials of the dot and the contacts. Writing the total energy of the dot at fixed N as $E_T(N) = \sum_{p=1}^{N} \varepsilon_p^0 + \frac{N}{2} e V_5(N)$, where ε_p^0 is the energy of the pth bare state, the chemical potential (at zero temperature) is simply:

$$\mu_{DOT} = \varepsilon_{N+1}^0 + \frac{N+1}{2} V_5(N+1) - \frac{N}{2} V_5(N) \tag{2}$$

We may use Eq. 1 to write $V_5(N) = \frac{1}{C_{55}}(Q_5(N) - \sum_{i \neq 5} C_{5i} V_i)$. Ignoring the level spacing in the contact we have the resonance tunneling condition as a function of, say, V_4 (which we now allow to vary) as $\mu_{DOT}(V_4, N) = E_F$, where E_F is the Fermi energy in the contact. Assuming the Fermi level in the contact changes little with gate voltage, the dot chemical potential at two succesive resonances must be equal,

$\mu_{DOT}(V_4, N) = \mu_{DOT}(V_4', N+1)$. This may then be used to obtain an expression for the resonance spacing with V_4. The result is:

$$e(V_4' - V_4) = \frac{e^2}{C_{45}} + \frac{C_{55}}{C_{45}}(\varepsilon_{N+1}^0 - \varepsilon_N^0) \tag{3}$$

A similar expression has been given by Kouwenhoven [7].

In contrast to the gate voltage spacing, the activation energy, i.e., the thermal energy needed to overcome the Coulomb barrier in the off-resonance condition when *all* gate voltages are fixed, is naturally related only to the self-capacity of the dot. A similar derivation yields:

$$E_{act} = \varepsilon_{N+1}^0 - \varepsilon_N^0 + \frac{e^2}{2C_{55}} \tag{4}$$

Since $C_{55} = \sum_{i=1}^{4} |C_{5i}|$, it is *manifestly* clear that when bare level spacings can be ignored (as is usually the case in SET) the activation energy will be less than the resonance voltage spacing, although this point seems to have provided some confusion in the literature.

Quantum Capacitance

For any object with a screening length comparable to its dimensions and, in particular, for a quantum dot, the voltage is not defined as a single number. Therefore the definition of capacity must be modified. The most general modification we can make is to define the capacity matrix as the kernel in an integral relation between the charge (density) of the ith object at point x and the voltage of the jth object at point $\mathbf{x}'$. In a semiconductor system however, the SC potential typically varies fairly smoothly and, due to screening, the influence of the voltage at one point on the charge density at a point more than a screening wavelength away is small. Hence for the self-capacity we write (for the dot in particular):

$$C_{55}(\mathbf{x}) \equiv \frac{\partial Q_5(\mathbf{x})}{\partial V_5(\mathbf{x})} \tag{5}$$

In the quantum dot system we are investigating, the principal other capacities affecting the dot come from metal gates whose voltage is well defined. Thus, these capacities are written as $C_{5j}(\mathbf{x}) \equiv \frac{\partial Q_5(\mathbf{x})}{\partial V_j}$. We will take the capacities which do not involve the dot to be the standard metallic ones (although we have not evaluated them). Note that the distinction between the i,j index and the spatial dependence is a matter of definition. Since all the capacitive elements are part of the same solid, the index merely designates an arbitrarily defined region of space. Given these local capacities of the dot, we integrate over the relevant region, in this case the dot, to obtain the total capacities.

Thus, for a quantum mechanical system, the capacity (or indeed the capacitance) measures the local self-consistent change in the charge distribution responding to a change in either an external potential or to a change in the local potential in the case where the total charge is varied.

For the contacts, we have done a separate calculation in the Thomas-Fermi approximation to estimate the screening by the charge which builds up in response to a change in the dot charge. We have written

$$\frac{1}{C_{55}} = \frac{1}{C_{55}^{0}} - \frac{1}{C_{55}^{sc}} \tag{6}$$

where here C_{55}^{sc} measures the change in the voltage at the dot brought about by a screening charge which the dot itself induces in the contact. We have determined for our geometry that $C_{55}^{sc}/C_{55}^{0} \sim 20$.

To calculate the capacities and the current-gate voltage characteristics we performed a SC electronic structure calculation for a resonant tunneling device of *proportions* equal to those described by Kouwenhoven [7] (our model is 30% smaller, however). The details of this calculation are similar to others in the literature [8]. Here we will note only one distinction, which has enabled us to perform fully quantum mechanical calculations for large numbers of electrons in the dot. Rather than solve Schrödinger's equation on a three dimensional grid, we expanded the Hartree potential in the direction perpendicular to the interface (z-direction) about the interface (at each point in x,y). We use the z-subband energies as an effective potential in x,y. In this adiabatic approximation we ignore gradients in the z-wavefunction with changes in x,y. We believe the approximation is quite good, but we plan to examine coupling terms between different z-subbands in the future.

For Poisson's equation we treated the gate potentials and the contacts via Dirichlet boundary conditions and used vanishing normal derivative elsewhere. We treated the electron charge density as though it were confined in a uniform layer 2 (effective) Bohr ≈ 200 Å thick. This possibly underestimates the electric repulsion, but the level spacings should be unaffected.

We show an example of the SC electrostatic potential in the 2DEG plane in Fig. 1. The point contacts and the gate configuration, 1000 Å above the 2DEG level, are labelled. In this device typically gate V_C ("center") is varied to change the electron number in the dot. The point contact gates are adjusted to pinch off the channel while still allowing significant tunneling. For typical N the charge distribution occupies predominantly the circularly symmetric region near the center.

In Fig. 2 we show the local self-capacity for the dot. The jagged nature of the figure is a result of the necessarily coarse grid. For a perfect *metal* the local capacities are zero everywhere but at the surfaces, where excess charge resides. For the *dot* we see that the capacity peaks at the perimeter, which expands as N is increased. Nonetheless, the capacity near the center is non-zero, reflecting in part the structure of the most recently filling wave function.

In the inset of Fig. 2 we show the *total* capacities C_{55} (top) and C_{54} as a function of N. Note that we cannot evaluate C_{54} at fixed N (it is zero). Thus, here we have written

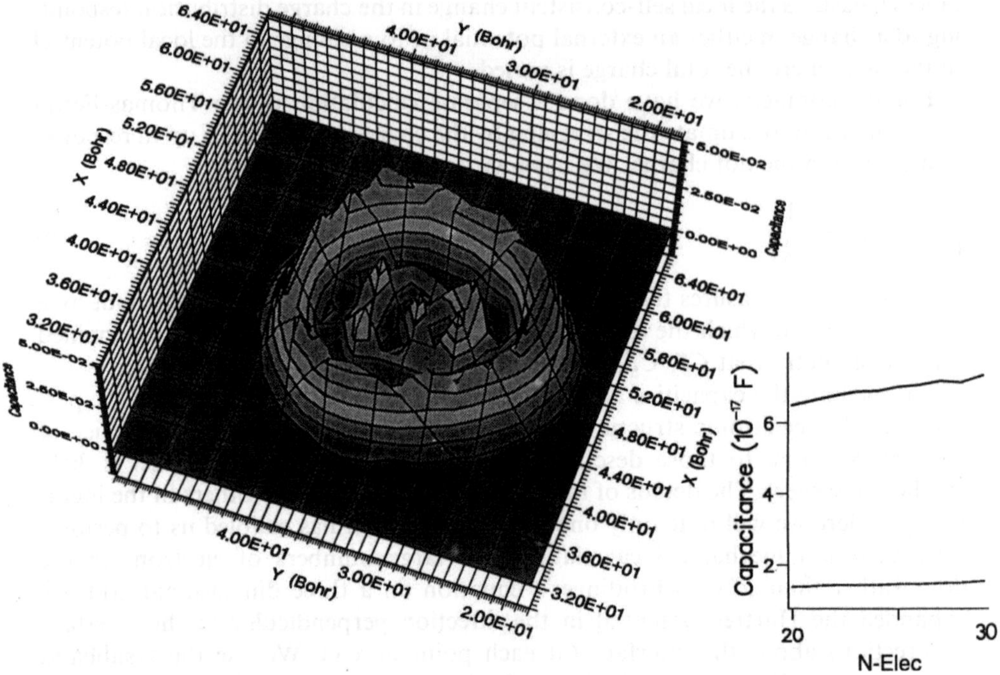

Fig. 2. Local self-capacity for a dot of width $\approx 1\mu m$ and 25 electrons. Capacitance units are arbitrary. *Inset*: total self-capacity C_{55} (*top*) and total "center" gate to dot capacity C_{54} as a function of electron number in the dot

$$C_{54} = \int dxdy \frac{\partial Q_{55}(x,y)}{\partial V_4} \tag{7}$$

$$= \int dxdy \frac{\partial Q_{55}(x,y)}{\partial V_5(x,y)} \frac{\partial V_5(x,y)}{\partial V_4} \tag{8}$$

$$= C_{55} \frac{\partial V_5^m}{\partial V_4} \tag{9}$$

where in the last step we approximate the local change in V_5 by the change in V_5^m, the minimum of the potential.

From the Fig. 2 inset it is clear that the capacity of the gate is only about 20% of the self-capacity. Further, C_{55} increases with N and hence with the dot area, while C_{54} is relatively constant. Naively, since charge is added principally to the perimeter, we would expect C_{55} to scale with the radius, although the current data is inconclusive. As a specific experimental prediction, since C_{55} governs the activation energy, it is clear that this quantity must decrease as electrons are added to the dot. Finally, from Eqs. 3 and 4, the average values of these capacities can be used to compute the resonance spacing, $\Delta V_4 \approx 0.023eV$, and the activation energy,

$E_{act} \approx 0.43 Ryd \approx 2.5 meV$ (1(*effective*)$Ryd \approx 5.83 meV$). We have assumed here a bare level spacing of 0.2 meV.

Self-Consistent Charging Energy

Based on a detailed balance argument, Beenakker [5] has derived the following formula for SET conductance in the low source-drain bias limit.

$$G(V) = \frac{e^2}{kT} \sum_p \sum_{\{n_i\}} \frac{\Gamma_p^l \Gamma_p^r}{\Gamma_p^l + \Gamma_p^r} P_{eq}(\{n_i\}) \delta_{n_p,0} f(E_p + U(N+1) - U(N) - E_F) \tag{10}$$

Where $\Gamma_p^{l(r)}$ is the tunneling coefficient from the left (right) contact to state p in the dot, $P_{eq}(\{n_i\})$ is the equilibrium occupation probability distribution for dot level occupancies $\{n_i\}$ in the grand canonical ensemble, $U(N)$ is the classical charging energy for N electrons in the dot, E_p are the bare dot levels, E_F is the contact Fermi energy, and f is the Fermi function. In a self-consistent formulation the charging energy is included in the levels ε_p and conservation of energy for tunneling to level p in the dot is given by:

$$E^{l(r)} + eV_1 = \frac{\partial}{\partial N} \sum_{p'}^{occupied} \varepsilon_{p'}(N, V_4) + \varepsilon_p(N+1, V_4) \tag{11}$$

(cf. Eq. 2). Note that the chemical potential of the dot is very different from the effective Fermi energy (highest filled level at $T = 0$) of the dot. When an electron is added to the dot, *all* the levels increase. It is the sum of all these level changes (plus the new level) which equals the initial energy. As a consequence *the effective Fermi level of the dot is far below the Fermi level of the contact* at resonance.

We have estimated the tunneling coefficients as follows. We take the *unperturbed* Hamiltonian for the contact to be a wide ($\approx 1 \mu m$) rectangular region of constant potential extending to infinity on the left, connected via a narrow rectangular channel to another wide region, of higher constant potential, extending to infinity on the right. Expanding in subbands in the three regions, we use a transfer matrix method to compute the decaying wavefunction on the right resulting from an incident wave in a given channel on the left. We compute the matrix element squared of: (the difference between the dot potential profile and the aforementioned unperturbed potential) between: (the righthand decaying state) and (the SC dot states). We sum over incident channels at fixed energy to obtain the tunneling rates. Using point contact parameters from the SC potential we find tunneling rates whose absolute magnitudes are ten orders of magnitude smaller than the dot level spacing. This is puzzlingly minute. In our calculation of $G(V)$ we normalized to the largest tunneling coefficient.

Our results for the conductance for a series of temperatures are shown in Fig. 3. The leftmost peak is affected at higher temperatures by the edge of our energy range sample. Earlier calculations based on a simplistic model [9] have examined the temperature dependence in the peaks as compared with the minima. Two

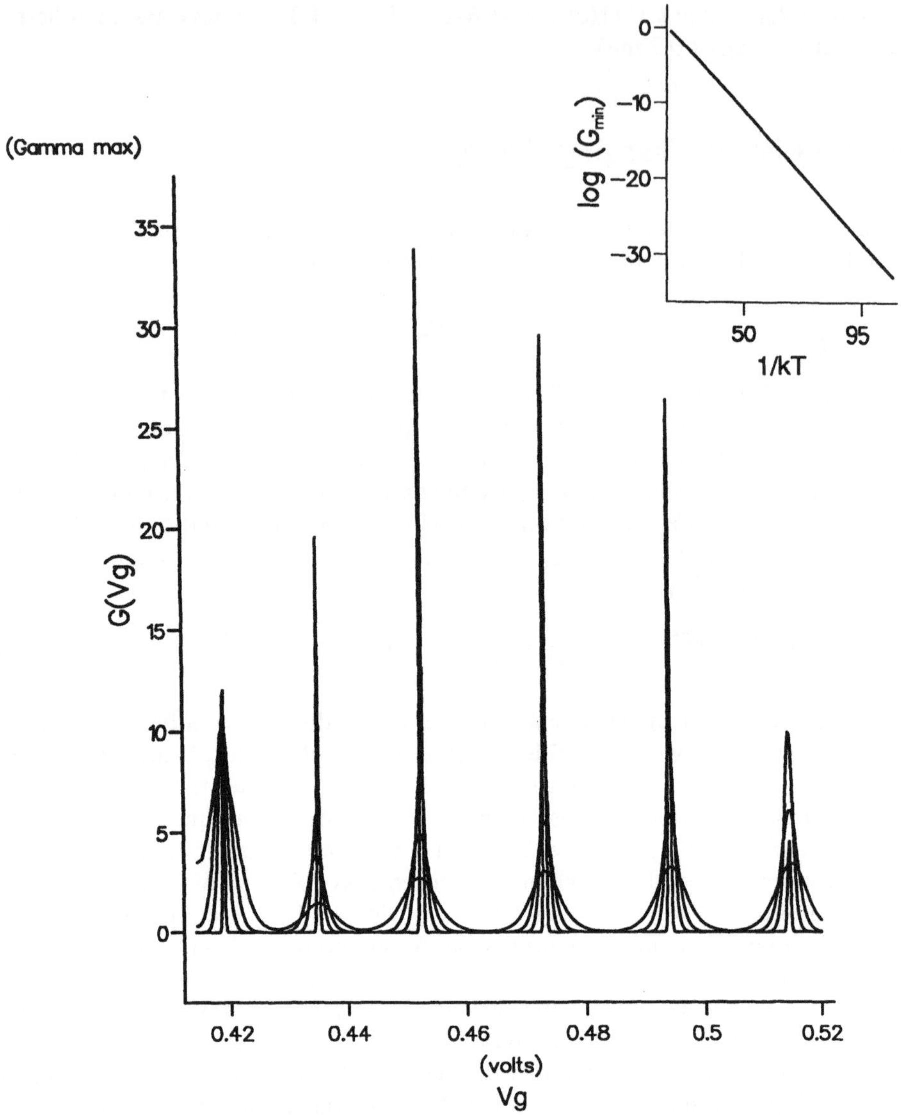

Fig. 3. Calculated conductance as a function of the center gate voltage of Fig. 1. Separate curves are for different temperatures, $T = 0.2$, 1.0, 2.0, and 4.0K. At the peaks the conductance increases with decreasing temperature, whereas the reverse is true at the minima. *Inset* shows log of the conductance at the minimum near 0.46 eV as a function of $1/kT$ in Ryd^{-1}. Conductance is normalized to value at $T = 4K$. The negative of the slope is the activation energy

features of our results are, however, completely new. First, the envelope structure of the peaks has frequently been observed experimentally. A closer examination of the calculation reveals that this structure arises from the geometric near-degeneracies of certain levels (rather than, say, random tunneling coefficient values as suggested by earlier researchers [7]). Second, the inset shows the log of the conductance in the minimum near 0.46 eV versus $1/kT$. The resulting activation energy (the slope) is, of course, different from the resonance spacing. This cannot in principle be derived from a calculation that treats the charging energy with a single capacitance parameter. The value we deduce for the activation energy is $E_{act} \approx 0.32Ryd \approx 1.9meV$, or about 25% smaller than that determined by the classical theory. Although the results are fairly close, the discrepancy is still somewhat outside the limits of our numerical accuracy. The resonance spacing, on the other hand, is $\approx 0.021eV$, which is in excellent agreement with our value determined from the capacities.

Conclusion

In conclusion, we have examined the nature of capacitance in quantum mechanical systems and in particular, in SET quantum dots. We introduced the term "local capacitance" to describe electrostatic effects in such a system. We described the charging energy in SET in terms of capacities, rather than in terms of the more traditional capacitances, and we showed that the resonance spacings in SET proceeded from the gate-dot capacity, while the activation energy was principally related to the self-capacity. Since the self-capacity is necessarily larger than that from the gate, this implies that the activation energy must always be smaller than the resonance spacing energy. Using the results from our self-consistent electronic structure calculation for such a structure, we evaluated these capacities (by integrating over the local capacities) and deduced E_{act} and the resonance spacing. We modified Beenakker's formula for use with self-consistent energy levels and, calculating tunneling coefficients, computed the conductance versus gate voltage of the device. We have shown that the resonance spacing and activation energy thus determined are in good agreement with our classically computed values.

Acknowledgments. The authors happily thank M. Ueda and A. Kumar for their helpful discussions. We also thank S. Tarucha, Y. Horikoshi, and T. Kimura for their continuing support.

References

1. Meirav U, Kastner MA, Wind SJ (1990) Phys Rev Lett 65:771
2. Kouwenhoven LP, Johnson AT, van der Vaart NC, Harmans CJPM, Foxon CT (1991) Phys Rev Lett 67:1630
3. Groshev A, Ivanov T, Valtchinov V (1991) Phys Rev Lett 66:1082

4. van Houten H, Beenakker CWJ, Staring AAM (1991) In: Grabert H, Devoret MH (eds) Single charge tunneling, NATO ASI Series B. Plenum, New York
5. Beenakker CWJ (1991) Phys Rev B44:1646
6. Jackson JD (1975) Classical electrodynamics. Wiley, New York
7. Kouwenhoven LP, van der Vaart NC, Johnson AT, Harmans CJPM, Williamson JG, Starting AAM, Foxon CT (1991) Proceedings of the German Physical Society Meeting. Festkörperprobleme/Advances in Solid State Physics, vol 31. 1991, Münster
8. Kumar A, Laux S, Stern F (1990) Phys Rev B42:5166
9. Meir Y, Winegreen N, Lee P (1991) Phys Rev Lett 66:3048

Many-Body Effects in Mesoscopic Double Junctions

32

MASAHITO UEDA[1] AND FRANCISCO GUINEA[2]

Summary. We predict novel many-body effects in mesoscopic double junctions which consist of a scanning-tunneling-microscope (STM) tip, a metallic micro-cluster, and a substrate. A tunneling event from the tip to the micro-cluster is shown to cause an infrared-divergent creation of electron-hole pairs in the substrate, thus making current and voltage steps outside the Coulomb gap unequal. It is also predicted that higher-order tunneling leads to a leakage current inside the Coulomb gap where tunneling is inhibited within the framework of the orthodox Coulomb blockade theory.

Introduction

Charging effects in small-capacitance tunnel junctions have opened up the possibility of manipulating electric charge at the level of single electrons. Possible applications for current standard and supersensitive electrometry have stimulated great experimental efforts with a rich variety of materials and configurations, and have motivated theoreticians to tackle controversial issues such as how far a tunneling electron probes its electrodynamic environment (for a collection of recent reviews, see [1]). While most experiments performed to date may well be described by the so-called orthodox Coulomb blockade theory [2], various efforts are now being made to extend this theory beyond simple one-body effects [1].

In this paper, we predict novel many-body effects in micro-cluster double junctions which consist of an STM tip/vacuum barrier/micro-cluster junction and a microcluster/oxide barrier/substrate junction, as illustrated schematically in Fig. 1. When an STM tip is negatively biased, an electron tunnels from the tip to the micro-cluster. Since the traversal time for tunneling is negligible for normal metal junctions, conduction electrons in the substrate see the charging process in the cluster as a sudden perturbation. If the perturbation may be regarded as local and static, then it causes the infrared-divergent creation of electron-hole pairs in the substrate [3], and thus the spectrum of subsequent tunneling from the micro-cluster to the substrate is expected to be renormalized by this shake-up.

[1] NTT Basic Research Laboratories, Midori-cho 3-9-11, Musashino, Tokyo, 180 Japan
[2] Instituto de Ciencia de Materiales (sede B), Consejo Superior de Investigaciones Científicas, Cantoblanco, E-28049 Madrid, Spain

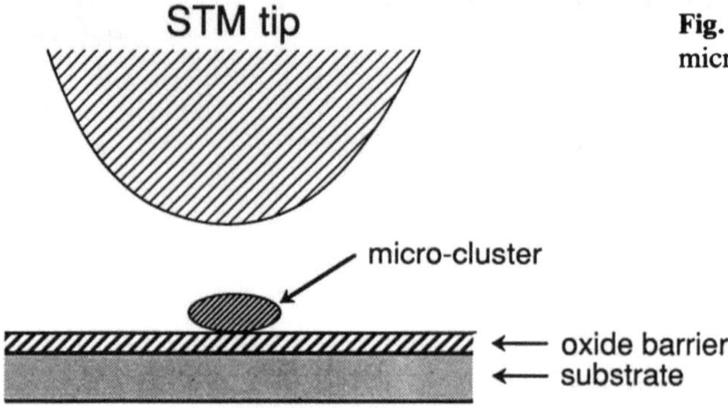

Fig. 1. Schematic illustration of a microcluster double junction

It is well-known that a high mobility of the charge disturbance smears out the singularities due to electron-hole pairs [4]. Also, the spontaneous fluctuations of junction charge due to the electromagnetic environment wash out the change in junction charge carried by tunneling [5–7]. By using a double junction configuration, we can circumvent these difficulties.

For the perturbation to be local on an atomic scale, the middle electrode should be as small as 10 Å. The other requirement is that the perturbation should be static. A finite temperature gives a cut-off time during which electron-hole pair excitation can affect the tunneling current spectrum. For the localized disturbance in the micro-cluster to be static, its lifetime, which is on the order of C_2R_2, must be larger than this cut-off. Hence, we have the following requirement:

$$\frac{\hbar}{k_B T} < C_2 R_2 \tag{1}$$

This condition is well satisfied for a typical experimental condition, where $k_B T \sim 1K$ and $C_2R_2 \sim 10^{-9}$ s. Therefore the localized disturbance caused by tunneling can be regarded as static within a cut-off time imposed by a finite temperature.

We will present some novel many-body effects associated with this shake-up. Our results are two-fold:

(1) According to the orthodox theory, the current-voltage characteristic of a voltage-biased double junction exhibits, under appropriate conditions, the Coulomb staircase with equal current and voltage steps. On the other hand, the shake-up makes current steps unequal.

(2) According to the orthodox theory, electron tunneling is inhibited inside the Coulomb gap. We show, however, that higher-order tunneling leads to a leakage current inside the Coulomb gap [8].

Orthodox Theory of Double Junctions

Before discussing the effects of electron-hole pairs, let us briefly summarize the orthodox theory for the current-voltage (I-V) characteristic of two normal tunnel junctions coupled in series [2].

Suppose that two normal tunnel junctions with capacitances C_i and tunnel resistances R_i (i = 1, 2) are driven by a voltage source V_{ex}. A tunneling event is allowed to occur if the free energy of the total system is lowered by it. Now suppose that n_1 and n_2 electrons tunneled through junctions 1 and 2, respectively, and that the amount of charge q was supplied from the voltage source to electrically equilibrate the double junctions. A simple circuit-theory calculation shows that the voltage across each junction is given by

$$V_1 = \frac{C_2}{C_1 + C_2} V_{ex} - \frac{e}{C_1 + C_2}(n_1 - n_2) \tag{2}$$

$$V_2 = \frac{C_1}{C_1 + C_2} V_{ex} + \frac{e}{C_1 + C_2}(n_1 - n_2) \tag{3}$$

The total free energy of the system $F(n_1, n_2)$ is therefore given by

$$\begin{aligned} F(n_1, n_2) &= \frac{1}{2} C_1 V_1^2 + \frac{1}{2} C_2 V_2^2 - qV_{ex} \\ &= \frac{e^2}{C_1 + C_2}\left[\frac{1}{2}(n_1 - n_2)^2 - n_1 \frac{C_2 V_{ex}}{e} - n_2 \frac{C_1 V_{ex}}{e}\right] + \text{constant} \end{aligned} \tag{4}$$

For the initial condition $n_1 = n_2 = n$, an electron can tunnel through junction 1 if the following condition is met:

$$F(n + 1, n) - F(n, n) < 0 \tag{5}$$

Substituting Eq. (4) into (5) yields the threshold voltage for junction 1.

$$V_{ex}^{th,1} = \frac{e}{2C_2} \tag{6}$$

Below this threshold, electron tunneling through junction 1 is inhibited due to Coulomb blockade. A similar calculation gives the threshold voltage for junction 2 as

$$V_{ex}^{th,2} = \frac{e}{2C_1} \tag{7}$$

For $C_1 \ll C_2$, Coulomb blockade is first lifted in junction 1 when the bias voltage exceeds $V_{ex}^{th,1}$. If an electron tunnels through junction 1, the subsequent tunneling through junction 2 becomes possible because then the inequality

$$F(n + 1, n + 1) - F(n + 1, n) < 0 \tag{8}$$

is automatically satisfied.

If $R_1 \ll R_2$, the middle electrode almost always has an excess charge corresponding to exactly one electron, since when this excess charge is released by tunneling through junction 2, another electron will immediately tunnel through junction 1. Assuming $k_B T \ll e^2/2C$, the tunneling current is thus given by

$$I(V_{ex}) = \frac{e}{2R_2(C_1 + C_2)}\left(1 + \frac{C_1 \Delta V_{ex}}{e}\right) \tag{9}$$

where $\Delta V_{ex} \equiv V_{ex} - V_{ex}^{th,1}$.

A middle electrode can have two excess electrons when V_{ex} is further increased to reach

$$V_{ex}^{th,2} = \frac{3e}{2C_2} \tag{10}$$

because above this point we have $F(n+2,n) - F(n+1,n) < 0$. In general, it can be shown that n electrons can stay in the middle electrode if $(n-1/2)e/C_2 < V_{ex} < (n+1/2)e/C_2$. This result, combined with Eq. (9), shows that the tunneling current is kept almost constant between voltage thresholds (that is, the I-V curve exhibits current plateaus) if $C_1 \ll C_2$. Thus we find that the I-V curve shows the so-called Coulomb staircase. The solid lines in Fig. 2 illustrate a typical Coulomb staircase for $C_1 \ll C_2$ and $R_1 \ll R_2$.

Conductance Anomalies Outside the Coulomb Gap

Now we return to the micro-cluster double junctions in Fig. 1. The only distinction from conventional double junctions is that the tunneling rate between the micro-cluster and the substrate should be replaced by [3]

$$I(V) = A\left(\frac{CV}{e} - \frac{1}{2}\right)^{1+g(V)} \theta\left(Q - \frac{e}{2}\right) \tag{11}$$

where $A = (\alpha^R \alpha^L / CR_T)(e^2/C\sqrt{\xi_0^R \xi_0^L})^g$. Since the infrared catastrophe occurs only in one electrode (substrate), the electron-hole symmetry does not hold, and hence the first-order phase shift remains in the anomalous power exponent,

$$g = \frac{\delta_R}{\pi} - \left(\frac{\delta_R}{\pi}\right)^2 \tag{12}$$

Thus we find that for the positive (negative) phase shift, the tunneling rate is suppressed (enhanced) in the first nonvanishing current plateau, but tends to be more and more enhanced (suppressed) in the following ones, as illustrated schematically in Fig. 2, where dashed and dotted-and-dashed lines correspond to positive and negative phase shifts.

Conductance Anomalies Inside the Coulomb Gap

According to the orthodox theory, tunneling is inhibited as long as the bias voltage falls inside the Coulomb gap. We will show, however, that a leakage current always exists inside the Coulomb gap because of higher-order tunneling [8].

In second-order perturbation theory, only one pair of virtual tunneling events foward and backward of lifetime $\hbar$/(the charging energy E_C) is allowed. Inside the Coulomb gap, this process does not contribute to the current-current correlation function, and therefore no leakage current exists. The orthodox theory is based on such lowest-order perturbation theory.

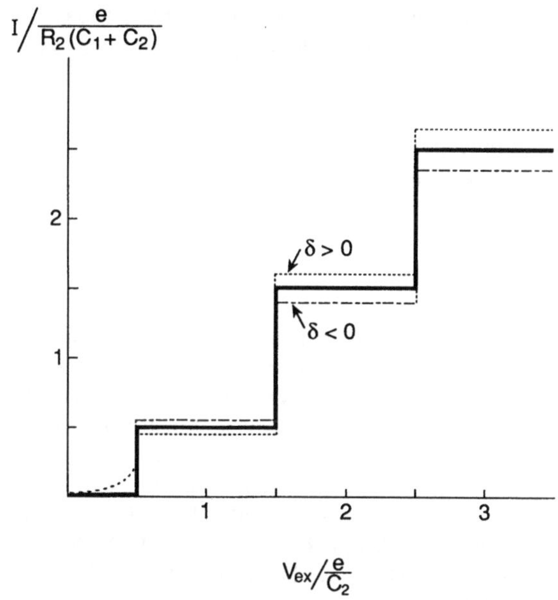

Fig. 2. Current-voltage characteristic of a voltage-biased double junctions, where *solid*, *dotted*, and *dotted-and-dashed* lines respectively correspond to zero (orthodox limit), positive, and negative phase shifts. A broken curve shows a leakage current due to higher-order tunneling

When the junction resistance is larger than the quantum resistance, higher-order tunneling can usually be neglected compared to second-order tunneling. Inside the Coulomb gap, however, the second-order contribution vanishes, and, therefore, fourth-order tunneling makes a dominant contribution.

The path which contributes most to the current-current correlation function is the one in which the system will be in the ground state for most of the time, with two excursions into the excited states for a short time $\hbar/E_C$. To evaluate the path, we can use the representation where kinks interact logarithmically with each other. The interaction between these charge dipoles therefore decays as t^{-4}. Taking the Fourier transform, we obtain a finite conductance at voltages inside the Coulomb gap:

$$\sigma(\omega) \sim \frac{e^2}{h} t^4 N_R(E_F)^2 N_L(E_F)^2 \left(\frac{\omega}{E_C}\right)^2 \sim \frac{e^2}{h}\left(\frac{R_Q}{R_T}\right)^2 \left(\frac{\omega}{E_C}\right)^2 \tag{13}$$

where $N_R(E_F)$ and $N_L(E_F)$ denote the density of states at the Fermi level in the right and left electrodes, respectively. This implies that a leakage current exists inside the Coulomb gap, whose magnitude is proportional to the third power of the voltage, as schematically shown in Fig. 2 (broken curve inside the Coulomb gap).

Conclusions

In this paper we predicted novel many-body effects in mesoscopic double junctions. An electron tunneling from an STM tip to a micro-cluster causes an infrared-divergent creation of electron-hole pairs in the substrate, thus renormalizing the tunneling current spectrum outside the Coulomb gap. Inside the Coulomb gap, we have shown that a leakage current always exists due to higher-order tunneling.

Acknowledgment. Author (M.U.) would like to thank Professor H. Fukuyama for his stimulating discussions.

References

1. Devoret M, Grabert H (eds) (1992) Proceedings of NATO Advanced Institute Study on Single Charge Tunneling. Plenum, New York
2. Averin DV, Likharev KK (1991) In: Altshuler BL, Lee PA, Webb RA (eds) Mesoscopic phenomena in solids. Elsevier, Amsterdam, p 173
3. Ueda M, Kurihara S (1991) In: Clark TD, Prance H, Prance RJ, Spiller TP (eds) Macroscopic quantum phenomena. World Scientific, Singapore, p 143
4. Müller-Hartmann E, Ramakrishnan TV, Toulouse G (1971) Phys Rev B3:1102
5. Nazarov Yu V (1989) Sov Phys JETP 68:561
6. Devoret MH, Esteve D, Grabert H, Ingold G-L, Pothier H, Urbina C (1990) Phys Rev Lett 64:1824
7. Girvin SM, Glazman LI, Jonson M, Penn DR, Stiles MD (1990) Phys Rev Lett 64:3183
8. Ueda M, Guinea F (1991) Z Phys B Condensed Matter 85:413

Anomalous Switching-Current Distributions to the Voltage State in a Small Josephson Junction

33

NORIYUKI HATAKENAKA AND SUSUMU KURIHARA[1]

Summary. Switching-current distribution for supercurrent-state decay is investigated in small Josephson junctions at zero temperature. Due to resonant features of the tunneling process reflecting the existence of quantized energy levels in Josephson potential, decay to the voltage state is enhanced or suppressed by changing bias current. As a result, the switching-current distribution has peculiar peaks in a low bias regime. Resonant processes are also controlled by the sweep rate of the junction.

Introduction

In *mesoscopic* Josephson junctions, the phase difference ϕ across the junction cannot in general be considered as a *classical* variable and must be treated as an operator that does not commute with the Cooper-pair-number operator n, i.e. $[\phi, n] = i$. Under current biases, the junction can be represented as a *quantum* particle moving in a one-dimensional tilted cosine potential. One aspect of the quantum nature of ϕ is *so-called* macroscopic quantum tunneling (MQT) [1–3], which has been studied to determine whether quantum mechanics is applicable on the macroscopic scale. MQT has been confirmed experimentally by finding the temperature-independent rate of decay to the voltage state at low temperatures from the switching-current distribution. Another aspect is the superposition of macroscopically distinct states, called macroscopic quantum coherence (MQC) [4, 5]. Although MQC is considered as the strongest evidence for quantum behavior on the macroscopic scale, it is very difficult to observe it with present techniques, due to dissipation.

Under these circumstances, macroscopic resonant tunneling (MRT) [6, 7] has been predicted on the basis of observations [8] of energy-level quantization. MRT can be regarded as intermediate between MQT and MQC in the sense that MRT is a tunneling between macroscopically different states, but is without coherence. Theoretical studies have shown some progress [9–12] and suggested several experi-

[1] NTT Basic Research Laboratories, 3-9-11 Midori-cho, Musashino, Tokyo, 180 Japan

ments, such as microwave-photon detection [6] and voltage-spike detection in the superconducting branch of current-voltage characteristics [9–11], but resonant tunneling phenomena have not yet been observed in Josephson systems. Therefore, the experimental observation of MRT remains as an important problem in Josephson systems on the way to understanding MQC.

Schmidt et al. [10] and the authors [13] have proposed another technique which uses the switching-current distribution as mentioned in MQT experiments. This technique is based on the fact that decay processes from the potential appear in switching phenomena from the supercurrent-carrying (zero voltage) state to the voltage state. The decay rate depends not only on barrier shape but also on the relation between the neighboring energy levels of Josephson potential in low bias currents. That is, in the *resonant* case, decay probabilities to the voltage state are sharply enhanced. This feature relates to the switching distribution. The switching distribution technique is much more useful for observation than previous techniques because it deals with statistical quantities. In this technique, Silvestrini et al. [14] pointed out that nonstationary conditions obtained by means of a fast sweep rate modify the shape of the switching distribution. Schmidt et al. have not discussed the effect of the rate of change of the potential barrier. In this paper we discuss the switching-current distribution in a small Josephson junction, taking into account MRT, as well as the sweep rate of bias currents. Both capacitance and resistance dependences of the distributions are also discussed.

Density Matrix Approach

Switching-current distribution has been discussed using the master equation developed by Larkin and Ovchinnikov (LO) [15]. According to Schmidt et al. [10], Josephson potential which has many wells can be reduced to a one-well potential. In the reduced potential scheme, resonant tunneling from the ground state of one well of the full Josephson potential to an excited state of the adjacent well is mapped onto transition within the single reduced well. MRT plays a role similar to the external radiation fields in the Martinis et al. [8] experiment, which shows the existence of quantized levels in the Josephson potential. The master equations of this system at an arbitrary bias current are

$$\frac{\partial \rho_j}{\partial t} = \sum_{k=1}^{N} (w_{jk}\rho_k - w_{kj}\rho_j) \qquad \text{for } j \leqslant n \tag{1}$$

$$\frac{\partial \rho_j}{\partial t} = \sum_{k=1}^{N} (w_{jk}\rho_k - w_{kj}\rho_j) - \gamma_j\rho_j \qquad \text{for } j > n \tag{2}$$

where ρ_j is the probability of finding the particle at the jth level, w_{jk} is the transition rate from the k th into the jth level within the potential, γ_j is the escape probability from the jth level, N is the total number of levels within the well, and n is the number of levels in the left well having energies lower than the top of the lower barrier in the right well (Fig. 1). The time evolution of the system is given

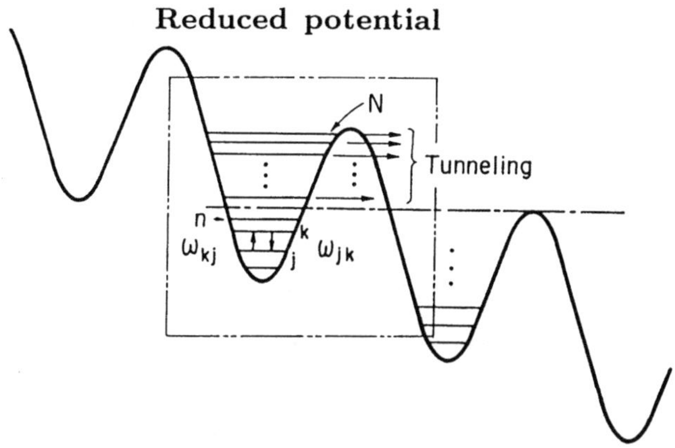

Fig. 1. Schematic representations of a current-biased Josephson potential

by solving these equations. Then, the distribution $P(I)$ of the switching-current values is

$$P(I) = -\frac{d\rho}{dI} \tag{3}$$

Simulation: Two-Level Case

Now let us consider a two-level system which gives some physical insight and presents a possible situation for experimental observations. We restrict ourselves to the case of weak dissipation and zero temperature. We also neglect the coherent tunneling or coherent oscillations of phase between wells. In this case, w_{01} and w_{10} are respectively replaced by the downward transition rate Γ_d and resonant tunneling rate Γ_t. These are estimated numerically; details are given in previous papers [9, 16]. The initial conditions for the system are set by $\rho_0(0) = 1$, $\rho_1(0) = 0$, for $t = 0$.

Figure 2 shows the sweep rate dependence of the switching distribution. We immediately find two kinds of peaks. The peaks located at $I/I_c \geq 0.18$ come from the usual MQT decay process and the peaks located near $I/I_c \sim 0.09$ originate from an indirect process that involves two consecutive processes in escaping from the well, i.e., an upward transition (MRT) and the direct process (MQT). In other words, the indirect process corresponds to successive tunneling in the extended scheme.

In the high-sweep rate regime, the indirect process cannot contribute to the decay process. This can be understood by comparing the decay time due to the indirect process with the inverse of the junction sweep rate normalized by the Josephson critical current I_c (the average duration at an arbitrary current). The indirect decay time can be roughly estimated by $(\Gamma_t + \Gamma_c)/\Gamma_t\Gamma_c$, where Γ_c is the tunneling rate from the excited state after MRT to the quasi-continuum states. In the Fig. 2f case, the inverse sweep rate is 10^{-8}s and the indirect decay time is 10^{-6}s.

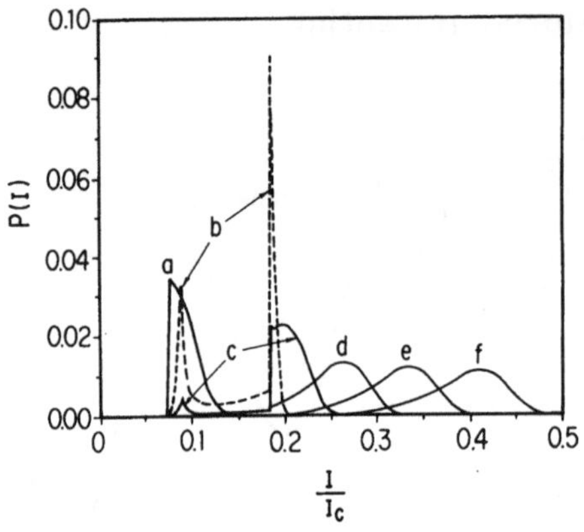

Fig. 2. Switching-current distribution with different sweep rates with C = 9fF and I_c = 100nA; *a*, 0.1mA/s; *b*, 1mA/s; *c*, 10mA/s; *d*, 100mA/s; *e*, 1A/s; *f*, 10A/s

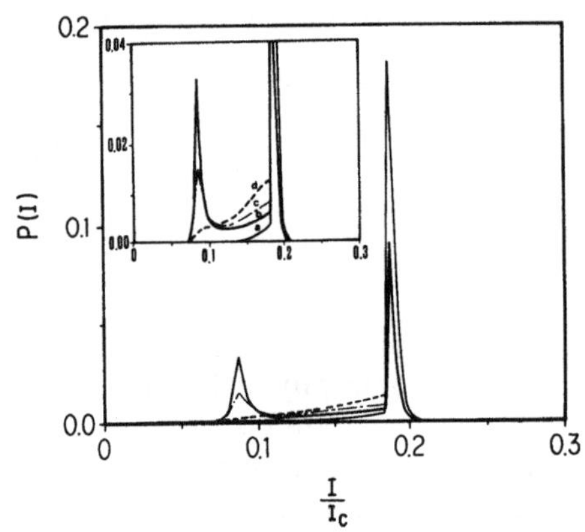

Fig. 3. Resistance dependence of switching-current distribution with a fixed sweep rate (1mA/s); *a*, 1kΩ; *b*, 100kΩ; *c*, 60kΩ; *d*, 20kΩ

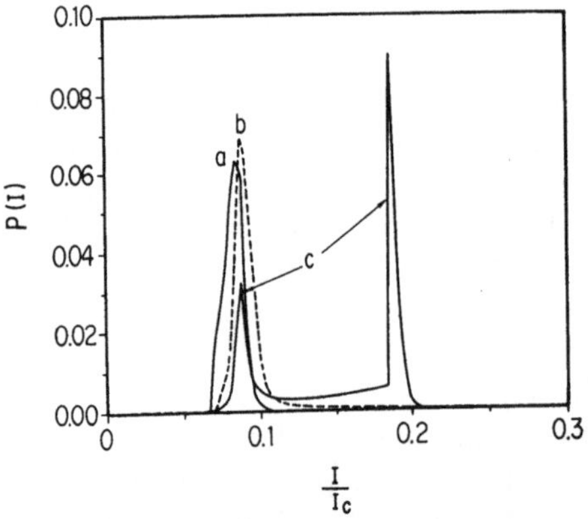

Fig. 4. Capacitance dependence of switching-current distribution with a fixed sweep rate (1mA/s); *a*, 7fF; *b*, 8fF; *c*, 9fF

This is not enough time to escape from the well. Therefore, the direct MQT process becomes the dominant process. We can find MQT peaks in the switching distribution. In the low-sweep rate regime, in contrast, the resonant decay process is dominant. These indirect transitions can be identified from their peak positions since resonant conditions are determined if the junction parameters are given. In the intermediate regime, we can find a coexistence of MRT and MQT peaks. MQT peaks are weaker, and crossover from MQT to MRT peaks occurs as the sweep rate is reduced.

Figures 3 and 4 show, respectively, resistance and capacitance dependence of switching-current distribution at a fixed sweep rate (1mA/s). The observed behaviors are similar to those sweep rate dependence. These behaviors arise from the relative time difference between the external driving force period and the period of internal motions of the particle in the potential. Changing the junction parameters corresponds to changing the characteristic decay times.

Discussion

We studied the switching-current distribution to the voltage states in small Josephson junctions and we found two kinds of peaks, i.e., MRT peaks and MQT peaks, that arose from the different processes with their characteristic time scale.

For the observation of resonant tunneling, it is necessary that switching experiments be performed at lower sweep rates so as to detect the peaks indicating this phenomenon. This is in contrast with the Silvestrini case, where a higher sweep rate is needed to produce nonstationary conditions. Resonant peaks are identified by bias current positions, which are estimated by the junction parameters. This approach has a further advantage for observing MRT. If the junction parameters have uncertainties in their values, we cannot identify a peak from the junction parameters. However, the MQT peak serves as a good reference for the resonant peak at intermediate sweep rates in our simulations.

Acknowledgments. We would like to thank Dr. Hideaki Takayanagi and Dr. Junsaku Nitta of NTT Basic Research Laboratories for their valuable discussions.

References

1. Caldeira AO, Leggett AJ (1981) Phys Rev Lett 46:211–214
2. Voss RF, Webb RA (1981) Phys Rev Lett 47:265–268
3. Jackel LD, Gordon JP, Hu EL, Howard RE, Fetter LA (1981) Phys Rev Lett 47:697–700
4. Chakravaty S (1982) Phys Rev Lett 49:681–684
5. Bray AJ, Moore MA (1982) Phys Rev Lett 49:1545–1549
6. Hatakenaka N, Kurihara S (1987) Solid State Commun 64:943–945
7. Hatakenaka N, Kurihara S (1988) Solid State Commun 68:131–134
8. Martinis JM, Devoret MH, Clarke J (1985) Phys Rev Lett 55:1543–1546

9. Hatakenaka N, Kurihara S, Takayanagi H (1990) Phys Rev B42:3987–3991
10. Schmidt JM, Cleland AN, Clarke J (1991) Phys Rev B43:229–238
11. Zhuravlev AV, Zorin AB, (1990) (in Russian). Sov J Low Temp Phys 16:102–105
12. Gurvitz SA, Marinov MS (1990) Phys Lett A149:173–178
13. Hatakenaka N, Kurihara S (1990) In: Clark TD (eds) Macroscopic quantum phenomena. World Scientific, Singapore, pp 95–96
14. Silvestrini P, Ovchinnikov YN, Cristiano R (1990) Phys Rev B41:7341–7344
15. Larkin AI, Ovchinnikov YN (1986) (in Russian) Zh Eksp Teor Fiz 91:318 [Sov Phys JETP 64:185–189]
16. Esteve D, Devoret MH, Martinis JM (1986) Phys Rev B34:158–163

Fabrication Technology

Silicon Nanoelectronics: The Road Less Travelled

34

THEOREN P. SMITH, III[1]

Summary. The study of nanometer-scale devices has attracted a great deal of attention in the past few years. A number of discoveries have heightened scientific interest in this area. There is also a great deal of interest from the technological point of view. This interest arises on both the primary (new device concepts) and the secondary (impact on existing devices) levels. Except in isolated instances, the unique potential of silicon-based materials has not been exploited to study mesoscopic phenomena. We will examine the key breakthroughs in mesoscopic physics and exploratory silicon technology that may provide extraordinary opportunities to investigate nanometer-scale electronic devices.

We would like to address the future work emerging in mesoscopic physics and nanoelectronics and its potential technological impact. In addressing this, several overriding assumptions must be made. The first issue is what technology base will launch nanoelectronics. The most obvious and most likely scenario is that silicon technology will continue to dominate computation, logic, and memory. Nanometer scale transistors and circuits employing devices with sub-100nm gate lengths have already been fabricated and studied in silicon. The leading edge of semiconductor product miniaturization is being driven by dynamic random access memory (DRAM) and the conflicting requirements of scaling and macroscopic charge storage may provide an opportunity to exploit new schemes such as single-electron circuits. In fact, a major breakthrough or innovation is probably needed for chips to store more than a gigabit of information.

Another important issue is the operating temperature for future circuits. While there are clear technological advantages associated with low-temperature (77K) operation of conventional complementary metal-oxide semiconductor (CMOS) circuits there are numerous questions regarding cooling and packaging. This is ominous, since the most optimistic projections for nanoelectronics require low temperatures and in some cases cryogenic (4.2K) temperatures. We must assume that nanoelectronics alone will not drive technology to low-temperature operation and CMOS must get there first.

The final issue is the type of physics that can be used effectively for technology.

[1] IBM Research Division, Thomas J. Watson Research Center Yorktown Heights, NY 10598, USA

We assume at this point that the leading contenders are charge quantization, spatial quantization, and resonant tunneling. There are a number of recent examples of semiconductor nanostructures that demonstrate these types of physics and might be precursors to a nanoelectronics technology: Quantum-dot capacitors, Coulomb-blockade transistors, and resonant-tunneling diodes. Each of these has been studied using III-V structures but has given only a hint of what might be possible for future silicon-based devices. Based on our assumptions concerning the future of nanoelectronics and the physics that emerges at these length scales it is critical to examine developments in silicon technology to determine the likelihood of translating mesoscopic physics into novel technology.

The single-transistor memory cell is beginning to face scaling limitations. The amount of charge required for the reliable storage of information cannot be reduced much beyond present levels (about 150 fC). This complicates further reduction of memory cell size since voltages continue to drop, demanding increasing capacitance. We have studied quantum-dot capacitors extensively to investigate spatial quantization. Quantum dots represent a new class of electronic structures where the electronic configuration and degree of confinement can be tuned at will. These structures mimic many atomic effects with a tremendous degree of flexibility. They can be tuned from isolated to coupled quantum dots simply by changing the bias on a patterned gate. Three distinct regimes can be examined in this way: A superlattice regime; a tight-binding regime; and an isolated-quantum-dots regime [1].

Quantum dots also have intriguing magnetic properties. The presence of an external magnetic field couples effects of the magnetic flux with the electrostatic potential depending on the strength of the potential modulation. In strongly-coupled quantum wires and quantum dots, the commensurability between the superlattice period and the cyclotron orbit is reflected in a modification of the envelope of the magnetocapacitance oscillations. In weakly-coupled quantum dots, a new structure is observed. Aperiodic oscillations in the density of states are related to fractal-like behavior in the energy dispersion relation first predicted by Hofstadter. In completely isolated quantum wires or quantum dots, the Landau-level magnetocapacitance oscillations are recovered with a Zeeman bifurcation of the states produced by the interplay of magnetic and spatial quantization [2].

Another important observation is that random impurity distributions and imperfections in large arrays of quantum dots do not obscure the underlying spatial quantization. This is still somewhat puzzling, but it demonstrates that spatial quantization of electrons can be detected and even averaged without losing the discrete nature of the process. While numerous questions still remain, one can envision schemes where the total charge required for reliable information storage could be reduced by exploiting discrete quantum-state effects.

Recent work by Meirav, Kastner, and Wind [3] on transport through quantum dots also elucidates some intriguing physics that may presage devices that exploit the energy associated with single electron charging of small capacitors. These experiments were conducted at cryogenic temperatures, and therefore translating such effects into a viable technology faces enormous challenges. Nonetheless, the fact that clear effects from the transport of single electrons can be measured opens the door to new areas of research and development. The potential for quantum dots in technology is no longer a question of physics but of fabrication.

At the same time, the fabrication of conventional silicon devices continues to demand greater and greater control of dopant profiles and increased vertical as well as lateral integration. In advanced CMOS the active dimensions of devices are becoming less than quarter-micron. This continued scaling of semiconductor devices poses problems for fabrication while offering enhanced and possibly unique performance. To meet this demand the overall thermal budget (the integrated time-temperature cycles for processing) has become progressively smaller and smaller. A number of powerful new processes have been developed to fabricate semiconductor devices at low temperatures. Molecular-beam epitaxy (MBE), atmospheric-pressure chemical vapor deposition (APCVD), ultra-high vacuum CVD (UHV CVD), and plasma-enhanced CVD (PECVD) are key examples. These processes allow band-gap engineering of novel structures in silicon-based heterostructures. Resonant tunneling diodes, heterojunction bipolar transistors, and multi-quantum-well photodetectors are notable examples and are the first of what may become a new phase of silicon solid state electronics that exploits the benefits of scaling and quantum-size effects.

We have observed two-dimensional quantization at an Si-SiO_2 interface where the oxide has been deposited rather than thermally grown. In these devices an accumulation layer is created in the undoped channel just below the interface but relatively close to a doped-substrate electrode. To minimize dopant diffusion and channel doping these devices were fabricated using low-temperature processes. Two series of samples were fabricated using molecular-beam epitaxy (MBE) and ultra-high vacuum chemical vapor deposition (UHV CVD) of silicon followed by plasma-enhanced chemical vapor deposition of SiO_2. The fabrication was particularly demanding since the thin (10-60nm) channel region was deposited directly on a heavily doped Si substrate. Capacitance and electrochemical potential measurements were used to probe the two-dimensional density of states of the electrons in accumulation layers. Our results demonstrate that silicon-based devices can incorporate these new materials and function effectively at low temperatures. Electron mobilities are approximately 4500 cm^2/V-s at 4.2K for both MBE and UHV CVD samples. These processes open the door to novel applications and quantum device concepts.

We have also made significant progress in understanding resonant tunneling in silicon-based materials. This is important, since resonant tunneling may have technological impact beyond the discrete device and high-frequency oscillator arena. There has been a great deal of work on silicon-based resonant tunneling diodes. Resonant tunneling has been observed in the conduction band as well as the valence band. While the effects in these devices are not nearly as large as in III–V structures, silicon devices can be integrated to a much higher degree. Applications of two-terminal devices are limited at this point, but one cannot completely rule out some novel logic or memory schemes.

We have directly explored the influence of strain on the valence band structure of Si-SiGe quantum wells, using angular-resolved magnetotunneling spectroscopy (ARMS). The in-plane dispersion relations of the resonant levels are probed by a magnetic field applied parallel to the interfaces. This method is unique, in that it measures the local band structure in k-space, in contrast to optical spectroscopy, which averages overall k-space. We detect large non-parabolicity and negative in-plane effective mass for some of the levels. By rotating the magnetic field the energy surface at constant $k_{\parallel}$ is measured, wherewith we see a large anisotropy of the energy levels, with the symmetry axes of light holes and heavy holes being rotated 45° with respect to each other. The valence band structure is less anisotropic in samples with unstrained SiGe quantum wells.

The fact that one can probe the details of SiGe bond structure is evidence of the level of control now possible in silicon-based heterojunctions and quantum-well structures. The additional ability to passivate structures with an stable oxide and form complex three-dimensional structures should provide unique leverage over III-V materials. In short, there are many challenges and opportunities ahead for nanoelectronics. However, the requirements are stringent and the prognosis is not clear. Research in this area must continue to address both fundamental and practical issues.

References

1. Ismail K, Smith, III TP, Masselink WT (1989) Appl Phys Lett 55:2766–2768
2. Hansen W, Smith, III TP, Lee KY, Brum JA, Knoedler CM, Hong JM, Kern DP (1989) Phys Rev Lett 62:2168–2171
3. Meirav U, Kastner MA, Wind SJ (1990) Phys Rev Lett 65:771–774

Properties of GaAs-AlGaAs Quantum Wire Superlattices Grown by Molecular Beam Epitaxy

35

PIERRE M. PETROFF[1]

Summary. A novel type of superlattice, the serpentine superlattice, with a *built in* quantum wire superlattice has been grown directly by molecular beam epitaxy. The uniformity issues and photoluminescence polarization anisotropy properties associated with the quantum wire superlattice are presented. Both transmission electron microscopy and photoluminescence measurements indicate that control of the growth kinetics still limits the quality of these structures.

Introduction

The well known quantum well structures directly produced by epitaxial growth have yielded an impressive number of new physical phenomena and devices associated with the quantization effects found with confinement dimension on the order of the electron de Broglie wavelength (<50nm).

Two (2D) and three-dimensional (3D) carrier confinement is achievable relatively easily when the confinement length scale is larger than 100nm. This is the dimensional range of the so-called mesoscopic devices. All the technologies which permit this mesoscopic regime are based on a lithography (electrons or ions) processing which will be used for producing a depletion layer [1, 2] or a lateral band gap modulation with a strain field [3]. Unfortunately, the interesting effects associated with the presence of a 2D superlattice and quantization are not found until the structure dimensions are well below the 50nm range. There are presently few processing techniques which will permit reaching these dimensions. Surprisingly, these are primarily based on direct crystal growth [4, 5] or regrowth [6].

The direct crystal growth of quantum wire superlattices is made possible through the growth of the so-called tilted superlattices (TSL) [4, 7, 8]. These superlattices are similar to the usual quantum well superlattices; however, the orientation of the quantum well with respect to the substrate surface, as well as their thickness, can be continuously changed during growth. The basic principle

[1] Materials Science Department, University of California, Santa Barbara, CA93106, USA

for this epitaxial growth method was developed and demonstrated several years ago [7]. It involves the alternate deposition of fractional monolayers m and n ($m, n < 1$) of two semiconductors (A)m-(B)n with different band gaps on a substrate comprising a well ordered array of steps. The ordered vicinal surface is made of monoatomic atomic steps with a periodic spacing L. L may be adjusted by choosing the substrate vicinal angle and the ordering of the steps by growth of a buffer layer in the step flow growth regime [9, 10].

The TSL growth method has been successfully demonstrated both by molecular beam epitaxy (MBE) [4] and by metal organic chemical vapor deposition (MOCVD) [8] for the GaAs-AlGaAs system. It has also been applied successfully to the GaSb-AlSb system [11] and with the appropriate growth conditions should be applicable to a wide variety of semiconductors or even metal or insulator systems.

This growth method has successfully been used to demonstrate 2D quantum confinement by optical methods [12–14]. It is extremely attractive because of its simplicity for producing two or three dimensional quantum wire arrays.

In this paper we discuss the crystal growth and uniformity of the structures as observed by transmission electron microscopy. The 2D confinement properties of these quantum wire superlattices are demonstrated by optical measurements.

Quantum Wire Uniformity and Composition

Control of the quantum wire dimensions ,and their uniformity, are prerequisites to achieving the most desirable properties of the 2D quantum confinement.

The TSL growth requires an extremely fine control over the $(A)_m$-$(B)_n$ fractional monolayers deposition. Indeed the vertical interface superlattice is only achieved for $m + n = 1$. A 1% deviation in the flux deposition will produce a 15° change in the TSL orientation for a GaAs (100) vicinal surface with a 2° misorientation. This required accuracy (better than 0.5%) in the flux control is presently impossible to achieve in MBE, as well as in MOCVD over, large substrate areas.

To relax these stringent growth requirements, we have introduced a novel superlattice, namely the serpentine superlattice (SSL) [15]. In the SSL, a "*built in*" 2D quantum wire confinement is present. As shown in the schematic in Fig. 1, the meandering interfaces of the quantum well, as well as their thickness, vary along the substrate normal direction. This meandering of the quantum wells in the SSL is achieved by varying the per cycle coverage during deposition. One coverage cycle corresponds to the deposition of $m < 1$ monolayer of $(AlAs)_m$ or $(AlGaAs)_m$ followed by $n < 1$ monolayer of $(GaAs)_n$. If the cycle coverage is repeated and kept constant during the growth a TSL is obtained [3, 6]. However, a linear variation in the per cycle coverage between two values p_A and p_B, such that $p_A < 1 < p_B$ is a function of distance along the growth direction, yields a superlattice of GaAs quantum wells with interfaces in the shape of parabola.

The parabola equation is given by:

$$x = Ky^2/2 \qquad (1)$$

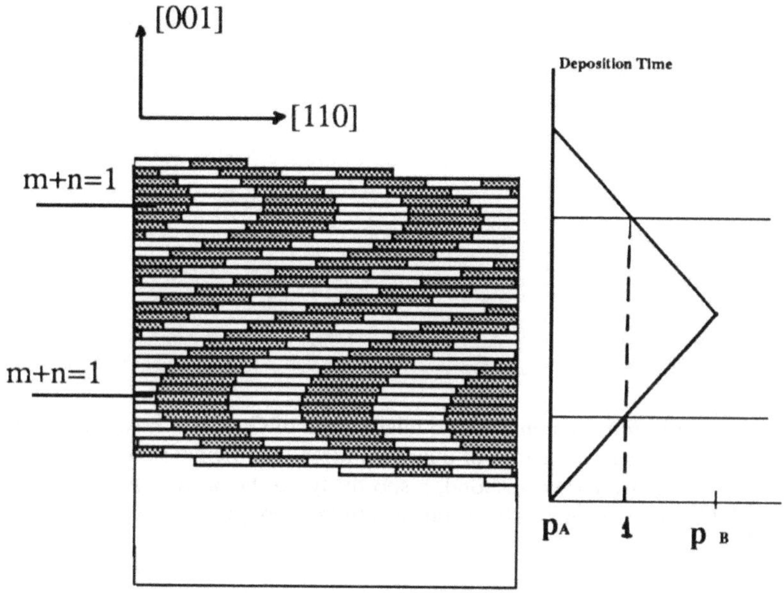

Fig. 1. Schematic of a serpentine superlattice produced by deposition of alternate fractional monolayers of $(AlAs)_m$ (*shaded areas*) and $(GaAs)_n$ (*unshaded areas*). The *graph* indicates the variations of the tilt parameter, $p(t) = m(t) + n(t)$ as a function of time during deposition. A linear variation of the coverage per cycle, $m(t) + n(t)$ as a function of time, yields a parabolic-shaped interface. The quantum wells are thickest for $p(t) = 1$, thus providing a quantum wire confinement in this region of the structure

Where the x and y axes are along the [110] and [001] directions, respectively, and K is the curvature.

We note that the apex of the parabola corresponds to the coverage $p = m + n = 1$ value and also corresponds to the point of maximum quantum well width W:

$$W = nd/tg\alpha \tag{2}$$

with the spacing between steps along the [001] direction = 2.68Å for GaAs and a substrate vicinal angle α.

The curvature K of the parabola is directly related to the quantum confinement along the growth direction, while the lateral confinement is determined by W. This curvature may be chosen at will since it is determined by the linear ramping rate of the per cycle coverage with

$$K = 2[p_B - 1]/tg\alpha \tag{3}$$

With this approach, one can afford to start ramping the per cycle coverage from $p_A = 0.9$ to a value $p_B = 1.1$ and thus make certain that variations in the deposition flux (2%–3%) will still insure the presence of regions in the film which contain identical quantum wires.

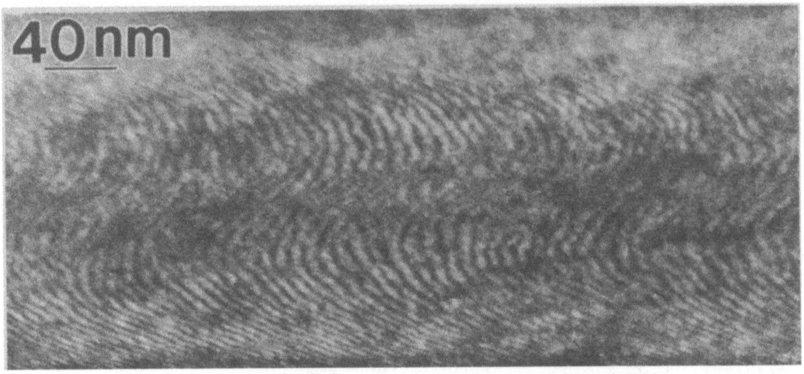

Fig. 2. Cross-section transmission electron micrograph of GaAs-AlGaAs serpentine superlattice containing two parallel arrays of quantum wires. The vicinal substrate orientation is 2°. The *white* and *dark areas* correspond, respectively, to the barrier (Al rich regions) and quantum well (Al poor regions). The distance between two quantum wires is 80Å

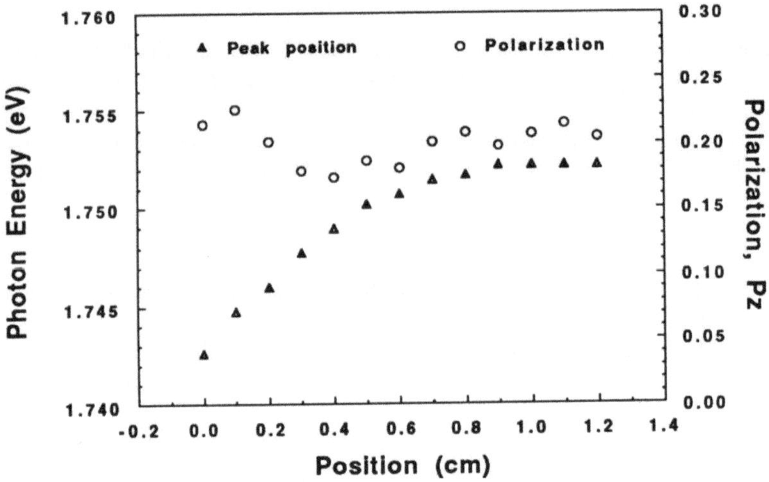

Fig. 3. Polarization photoluminescence as a function of position for the serpentine superlattice shown in Fig. 2

We will see in the next section that one of the characteristics of the quantum wire confinement is the photoluminescence (PL) polarization for the electric vector parallel or perpendicular to the wire axis [16, 17]. As a proof of the quantum wire superlattice existence over large wafer areas, we show (in Fig. 3) a plot of the polarization as a function of position for a quantum wire superlattice imbedded in the SSL structure shown in Fig. 2. The small variation in this ratio (2%–3%) for this wafer is similar to that observed in the ground state energy measured by PL for a quantum well included in the same wafer as the SSL. These variations can be

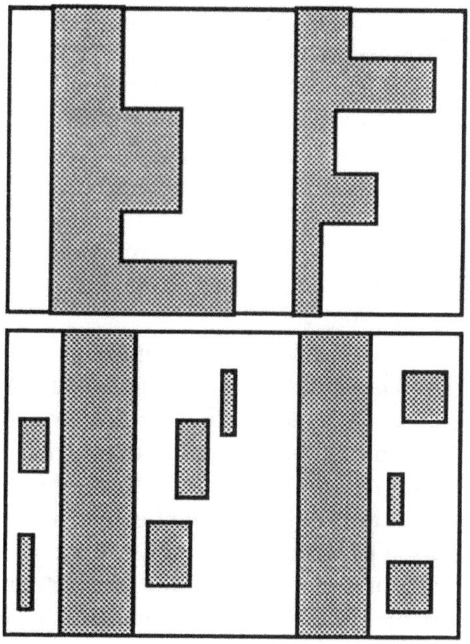

a

b

Fig. 4a, b. Schematic of the two possible growth processes that may yield quantum wires with low lateral barrier confinement. **a** Superkinks (K) produced by surface reconstruction and growth kinetics conditions. **b** Islands related to the poor segregation of the barrier material at the step edges

traced to perhaps temperature variations across the wafer or to the Al content in the barrier material. The important point is that this type of uniformity could not be achieved for quantum wires made using the TSL method.

The cross-section transmission electron micrograph shown in Fig. 2 for a double crescent serpentine superlattice shows the parabolic shaped quantum wells and barriers. The chemical contrast imaging mode used in this micrograph indicates that the programmed compositional variations between the $Ga_{.5}Al_{.5}As$ barriers and the GaAs quantum well regions are not present in the structure. In fact, this micrograph indicates that some of the Al that was supposed to be present in the barrier region has been trapped during growth in the quantum well and quantum wire regions of the structure.

The question of the Al distribution during the growth needs to be addressed. Figure 4 shows a schematic of two possible processes that play a role in the apparent poor Al segregation at step edges.

In Fig. 4a, we show the effects of surface reconstruction on the structure of step edges. The step edges will be formed of (2×4) reconstructed surface unit cells. As a result, a reconstructed unit cell with the long dimension oriented perpendicular to the step edge will protrude by as much as 16Å. This superkink, when covered by the next layer, will stay buried and give rise to a rough step edge with a roughness of up to 16Å. If the barrier separation is 40Å for an ideal vicinal substrate, the reconstruction at the step edges will narrow this distance by as much as 32Å!. The scanning tunneling micrographs we have obtained of steps on vicinal GaAs (100) surfaces as well as those published by Pashley et al. [18] support this interpretation.

In Fig. 4b, we show the second process for the inclusion of Al in the quantum wire structure. The formation of Al islands due to incomplete incorporation at step edges and their trapping in the structure in the next coverage cycle will produce an Al-containing quantum wire. The absence of RHEED oscillations during the GaAs growth cycle for the SSL or TSL deposition indicates a growth mode which is dominated by step flow. However, we have seen in earlier growth studies [9, 10] that step flow growth conditions are very different for AlAs growth (about 100°C lower or higher substrate temperatures than those used for GaAs are required). Unfortunately a temperature cycling of 100°C would be required to insure a good step flow mode for both AlAs and GaAs and this has not been implemented yet.

A new and promising approach has recently been tried by us: The growth of AlAs-GaAs TSL on {110} vicinal surfaces. The advantage of this surface resides in the fact that there is no surface reconstruction that will induce large dimension superkinks. In addition, there is a natural spinodal decomposition that is present for the AlGaAs system for this substrate orientation [19]. Thus, for this surface orientation, both the kinetics and thermodynamics should favor the segregation and self ordering of the AlAs at the step edges.

The problem of the poor Al segregation at step edges has also been found to exist in MOCVD grown quantum wire structures made using the TSL method [14].

Measurements of the optical properties of quantum wire arrays in the SSL structure, in conjunction with theoretical calculations, have allowed an estimation of the Al compositional fluctuation in the structure.

Quantum Wire Superlattice Optical Properties

The photoluminescence (PL) and resonantly excited photoluminescence (PLE) of the quantum wire superlattice in the SSL structures have been measured [16, 17] at low temperature (1.4°K). The linear polarization effects on the PL emission or on the exciting beam in PLE are detected at low temperature using a photoelastic modulation technique. For an observation direction parallel to the sample surface normal, the SSL emission shows linear polarization along the wire direction (up to 30%) due to lateral confinement [16, 17]. The PL spectra of an SSL structure for PL emission polarized parallel or perpendicular to the wire axis are shown in Fig. 5. This sample also contained witness quantum wells (thickness: 60Å and 80Å) that did not show any polarization anisotropy. The polarization is defined as:

$$P_z = (I_x - I_y)/(I_x + I_y) \qquad (4)$$

The subscript indices refer to the directions of the emission intensity (I) or polarization (P) along the axis shown in the inset of Fig. 5. The polarization of the light emitted in the perpendicular directions is, respectively, $P_x = 36\%$ and $P_y = 57\%$, where cyclically permuted indices define the other polarization ratios. The polarization anisotropy ratios for the 60Å quantum well were, respectively, $P_z = 0\%$, $P_x = 67\%$, and $P_y = 76\%$.

The lifetime for the quantum wire superlattice was 379ps while those of the 80Å

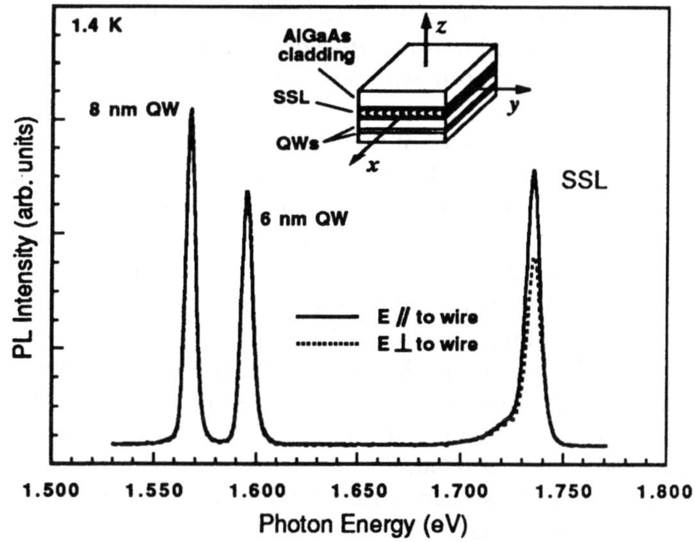

Fig. 5. Polarization-dependent PL spectra for a single quantum wire superlattice [16]. The observation direction is along the z axis, as defined in the *inset*, which also shows the SSL position and the quantum wells (*QWs*) in the sample. The *solid* and *dashed lines* are for light polarized parallel and perpendicular to the quantum wires, respectively. The quantum wells built in this same structure emit unpolarized light in this direction while the SSL displays a polarization of 23%

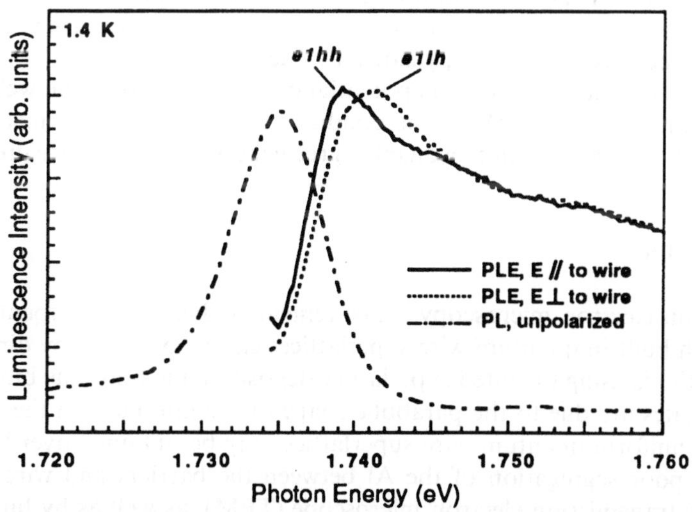

Fig. 6. Polarization-dependent PLE spectra of light emitted normal to the sample in the z direction, as shown in Fig. 5. The *solid* and *dashed lines* are for the exciting light polarized parallel and perpendicular to the quantum wire axis, respectively. The *dash-dotted line* is the unpolarized PL

and 60Å quantum wells in the structures were 250ps. The lifetimes measured here include both the radiative and non-radiative decay times. Systematically, we have observed that lifetimes in quantum wire superlattices are 1.5 to 2 times larger than those of quantum wells [17].

The intrinsic excitonic character of the emitted light has also been observed by following the quantum wire superlattice PL emitted intensity variations as a function of the laser pump power. The SSL emission is observed to vary linearly over three orders of magnitude while the quantum wells have an initial increase that is less than linear and becomes linear at higher pump power intensities.

The PLE spectra in Fig. 6 were observed by detecting the luminescence in a direction normal to the sample surface. The detection energy was set on the low energy side of the SSL emission peak. The exciting beam was polarized either in a direction parallel or perpendicular to the wires. A Stokes shift of 4.6meV from the detection energy is observed for the parallel polarization direction of the exciting light. The Stokes shift is 6.6meV for the perpendicular direction. The 60Å quantum well does not show such anisotropy effects. The heavy and light hole transitions are indicated in Fig. 6.

The imperfect segregation of the Al has been estimated by comparing the polarization measured in the x, y, and z directions with that computed in the framework of an envelope function approximation [20]. The conduction band and four coupled valence band states, and the optical polarization of the light emitted, were calculated. The Al concentration in the barrier, x_B varied from 0 to 0.33. A value of $P_z = 23\%$ yielded a value of $x_B = 0.19$. The other measured ratios, P_x and P_y, yielded values of $x_B = 0.21$ and 0.22, respectively. Thus the quantum wires contain about 10% Al. This value is in good agreement with that obtained from the PLE data which show a 2 meV shift in the PLE emission with parallel and perpendicular polarization excitation [16]. The calculation, assuming a uniform redistribution of the Al in the wire and barrier, yielded a value of $x_B = 0.22$.

From the PL measurements, the potential difference between the well and barrier was estimated at 70meV in the conduction band. The PLE measurements yielded a value of 90meV, in reasonable agreement with the PL measurements.

Conclusions

Transmission electron microscopy measurements indicate that serpentine superlattices with built in quantum wire superlattices can be obtained by direct crystal growth methods using the tilted superlattice deposition method. The built in quantum wire states are due to the parabolic shape of the interfaces in the SSL. With the SSL, a uniform quantum wire superlattice may be obtained over large wafer areas. The poor segregation of the Al between the barriers and wire regions is detected by transmission electron microscope (TEM), as well as by luminescence experiments. We find that roughly 30% of the Al intended for the barrier stays trapped in the wire region. The growth kinetic conditions and surface reconstruction at steps during deposition are both probably responsible for this poor segregation. Using deposition conditions which favor a step flow mode and segregation for both Al and Ga at step edges should dramatically improve the carrier confinement in these wire structures. Optimal deposition conditions should make the SSL approach for room temperature quantum wire devices and structures a reality.

Acknowledgments. This work was supported by grants from the AFOSR (No. 88-0334) and QUEST, an NSF Science and Technology Center.

References

1. Kotthaus J, Hansen W, Pohlman H, Wassermeier M, Ploog K (1988) Surf Sci 196:600
2. Kohl M, Heitman D, Grambow P, Ploog K (1989) Phys Rev Lett 63:2124
3. Kash K, Vander Waag BP, Mahoney DD, Gozdz AS, Florez LT, Harbison JP, Sturge DM (1991) Phys Rev Lett 67:1326
4. Gaines JM, Petroff PM, Kroemer H, Simes RJ, Geels RS, English JH (1988) J Vac Sci Technol B6:1378
5. Tokura Y, Saito H, Fukui T (1989) J Crystal Growth 94:46
6. Kapon E, Hwang DM, Bhat R (1989) Phys Rev Lett 63:430
7. Petroff PM, Gossard AC, Wiegman W (1984) Appl Phys Lett 45:620
8. Fukui T, Saito H (1988) J Vac Sci Technol B6:1373
9. Chalmers SA, Gossard AC, Petroff PM, Gaines JM, Kroemer H (1989) J Vac Sci Technol B7:1357
10. Chalmers SA, Gossard AC, Petroff PM, Kroemer H (1990) J Vac Sci Technol B8:431
11. Chalmers S, Gossard AC, Kroemer H (1990) J Cryst Growth 111:647
12. Tsuchiya M, Gaines JM, Yan RH, Simes RJ, Holtz PO, Coldren LA, Petroff PM (1989) Phys Rev Lett 62:466
13. Tanaka M, Sakaki H (1989) Appl Phys Lett 54:1326
14. Kasu M, Ando H, Saito H, Fukui T (1991) Appl Phys Lett 59:301
15. Miller MS, Pryor CE, Wehman H, Samoska LA, Kroemer H, Petroff PM (1991) J Cryst Growth 111:323
16. Miller MS, Weman H, Pryor CE, Krishnamurty M, Petroff PM, Kroemer H, Merz JL (1992) Phys Rev Lett 68:3464
17. Weman H, Miller MS, Pryor CE, Krishnamurty M, Petroff PM, Kroemer H, Merz JL (1992) Phys Rev B
18. Pashley MD, Haberern KW, Gaines JM (1991) Appl Phys Lett 58:406
19. Petroff PM, Logan RA, Cho AY, Reinhart FK, Gossard AC, Wiegmann W (1982) Phys Rev Lett 48:170
20. Pryor CE (1991) Phys Rev B44:12912

Nanostructure Technology Developed Through Electron-Beam-Induced Surface Reaction

36

Shinji Matsui, Toshinori Ichihashi, Yukinori Ochiai, Masakazu Baba, Heiji Watanabe, and Akinobu Sato[1]

Summary. Nanostructure technology has been studied by using electron beam-induced surface reactions. Ten-nm linewidth polymethylmethacrylate (PMMA) resist patterns and 14-nm diameter carbon dot patterns have been demonstrated by 50 kV electron beam with a gas introduction line. It has been confirmed that damage induced by electron beam assisted dry etching is less than that induced by reactive ion beam etching. Atomic layer etching for GaAs has been achieved by scanning tunneling microscope (STM) assisted etching using a Cl_2 adlayer. A side-wall roughness of GaAs patterns etched by reactive ion beam etching has been successfully measured by using a new STM tip made by electron beam deposition.

Introduction

The objective of research in nanostructure technology is to explore the basic physics, technology, and applications of ultra-small structures and devices with dimensions in the sub-100 nm regime. Today, the minimum size of production devices is down to 0.5 μm or less. Nanostructure devices are fabricated in many laboratories to explore the effects created by the downscaling of existing production devices, such as field-effect transistors (FETs) in silicon or III-V compounds, or to explore new principles, such as ballistic and quantum effects in semiconductors, or tunneling effects in superconductors, etc. In addition, new phenomena are being explored in an attempt to build switching devices with dimensions down to the molecular level.

Electron beam-induced selective etching and deposition (EBISED) technology is a new method for the nanofabrication of semiconductors and other devices [1, 2]. This technique uses electron beam (EB)-induced surface reaction at a gas-substrate interface allowing etching and deposition to occur. EBISED technology is classified into several types, as shown in Table 1. To date, we have demonstrated nanofabrication using EBISED technology with a scanning transmission electron microscope (STEM) [3]. and a scanning tunneling microscope (STM) [4], both using focused EB. Both a tungsten rod pattern with 15-nm diameter and carbon dot patterns with 10-nm diameters have been fabricated by STEM and STM-

[1] NEC Corporation, 34 Miyukigaoka, Tsukuba, Ibaraki, 305 Japan

Table 1. EBISED Technology

Analog process	Focused EB	SEM EBISED
		STEM EBISED
		STM EBISED
	Showered EB	–Masked EBISED
Digital process (ALE)	Focused EB	SEM EBISED
		STEM EBISED
		STM EBISED
	Showered EB	–Masked EBISED

EBISED technology, respectively, as shown in Fig. 1a and b. As a result, direct writing at nanometer dimensions has already been achieved with EBISED technology using focused EB.

This paper describes nanostructure technology developed through EBISED, such as 10-nm resolution lithography, low damage dry etching, atomic layer etching, and pattern evaluation of sidewall roughness.

Nanolithography by 50 kV SEM-EBISED

Figure 2 shows a photograph of a 50 kV electron beam lithography system developed by modifying the JEOL-5FE 25 kV EB system [5], and the design target is shown in Table 2 [6]. The major modified points of this electron beam system are acceleration voltage and a gas feed system. The acceleration voltage has been changed from 25 kV to 50 kV. The column design with respect to the electron gun chamber is modified to withstand the high voltage in order to avoid the field

Table 2. Design of 50 kV electron beam direct writing system using a thermal field emitter gun

Acceleration voltage:	50 kv
Electron beam:	point beam
Electron source:	Zr/O/W thermal field emitter
Beam diameter:	5 nm
Scanning field:	80 μm
Minimum step:	2.5 nm
Scanning speed:	0.167 μsec (6MHz)
Stitching accuracy:	40 nm at 2σ
Overlay accuracy:	40 nm at 2σ
Wafer size:	6 inch (max.)
Wafer movement:	step and repeat
Gas injection:	nozzle

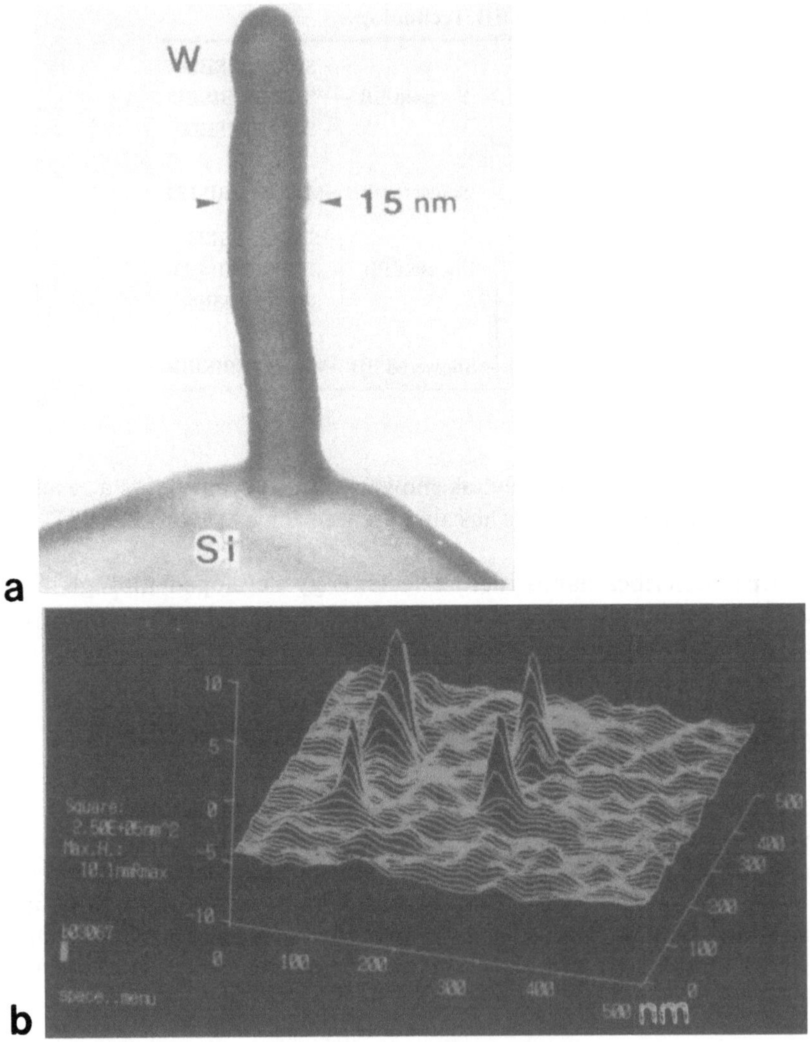

Fig. 1. a W rod with 15-nm diameter fabricated by a 120kV EB with 3-nm diameter **b** Carbon dot patterns with 10-nm diameter fabricated by STM

emission from the electrode surface. Because the Zr/O/W thermal field emission (TFE) electron gun is used in this system, the vacuum in the electron gun chamber must be maintained under 10^{-9} Torr. The TFE gun has a small virtual source size and a high angular current intensity and is operated at 1800 K. This system is fully controlled by a PDP-11 computer. Various types of pattern format, such as JEOL01, PG-3000, and PG-3600 can be used for the lithography.

Various gas species can be introduced into this specimen chamber to investigate the electron-beam-induced surface reactions. The gas feed system is shown in Fig. 3a, b. The 50-cc capacity gas cylinder contains the reaction gas. The gas is intro-

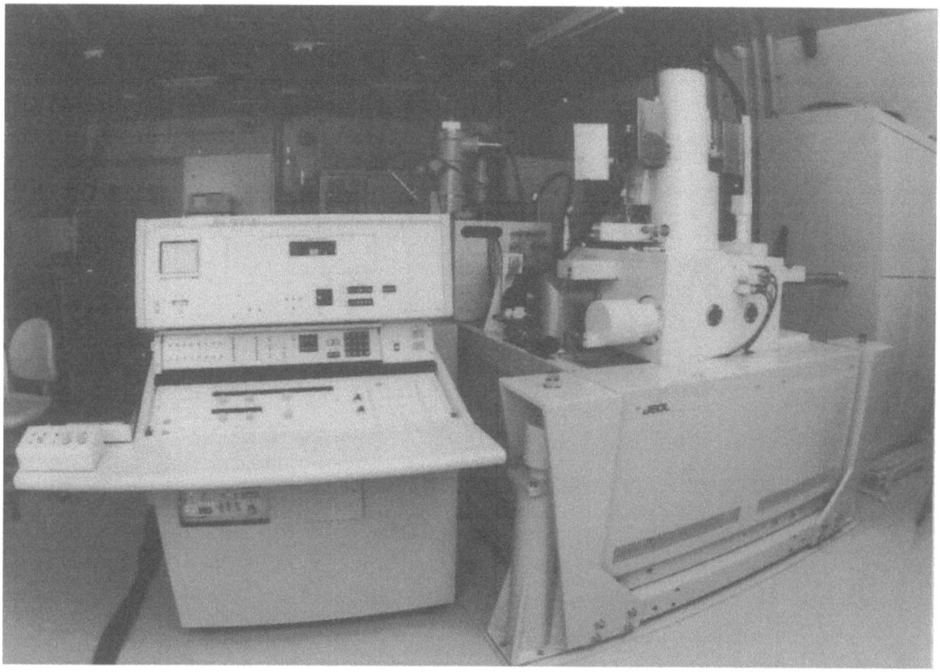

Fig. 2. Photograph of 50 kV electron beam direct writing system

duced through a variable leak valve and a fine nozzle with an inner diameter of 200 μm. The gas line between the variable leak valve and the nozzle is evacuated through a bypass line. The nozzle is placed at a distance of 3 mm apart from the wafer.

The electron beam diameter was measured by using a knife edge method. The edge of Ni mesh was used as a knife edge. The acceleration voltage was 50 kV and the beam current used was 100 pA for a 40 μm diameter objective aperture. The beam current change was measured by a Faraday cup when the beam was scanned over the knife edge. The beam diameter was determined to be the scan distance which corresponded to 70% of the total current change (that is, between 15% and 85%). The results of the measurement for the x and y directions are shown in Fig. 4. It was found that the beam diameter for both the x and y directions was less than 5 nm and there was a small amount of astigmatism. The results are in reasonable agreement with the calculated values. Overlay and stitching accuracy were evaluated by exposing 16 chips (4 × 4 in array) with a size of 80 × 80 μm in PMMA resist at 50 kV. The stitching accuracy and overlay accuracy were 0.021 and 0.016 μm at 2 σ, respectively. These results sufficiently satisfied the target.

PMMA resist exposure was carried out to evaluate the fine pattern exposure characteristics using this electron beam system. PMMA is known as the positive resist with the highest resolution. Thirty-nanometer-thick PMMA was spin-coated on a bare thick Si wafer. After PMMA was prebaked at 170°C for 20 minutes,

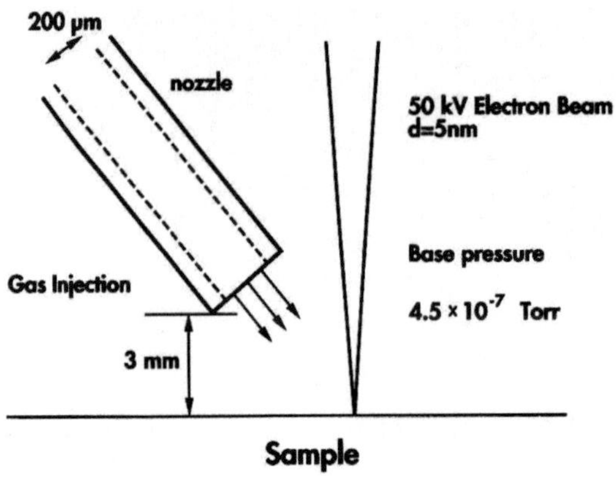

a

Fig. 3. **a** Schematic diagram of gas introduction system. A fine nozzle with an inner diameter of 200 μm is used. **b** Photograph of gas introduction system. Gas contained in a 50-cc gas cylinder is introduced into a chamber through a variable leak valve

b

electron beam exposure was carried out. The line dose was 0.8 nC/cm for both periods. The PMMA was developed in a mixture of MIBK : IPA = 1 : 3 for 1 min and was then rinsed in IPA for 1 min. The point electron beam was line-scanned with spatial periods of 50 nm, as shown in Fig. 5a, and 100 nm, as shown in Fig. 5b. Ten-nm-thick Au-Pd (60%–40%wt) metal was deposited on the resist to avoid a charge-up effect. Ten-nm width line patterns in PMMA resist were obtained for both a and b cases in Fig. 5. This linewidth is the same as others that have been reported and carried out in PMMA coated on a thick substrate [7]. It was observed that there were few proximity effects due to the linewidth being the same for

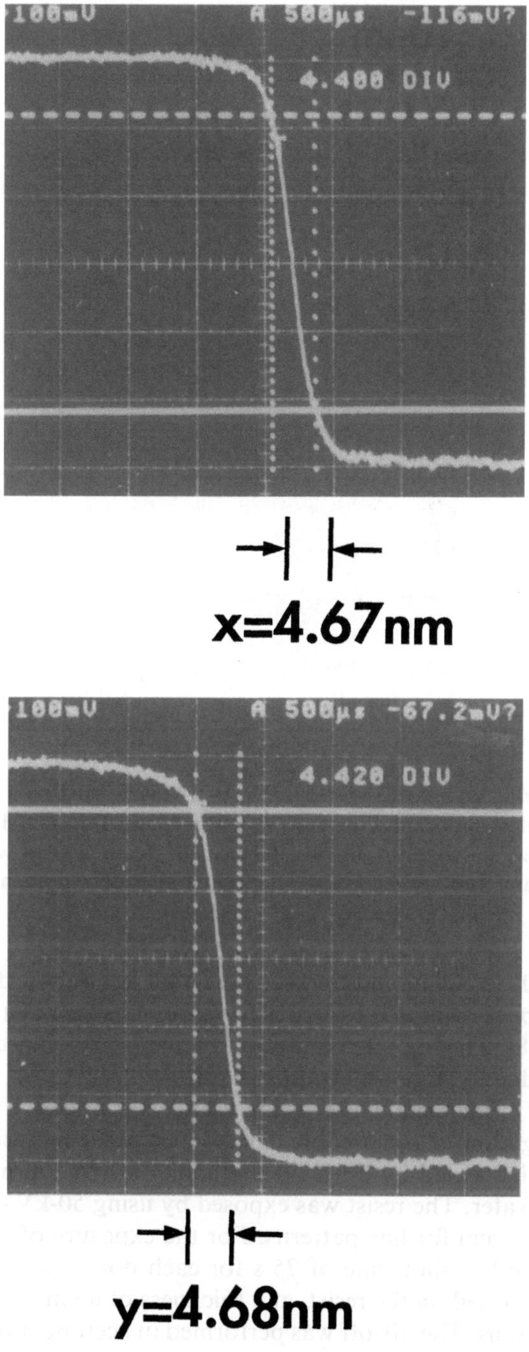

Fig. 4. Electron probe size measurements by a knife edge method. The sizes in x and y directions are 4.67 nm and 4.68 nm, respectively

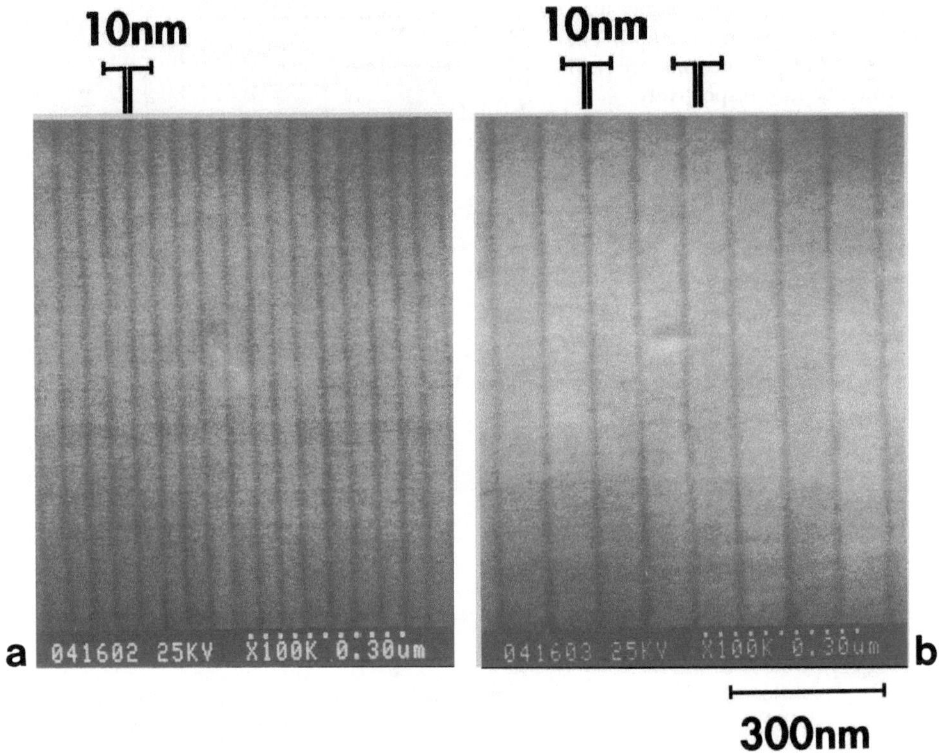

Fig. 5a, b. SEM photographs of PMMA resist on a thick Si wafer exposed by using 50 kV electron beam. Line dose was 0.8 nC/cm. Ten-nm width line patterns with periods of 50 nm (**a**) and 100 nm (**b**) were fabricated. Beam current was 100 pA and resist thickness was 50 nm. Ten-nm Au-Pd was deposited after development

different periods, since there was less forward scattering in the resist and fewer back-scattered electrons from the substrate at a high acceleration voltage, of 50 kV [8]. The high acceleration voltage is also expected to reduce the charge up effect due to large penetration depth.

Fine metal patterns are useful as conductive wires or gate metals for the investigation of mesoscopic devices and other nanostructure physics. Au-Pd metal was delineated by a lift-off method. PMMA (30-nm-thick) was coated on a thick Si wafer. The resist was exposed by using 50-kV electron beam at a line dose of 1.44 nC/cm for line patterns. For the exposure of dot patterns, the resist was exposed with a shot time of 75 s for each dot. After development, Au-Pd metal was deposited on the resist, at a thickness of 6 nm for line patterns and 3 nm for dot patterns. The lift-off was performed in acetone. Figure 6 shows the SEM photographs of Au-Pd metal line and dot patterns on a Si wafer. Twenty-nanometer-width line and 20-nm-diameter dot patterns with a period of 100 nm were successfully fabricated.

In-situ processes such as beam-induced chemistry are promising technologies to

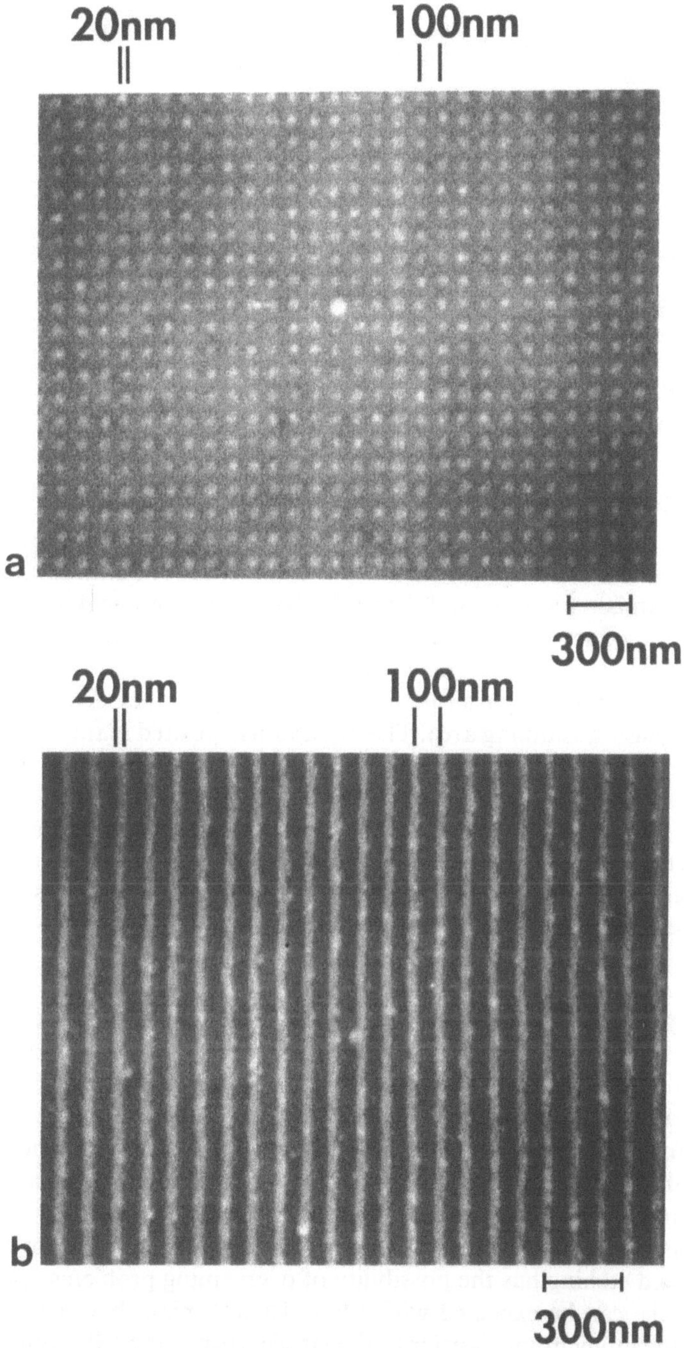

Fig. 6a, b. SEM photographs of Au-Pd metal line and dot patterns on Si wafer made by a lift-off method using PMMA. Au-Pd thickness was 3 nm for dots (**a**) and 6 nm for lines (**b**)

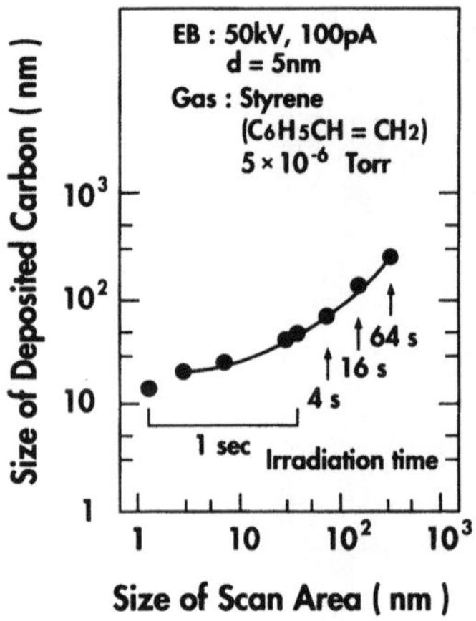

Fig. 7. Size of deposited carbon versus size of scan area using styrene gas as a source gas

fabricate ultimate fine patterns and to reduce process steps. Styrene ($C_6H_5CH = CH_2$) gas contained in a cylinder was used as the source gas in this experiment. A bare Si wafer was used as a substrate. The gas was introduced until the vacuum in the specimen chamber was 5×10^{-6} Torr; the base pressure was 4.5×10^{-7} Torr before the gas introduction. The vacuum in the gun chamber had changed from 2×10^{-9} Torr to 3×10^{-9} Torr due to gas introduction.

The sizes of the deposited patterns were plotted (Fig. 7); they decreased considerably with decreasing scanning area. The minimum deposited material size was 14 nm; this was obtained without scanning the beam during deposition. This size is three times larger than the beam spot size and is smaller than lift-off Au-Pd metal lines and dots. Figure 8 shows the carbon "NEC" pattern deposited at a dose of 10 C/cm^2. Carbon has a high etching selectivity for various semiconductors [7, 8]. The carbon deposited film can be used as a mask for etching to transfer the fine patterns into semiconductors and other materials.

Low Damage Dry Etching by Showered EB Assisted Etching

Dry etching is a very important technology that enables the transfer of patterns where mask patterns are transfered during device processes. Usually, ion beams are applied during the dry etching process. Ion bombardment enhances surface chemical reactions to produce volatile products. Ion beam etching has a serious disadvantage in that ion beam bombardment causes such radiation damage in semiconductors as point defects, dislocation, and amorphization. On the other hand, EB assisted etching has the possibility of overcoming problems due to damage. Less damage can be expected with EB assisted etching than with ion beam etching, since the momentum transfer to the lattice when using EB is much smaller than when ion beams are used. Also, only chemical reactions are introduced in EB assisted etching. Akita et. al. reported that 10 keV focused EB assisted etching using Cl_2 gas caused less damage to a GaAs sample than when Ar plasma etching was used [9]. It is expected that dry etching using showered EB assisted etching will

Fig. 8. Carbon pattern formed by EB deposition using styrene gas

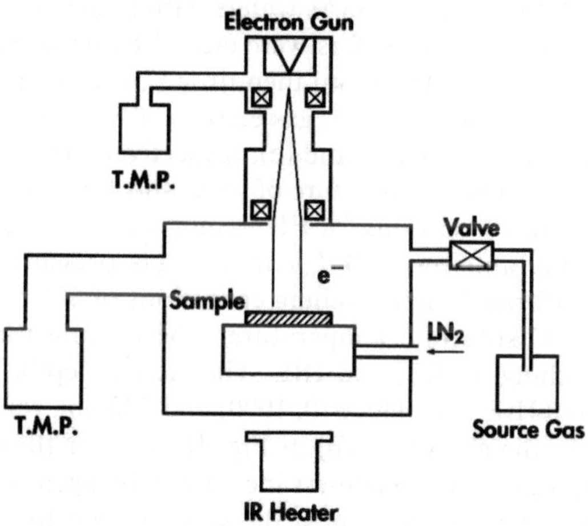

Fig. 9. Experimental arrangement for showered EB assisted etching apparatus. *T.M.P.*, turbo-molecular pump

overcome this problem of damage in ion beam dry etching. Dry etching is required in device fabrication to limit damage, but also to produce stronger anisotropy and higher selectivity.

Figure 9 schematically illustrates the experimental apparatus. The system was evacuated with a turbo-molecular pump. The base vaccum was 1×10^{-7} Torr. The chamber contains a gas introduction system, an EB gun with a differential

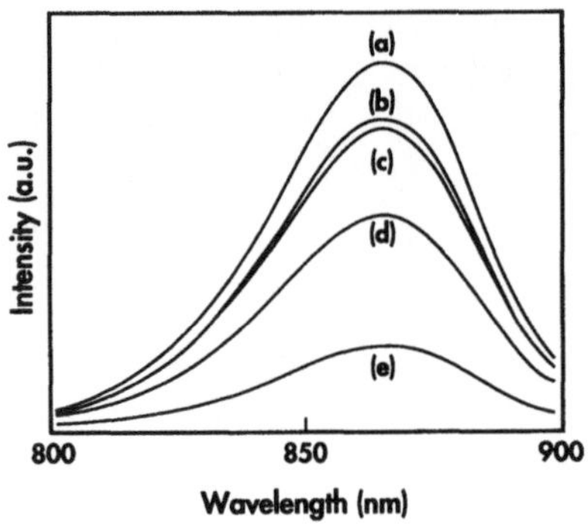

Fig. 10. PL spectra at the etched surface of GaAs. (*a*) Reference, (*b*) Cl_2 gas etching, (*c*) EB assisted etching with Cl_2 gas (*d*) Cl_2 RIBE, (*e*) ArIBE

pumping system, and a sample holder. A GaAs sample can be heated using an infrared (IR) heater and cooled by liquid nitrogen and wafer circulation. A reflection high energy electron diffraction gun was used as the showered EB source in this experiment. The beam voltage, current density, and diameter for the sample were 10 kV, 1.4 $\mu A/cm^2$ and 40 mm, respectively. The calculated temperature increase on the sample surface is less than 10°C for GaAs substrate.

The damage induced by EB assisted etching was evaluated through photoluminescence (PL) measurements and compared with the damage induced by reactive ion beam etching (RIBE) and ion beam etching (IBE) [10]. The PL was measured at room temperature in air. These experiments were performed using HB Si doped n^+ GaAs wafers with a carrier concentration of $1 \times 10^{18} cm^{-3}$. Prior to etching, the wafers were etched by immersion in HCl for 1 min, to remove the native oxide layer and then rinsed by dipping for 5 min. in H_2O. PL measurement spectra for reference, gas etching, EB etching, RIBE, and IBE samples are shown as (a), (b), (c), (d), and (e), respectively, in Fig. 10. Gas etching was carried out at a substrate temperature of 75°C and 1.5×10^{-4} Torr Cl_2 gas pressure. The experimental conditions for EB etching were 10 kV, 1.4 $\mu A/cm^2$, 75°C substrate temperature and 1.5×10^{-4} Torr Cl_2 gas pressure. Both Cl_2 RIBE and Ar IBE were performed under etching conditions of 500 V and 1.5×10^{-4} Torr gas pressure at 75°C substrate temperature. The etching time was 30 min for gas etching, EB etching, RIBE, and IBE. The etched depths for gas etching, EB etching, RIBE, and IBE were 440, 670, 1000, and 430 nm, respectively. The PL intensity of the gas etching shown by (b) in Fig. 10 is lower than that of the reference shown by (a), since surface roughness increases when gas etching is used. The PL intensity of the EB etching was almost the same as that for gas etching, as shown by (b) and (c) in Fig. 10. On the other hand, the PL intensities of RIBE and IBE were much smaller than that for EB etching, as shown by (d) and (e) in Fig. 10. These results indicate that showered EB assisted etching rather than RIBE and IBE, is a promising method for achieving damage-free dry etching.

Figure 11 shows the etching depth for EB assisted and Cl_2 gas etching under various substrate temperatures, from −170°C to 100°C over 60 min [11]. At 75°C substrate temperature, the etching rate of EB assisted etching is twice that of Cl_2 gas etching. The increase in etching rate due to EB irradiation was observed even

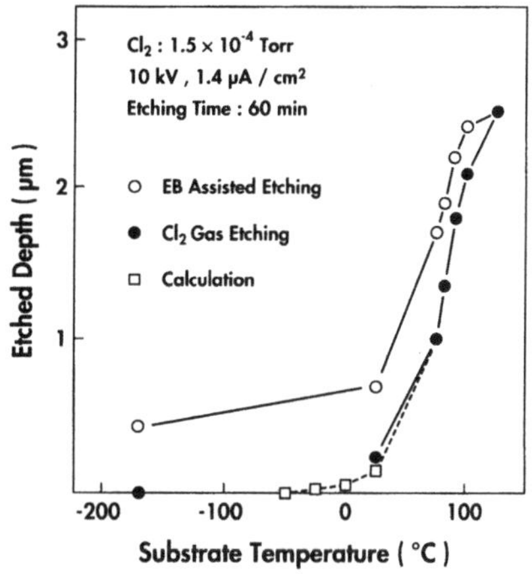

Fig. 11. Etching depth for EB assisted etching and Cl_2 gas etching under various substrate temperatures

at −170°C substrate temperature. This Fig. shows the calculated etching depth, using Cl_2 below a temperature of 25°C, estimated from the experimental data at a range of 75°C to 100°C. The Cl_2 gas etching depth at 25°C correlates well with calculated and experimental results. Cl_2 gas etching does not occur at −170°C because GaClx and AsClx are nonvolatile materials at this temperature. In any case, etching rate ratio can be improved by reducing substrate temperature. These results indicate that EB irradiation enhances not only the chemical reaction between Cl_2 gas and the GaAs substrate, but also the desorption of products, that is, electron-stimulated desorption. The etching rate and etching yield were estimated to be 6.5 nm/min and 55 atoms/electron from the dependence of low temperature EB assisted etching depth on etching time.

Figure 12 shows the cross-section of a 0.4 μm linewidth GaAs pattern with 0.7 μm depth formed by EB assisted etching at a substrate temperature of −170°C, using a resist mask with a linewidth of 0.5 μm. This result indicates that high-precision pattern transfer onto a GaAs substrate can be achieved using this method.

Atomic Layer Etching by STM-EBISED

The STM microfabrication system was constructed by modifying a conventional STM apparatus (USM-201; Unisoku Corp.), as shown in Fig. 13. The STM unit is built within a vacuum chamber into which gases can be introduced through a variable leak valve. This chamber was evacuated by a turbo-molecular pump, and a base chamber pressure of 1.0×10^{-7} Torr was established. Scanning areas from $50 \times 50 nm^2$ to $3000 \times 3000 nm^2$ can be selected. Pt mechanical tips were used as tips. The GaAs(100) substrates were held in the STM head. At this base pressure, the surface of the sample is covered by an oxide layer; this layer can be easily removed with Cl_2 gas by heating up to 65°C. The samples were heated to more

Fig. 12. SEM micrograph of GaAs etching pattern formed by low temperature EB assisted etching

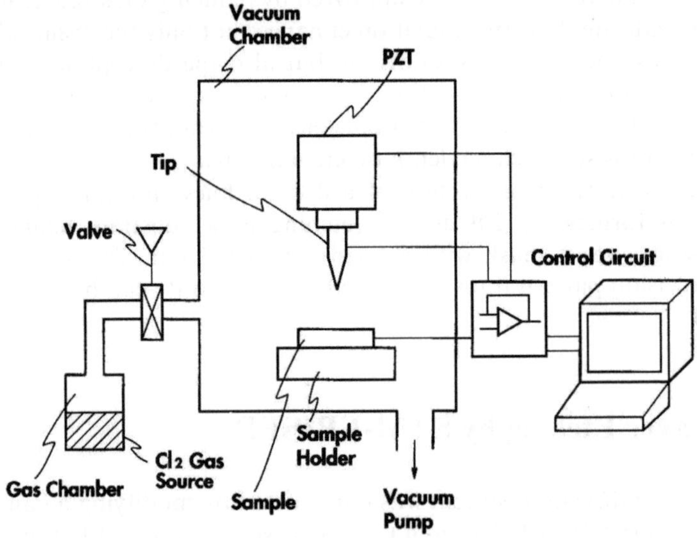

Fig. 13. STM microfabrication system

Fig. 14. Experimental procedures for GaAs atomic layer etching by STM-EBISED

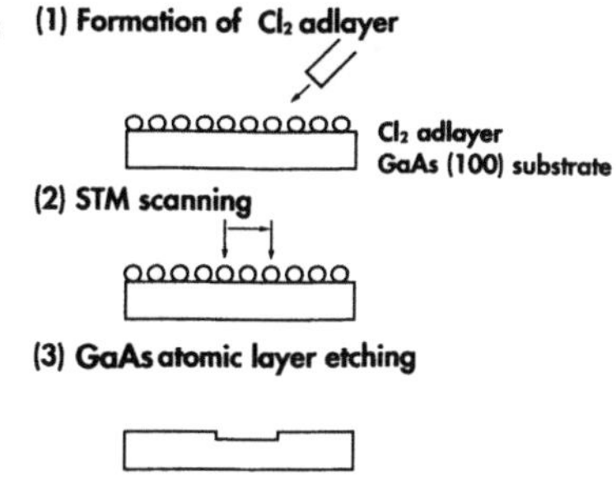

than 65°C to remove the oxide layer on the GaAs surface in this experiment. Imaging was typically performed with a steady-state sample voltage of -1V and a constant current of 1 nA. Figure 14 shows the experimental procedures for GaAs atomic layer etching [12]. First, the Cl_2 gas was allowed to flow for 5 min with a pressure of 1×10^{-5} Torr and the residual Cl_2 gas was evacuated outside. As a result, an adlayer of Cl gas molecules was formed on the GaAs surface. The STM tip was then approached to the surface of sample. The tip was moved within a $500 \times 500 nm^2$ square by applying sample voltage (-1V) and constant current (1nA). Finally, the etched surface was found by measuring in a $1000 \times 1000 nm^2$ square. Figure 15a shows the etching pattern in this procedure.

The GaAs etched depth was about 0.3–0.6 nm, as shown in Fig. 15b. One layer height of GaAs(100) is 0.246 nm. Therefore, the etched depth corresponds to almost the value for one or two layers. These results indicate that the tunneling electrons cause an assist-etching reaction on a GaAs surface and the GaAs atomic layer etching can be achieved by STM-EBISED using a Cl_2 adlayer.

Measurement of Sidewall Roughness by STM

Akama et al. have already reported that a new STM tip fabricated by EB deposition not only had an atomic scale resolution, but also had the capability of correctly imaging deeply structured patterns [13].

It is impossible to measure sidewall roughness, even using an EBD tip, since the tip may or not employ a well-found interaction to the sidewall surface through the tunneling current, as shown in Fig. 16a. That is, the tunneling current does not always flow through the tip apex in the vicinity of a steep step. If a STM tip has a flat surface part and a sidewall part, as shown in Fig. 16b, the horizontal tip apex has well-found interaction to the sidewall surface, and the tip can be driven along the sidewall. A tip of this sort is termed a "branched EBD tip"; Fig. 17 shows a scanning electron micrograph of such a tip [14]. The tip has two branches. One part is perpendicular for measuring a flat surface, and the other is horizontal, for tracing a sidewall surface. The tip was prepared as follows: An electrochemically sharpened tungsten tip was set into the scanning electron microscope. An electron-beam was then focused and fixed on the tip apex for about 1h. A lightning conductor-like contamination tip grew with EB irradiation time. The deposition rate was

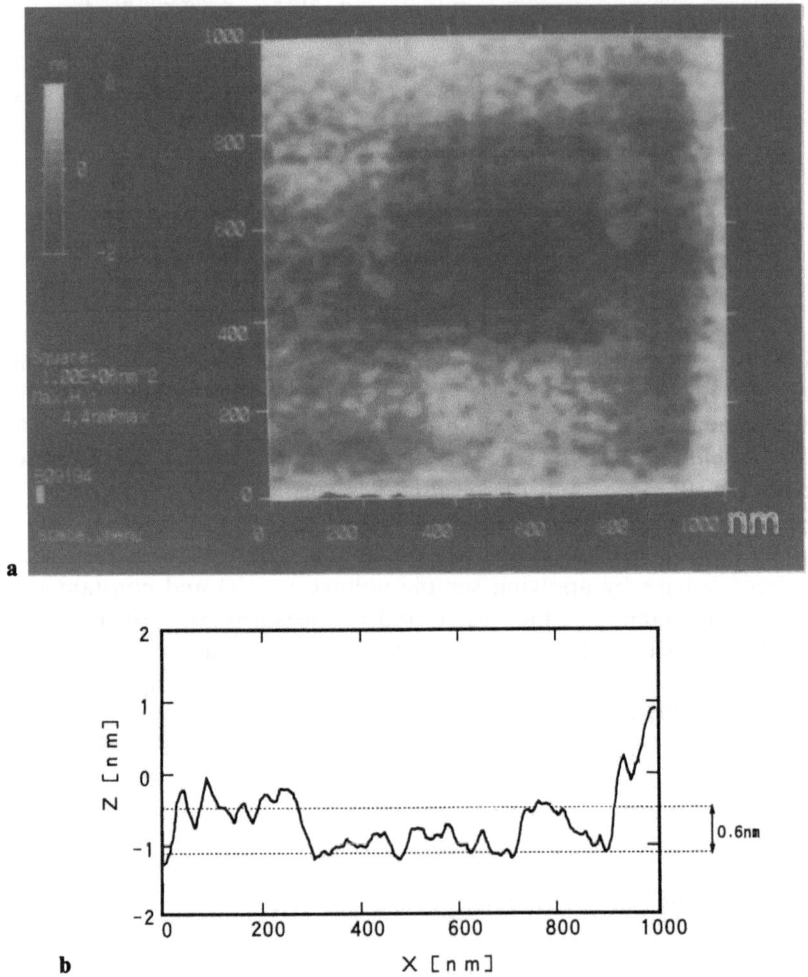

Fig. 15a, b. Atomic layer etched GaAs pattern. **a** STM image of GaAs etching pattern. **b** Cross-section of **a**

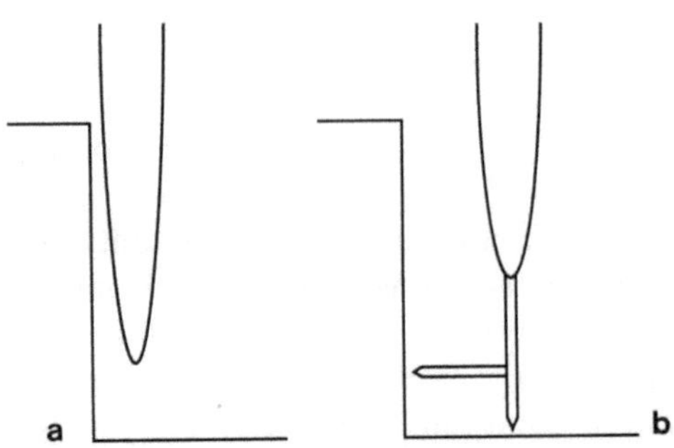

Fig. 16a, b. Illustration of sidewall roughness measurement **a** Conventional STM, **b** new STM using EBD tip

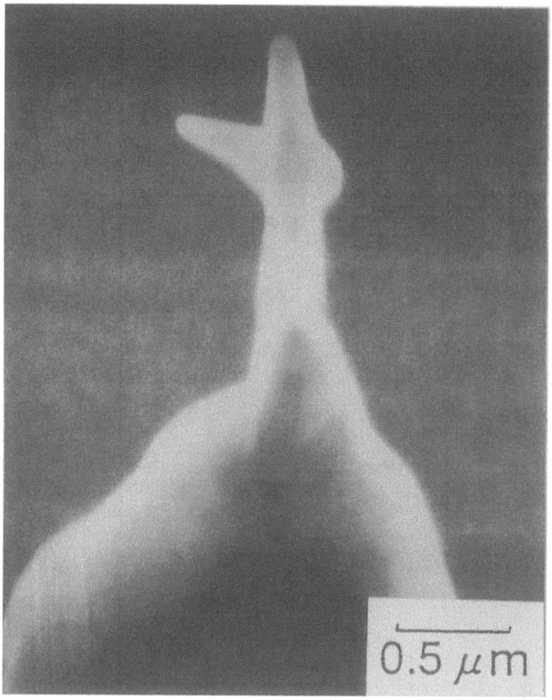

Fig. 17. SEM micrograph of a branched EBD tip

about 1 μm/h at an acceleration voltage of 25 kV. A 50-nm-thick Au film was finally coated onto the tip surface by radio frequency (RF) sputtering to increase its own electrical conductivity.

In a conventional STM, a one-dimensional servo system is adopted, where only a z-piezoelectric actuator is controlled by feedback servo, and both x- and y-piezoelectric actuators are driven at a given scanning speed. The tip is apt to collide with a large protrusion and a steep step on a sample surface. Namely, the tip cannot trace along sidewall unevenness. A newly-developed two-dimensional, at least x- and z-axis, servo system is required for sidewall roughness measurement. The two-dimensional servo system is based on partition of the x-axis and z-axis tunneling currents. The tunneling current is resolved into the x-axis and z-axis components by a subnanovibration mode. In tunneling current measurements, Δx and Δz vibrations were applied by x- and z-piezoelectric servo control actuators, respectively, in order to separate the x-axis and z-axis tunneling currents. In this case, Δx and Δz should be less than 1 nm to avoid a collision between tip and sample. Where ΔIz is larger than ΔIx, the tip is located as shown in Fig. 18a; where ΔIz is nearly equal to ΔIx, the tip is located at the corner, as shown in Fig. 18b; and where ΔIz is smaller than ΔIx, the tip is located at a sidewall, as shown in Fig. 18c. X- and z-tunneling current components were employed as feedback signals to x- and z-piezoelectric actuators, respectively.

The sidewall roughness measurement for line and space patterns was carried out as follows:

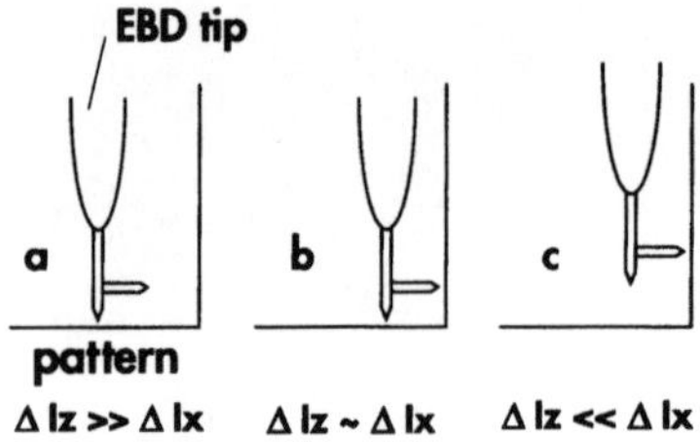

Fig. 18a–c. Relationship between EBD tip position and tunneling current for subnanovibration

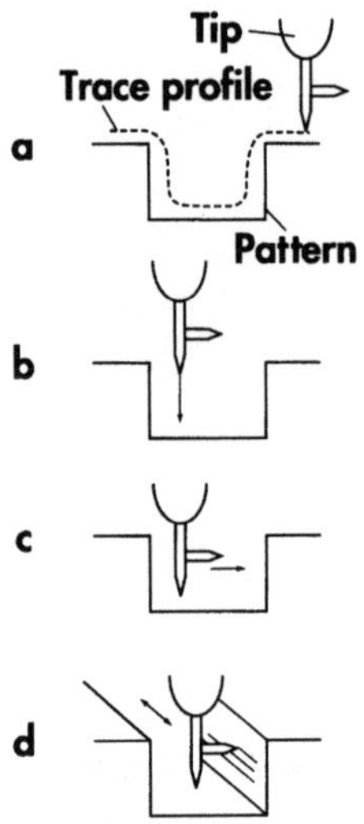

Fig. 19a–d. Illustration of sidewall roughness measurement (see text for explanation)

1. A line and space pattern was roughly recognized by using a two-dimensional servo system, as shown in Fig. 19a.
2. The tip was driven to a groove region in the line and space pattern, as shown in Fig. 19b.
3. The tip was approached to the sidewall until the tunneling current began to flow between the horizontal part of the branched EBD tip and the sidewall surface, as shown in Fig. 19c.
4. Finally, the tip scanned along the sidewall surface rising in the z-direction.

The sidewall roughness was measured, as shown in Fig. 19d.

Figure 20a, and b, shows the measured sidewall roughness values for the SAL601-ER7 resist pattern fabricated by EB and the GaAs pattern etched by Cl_2 RIBE under etching conditions of 300 V and 1.5×10^{-4} Torr gas pressure, respectively. The average measured sidewall surface roughness for the resist and GaAs etched patterns was 3.49 nm and 3.05 nm, respectively. It has been clearly shown that the newly developed STM system is a useful tool for the characterization of sidewall roughness.

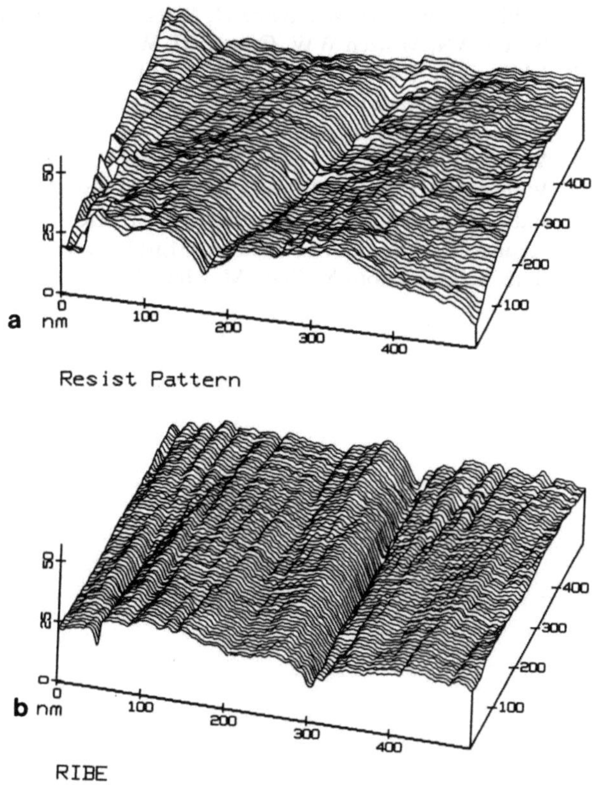

Fig. 20a, b. Sidewall roughness distribution **a** SAL601-ER7 resist pattern fabricated by EB, average roughness = 3.49 nm. **b** GaAs pattern etched by Cl_2 RIBE, average roughness = 3.05 nm

Conclusion

Nanostructure technology has been studied by using electron beam-induced surface reactions. Ten-nm linewidth PMMA resist patterns and 14-nm diameter carbon dot patterns were demonstrated by a 50 kV electron beam system with a gas introduction line. It has been confirmed that the damage induced by EB assisted dry etching is less than that induced by reactive ion beam etching. Atomic layer etching for GaAs was achieved by STM assisted etching using a Cl_2 adlayer. The side-wall roughness of GaAs patterns etched by Cl_2 RIBE has been successfully measured by using a new STM tip made by electron beam deposition.

References

1. Matsui S, Ichihashi T, Mito M (1989) J Vac Sci Technol B7:1182
2. Matsui S, Ichihashi T, Baba M, Satoh A (1990) Superlattices and Microstructures 7:295
3. Matsui S, Ichihashi T (1988) Appl Phys Lett 53:842
4. Baba M, Matsui S (1990) Jpn J Appl Phys 29:L2854
5. Nakazawa H, Takemura H, Isobe M (1988) J Vac Sci Technol B6:2019

6. Ochiai Y, Baba M, Watanabe H, Matsui S (1991) Jpn J Appl Phys 30:3266
7. Broers AN, Molzen WW, Cuomo JJ, Wittels ND (1976) Appl Phys Lett 29:596
8. Ishibashi A, Funato K, Mori Y (1988) Electron Lett 24:1035
9. Akita K, Sugimoto Y, Taneya M, Hiratani Y, Ohki Y, Kawanishi H, Katayama Y (1990) Proc SPIE 1392:576
10. Watanabe H, Matsui S (1991) Jpn J Appl Phys 30:3190
11. Watanabe H, Matsui S (1992) Jpn J Appl Phys Lett 31:L810
12. Baba M, Matsui S (1992) Proceedings of MRS 1991 Fall Meeting, 236:159
13. Akama Y, Nishimura E, Sakai A, Murakami H (1990) J Vac Sci Technol A8:429
14. Sato A, Tsukamoto Y, Baba M, Matsui S (1991) Jpn J Appl Phys 30:3298

MOCVD Methods for Fabricating Semiconductor Nano-Structures

37

TAKASHI FUKUI[1], SEIGO ANDO, AND HISAO SAITO[2]

Summary. Methods of fabricating quantum well wires and dots by metalorganic chemical vapor deposition (MOCVD) are reviewed. There are two methods of fabricating very narrow as-grown quantum wires and dots by crystal growth. One, using crystallographic facets grown by selective area MOCVD, produces facet quantum wires and quantum dots. The other using fractional-layer superlattices on GaAs vicinal surfaces, grows quantum wire arrays. The advantage of these quantum well wire and dot fabrication methods is that wire and dot widths as narrow as 10 nanometers can be obtained without any processing damage.

Introduction

Semiconductor quantum wires and dots have recently been actively studied to investigate their theoretical properties, and to apply them to high-speed, quantum interference and opto-electronics devices [1–3]. There have been several attempts to fabricate quantum structures by simply restricting the lateral dimensions of quantum well or modulation doped wafers using physical or chemical processing. Processing technologies such as reactive ion etching combined with electron beam lithograpy have produced quantum wires and ring structures on the 30–100nm scale [4, 5].

Recently, we have reported new quantum wire fabrication methods based on crystal growth, in which electron channel regions or active regions are surrounded two- or three-dimensionally by cladding materials. These methods, compared to other processing techniques, have the advantage of forming damage-free interfaces.

This paper reviews MOCVD methods for fabricating quantum wires and dots based on our recent study. Facet quantum wires and dots can be fabricated, using crystallographic facets grown by selective area growth [6–9], and quantum wires can also be fabricated using fractional-layer superlattices on vicinal surfaces [10, 11].

[1] Research Center for Interface Quantum Electronics, Hokkaido University, N13, W8, Sapporo, 060 Japan
[2] NTT Basic Research Laboratories, Midori-cho 3-9-11, Musashino Tokyo, 180 Japan

Quantum Wires and Dots by Facet Growth

In MOCVD, crystallographic facets appear during selective area growth. The facet orientations are usually low index crystal planes, such as (111)A, (111)B, and (110). On these crystal planes, crystal growth rates are slow and they depend strongly on the growth conditions. Using these facet properties, we fabricated three kinds of quantum confinement structures: facet quantum well wires [6, 7], lateral quantum well wires [8], and tetrahedral quantum dots [9].

Facet Quantum Well Wire

GaAs quantum well wires were fabricated on (111)B crystallographic facets by selective area growth. Figure 1 shows a schematic diagram and a scanning electron microscope (SEM) photograph of the cross-section of an $Al_{0.3}Ga_{0.7}As$/GaAs double-hetero (DH) structure facet quantum wire. The substrates were GaAs (001) wafers with SiO_2 stripe masks (2μm-wide lines and spaces). The stripes were oriented to the (110) direction . First, undoped $Al_{0.3}Ga_{0.7}As$/GaAs DH structures were selectively grown on SiO_2 stripe windows with high AsH_3 partial pressure and a low growth temperature (650°C). The selectively grown DH layers were trapezoidal with (111)B facets as the sidewalls. The results suggest that no crystal growth occurs on the SiO_2 mask region, or even on the (111)B sidewalls. The group III source materials diffuse through the (111)B sidewalls to the (001) top surface.

Next, undoped and Si-doped $Al_{0.3}Ga_{0.7}As$ were grown over the (111)B and (001) facets at 800°C. At higher growth temperatures, crystal growth occurs both on (111)B and (001) surfaces. Modulation doped electrons accumulate in the GaAs at the (111)B sidewall interface, and the GaAs channel width can be controlled by the growth thickness of GaAs. The GaAs channel width in Fig. 1 is 0.1μm. The existence of a 2-dimensional electron gas (2DEG) on (111)B facets was confirmed by the dependence of Shubnikov-de Haas (SdH) oscillations on magnetic field direction.

Similar facet quantum wires with wider channel widths (0.3–1.0μm) were used to investigate quantum interference properties, such as universal conductance fluctuations and negative magnetic resistance. We also observed the temperature dependence of the inelastic scattering lengths in both 1-dimensional and 2-dimensional electron cases [12].

Compared with other techniques for fabricating quantum well wires, the advantage of this technique is that there is no size fluctuation in the quantum wire width along the wire direction, because the wire width is fundamentally determined by the growth thickness of the trapezoidal GaAs layer.

Lateral Quantum Well Wires

In selective area growth, it is very convenient if we can obtain arbitrarily-shaped quantum structures by independently controlling the vertical and lateral growth by changing the growth conditions. In MOCVD, (111)B and (110) are extremely

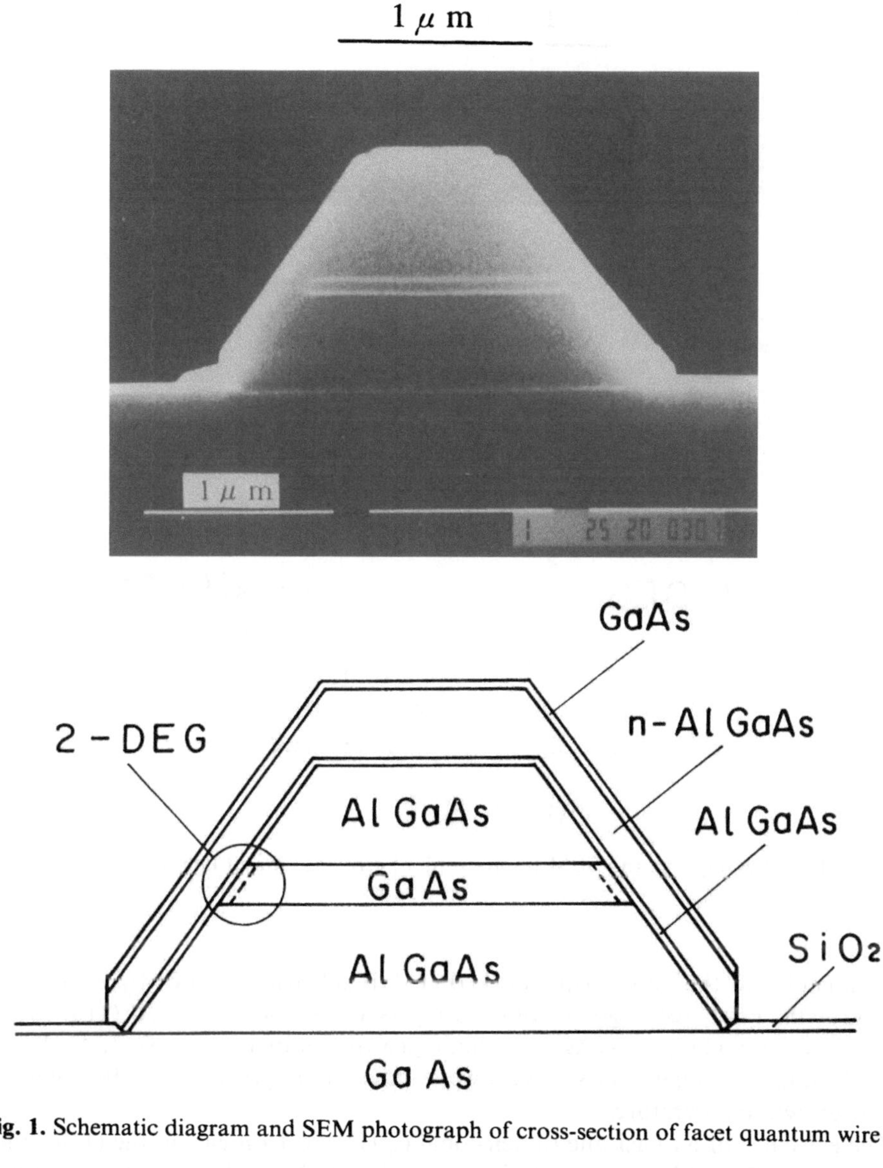

Fig. 1. Schematic diagram and SEM photograph of cross-section of facet quantum wire

contrasting surfaces, because crystal growth only occurs at a high growth temperature on (111)B surfaces, but at a low growth temperature on (110) surfaces. Therefore, if we use (111)B substrates and (110) sidewalls perpendicular to the surfaces in selective area MOCVD growth, we can control the growth on the (111)B top surface and (110) sidewalls independently by changing the growth conditions. Furthermore, we can obtain various lateral structures, such as lateral quantum wires and lateral superlattices.

Figure 2 shows typical GaAs lateral quantum wires (LQWs) on (110) sidewalls

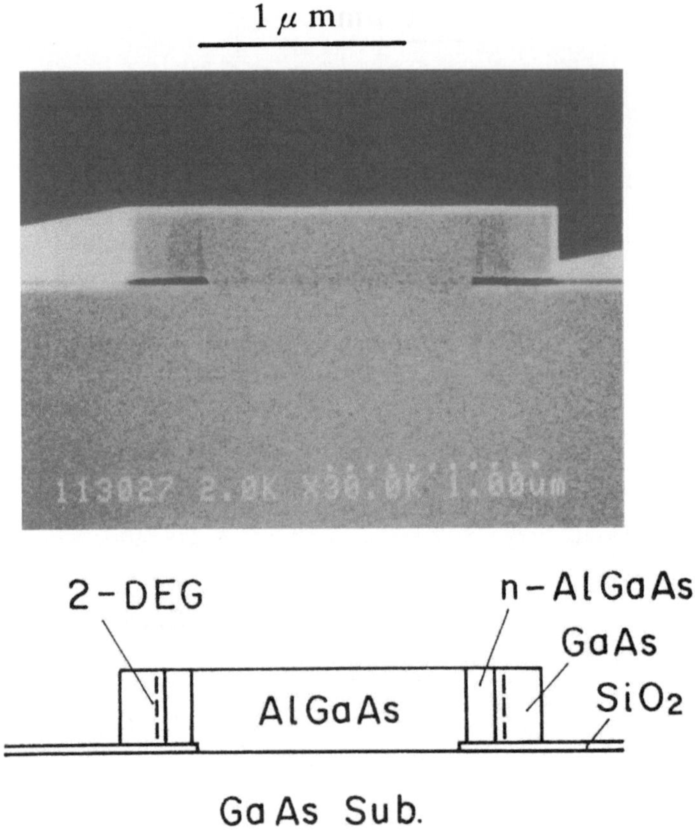

Fig. 2. Schematic diagram and SEM photograph of cross-section of lateral quantum wire

perpendicular to the SiO_2 stripe masked (111)B substrates grown by using two-step growth. First, rectangular AlGaAs layers whose sidewalls had (110) facets were grown on (111)B GaAs at a high growth temperature. Next, Si-doped AlGaAs and non-doped GaAs layers were sequentially grown on (110) sidewalls at low growth temperatures.

The existence of a quasi-one-dimensional (Q1D) electron gas on the (110) sidewalls was confirmed by the orientation-dependent SdH oscillations. For the sample shown in Fig. 2, the GaAs channel width measured by SEM is 0.35μm. The actual channel widths of the Q1Ds on (110) sidewalls were determined by fitting the magnetoconductance in a weak magnetic field using a one-dimensional weak localization theory. The width of the electron channel is about 160nm, which is smaller than the geometrical width. This indicates that there is a vertical depletion length of about 100nm at each channel edge.

There are many applications of this alternate vertical and lateral growth. If DH semiconductor laser structures are grown on SiO_2 stripe-masked GaAs (111)B substrates with facets several hundred nanometers wide, (110) laser cavities are automatically formed. Furthermore, AlGaAs/GaAs lateral superlattices can be formed on (110) sidewalls.

Tetrahedral Quantum Dots

New GaAs quantum dot structures, called tetrahedral quantum dots (TQDs) are one way to make a zero-dimensional electron-hole system. They can be fabricated using procedures similar to those used for lateral quantum well wires. The substrates were SiO_2 masked (111)B GaAs partially etched to remove a triangle of SiO_2. Figure 3 shows the fabrication procedure for the tetrahedral quantum dot structures on GaAs (111)B substrate. First, thick AlGaAs buffer layers were grown on the substrate. Next, only near the top of the tetrahedral structure, GaAs dots were grown. Finally, AlGaAs was over-grown on the whole tetrahedral structure at lower growth temperature. Under these growth conditions, crystal growth occurs on (110) facets.

The preliminary fabrications of TQDs were studied using low-pressure MOCVD. Growth temperature was 800°C, and sidewalls showed (110) facets. Figure 4 shows a schematic view and SEM image of a cross-section of GaAs tetrahedral structures buried in AlGaAs. Although the pattern size of substrate was 2μm in this trial, it is easy to obtain sub-micron sized tetrahedrons by the same procedure.

The energy sublevel structures of 0-dimensional electrons in a GaAs TQD were also calculated. Large quantum size effects were obtained compared with a single quantum well, since electrons were confined three-dimensionally. Furthermore, for electrons, the effective size of the TQD is small: close to the diameter of the inscribed sphere, rather than the length of the side of the tetrahedron. The ground

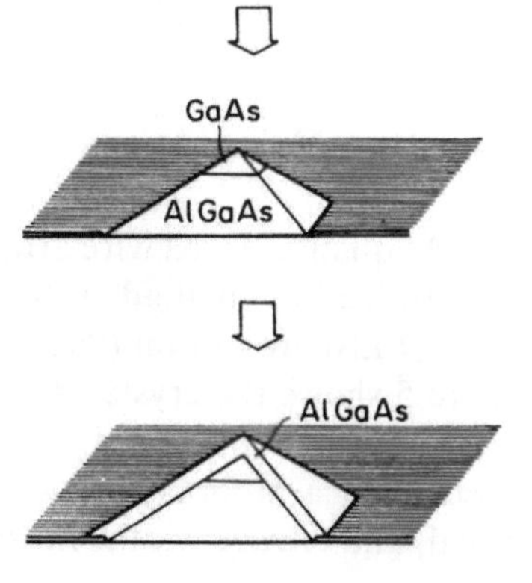

Fig. 3. Fabrication procedure for tetrahedral quantum dot structures on GaAs (111)B substrate

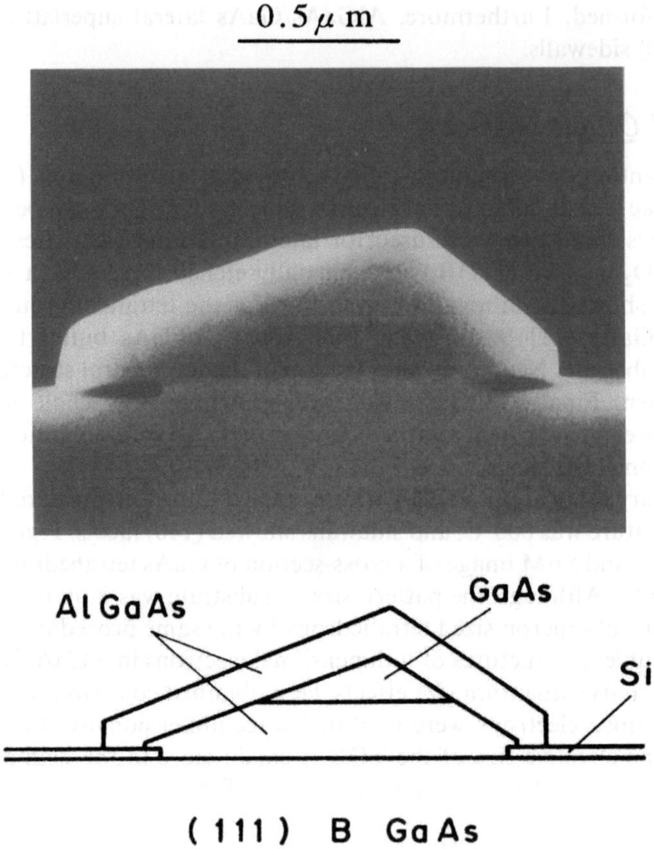

Fig. 4. Schematic view and SEM image of cross-section of GaAs tetrahedral structures buried in AlGaAs

sublevel E_o is 20 meV above the bottom of the conduction band E_c for 100-nm TQDs.

TQDs are interesting structures for investigating 0-dimensional excitons and nonlinear optical properties, and for applying to quantum dot lasers.

Quantum Wire Array on GaAs Vicinal Surface

A GaAs quantum well wire array less than 10nm wide was fabricated by MOCVD. The fabrication method is based on $(AlAs)_{1/2}(GaAs)_{1/2}$ fractional-layer superlattice (FLS) growth on (001) GaAs substrates slightly misoriented toward (110). Figure 5 shows the crystal growth procedure for a FLS grown on a (001) vicinal surface, as first proposed by P.M. Petroff [13]. Growth must be controlled to exactly half a monolayer, two-dimensional growth on the terrace must be prevented, and terrace widths must be equal.

Fig. 5. Crystal growth procedure for a FLS grown on a (001) vicinal surface

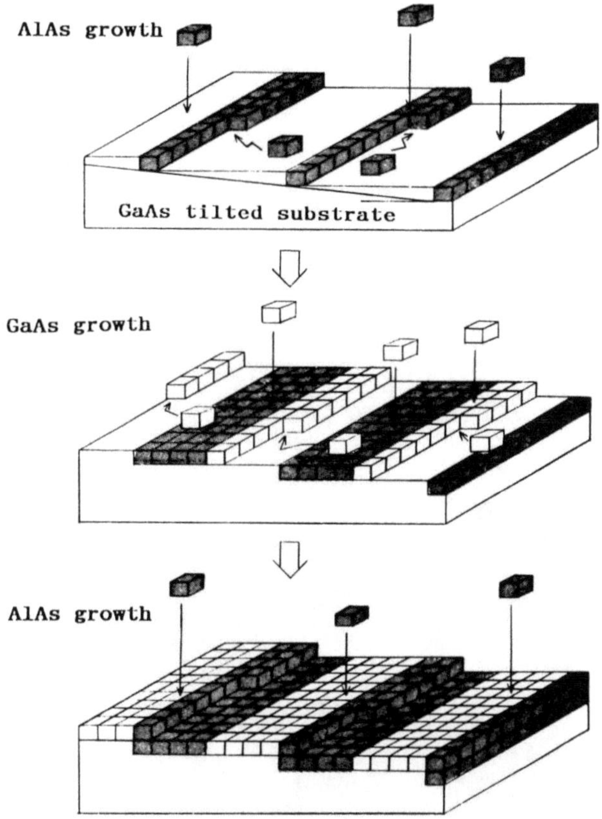

Very thin epitaxial layer growth, e.g., alternate half-monolayer growth, was achieved at a low growth rate and rapid change in the gas composition inside the reactor. The growth temperature was 600°C. The substrate misorientation angles were 1.0°–3.5°, corresponding to superlattice periods of 16–5 nm. Figure 6 shows transmission electron microscope (TEM) photographs of the top of an FLS. The clear contrasts between the AlAs (bright region) and GaAs (dark region) have an 8-nm period, which is equal to the mean terrace width on a 2-degree-tilted substrate. Furthermore, these TEM images prove that each superlattice period has a uniform width.

Polarization-dependent optical absorption and photoluminescence in FLS shows the absorption anisotropy at the band edge caused by the carrier confinement effects in the parallel direction of superlattices [14, 15]. The observed separation in the band edge wavelengths corresponds to the energy difference between the heavy- and light-hole related transitions.

We have also fabricated quantum wire transistors and electron wave interference transistors, in which we observed mobility modulation and electron wave interference effects [16, 17]. The FLS structure is very promising for new types of low-dimensional electron and optical devices.

Fig. 6. TEM photographs of the top of a FLS. Substrate misorientation angle is 2°

To summarize, quantum well wires and dots were fabricated by MOCVD. Facet quantum wires, lateral quantum wires and tetrahedral quantum dots were grown using selective area growth. The fundamental characteristics of low-dimensional electron gas were discussed. We also fabricated quantum well wire arrays on a GaAs (001) vicinal surface, using fractional-layer superlattices.

Acknowledgments. The authors would like to thank H. Asai, S. Yamada, Y.K. Fukai, K. Tsubaki, Y. Tokura , H. Ando, M. Kasu, and H. Kanbe for their fruitful discussions.

References

1. Sakaki H (1980) Jpn J Appl Phys 19:L735
2. Tokura Y, Tsubaki K (1988) Appl Phys Lett 53:1807
3. Arakawa Y, Sakaki H (1982) Appl Phys Lett 40:939
4. Washburn S, Umbach CP, Laibowitz RB, Webb RA (1985) Phys Rev Lett B32:4789
5. Roukes ML, Scherer A, Allen SJ Jr, Craighead HG, Ruthen RM, Beebe ED, Harbison JP (1987) Phys Rev Lett 59:3011
6. Asai H, Yamada S, Fukui T (1987) Appl Phys Lett 51:1518
7. Fukui T, Ando S (1989) Electron Lett 25:410
8. Fukui T, Ando S, Fukai Y (1990) Appl Phys Lett 57:1209
9. Fukui T, Ando S, Tokura Y, Toriyama T (1990) Extended abstracts of the 22nd conference on solid state developed matter, Sendai, August 22–24, p 99
10. Fukui T, Saito H (1987) Appl Phys Lett 50:824
11. Fukui T, Saito H (1988) J Vac Sci Technol B6:1373
12. Yamada S, Asai H, Fukai Y, Fukui T (1989) Phys Rev B15:11 199
13. Petroff PM, Gossard AC, Wiegmann W (1984) Appl Phys Lett 45:620
14. Ando H, Saito H, Fukui T (1990) Extended abstracts 22nd conference on solid state developed matter, Sendai, August 22–24, p 123
15. Kasu M, Ando H, Saito H, Fukui T (1991) Appl Phys Lett 59:301
16. Tsubaki K, Fukui T, Tokura Y, Saito H, Susa N (1988) Electron Lett 24:1267
17. Tsubaki K, Tokura Y, Fukui T, Saito H, Susa N (1989) Electron Lett 25:728

Ultra-High Resolution Electron Beam Patterning of SiO_2: A Review

38

David R. Allee,[1] Xiao Dan Pan, Alec N. Broers,[2] and Corwin P. Umbach[3]

Summary. Arrays of lines with periods down to 15nm have been directly patterned in SiO_2 with electron irradiation. Two methods have been developed to eliminate the surface contamination and enable the subsequent development in HF based etches: 1) exposing the oxide through a sacrificial layer and 2) oxygen reactive ion etching. This period is three times better than is possible with polymethylmethacrylate (PMMA), the most widely used high resolution electron beam resist. Preliminary pattern transfer experiments using chlorine reactive ion etching have transferred feature sizes down to 10nm into Si. Because patterned SiO_2 is itself extremely useful as a key component of many electronic devices, the ability to directly pattern SiO_2 with nanometer scale resolution, albeit with a high dose, will enhance the techniques available to prototype both conventional and quantum effect ultra-small devices.

Introduction

Numerous laboratories in recent years have investigated various ultra-small conventional and quantum effect devices [1–9]. Most of these devices incorporate a nanometer scale (sub-100nm) gate electrode(s). The nanometer scale electrodes have been fabricated almost exclusively with electron beam lithography, although there are several other ultra-high resolution lithographic techniques in development, including focused ion beam [10], scanning x-ray [11], and scanning tunneling lithography [12]. Of these alternative lithographic technologies, only scanning tunneling lithography has an ultimate resolution equal to or greater than electron beam lithography [13, 14]. Hence, for the present, the limits of ultra-small device fabrication are the limits associated with electron beam lithography.

For nanolithography, the minimum achievable electron beam diameter is set by spherical aberration, axial chromatic aberration, diffraction, and the gaussian spot

[1] Department of Electrical Engineering, Arizona State University, Tempe, AZ 85287-5706, USA
[2] Department of Engineering, Cambridge University, Cambridge, CB2 1PZ, UK
[3] IBM Research Division, Thomas J. Watson Research Center, Yorktown Heights, NY 10598, USA

diameter. Spherical and chromatic aberration are proportional to α and α^3, respectively, where α is the semi-angle of convergence at the image plane. Diffraction and the gaussian spot diameter, on the other hand, are inversely proportional to α. The optimum semi-angle of convergence is typically a few mrad. Since spherical and chromatic aberrations scale with the lens focal length, they are fundamentally limited by the saturation flux density of iron, 1.8T. The gaussian spot diameter, which is the minimum beam diameter neglecting aberrations necessary to contain a useful current, is limited by the brightness of the electron source. Increasing the beam potential both increases the source brightness, thereby reducing the gaussian spot diameter, and also reduces the effects of diffraction. Hence the highest resolution electron beam lithography columns have a beam potential as high as is technologically practical. Most of the experiments reviewed in this paper were done on a modified transmission electron microscope that operates up to 400kV. Using a lanthanum boride cathode, 0.4pA was obtainable in a 0.4nm beam diameter [15]. This beam diameter is equal to the smallest diameters achieved at several laboratories. The use of a higher brightness source such as a field emitter would increase beam current substantially but the minimum spot would not be altered.

Although the minimum beam diameter, 0.4nm, is atomic in size, the limits of nanofabrication are one or two orders of magnitude larger, due to beam substrate interactions and resist limits. When the beam penetrates the substrate, the electrons occasionally collide with an atomic core, resulting in large angle inelastic scattering, and interact with the bonding electrons, resulting in small angle inelastic scattering. The latter generates numerous low energy secondary electrons along the primary electron's path. From the lithographer's point of view, the beam substrate interactions have two effects. The broadening or forward scattering of the beam as it penetrates the resist directly limits the minimum feature size. Primary electrons that reemerge from the substrate surface or backscattered electrons result in a low background exposure of the resist over an area extending many microns from the incident beam. This background exposure reduces image contrast and can indirectly reduce resolution. Forward scattering can be effectively eliminated by using a thin resist. Although not usually consistent with device fabrication, backscattering can be effectively eliminated by exposing a resist on a free standing thin membrane.

Usually most serious are the resolution limits associated with the resist material itself. For nanometer scale patterning, poly-methylmethacrylate or PMMA is the most widely used and the highest resolution conventional electron beam resist. As with most positive e-beam resists, exposure to electrons breaks the polymer chains, enhancing the solubility in developer. Structures have been fabricated with PMMA down to approximately 10nm, and the densest structures fabricated have had 50nm periods [16, 17]. This minimum feature size and period agree with an estimation made from measurements of the point spread function for PMMA [18]. A thorough understanding of the exposure mechanism and an explanation as to why the resolution limit of PMMA is more than an order of magnitude larger than the electron beam diameter are still lacking. It has been suggested that the resolution limit is due to the lateral range of the low energy secondary electrons, which can break bonds with energies as low as 5eV [18]. Alternative explanations include the size of a PMMA molecule, gas evolution and pore formation upon exposure, and the dynamics of development [19].

Another ultra-high resolution electron beam resist that has been used to fabricate several devices is the negative acting "contamination resist". A finely focused, high current density electron beam polymerizes residual hydrocarbons on the specimen surface, forming a contamination mask for a wet chemical etch or ion milling. The latter approach was used to fabricate a 5nm AuPd line, and a AuPd grating has been formed on a Si substrate with 25nm linewidths on a 65nm period [15]. Again, the resolution is an order of magnitude larger than the electron beam diameter.

Several inorganic resists have resolutions that more closely approach the beam diameter. Features down to 1–2nm were achieved in AlF_3, LiF, MgF_2, NaCl [20–23], and metalβ-aluminas [24]. The densest structures have had periods of 3–4nm. These resists are self developing; a common denominator in the exposure mechanism appears to be electron stimulated desorption of the more electronegative element. Pattern transfer processes, however, that replicate this 1nm resolution have yet to be developed and these resists have therefore not been used to fabricate devices. In general, the metal halides do not hold up well to wet and dry etches. The smallest features transferred to date range between 8 and 20nm and have been fabricated in Si, Au, or Si_3N_4 with reactive ion etching or ion milling [25, 26].

In this paper, we review our work in directly patterning SiO_2 with nanometer scale resolution by electron beam exposure and development in various wet etches. Lines and spaces with a period down to 15nm have been achieved in SiO_2 membranes imaged in a transmission electron microscope. Of particular importance is the fact that SiO_2 can be directly incorporated in many devices, eliminating the difficult task of developing a pattern transfer process with nanometer scale resolution.

In 1967, O'Keeffe and Handy demonstrated that bombardment of thermally grown SiO_2 with 1 to 15kV electrons would enhance the etch rate in p-etch (15:10:300 $HF:HNO_3:H_2O$) by 3 to 4 times [27]. The exposed oxide required a dose of approximately $1C/cm^2$, and the exposure could be removed with an 800C anneal for 30min. The major difficulty with repeating these experiments has been the build-up of contamination on the surface of the oxide in the presence of the electron beam. Contamination build-up is even more rapid in the presence of finely focused electron beams. This polymerization of residual surface hydrocarbons has itself been used as a nanostructure ion milling mask [28]. However, in this case, even trace amounts of contamination in the region of exposure is deleterious and blocks the wet etch. To the authors' knowledge the smallest feature size achieved was the 0.6μm trench reported by O'Keeffe and Handy [27].

We have overcome this contamination problem with two alternative methods. The first method exposes the oxide through a sacrificial layer [29]. The contamination builds up on the surface of the sacrificial layer, which is subsequently stripped either mechanically or chemically before etching the SiO_2 in p-etch, aqueous HF, or aqueous NaOH. Both PMMA and aluminum have been successfully used as sacrificial layers. The second method uses an oxygen reactive ion etch (RIE) to remove the contamination directly from the oxide surface, eliminating the need for a sacrificial layer [30]. The latter approach improves resolution by minimizing the forward scattering of the electron beam.

Experiment

The initial experiments were done on bulk silicon substrates using a sacrificial layer to remove the contamination. Oxide was thermally grown on p-type 1-0-0 silicon in wet O_2 at 900C. The thickness was measured with an ellipsometer to be 253nm. A 100nm layer of PMMA was spun on as a sacrificial layer. Thin membranes, producing minimal lateral electron scattering, were required for studies of the ultimate resolution of this process. Substrates consisting of thin SiO_2 membranes exhibit buckling due to the compressive stress in thermally grown SiO_2. This buckling was avoided by using bilayer Si_3N_4/SiO_2 membranes, where the tensile stress in the Si_3N_4 counteracted the compressive stress in the SiO_2. The bilayers were formed by chemical vapor deposition of 250nm of Si_3N_4 onto a 250nm-thick SiO_2 layer thermally grown on a Si wafer. Chemical backetching of the Si up to the SiO_2/Si interface resulted in free standing, unbuckled, Si_3N_4/SiO_2 membranes. The thickness of the SiO_2 in the completed membranes was not under tight control due to the lack of etch selectivity between the Si and SiO_2. For the thin membrane samples, the contamination was removed with either an aluminium sacrificial layer or with an O_2 reactive ion etch.

SiO_2 samples were exposed at 100kV and 300kV in a modified JEOL 4000EX transmission electron microscope (capable of patterning bulk substrates as well as transparent membranes), and at 50kV in a high resolution scanning transmission electron microscope. With a current of 0.4 pA, the minimum beam diameter in the transmission electron microscope is 0.4nm and can be deflected under computer control with an accuracy of 1nm. Because of the relative insensitivity of SiO_2 the beam current for these experiments was increased. The highest resolution patterns on the thin membranes exposed at 300kV were obtained with higher beam current and a 3nm beam diameter as imaged directly under high magnification on the phosphor screen without a sample in the beam path. The 50kV exposures were performed using a beam current of 50pA and a beam diameter of 2–3nm. To avoid charging of the oxide/nitride bilayer membrane during exposure, a thin Al layer was thermally evaporated on the Si_3N_4 side of the membrane where the beam exited the bilayer.

After exposure of the oxide on the bulk Si substrate, the PMMA sacrificial layer was removed with ultrasonic acetone. Some polymerized PMMA remained in the region of the exposures and was removed by rubbing with a q-tip making PMMA unsuitable for the membrane samples. After exposure of the oxide on the membrane samples with an aluminium sacrificial layer, the sacrifical layer was removed with standard Al etch. The contamination build-up on other membrane samples was removed with two 15min oxygen reactive ion etches, once for each side of the bilayer. The oxygen pressure was 100mTorr. A low power of 90watts maintained the sheath voltage less than 30volts, minimizing damage to the oxide. The anti-charging layer on the Si_3N_4 side was then removed with Al etch. For the large area exposures that could be seen at 2000x magnification in an optical microscope, there was no visual difference in the exposed and unexposed regions of the oxide.

Exposed samples were then developed with either p-etch (6 min), aqueous HF (room temperature, 48%, 8.7%, and 4.6% for 5.60, and 180 s respectively), or aqueous NaOH (75C, 1.3M for 15min). The etch time for the p-etch was calibrated to remove approximately 75nm or 100nm of the unexposed oxide. The exposed patterns developed at a faster rate, resulting in trenches in the oxide. The resulting resist profiles were shadowed with AuPd at 45 degrees.

Results and Discussion

Using a 100nm PMMA sacrificial layer, lines and spaces with a period down to 21nm were written in SiO_2 on bulk substrates (Fig. 1). There are five arrays of lines decreasing in period from 340nm to 21nm by factors of two. The line dose was 7.5μC/cm. The smallest period, 21nm, is clearly discernible, but the trench width is obscured by the grain size of the metal shadowing and by the fundamental resolution limit of secondary electron images set by the lateral range (several nanometers) of the secondary electrons.

Transmission imaging, however, is based on diffraction and scattering of the high energy electrons and has an imaging resolution of a few tenths of a nanometer. Using an O_2 reactive ion etch and a thin membrane sample, an array of lines in SiO_2 with decreasing period was fabricated with 300kV electrons (Fig. 2). This array of lines was shadowed with AuPd along the diagonal. The line dose for all lines was 7.5μC/cm. There are five periods with five lines each. Every fifth line is longer, marking the change in period. The largest four periods, of 25, 22, 19, and 15nm, are clearly discernible. The smallest period, of 12nm, is possibly present but is partially obscured by the granularity of the AuPd shadowing. In addition, at these dimensions, the forward scattering of the beam may be limiting the pattern resolution. At 300kV, the forward scattering in the 250nm-thick oxide was estimated with a Monte Carlo program [31] to be 12nm (90% diameter) at the base of the oxide.

Our experiments with shadowing large area exposures confirm O'Keefe and Handy's [27] observation that the maximum etch rate enhancement is three to four times. This relatively low etch rate enhancement results in trenches which become progressively wider near the surface of the oxide where there has been more lateral etching of the unexposed material. By superposition, when dense arrays of lines are written, the sloping sidewalls overlap, decreasing the amplitude of the oxide between the trenches. Assuming an etch rate enhancement of 3 and an exposed linewidth of 5nm, a profile of an isolated line and closely spaced lines with three different periods was calculated (Fig. 3a). The etch depth of the unexposed oxide was set at 75nm. The dashed line represents the oxide surface before development. This profile in qualitative agreement with what we have observed. Isolated lines appear substantially broader than the lines in dense arrays. Direct evidence of the sloping sidewalls has been obtained by terminating line exposures with area expo-

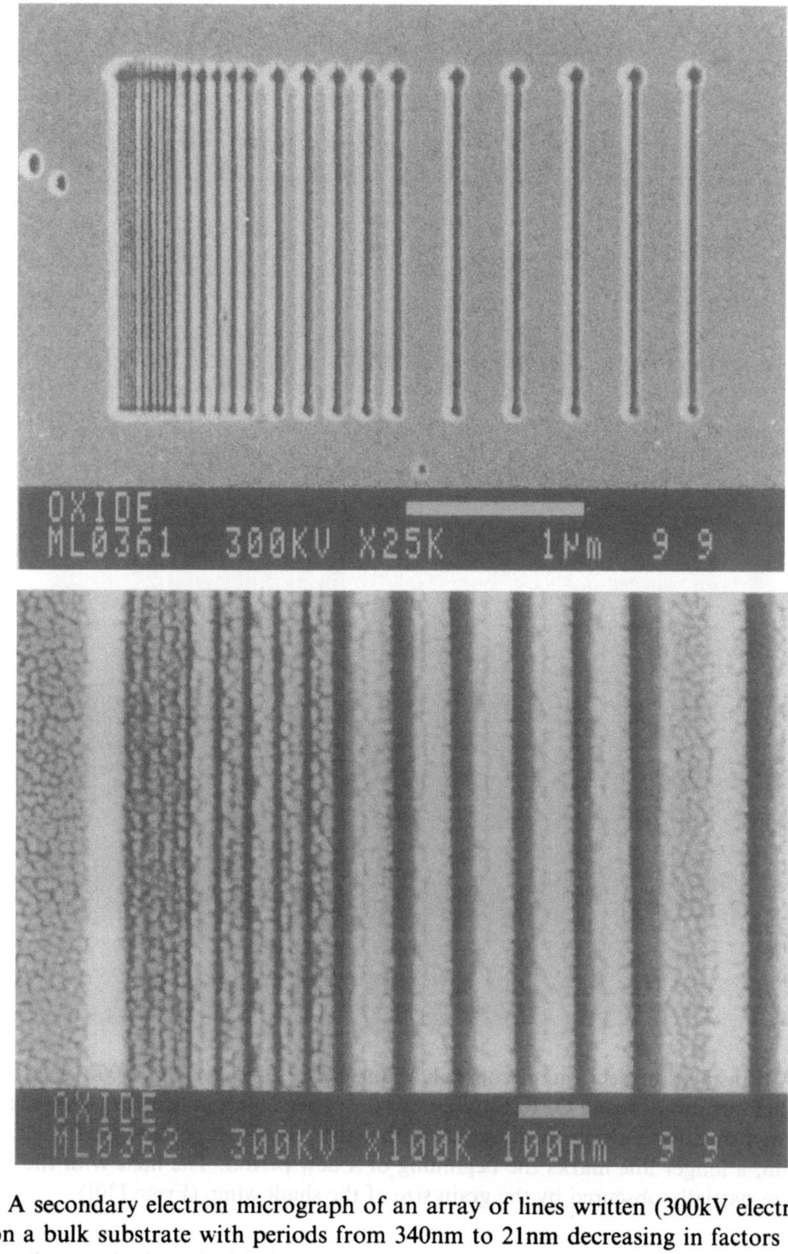

Fig. 1. A secondary electron micrograph of an array of lines written (300kV electrons) in SiO_2 on a bulk substrate with periods from 340nm to 21nm decreasing in factors of two. The sample was shadowed with Au from the right (From [29])

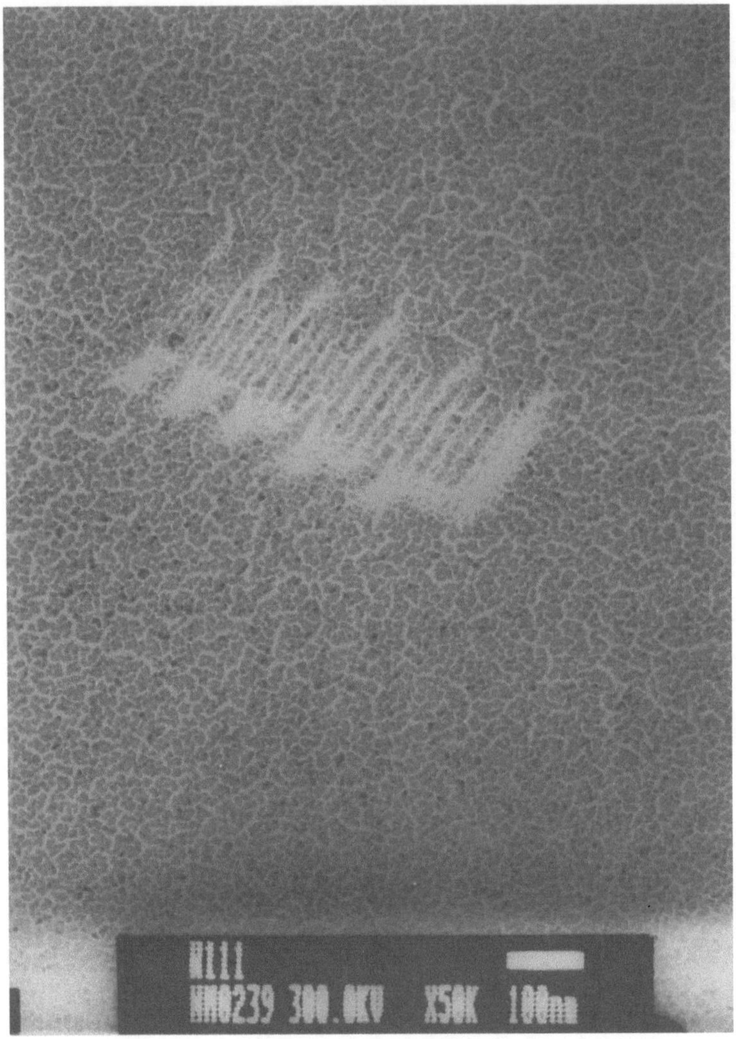

Fig. 2. A transmission electron micrograph of an array of lines written (300kV electrons) in a SiO_2 thin membrane using O_2 RIE to remove the contamination before development. The sample was shadowed along the diagonal with AuPd. The five periods are 25, 22, 19, 15, and 12nm; a longer line marks the beginning of a new period. The lines with the smallest period are partially obscured by the grain size of the shadowing. (From [30])

sures, developing, and then shadowing AuPd from a direction 45 degrees to the substrate and parallel to the lines into the developed areas. More recently (and presented here for the first time), a scanning force microscope has been used to profile an array of lines in SiO_2 similar to the patterns in Fig. 1 (Fig. 3b). The decreasing oxide amplitude due to the overlapping sidewalls is evident. Clearly, there is a need to find a more selective developer to improve the aspect ratio of the

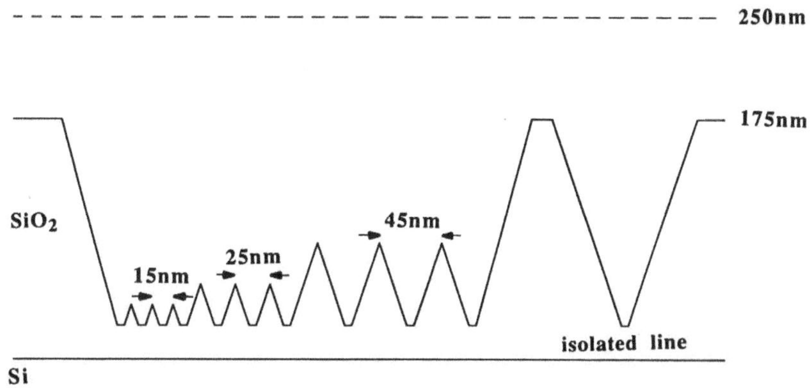

Fig. 3a. The predicted profile of lines written in SiO_2 assuming the exposed lines are 5nm wide and have an etch rate 3x faster than the unexposed regions. Forward scattering of the electron beam is neglected in this simple model. The *dashed line* represents the surface of the oxide before development. The relatively small difference in etch rates between the exposed and unexposed regions results in trenches with sloping sidewalls. By superposition, closely spaced lines result in a decreased oxide amplitude. An isolated line and three periods, 45, 25, and 15nm, are shown. (From [30])

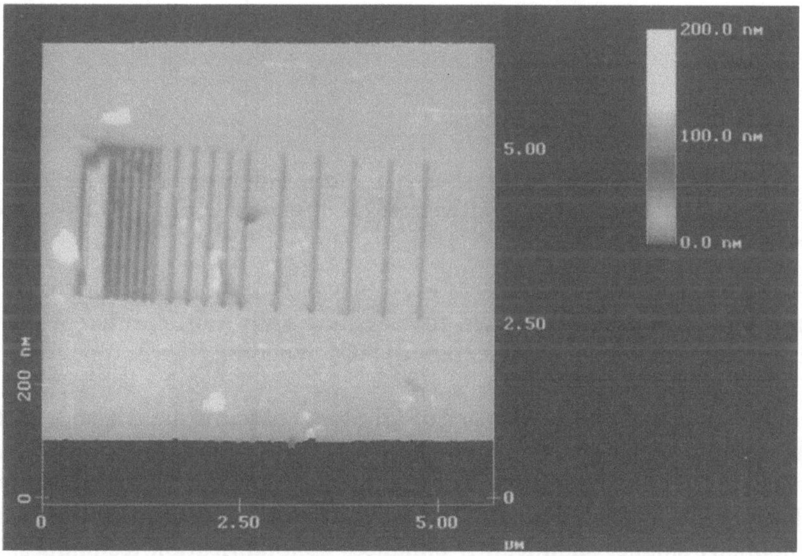

Fig. 3b. A scanning force microscope image of an array of lines in oxide on a bulk Si substrate similar to that in Fig. 1. The decreasing oxide amplitude due to the overlapping sidewalls can be clearly seen

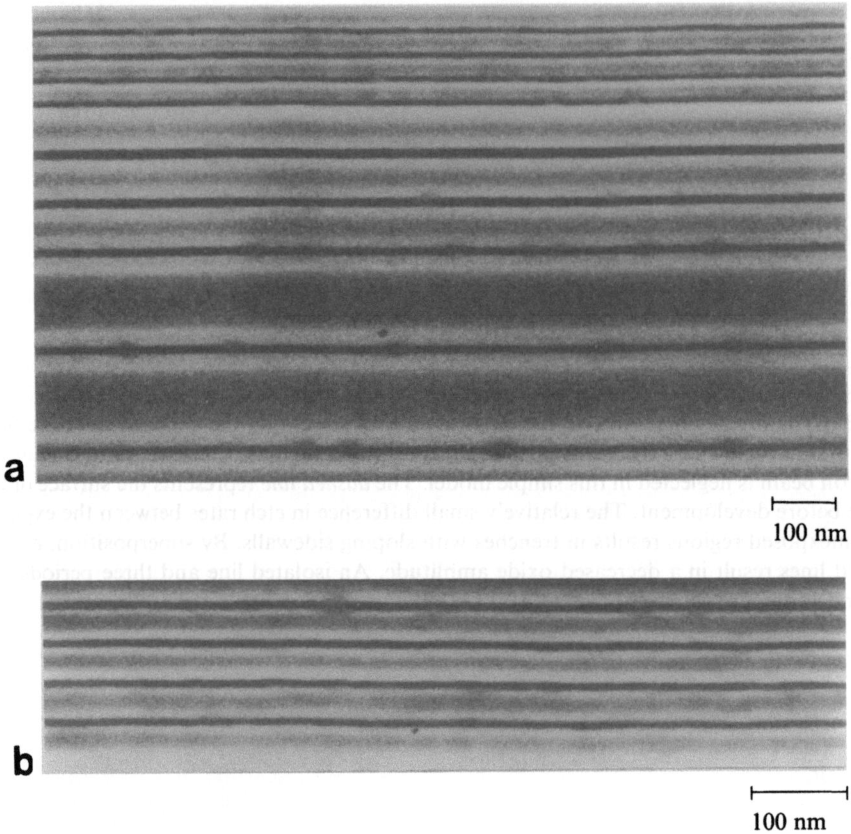

Fig. 4a, b. Replicated trenches in single crystal silicon with chlorine reactive ion etching **a** before removal of oxide mask, **b** after removal of the oxide mask. The width of the trenches in 10nm. (From [32])

patterns. All of the developers used to date have shown little difference in plots of the normalized thickness of exposed SiO_2 remaining as a function of log dose.

In conjunction with the University of Glasgow, patterns have been transferred into the underlying silicon using the patterned SiO_2 as a mask for chlorine reactive ion etching [32]. In these preliminary experiments, transferred features as small as 10nm have been achieved (Fig. 4). The ability to transfer features on this size scale and potentially smaller significantly enhances nanofabrication techniques. However, it is important to note that nanometer scale patterned SiO_2 is particularly useful itself as a key component of many devices.

The mechanism of this electron beam exposure of SiO_2 is not clearly understood. In light of O'Keefe and Handy's experiments at 1-15kV, which is well below the nuclear damage threshold [33], it is very likely that processes associated with direct atom displacement do not dominate the exposure mechanism(s). The enhancement of the SiO_2 etch rate in HF based etches may be due to the effects of ionizing radiation. It is known that electrons beam irradiation of amorphous SiO_2 can cause oxygen vacancies, peroxy radicals, non-bridging oxygens [34], and neutral traps that can subsequently fill with either electrons or holes [35]. Defect creation in amorphous SiO_2 has often been explained in terms of breaking strained Si-O-Si bonds [34] or, more recently, in terms of interatomic Auger processes [36]. Of particular importance is whether or not the minimum energy required for an exposure event is above the energy of the secondary electrons (several tens of eV). There is some evidence that the delocalized exposure due to secondary electrons is the source of the resolution limit of PMMA [18]. The mechanisms associated with defect formation due to interatomic Auger processes have energy thresholds of 30 to 40eV [37]. The mechanism of exposure of SiO_2 may be related to the density and stress of the oxide film. The density and stress are known to be sensitive functions of the oxide growth conditions and correlate well with the refractive index [38]. Furthermore, the etch rates of HF based SiO_2 etches, in particular p-etch, are known to be sensitive to oxide density, bond strain, etc [39].

Conclusions

In conclusion, we have overcome the problem of surface contamination by using either a sacrificial layer or an O_2 reactive ion etch, and we have demonstrated that SiO_2 can be directly patterned with nanometer scale resolution. The resolution for dense structures, lines and spaces with a 12 to 15nm period, is 3 to 4X better than PMMA, the most widely used high resolution electron beam resist. Preliminary pattern transfer experiments using chlorine reactive ion etching have resulted in 10nm features in Si. Clearly, much work remains to be done to fully develop the potential of this ultra-high resolution material, including a better understanding of the exposure mechanism. Because patterned SiO_2 is itself extremely useful as a key component of many electronic devices, the ability to directly pattern SiO_2 with nanometer scale resolution, albeit with a high dose, will enhance the techniques available to prototype both conventional and quantum effect ultra-small devices.

Acknowledgment. We would like to thank David Joy at the University of Tennessee for the use of his Monte Carlo programs to model the forward scattering and C.P. D'Emic for processing the Si_3N_4/SiO_2 membranes. We would also like to thank Les Peters for film processing.

References

1. Chao PC, Shur MS, Tiberio RC, Duh KHG, Smith PM, Ballingall JM, Ho P, Jabra AA (1989) IEEE Trans Electron Dev 36(3):461–471
2. Patrick W, Mackie WS, Beaumont SP, Wilkinson CDW, Oxley CH (1985) IEEE Electron Dev Lett 6(9):471–472
3. Han J, Ferry DK, Newman P (1990) IEEE Electron Dev Lett 11(5):209–211
4. Allee DR, de la Houssaye PR, Schlom DG, Harrris JS Jr, Pease RFW (1988) J Vac Sci Technol B6(1):328–332

5. Umbach CP, Palevski A, Heiblum M, Sivan U (1989) J Vac Sci Technol B7(6):2003–2006
6. Ismail K, Chu W, Antoniadis DA, Smith HI (1989) Appl Phys Lett 54:460–462
7. Ismail K, Chu W, Tiberio R, Yen A, Lezec HJ, Shepard MI, Musil CR, Melngailis J, Antoniadis DA, Smith HI (1989) J Vac Sci Technol B7(6):2025–2029
8. Allee DR, Chou SY, Harris JS Jr, Pease RFW (1989) J Vac Sci Technol B7(6):2015–2019
9. Bernstein G, Ferry DK (1987) J Vac Sci Technol B5(4):964–966
10. Kubena RL, Joyce RJ, Ward JW, Garvin HL, Sratton FP, Brault RG (1988) J Vac Sci Technol B6:353–356
11. Rarback H, Shu D, Feng SC, Ade H, Kirz J, McNulty I, Kern DP, Chang THP, Vladimirsky Y, Iskander N, Attwood D, McQuaid K, Rothman S (1988) Rev Sci Instrum 59:52–59
12. McCord MA, Pease RFW (1986) J Vac Sci Technol B4:86–88
13. McCord MA, Pease RFW (1985) J Vac Sci Technol B3:198–201
14. Eigler DM, Schweizer EK (1990) Nature 344:524–526
15. Broers AN, Timbs AE (1989) Microelectron Engin 9:187–190
16. Craighead HG, Howard RE, Jackel LD, Mankiewich PM (1983) Appl Phys Lett 42(1):38–40
17. Broers AN, Harper JME, Molzen WW (1978) Appl Phys Lett 33(5):392–394
18. Broers AN (1981) J Electrochem Soc 128(1):166–170
19. Ouano AC, Gipstein E, Johnston D, Need OU III (1977) Polymer Eng Sci 17:396–401
20. Broers AN, Cuomo JJ, Harper J, Molzen W, Laibowitz RB, Pomerantz M (1978) Electron Micros III:343–354
21. Broers AN, Cuomo JJ, Krakow W (1981) IBM Tech Disclosure Bull 24:1534
22. Isaacson M, Muray A (1981) J Vac Sci Technol 19:1117–1120
23. Muray A, Scheinfein M, Isaacson M, Adesida I (1985) J Vac Sci Technol B3(1):367–372
24. Mochel ME, Humphreys CJ, Eades JA, Mochel JM, Petford AM (1983) Appl Phys Lett 42(4):392–394
25. Muray A, Isaacson M (1983) J Vac Sci Technol B1(4):1091–1095
26. Kratschmer E, Isaacson M (1986) J Vac Sci Technol B4(1):361–364
27. O'Keeffe TW, Handy RM (1968) Solid state electronics, vol 11. Pergamon, London, pp 261–266
28. Broers AN, Molzen WW, Cuomo JJ, Wittels ND (1976) Appl Phys Lett 29(9):596–598
29. Allee DR, Broers AN (1990) Appl Phys Lett 57(21):2271–2273
30. Allee DR, Umbach CP, Broers AN (1991) J Vac Sci Technol B9(6):2839–2843
31. Joy DC (1988) In: Craven AJ, Elder H (eds) Proc European meeting on electron microscopy 1988. Inst Phys Conf Ser 93(1):23
32. Pan X, Allee DR, Broers AN, Tang YS, Wilkinson CW (1991) Appl Phys Lett 59(24):3157–3158
33. Vasilov VS, Kiv AE, Niyazova OR (1975) Phys Sol (a) 32:11
34. Beall Fowler W (1983) Semicond Insul 5:583
35. Aitken JM (1980) J Noncryst Sol 40:31
36. Knotek ML (1983) Semicond Insul 5:361
37. Feibelman PJ, Knotek ML (1978) Phys Rev B18(12):6531–6539
38. Taft EA (1978) J Electrochem Soc 125:968
39. Pliskin WA (1977) J Vac Sci Technol 14(5):1064–1081

Ultrafine AlGaAs/GaAs Quantum-Well Wire Fabrication by Combining Electron Beam Lithography and Two-Step Wet Chemical Etching

39

EIICHI ISHIKAWA[1], SUSUMU FUKATSU[2], KENTARO ONABE[1], YASUHIRO SHIRAKI[2], AND RYOICHI ITO[2]

Summary. Ultrafine AlGaAs/GaAs quantum well wire structures as small as 35nm, with smooth sidewalls and low damage, which would otherwise be difficult to produce, e.g., by conventional dry etching, have been successfully fabricated by combining electron beam lithography and two-step wet chemical etching. Photoluminescence measurement has revealed that the thickness of the nonradiative "dead layer" of sidewalls formed during etching is remarkably reduced, even for the finest wires in the quantum regime (35nm), and the associated blue shift of emission due to the two-dimensional quantum confinement of the carriers has been clearly observed.

Introduction

There has been considerable interest in reducing the physical dimensions (or feature sizes) of semiconductor structures to elucidate the long-sought quantum effects for optical and electronic applications [1]. Various schemes have been proposed [2] and exercised [3–6] to achieve quantum mechanical confinement of electrons. "Vertical" confinement has been demonstrated in quantum well structures owing to the advance of *state-of-the-art* epitaxial growth techniques, such as molecular beam epitaxy (MBE) and metal organic chemical vapor deposition (MOCVD). Two-dimensional and three-dimensional confinement of carriers, i.e., wires and dots, respectively, can be attained by a naive epitaxial growth on tilted or patterned substrates, and by sophisticated lithography techniques like electron beam lithography (EB) and focused ion beam lithography (FIB). The former method has eventually provided very small structures [4], but the pattern feature design is undesirably limited. The latter method, in contrast, is believed to offer one much flexibility in generating patterns. EB lithography, in particular, has the potential for creating the finest features where quantum confinement effects are most likely to be manifest.

[1] Department of Applied Physics, The University of Tokyo, 7-3-1 Hongo, Bunkyo-ku, Tokyo, 113 Japan
[2] Research Center for Advanced Science and Technology (RCAST), The University of Tokyo, 4-6-1 Komaba, Meguro-ku, Tokyo, 153 Japan

The pattern transfer from the resist to the semiconductor substrate comprises the final stage of small feature fabrication and has generally been carried out by dry etching represented by reactive ion etching (RIE). However, the radiation-induced damage often leads to undesirable sidewall degradation, including optically inactive layer formation, in the smallest structures that would otherwise exhibit quantum mechanical confinement effects. Obviously, it is this etching-induced damage that principally determines the optical quality of either one- or zero-dimensional structures fabricated by ultrafine lateral etching. In fact, photoluminescence in the quantum confinement regime has rarely been observed in open wire structures of AlGaAs/GaAs systems fabricated by dry etching [7]. Unfortunately, the full characterization of etching-induced damage in AlGaAs/GaAs ultrasmall structures is still missing. The optically inactive "dead layer" width was only roughly estimated to be 65 nm in earlier reports [8]. This result naturally leads us to conclude that the wire structure is totally optically inactive when the wire feature size is smaller than 100nm in width.

In order to circumvent these problems, we previously proposed a novel wet chemical etching technique, "two-step wet chemical etching" [9], which is particularly useful for fine pattern definition. The whole process basically consists of two kinds of etching using different etchants. A H_2SO_4-H_2O_2-H_2O system [10] is used as the first etchant, followed by the second etchant, a Br_2-CH_3OH system [11]. The former etchant is insensitive to surface contamination like carbon hydrides, but heavily erodes the GaAs, offering the advantage of coarse feature patterning. The latter, however, is favorable and is essential for the smoothing of clean surfaces, though it is sensitive to surface contamination.

In the present study, we applied this two-step wet chemical etching to the fabrication of wire structures with the wire axis running in the (100) direction on GaAs/AlGaAs quantum wells. Ultrafine (35nm) wires of smooth sidewalls with slight etch-induced damage were obtained, which would not be obtained by conventional dry etching. Photoluminescence (PL) measurements revealed a marked reduction in the thickness of the non-radiative surface layer, and the realization of the two-dimensional confinement of photogenerated carriers was confirmed.

Experimental Results and Discussion

An AlGaAs/GaAs double quantum well structure grown by MBE on GaAs(001) substrate was used as a starting material. The structure consisted of a GaAs buffer layer, a 50-nm-thick $Al_{0.3}Ga_{0.7}As$ barrier, a 20-nm-thick GaAs quantum well, a 180-nm-thick $Al_{0.3}Ga_{0.7}As$ barrier, a 10-nm-thick GaAs quantum well, and a 10-nm cap layer of GaAs. Only the 10-nm quantum well layer was patterned into wire structures, leaving the 20-nm quantum well intact as the standard for PL measurement.

The line-and-space features were created on a substrate, coated with a 50-nm-thick poly-methylmethacrylate (PMMA) positive resist, using a high-resolution electron beam lithography system (JEOL JBX-5DII) at an accelerating voltage of 50kV. Orientation of the patterns was chosen so that the wire was tilted 45°

from the cleavage direction ⟨110⟩, i.e., running along ⟨100⟩. This orientation was found to be critical for the fabrication of wire structures with favorable etched surface characteristics. It is generally known that ordinary mesa structures are obtained if the wires run parallel to the ⟨110⟩ direction and are chemically transferred-and-etched on a (001) GaAs surface. On the other hand, reversed-mesa structures are obtained after chemical etching by letting the wire direction run along ⟨$1\bar{1}0$⟩. It is therefore expected that almost perpendicular sidewalls could be obtained with the wire axis twisted by 45° with respect to ⟨110⟩. This is supported by a previous study [10], where symmetric sidewalls with surfaces normal along ⟨010⟩ [⟨100⟩] were shown to be obtained by rendering the rectangular window edge aligned with ⟨100⟩ [⟨010⟩].

Using appropriate conditions for resist development, resist patterns were produced which directly served as a mask for the subsequent wet etching. It should be noted that resist postbake is unfavorable for such a fine pattern creation since an elevated temperature environment rapidly degrades the structural integrity, due to the thermal flow of the resist [9]. Hence we did not undertake resist postbake. For pattern transfer to the quantum well substrates, two-step wet chemical etching was used. The first etching was performed in a mixture of H_2SO_4-H_2O_2-H_2O, with a volume ratio of 8:1:20, for 30s at temperature of 3.5°C, which provided coarse erosion of the contaminated GaAs surface. After a CH_3OH dip, the second etching was carried out in a Br_2:CH_3OH (1:1000) solution for 20s at 20°C, with resultant ultrafine wires with truly flat sidewalls and narrow width down to 35nm. The resist masks were removed with boiled methylethylketone after etching. A cross-sectional view of the array of the 35nm quantum wires is shown in Fig. 1. It can be seen that the (110) and ($1\bar{1}0$) facets are preferentially eroded in the first

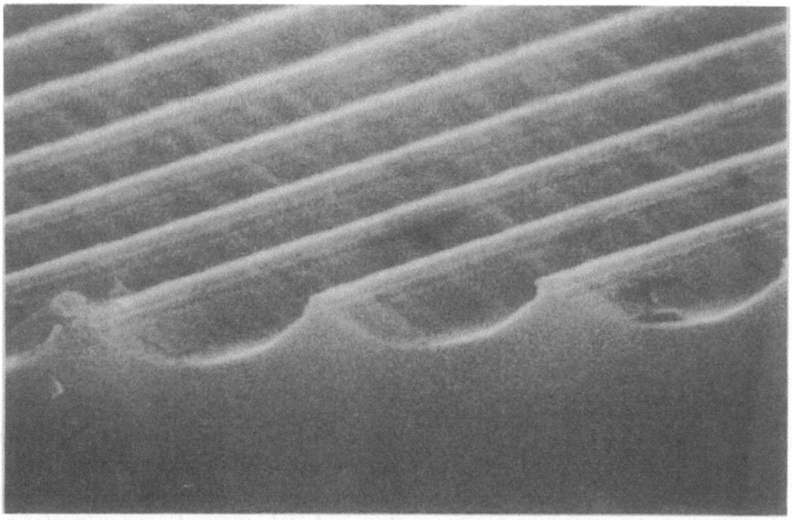

Fig. 1. Cross-sectional scanning electron micrograph of (100) oriented wire structures fabricated by two-step wet chemical etching

etching, irrespective of the resist mask orientation. Thus many square hole features develop along (100) direction, with their sides aligned with (110) and (1$\bar{1}$0), giving the fluctuation of the wire width. The Br_2-CH_3OH system etchant was found to be effective for smoothing the sidewalls, very flat sidewalls being obtained by a proper choice of etching time. Detailed inspection revealed that the undercut length beneath the mask, provided by lateral etching, was comparable to the etch depth, indicating the isotropic character of the etching. It is readily seen that the sidewalls are nearly perpendicular to the (001) plane, with a rounded entrailing edge, as shown in Fig. 1.

Figure 2 shows the well width dependence of the normalized PL intensity for wet chemically etched wires of feature sizes down to 35nm. PL experiments were carried out at 20K with the 514.5nm line of an Ar^+ ion laser as the excitation source. The PL intensity was scaled and normalized, taking account of the relationship of the filling factor, to the intensity of the 20nm quantum well reference. For comparison, results on the dry-etched samples obtained by Maile et al. [12] are also shown (open triangles). It can be seen that, for both the wet-etched and dry-etched wires, luminescence intensity decreases by more than five orders of magnitude as the feature size is reduced. This strong feature size dependence of the luminescence intensity is most likely due to the surface recombination of carriers at the wire sidewalls. It should be pointed out that the luminescence intensity is consistently larger in the wet-etch samples than in the dry-etched wires. In particular, the decay slope of the wet-etched result is significantly smaller than that of the dry-etched result.

The feature size variation of the luminescence is discussed based upon a simple model [13], incorporating the carrier diffusion and recombination in an optically inactive "dead layer" at the sidewalls. Assuming the luminescence intensity to be

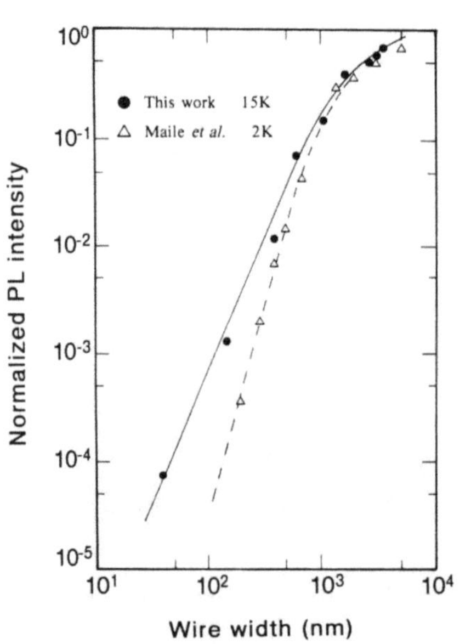

Fig. 2. Normalized PL intensity as a function of feature size. *Solid circles* represent the experimental data for wires etched by two-step wet chemical etching, and *open triangles* represent the data for dry-etched wires. (After Maile et al. [12])

proportional to the square of the photogenerated carrier density, i.e., bimolecular recombination, normalized luminescence intensity can be described by a simple equation,

$$\eta = \left[1 - \frac{2S\tau}{w - 2d + 2S\tau}\right]^2 \tag{1}$$

where S is the nonradiative surface recombination velocity, τ is the carrier life time, w is the geometrical wire width, i.e., the feature size, and d is the "dead layer" width on both sides of the wire structure. Effective wire width which is optically active for light emission is therefore given by (w-2d).

By proper fitting to the experimental data, $2S\tau = 250$nm and $d = 6$nm were obtained by the least squares method. This dead layer width (6nm) is significantly small compared with that of the dry-etched wires reported previously (50–60nm) [8]. Therefore it might be concluded that the present wet etching scheme offers a greater advantage for nanofabrication, requiring less damage on the surface, over conventional methods reported to date. Surface recombination velocity, S, of 1.3×10^5cm/s was obtained at 20K, assuming the radiative decay lifetime to be $\approx$1nsec. This is of the same order as the decay time of dry-etched samples [12].

The filled circles in Fig. 3 represent the PL peak energies as a function of the actual feature size. PL peak energy is likely to change as the excitation beam spot moves across different positions of the surface, since the well thickness homogeneity of the present sample might be rather poor due to growth-related origin (beam flux nonhomogeneity). The PL peak energy variation was investigated by mapping the entire surface, which is shown by the region bounded by broken lines in Fig. 3. The solid line in this figure, representing the peak energy averaged over the surface of the as-grown quantum well, was drawn to unambiguously indicate the blue-shift of PL energy. In all wires for $w \geqslant 150$nm, the peak energy was found to lie within the bound region. In contrast, the peak energy of the finest wire (35nm) was clearly higher, giving a finite blue-shift. Thus it can be plausibly stated that the two-dimensional confinement of electrons was realized in the finest wire, 35nm. Though the lateral confinement energy was calculated to be 5.2meV in 35-nm-wide

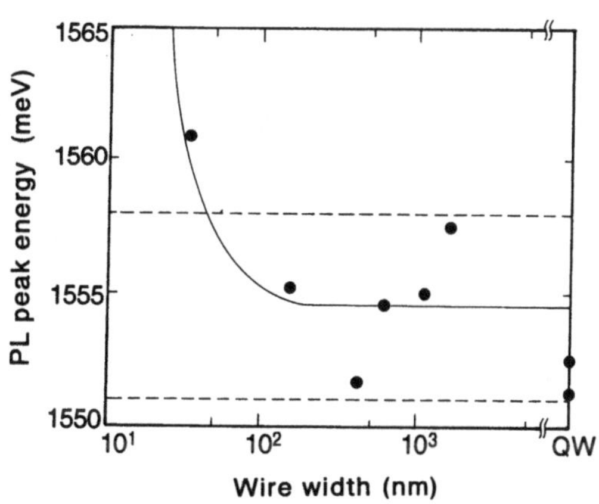

Fig. 3. PL peak energy positions as a function of wire width. The *broken lines* bound the energy fluctuation range in the as-grown quantum well. The *solid line* represents the calculated wire width dependence of the peak energy

wires, the amount of the blue-shift was not precisely determined solely from this. Further study is now in progress.

In conclusion, we have successfully fabricated ultrafine quantum well wire structures as small as 35nm by combining EB exposure and two-step wet chemical etching. A dramatic reduction in the optically inactive surface "dead layer" on the sidewalls was demonstrated, in marked comparison with dry-etched samples. For wires with a lateral dimension of 35nm, we observed a substantial spectral blue-shift of the excitonic emission due to two-dimensional quantum mechanical confinement of electrons.

Acknowledgments. The authors acknowledge the support of M. Morioka, A. Fujiwara, K. Momma, and S. Ohtake. This work was supported, in part, by a Grant-in-Aid for Scientific Research on the Priority Area, "Electron Wave Interference Effects in Mesoscopic Structures", from the Ministry of Education, Science, and Culture of Japan.

References

1. Arakawa Y, Sakaki H (1982) Appl Phys Lett 40:939–941
2. Sakaki H (1980) Jpn J Appl Phys 19:L735–L738
3. Petroff PM, Gossard AC, Logan RA, Wiegmann W (1982) Appl Phys Lett 41:635–638
4. Tsuchiya M, Gaines JM, Yan RH, Simes RJ, Holtz PO, Molden LA, Petroff PM (1991) Phys Rev Lett 62:466–469
5. Kapon E, Hwang DM, Bhat R (1989) Phys Rev Lett 63:430–433
6. Hirayama Y, Tarucha S, Suzuki Y, Okamoto H (1988) Phys Rev B37:2774–2777
7. Izrael A, Sermage B, Marzin JY, Ougazzaden A, Azoulay R, Etrillard J, Uhierry-Mieg V, Henry L (1990) Appl Phys Lett 56:830–832
8. Maile BE, Forchel A, Germann R, Straka J, Korte L, Thanner C (1990) Appl Phys Lett 57:807–809
9. Katoh T, Nagamune Y, Li GP, Fukatsu S, Shiraki Y, Ito R (1990) Appl Phys Lett 57:1212–1214
10. Iida S, Ito K (1971) J Electrochem Soc 118:768–771
11. Tarui Y, Komiya Y, Harada Y (1971) J Electrochem Soc 118:118–122
12. Maile BE, Forchel A, Germann R, Menshig A, Meier HP, Grutzmacher D (1988) J Vac Sci Technol B6:2308–2311
13. Maile BE, Forchel A, Germann R, Grutzmacher D (1989) Appl Phys Lett 54:1552–1554

Mesoscopic Structure in Lattice-Mismatched Heteroepitaxial Interface Layers

40

MASAO TABUCHI, SUSUMU NODA, AND AKIO SASAKI[1]

Summary. Low dimensional mesoscopic structures are important for the realization of electron wave and quantum optoelectronic devices. In this study, we propose to utilize three-dimensional (3D) growth, which occurs at the initial stage of the heteroepitaxial growth of a lattice-mismatched system, for the realization of a mesoscopic structure. InAs layers grown 3-dimensionally on (001) just-oriented and on 3.5° off-angled GaAs substrates were investigated by photoluminescence (PL) measurement and by transmission electron microscopy. A strong PL emission was observed from the 3D grown InAs samples. Both TEM observation and PL measurement suggest that the strong PL emission would be caused by the quantum effect of the 3D InAs layer sandwiched by the GaAs layers.

Introduction

Low dimensional mesoscopic structures are attracting much interest in the area of electron wave and quantum optoelectronic devices; these structures have been fabricated by, for example, electron beam lithography and focused ion beam techniques [1]. Lateral growth on vicinal substrates has also been utilized to fabricate quantum wires and boxes [2].

Thus far, we have investigated heteroepitaxial growth and heterointerface structures of InGaAs layers grown by molecular beam epitaxy (MBE) on various substrates, such as GaAs, InAs, and InP [3, 4]. We have found that the crystalline quality of lattice-mismatched heteroepitaxial layers depends not only on the degree of lattice mismatching, but also on other factors. The initial 3D growth process plays a particularly important role in the improvement of the crystalline quality, even at a large lattice-mismatching. From transmission electron microscope (TEM) observations, we have found that 3D-grown layers show some periodical structure at the interface. We propose here to utilize the 3D growth process for the realization of a mesoscopic structure.

In this paper, we report the results of our investigation, using photoluminescence (PL) measurements and TEM observations, into the possibility of utilizing a periodical 3D structure as a low dimensional mesoscopic structure.

[1] Department of Electrical Engineering, Kyoto University, Kyoto, 606 Japan

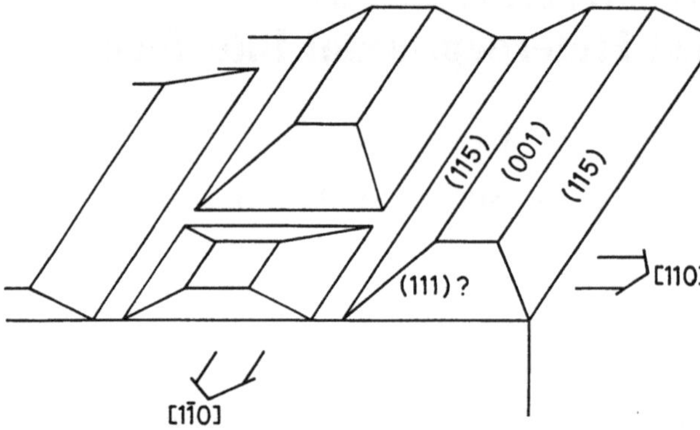

Fig. 1. Schematic structure of three-dimensionally grown InAs islands on GaAs substrate

Experiment

The GaAs/InAs/GaAs heterostructure was grown by MBE on (001) GaAs substrate, where the degree of lattice-mismatching between GaAs and InAs is about 7%. The thickness of the InAs layer was 1, 2, 4, or 10 molecular layers (ML), and the thickness of the GaAs cap layer was 100Å. Both the InAs layer and the GaAs cap layer were grown at 480°C. These samples were investigated by TEM observations and their optical properties were investigated by PL measurement at 77K.

As reported previously, we have already investigated InAs growth on GaAs substrate by reflection high-energy diffraction electron (RHEED) observations; three steps were observed during this growth. In the first step, InAs grows two-dimensionally (2D) on GaAs substrate for about 1ML. In the second step, the growth mode changes to the 3D growth. Figure 1 shows a schematic image of the 3D-grown InAs islands on GaAs substrate based on the knowledge obtained from the RHEED pattern. Finally, the growth mode recovers to the 2D growth when the InAs grows thick enough (~10ML), since the lattice-mismatching is well accommodated.

Figure 2 shows the TEM images of the heterostructure observed from the [001] direction normal to the substrate surface. The image of the sample with the 1ML-InAs layer shows no particular contrast. It agrees well with the result of RHEED observations, which showed that 1ML-InAs grew two-dimensionally. On the other hand, some characteristic contrasts are observed in the image of the sample with the 2ML-InAs layer. The strong oval contrasts (marked A in Fig. 2b) observed in this image are thought to indicate the misfit dislocation. The weak dotted image (marked B in Fig. 2b) is thought to be the image of the 3D-grown InAs island observed by RHEED. The sample with the 4ML-InAs layer shows both of the contrasts: the 3D-grown islands and the misfit dislocations. The images of dislocations observed in the 4ML-InAs samples were longer and of higher concentration than those of the 2ML-InAs samples. In the sample with the 10ML-InAs layer, only the image which indicates dislocation was observed, while the image of the

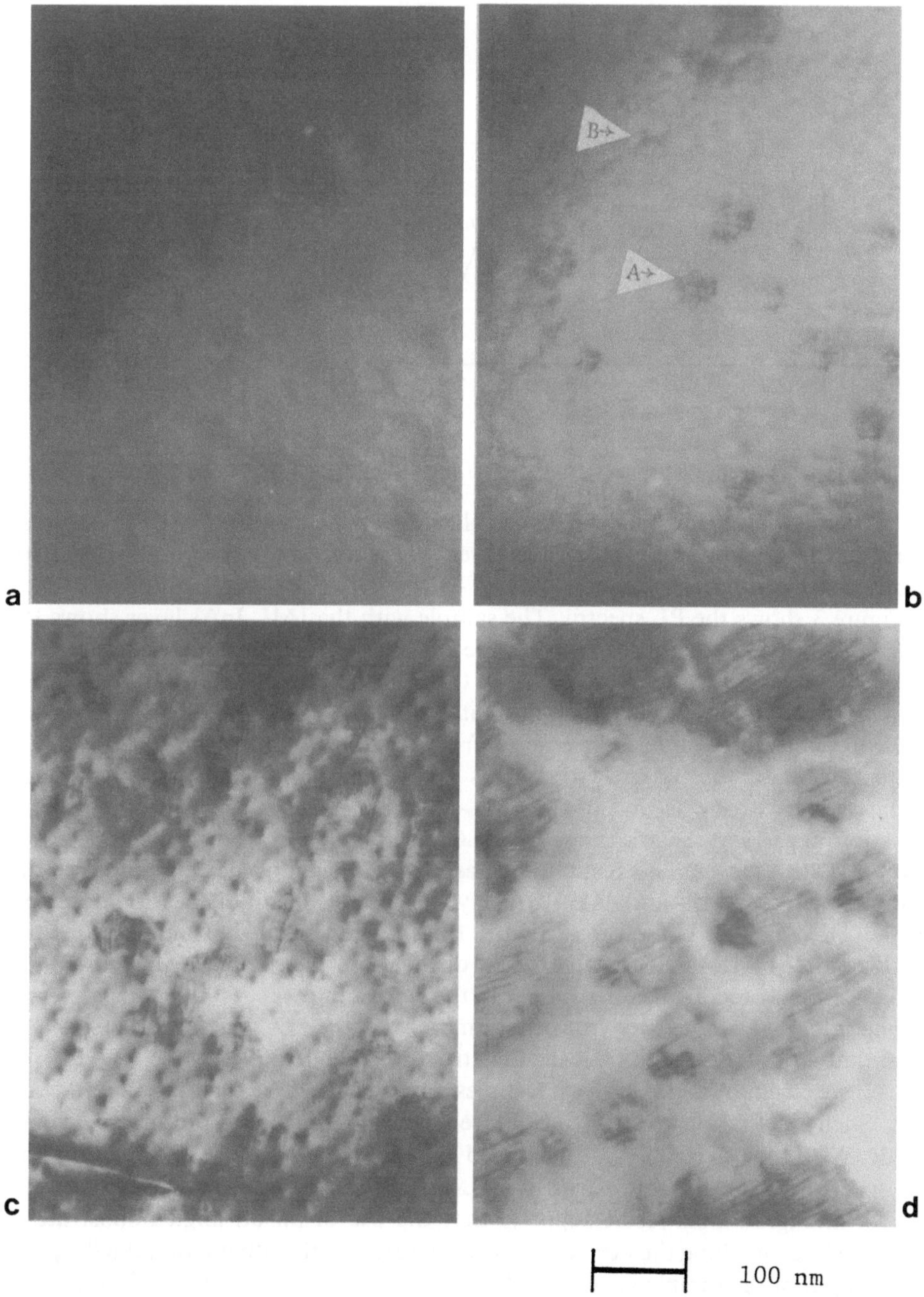

Fig. 2a–d. TEM images of InAs on GaAs, where the thickness of the InAs layer is **a** 1ML, **b** 2ML, **c** 4ML, and **d** 10ML, respectively. Note strong oval contrasts *A* and weak dotted image *B* in **b** (see text)

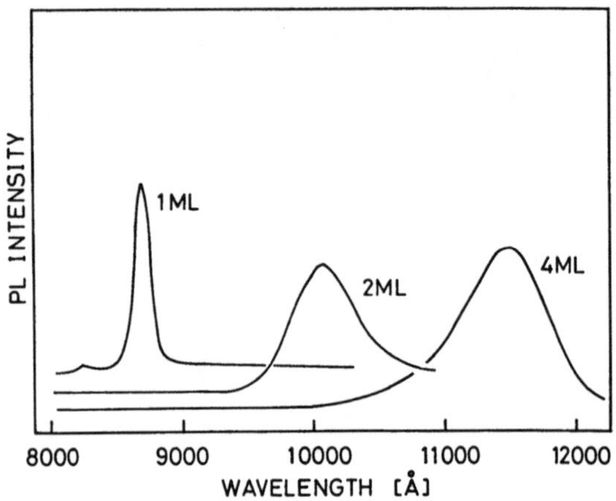

Fig. 3. PL spectra of 1ML-, 2ML-, and 4ML-InAs grown on GaAs (001) just-oriented substrate

InAs islands was negligible. The results also agree well with the RHEED observations during growth, in that the InAs layer that is thicker than several MLs again shows a streaky pattern.

Figure 3 shows the PL spectra. The sample with the 1ML-InAs layer shows PL emission which is thought to be from the 2D-grown quantum well structure. The full width at half-maximum (FWHM) of the PL spectrum of the 2D grown sample is as narrow as 20meV. The samples with the 2ML- and 4ML-InAs well layers show PL which is thought to be emitted from the 3D structure. The FWHM of the PL spectrum of the 3D-grown samples are wider than 100meV, but the peak intensity is as strong as that of the 1ML-grown InAs. However, we observed no emission in the wavelength range from 800nm to 1600nm from the samples with the 10ML-InAs layer. This result indicates that the origin of the emission from the samples with the 2ML- and 4ML-InAs layers is not the misfit dislocation observed in the TEM image. If the origin of the PL emission from the samples with the 2ML- and 4ML-InAs layers were misfit dislocation, the samples with the 10ML-InAs islands would also emit PL light in the same range.

The integral PL intensity from the sample with the 3D structure was stronger than that from the sample with the 2D structure. The relatively wide FWHMs of the PL spectrum of the 3D-grown samples may be caused by fluctuation in the 3D size. Such fluctuation can possibly be reduced by controlling the initial growth mode. With the use of the 3.5° off-angled substrate toward the [110] direction, the PL spectrum from the sample with 2ML-InAs becomes as narrow as 40meV, as shown in Fig. 4. This result suggests that the use of an off-angled substrate is effective in controlling the periodical 3D structure. On the other hand, the PL peak wavelength of the sample grown on the 3.5° off-angled substrate was shorter than that of the sample grown on (001) just-oriented substrate. The blue shift of the PL peak is thought to be related to the TEM finding that no images of dislocation were observed in the sample on the 3.5° off-angled substrate (Fig. 5). The lattice-mismatching between InAs and GaAs is about 7%. If the lattice-mismatching is

Fig. 4a, b. PL spectra of **a** 2ML- and **b** 4ML-InAs grown on GaAs (001) just-oriented, 3.5° off-angled, and 2° off-angled substrates

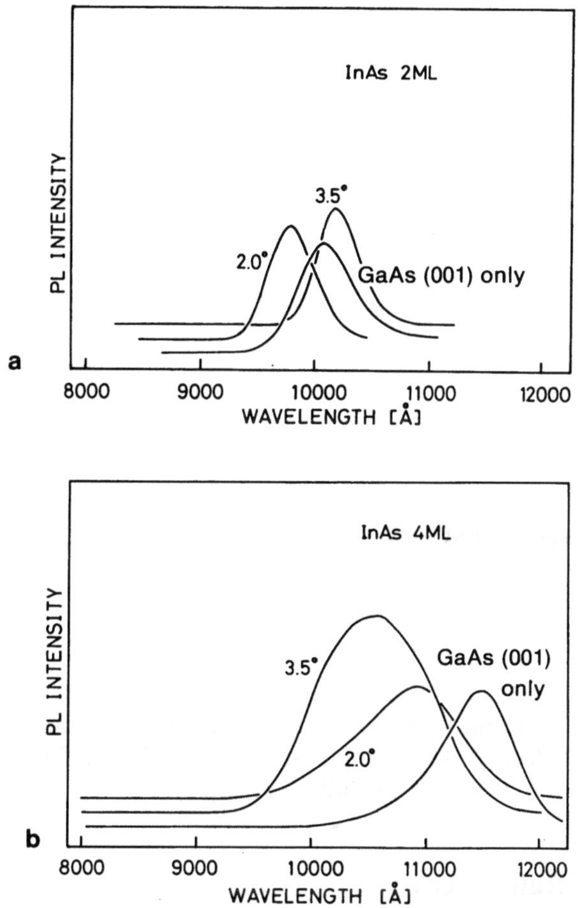

not accommodated by misfit dislocation, the compressive lattice strain which remains in the heteroepitaxial grown layer increases the width of the band gap. This result also indicates that the origin of the PL emission observed in these samples would not be misfit dislocation, but would be due to the quantum effect of the GaAs/InAs/GaAs confinement structure.

Conclusions

We carried out PL and TEM investigations of InAs layers grown 3-dimensionally on GaAs just-oriented (001) and on 3.5° off-angled GaAs substrates. Strong PL emission was observed from the 3D-grown InAs samples. TEM and RHEED observations suggest that the origin of the PL emission would be due to the quantum effect of GaAs/3D-InAs/GaAs structure.

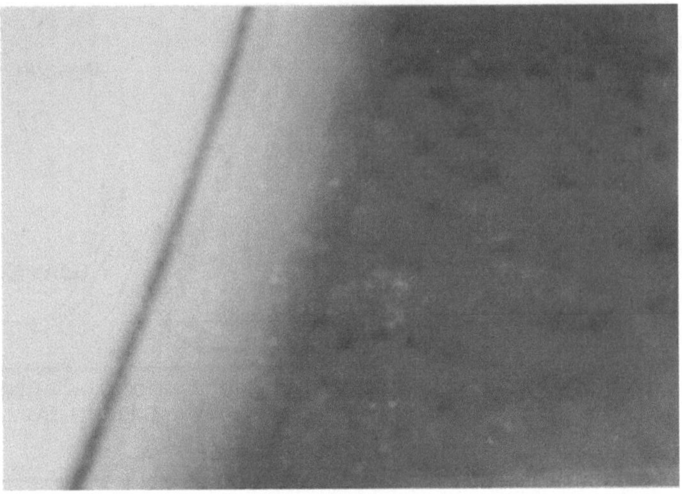

Fig. 5. TEM image of 2ML-InAs on 3.5° off-angled GaAs substrate observed from [001] direction

Acknowledgments. The authors wish to thank Prof. Y. Takeda and Dr. A. Wakahara for their helpful discussions. This work was supported, in part, by a Grant-in-Aid for Scientific Research on a Priority Area, "Electron Wave Interface Effects in Mesoscopic Structures," (No. 03237103) from the Ministry of Education, Science and Culture of Japan.

References

1. Ando S, Fukui T (1989) J Cryst Growth 98:646
2. Fukui T, Saito H (1988) J Vac Sci Technol B4:1373
3. Fujita S, Nakaoka Y, Uemura T, Tabuchi M, Noda S, Takeda Y, Sasaki A (1989) J Cryst Growth 95:224
4. Tabuchi M, Noda S, Sasaki A (1990) J Cryst Growth 99:315
5. Tabuchi M, Noda S, Sasaki A (1990) In: Proceedings of the First International Conference and Exhibition of Computer Applications, Science and Engineering. North-Holland, Amsterdam, p 711 Tokyo, Japan, PII-24

Luminescence Properties of Disordered Superlattices

41

AKIO SASAKI, KAZUYUKI UNO, AND SUSUMU NODA[1]

Summary. A disordered superlattice, which can be considered as an example of a disordered crystalline semiconductor, was fabricated, and its luminescence properties were investigated. It has been experimentally confirmed that the AlAs/$Al_xGa_{1-x}As$ disordered superlattice exhibits remarkable luminescence capability. Based on the investigation of the optical absorption, the time-resolved photoluminescence, and the luminescence temperature dependence, it was concluded that the unusual luminescence capability could be interpreted in terms of the Lifshitz model of localization created by artificial disordering in the physical thickness of the superlattice.

Introduction

Amorphous materials can radiate an intensified luminescence as compared with single-crystal materials, due to the recombination process through localized states [1], such as a-Si:H. However, epitaxial growth is impossible on an amorphous substrate.

A disordered superlattice is a new semiconductor material in which the chemical composition is disordered enough to create localized states but in which the physical alignment is sufficiently ordered for epitaxial growth to be possible [2]. Examples of disordering are shown in Figs. 1a–d. The temperature dependence of luminescence intensity, the optical absorption, and the decay time of photoluminescence are presented. Based on these experimental results, a possible recombination process is suggested.

Experimental Procedure

Three types of samples, i.e., $Al_{0.5}Ga_{0.5}As$ bulk alloy (b-AL), $(AlAs)_2(GaAs)_2$ ordered superlattice (o-SL), and $(AlAs)_m$ $(GaAs)_n$ disordered superlattice (d-SL) with m,n = 1,2,3, were grown by molecular beam epitaxy (MBE) at about 600°C,

[1] Department of Electrical Engineering, Kyoto University, Kyoto, 606 Japan

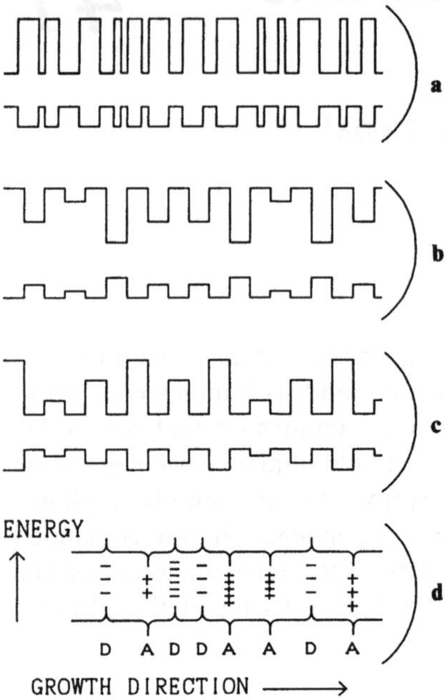

Fig. 1a–d. Examples of disordered crystalline semiconductor: disordered superlattices. Disorder variation which can be realized **a** in the individual layer thicknesses, **b** in the composition of well layers, **c** in the composition of barrier layers, and **d** in the period and level of δ-doping

on (001)-oriented semi-insulated GaAs substrates. A 0.2 to 0.3μm-thick GaAs buffer layer was grown on the substrate, and a smoothly grown surface was confirmed by reflection high-energy electron diffraction (RHEED) pattern. The thickness of the $Al_{0.5}Ga_{0.5}As$ b-AL was 1.0 μm, whereas the total thicknesses of the o-SL and the d-SL layers were 0.34 μm (1200MLs).

The $(AlAs)_m(GaAs)_n$ d-SL with m,n = 1,2,3 consists of 1, 2, and 3MLs irregularly distributed, as determined with the linear congruential method used widely in computers. However, the appearance probabilities were maintained equal, i.e., P(1) = P(2) = P(3) = 1/3. The macroscopic compositions of $Al_{0.5}Ga_{0.5}As$ b-AL, $(AlAs)_2(GaAs)_2$ o-SL, and $(AlAs)_m(GaAs)_n$ d-SL with m,n = 1,2,3 are all the same, but their microscopic structures are different.

Besides these samples, the $(AlAs)_4(Al_xGa_{1-x}As)_4$ o-SL and the $(AlAs)_m(Al_xGa_{1-x}As)_n$ d-SL with m,n = 2,4,6 were fabricated to reduce the influence from the interface roughness and to obtain a shorter luminescence wavelength. Here, the Al composition x was chosen as 0.0, 0.1, and 0.25.

The X-ray diffraction was observed to determine the structural differences among b-AL, o-SL, and d-SL by using Cu-α line, and the microscopic structure was investigated by measuring satellite peaks [3].

The photoluminescence (PL) was measured in the temperature range 4.2K–300K, and an Ar ion laser with multiple spectral lines. The typical laser power was about 0.5W. The time-resolved PL was also measured, with the Horiba NAES-700L system, whose time response is 70ps. The samples immersed in liquid nitrogen were excited with 0.3μJ/cm^2 from the dye laser.

Fig. 2. The PL intensities for the $Al_{0.5}Ga_{0.5}As$ b-AL and $(AlAs)_m(GaAs)_n$ o-SL with (m,n) = (2,2), (3,3), and (9,3), and the $(AlAs)_m(GaAs)_n$ d-SL with m,n = 1,2,3 and P(1) = P(2) = P(3) = 1/3

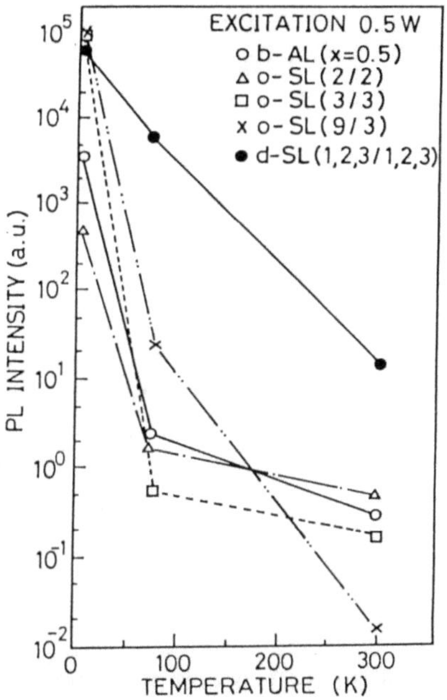

Experimental Results

Photoluminescence Intensity

The PL intensities of b-AL, o-SLs, and d-SL were compared; they are shown in Fig. 2 [3]. At 77K, the PL intensity of the d-SL becomes remarkably strong, i.e., the intensity ratios of the d-SL to the b-AL and the $(AlAs)_2(GaAs)_2$ o-SL were 2×10^3 and 3×10^3, respectively.

The main peaks of the PL were $\lambda = 0.605\mu m$, $0.617\mu m$, and $0.633\mu m$ for the b-AL, the $(AlAs)_2(GaAs)_2$ o-SL, and the d-SL, respectively. The PL wavelength from the d-SL exhibited a red shift relative to the wavelength from the b-AL and the $(AlAs)_2(GaAs)_2$ o-SL.

The temperature dependence of the PL intensity was measured in more detail between 6K and 100K. The samples used for the measurements were the $(AlAs)_4(Al_xGa_{1-x}As)_4$ o-SLs and the $(AlAs)_m(Al_xGa_{1-x}As)_n$ d-SLs with m,n = 2,4,6; the results are shown in Fig. 3. A noticeable difference in the temperature dependence can be seen between the o-SLs and the d-SLs. The linear portions of the decrease shown in Fig. 3 are approximated as $exp(-T/18)$, $\sim exp(-T/28)$, and $\sim exp(-T/6.5)$, respectively.

Photoluminescence Spectra

The PL spectra of the $(AlAs)_4(GaAs)_4$ o-SL and the $(AlAs)_m(GaAs)_n$ d-SL with m,n = 2,4,6 are shown in Fig. 4. The spectrum of the o-SL is relatively simple, but

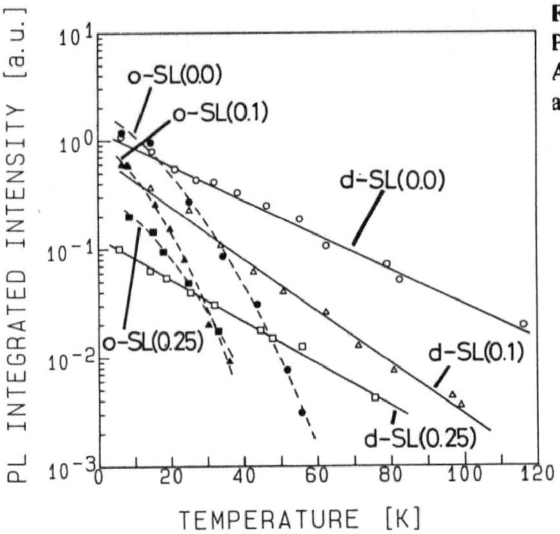

Fig. 3. Temperature dependence of PL integrated intensities of AlAs/$Al_xGa_{1-x}As$; x = 0.0, 0.1, 0.25 o-SLs and d-SLs

that of the d-SL is complex. In addition, the temperature dependence of the d-SL is also complex.

Optical Absorption

The absorption measurements of the $Al_{0.5}Ga_{0.5}As$ b-AL, the $(AlAs)_2(GaAs)_2$ o-SL, and the $(AlAs)_m(GaAs)_n$ d-SL were carried out at around 295K, 77K, and 7K [3, 4]. The results obtained at 7K are shown in Fig. 5, in which the PL spectra of the b-AL and the d-SL are superimposed. With regard to the optical absorption near the band edge, the o-SL followed the characteristics of an indirect band, but the d-SL was characteristic of neither an indirect nor a direct band.

Photoluminescence Decay Time

The PL decay times were derived from the time-resolved PL measurements [5]. The decay times of the $Al_{0.5}Ga_{0.5}As$ b-AL, the $(AlAs)_2(GaAs)_2$ o-SL, and the $(AlAs)_m(GaAs)_n$ d-SL with m,n = 1,2,3 were 2.6ns, 2.8ns, and 0.25ns, respectively, with the excitation intensity of 0.3 $\mu J/cm^2$. The decay time of the d-SL was shorter than those of the b-AL with an indirect band gap and the o-SL, and was longer than those of the b-AL with a direct band gap.

Discussion

The prominent PL capability of a d-SL is discussed in connection with a wider quantum well which does not appear in the o-SL, but appears in the d-SL. In a wider well, excited carriers are easily trapped and luminescence wavelength is

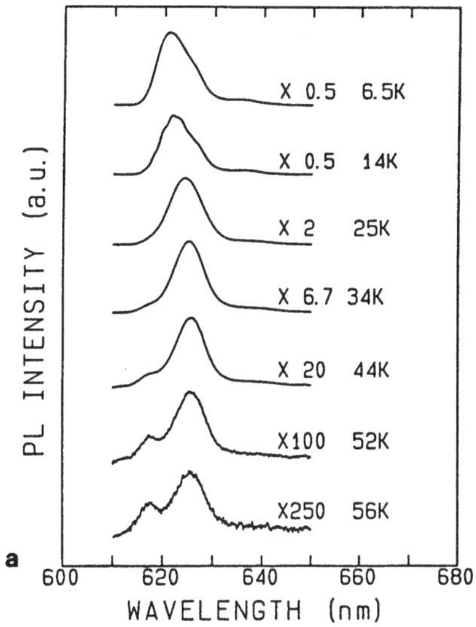

o-SL(0.0)

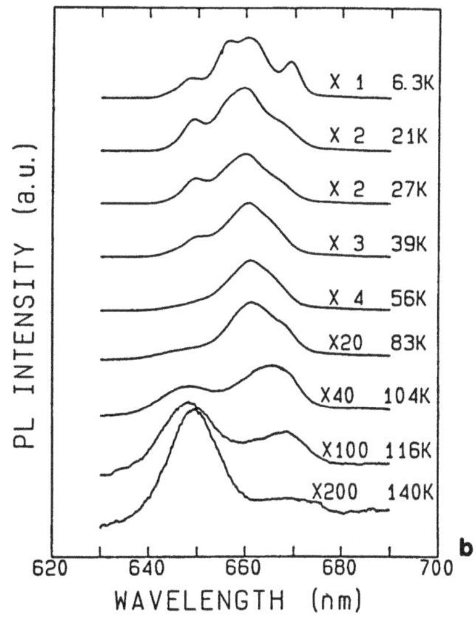

d-SL(0.0)

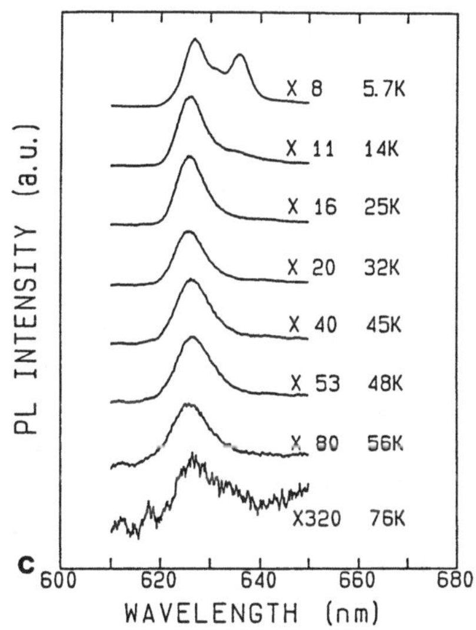

d-SL(0.25)

Fig. 4a–c. The PL spectra of the $(AlAs)_4(GaAs)_4$ o-SL and the $(AlAs)_m(Al_xGa_{1-x}As)_n$ d-SL with m,n = 2,4,6. The Al composition x here is 0.0 and 0.25

longer than in a narrow well. The PL wavelengths of the $(AlAs)_3(GaAs)_3$ and $(AlAs)_9(GaAs)_3$ o-SLs are shorter than that of d-SL, and the PL intensity of the o-SLs is weaker than that of d-SL. Additionally, as shown in Fig. 3, the temperature dependence of the d-SLs is different to that of the o-SLs. These results show that the PL enhancement in the d-SL would not be caused by the carrier trap and confinement in the wider GaAs-well layer.

The PL peak energy of the $Al_{0.5}Ga_{0.5}As$ b-AL (Fig. 5) shows that the phonon

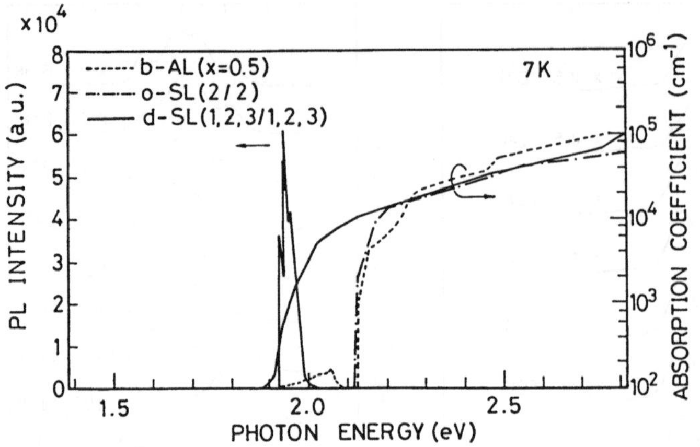

Fig. 5. Optical absorption spectra of the b-AL (*broken line*), the o-SL (*dash-dotted line*), and the d-SL (*solid line*) at 7K and PL spectra at 4.2K

emission is required in the recombination process. On the other hand, the PL peak of the d-SL appears in the extended energy range. This suggests that the phonon assistance is not necessary for the recombination process in a d-SL, and its suggestion is supported by the result of time-resolved PL measurement.

The intensity comparison of PL at equal wavelengths is interesting from the point of view of application. The luminescence wavelengths of the o-SL (0.0) and the d-SL (0.25) are almost the same, 0.625μm. The intensity of the d-SL (0.25) becomes stronger beyond 50K, as shown in Figs. 3 and 4.

The possible luminescence mechanism of a d-SL can be considered to be that of a recombination process that occurs in the Lifshitz model of localized states [6], which are created by disordering in the layer thickness of the superlattice; the momentum conservation being relaxed in the recombination process. This mechanism is supported by the following three experimental results: (a) the temperature dependence of the PL of the d-SL in Fig. 3 can be fitted by an equation which is interpreted in terms of the existence of localized states in an amorphous semiconductor [7]; (b) the PL peak energy and the absorption edge of the d-SL do not show interactions between carrier and phonon, and (c) the PL intensity of the d-SL decays faster than is the case in samples with an indirect band.

Further experimental and theoretical studies are continuing with the goal of thoroughly elucidating the PL characteristics of a d-SL.

Acknowledgments. The authors would like to thank Drs. Y. Takeda (now at Nagoya University) and A. Wakahara, and Messrs. M. Kasu (now at NTT) and T. Yamamoto (now at Murata Mfg. Co. Ltd.). Appreciation is extended to Horiba Ltd. for time-resolved photoluminescence measurement. This work was supported, in part, by Grant-in-Aid No. 0142006 for Scientific Research from the Ministry of Education, Science and Culture of Japan.

References

1. Cody GD (1984) In: Pankov JI (ed) Semiconductors and semimetals vol 21, Part B. Academic, London, pp 11–82
2. Sasaki A, Kasu M, Yamamoto T, Noda S (1989) Jpn J Appl Phys 28:L1249
3. Kasu M, Yamamoto T, Noda S, Sasaki A (1990) Jpn J Appl Phys 29:828
4. Yamamoto T, Kasu M, Noda S, Sasaki A (1990) J Appl Phys 68:5318
5. Kasu M, Yamamoto T, Noda S, Sasaki A (1991) Appl Phys Lett 59:800
6. Lifshitz IM (1965) Phys Uspekhi (in Russian) 7:549
7. Street RA, Searle TM, Austin IG (1974) In: Stuke J, Brenig W (eds) Amorphous and liquid semiconductors. Taylor and Francis, London, p 953

Photoluminescence Spectra of GaAs Quantum Wires in High Magnetic Fields

42

YASUSHI NAGAMUNE, YASUHIKO ARAKAWA, SHIRO TSUKAMOTO[1], MASAO NISHIOKA[2], SATOSHI SASAKI, AND NOBORU MIURA[3]

Summary. We investigated the photoluminescence (PL) spectra of GaAs quantum wires in high magnetic fields up to 40 T. The observed PL peak shift of the quantum wires with the increase of the applied magnetic field was much smaller than that of the bulk. In addition, it was found that the PL peak shift of the quantum wires was strongly dependent on the direction of the magnetic field. These results demonstrate the existence of the two-dimensional confinement effect in quantum wires.

Introduction

The fabrication of low dimensional semiconductor structures such as quantum wires has been intensively investigated, it is expected that since newly discovered physical phenomena will be applied to semiconductor lasers or other functional devices [1, 2]. However, conventional fabrication techniques involving the use of wet chemical etching [3], reactive ion etching [4], ion beam implantation [5], or ion beam milling [6, 7] on samples with quantum wells have problems of surface recombination and damage or disordering of the samples generated by ions with large energy. Although regrowth has been carried out to eliminate surface recombination at the free surface, problems of the mass-transport effect [8] and contamination at the growth interface still remain.

To avoid these problems, selective growth on masked substrates [9, 10], nonplanar or misoriented substrates [8, 11–13], or substrates with reverse-mesa stripes [14] has also been investigated, by employing metal-organic chemical vapor deposition (MOCVD) techniques or molecular beam epitaxy (MBE). Recently, by using an electron beam (EB) lithography technique and MOCVD selective growth on masked substrates [15], we obtained triangular-shaped GaAs quantum wires,

[1] Research Center for Advanced Science and Technology, University of Tokyo, 4-6-1 Komaba, Meguro-ku, Tokyo, 153 Japan
[2] Institute of Industrial Science, University of Tokyo, 7-22-1 Roppongi, Minato-ku, Tokyo, 106 Japan
[3] Institute for Solid State Physics, University of Tokyo, 7-22-1 Roppongi, Minato-ku, Tokyo, 106 Japan

with a lateral dimension of 20 nm, which showed a clear blue-shifted photoluminescence (PL) peak with high intensity due to two-dimensional confinement.

PL spectra and PL excitation spectra have been studied to obtain evidence of two-dimensional confinement effects in quantum wires [16, 17], while magneto-PL measurement of quantum wires has also been carried out [18, 19]. However, in order to clarify the two-dimensional confinement effects more clearly, we need to investigate the magneto-PL effects through varying the direction of the applied magnetic fields three-dimensionally.

In this study, we measured the magneto-PL spectra of excitons in our GaAs quantum wires in high magnetic fields up to 40 T; the orientation of the applied magnetic fields to the quantum wires was varied in three orthogonal directions in order to investigate the two-dimensional confinement effects. The results indicate the clear dependence of the PL spectra on both the magnitude and the direction of the magnetic fields, which result demonstrates the existence of two-dimensional confinement effects in quantum wires.

Experimental Procedure

The fabrication procedure for GaAs quantum wires was as follows: Before MOCVD growth, a 20-nm-thick SiO_2 layer was formed by plasma chemical vapor deposition on a semi-insulating (100) GaAs substrate. A polymethylmethacrylate (PMMA) resist, 50-nm thick, and with 100 nm line and spaces parallel to $\langle 01\bar{1} \rangle$ direction, was lithographically defined in an area of 3×3 mm^2 on the SiO_2 layer by an EB lithography technique, followed by wet chemical etching of the SiO_2 layer with buffered HF.

After removing the PMMA resist, GaAs triangular prisms with (111)A facet sidewalls were selectively grown in the windows of the SiO_2 mask on the substrate by MOCVD growth. Further continuation of the growth, which leads to smoothing of the (111)A facet sidewalls, makes the dimensions of the triangular prisms more uniform, and, at the same time, growth of the (111)A facet sidewalls occurs laterally above the SiO_2 mask, making the space between the triangular prisms smaller, and then a thin $Al_{0.4}Ga_{0.6}As$ layer was grown. When the triangular prisms connected together, very sharp $Al_{0.4}Ga_{0.6}As$ V-grooves were formed between them; these were much sharper than those made by a wet chemical etching technique [10]. GaAs wires were grown at the bottom of the V-grooves, and an $Al_{0.4}Ga_{0.6}As$ layer was grown again to embed the GaAs wires. After this procedure was repeated three times, three vertically stacked GaAs quantum wires were formed without being exposed to the air.

Here, as for MOCVD growth conditions, the partial pressures of trimethylgallium (TMG), trimethylaluminum (TMA), and AsH_3 were 4.4×10^{-6}, 1.5×10^{-6}, and 4.4×10^{-4} atm, respectively, the total pressure, including H_2 carrier gas (6 L/min) was 100 torr, the group V to III ratio was about 100, and the growth temperature was 700°C. In the whole growth process, the growth conditions were not changed [15].

Figure 1, parts a and b, respectively, show a schematic illustration of a cross

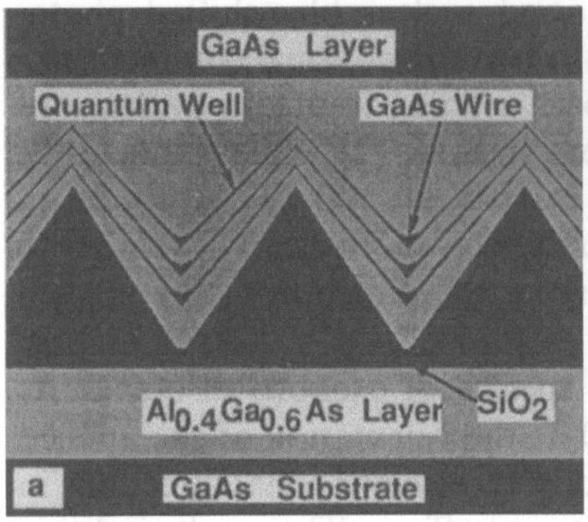

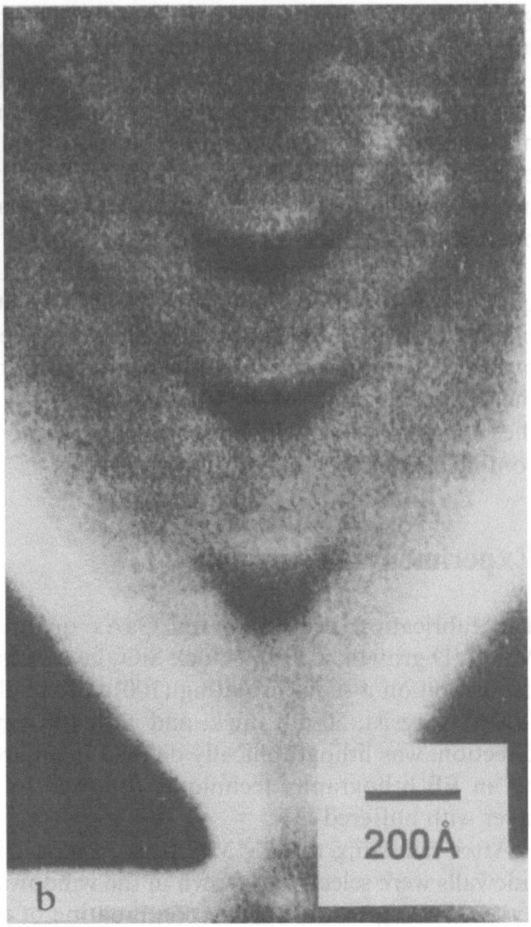

Fig. 1. **a** Schematic illustration of a cross section of vertically stacked multiple quantum wires. **b** High resolution secondary electron image of the region near the quantum wires

section of the vertically stacked multiple quantum wires and a high-resolution secondary electron image of the region near the quantum wires fabricated by the present growth process, where the sample was stain-etched by an NH_4OH and H_2O_2 system. As shown in Fig. 1b, triangular-shaped GaAs quantum wires, of base size and height 20 nm and 10 nm, respectively, were fabricated. The quantum wires were connected to thin GaAs quantum wells because of the finite growth rate of GaAs at (111)A plane.

The magnetic fields were generated with a pulse magnet, and the PL spectra from the quantum wires at 4.2 K were detected with an optical multi-channel analyzer (OMA) system installed at the exit of the spectrometer through the optical fiber [20]. The pulse duration of the magnetic field was 10 ms and the maximum field was about 40 T. The variation of magnetic field during the measurement was within $\pm 3\%$ at the top of the pulsed magnetic field. The spectral resolution determined by the entrance slit width of the spectrometer was about 2.4 meV around the measured wavelength range. The relationship between the direction of the

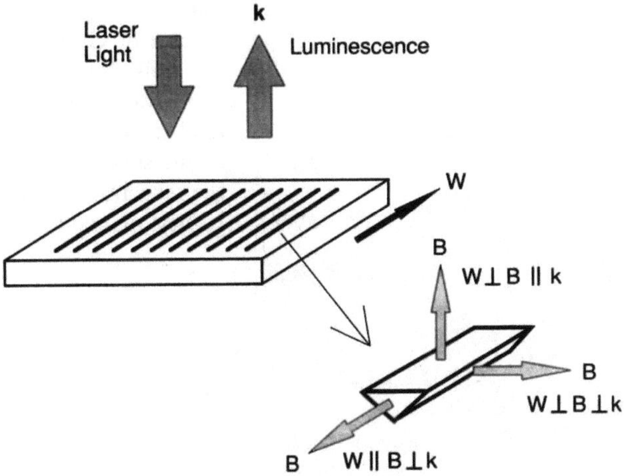

Fig. 2. Relationship between the direction of the quantum wires ***W***, that of the applied magnetic field ***B*** and that perpendicular to the substrate ***k***

quantum wires ***W***, the applied magnetic field vector ***B***, and the direction perpendicular to the substrate ***k***, is shown in Fig. 2. As shown in this figure, three kinds of configurations were taken for the present magneto-PL measurement, and it should be noted that $\boldsymbol{W} \perp \boldsymbol{B} // \boldsymbol{k}$ and $\boldsymbol{W} \perp \boldsymbol{B} \perp \boldsymbol{k}$ are different configurations because of the triangular-shaped quantum wires.

Results and Discussion

The PL spectra of the GaAs quantum wires clearly showed a blue-shifted PL peak which corresponded to the first subband state of these quantum wires. The energy difference between the PL peak of the quantum wire and that of the GaAs bulk was 46 meV, which is in good agreement with the calculated quantized energy level of the present quantum wires. Note here that the present triangular-shaped quantum wires are almost equivalent to rectangular-shaped quantum wires with a dimension of 14 nm with regard to the quantized energy level [21].

Figure 3 shows the magneto-PL spectra of the quantum wires with various magnetic fields. Each spectrum was measured at the magnetic field indicated on the right-hand side of the spectra, and at a temperature of 4.2 K with the configuration of $\boldsymbol{W} \perp \boldsymbol{B} // \boldsymbol{k}$. In this Figure, the spectra for $\boldsymbol{W} // \boldsymbol{B} \perp \boldsymbol{k}$ are also shown for the highest magnetic field of ~40 T (dotted line) for comparison. Here, the PL peak at zero magnetic field around 1.51 eV corresponds to the bulk transition, and that at 1.56 eV corresponds to the quantum wires. The PL peak related to the quantum wells, which is not shown here, is placed around 1.74 eV. The energy shift of the main bulk peak with the highest intensity, which may arise from the transition from conduction electrons to neutral donors is almost equally changed independent of the configurations, even in the high magnetic field region. On the other

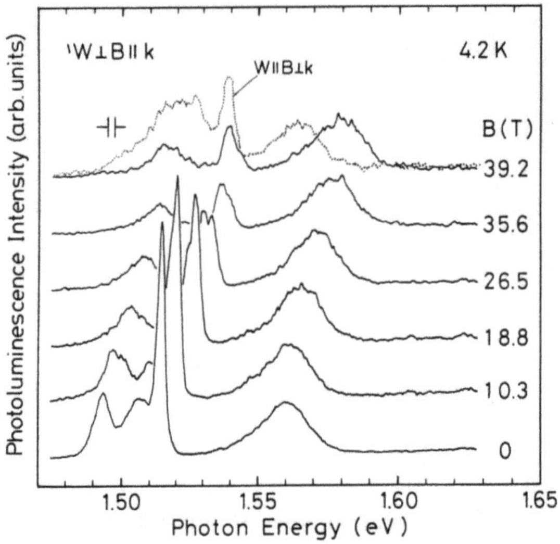

Fig. 3. Magneto-photoluminescence spectra of GaAs quantum wires and the bulk for $\boldsymbol{W} \perp \boldsymbol{B} // \boldsymbol{k}$. Each spectrum was measured at the magnetic field indicated on the *right-hand side* of the spectra. The *dotted line* shows the spectra for $\boldsymbol{W} // \boldsymbol{B} \perp \boldsymbol{k}$ for the highest magnetic field

hand, the energy shift of the PL peak position of the quantum wires is strongly dependent on the configurations. Namely, the energy shift for $\boldsymbol{W} \perp \boldsymbol{B} // \boldsymbol{k}$ is much larger than that for $\boldsymbol{W} // \boldsymbol{B} \perp \boldsymbol{k}$.

In Fig. 4, the PL peak positions from the quantum wires and the bulk at various configurations are plotted as a function of the applied magnetic fields. In the low magnetic field region, all the PL peak positions change diamagnetically, while in the higher magnetic field region they change almost linearly against the magnetic fields corresponding to the Landau level energy shift.

However, the behavior of the PL peak shift is different in the bulk and the quantum wires. The bulk PL peak shifts at various magnetic fields are almost equal, independent of the configurations in all the magnetic field regions. In contrast, the PL peak shift of the quantum wires clearly depends on the configuration. The ratio of the energy shifts of the quantum wires in the high magnetic field region is about 4 : 2 : 1 against the configurations of $\boldsymbol{W} \perp \boldsymbol{B} // \boldsymbol{k}$, $\boldsymbol{W} \perp \boldsymbol{B} \perp \boldsymbol{k}$, and $\boldsymbol{W} // \boldsymbol{B} \perp \boldsymbol{k}$, respectively. This tendency is qualitatively in agreement with the cross section sizes of the quantum wires perpendicular to the direction of the applied magnetic fields.

These magnetic field dependences show that quasi-one-dimensional excitons are realized by the two-dimensional confinement effect in quantum wires. The ratio of the energy shift of the PL peak position to the square of the magnetic field in the low magnetic field region is almost in proportion to the cross section size of quasi-one-dimensional excitons in quantum wires. On the other hand, in the high magnetic field region, where Coulomb interaction can be ignored, the magnetic field dependence of the PL peak position is expected to be analyzed in terms of the interruption of the cyclotron motion of electrons and holes by two-dimensional confinement in triangular-shaped GaAs quantum wires.

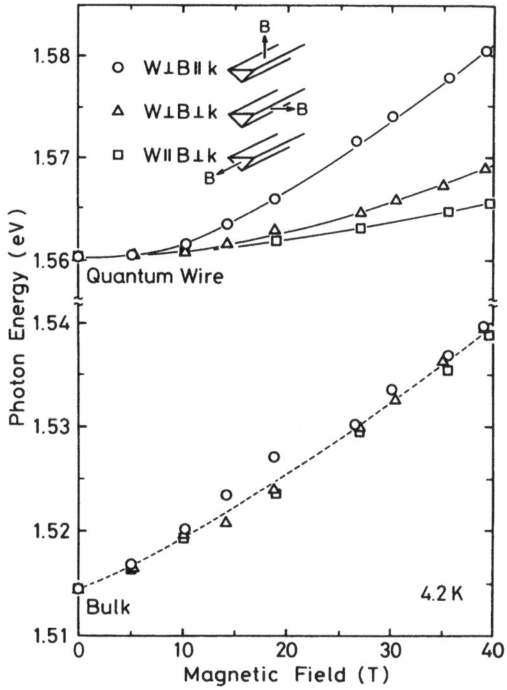

Fig. 4. PL peak positions from the quantum wires and the bulk as a function of the applied magnetic field for various configurations. *Circles*, *triangles*, and *squares* represent the data for $W \perp B // k$, $W \perp B \perp k$ and $W // B \perp k$, respectively

Conclusion

We investigated the PL spectra of GaAs quantum wires, fabricated by the MOCVD selective growth technique, in high magnetic fields up to 40 T. The difference between the PL peak shift of the GaAs quantum wires and that of the GaAs bulk against various configurations clearly demonstrated that the two-dimensional confinement effect *really* exists in GaAs quantum wires.

References

1. Arakawa Y, Sakaki H (1982) Appl Phys Lett 40:939
2. Arakawa Y, Vahala K, Yariv A (1984) Appl Phys Lett 45:950
3. Miller BI, Shahar A, Koren U, Corvini PJ (1989) Appl Phys Lett 54:188
4. Kash K, Scherer A, Worlock JM, Craighead HG, Tamargo MC (1986) Appl Phys Lett 49:1043
5. Cibert J, Petroff PM, Dolan GJ, Pearton SJ, Gossard AC, English JH (1986) Appl Phys Lett 49:1275
6. Temkin H, Dolan GL, Panish MB, Chu SNG (1987) Appl Phys Lett 50:413
7. Gershoni D, Temkin H, Dolan GL, Dunsmuir J, Chu SNG, Panish MB (1988) Appl Phys Lett 53:995

8. Colas E, Simhony S, Kapon E, Bhat R, Hwang DM, Lin PSD (1990) Appl Phys Lett 57:914
9. Lebens JA, Tsai CS, Vahala KJ (1990) Appl Phys Lett 56:2642
10. Fukui T, Ando S, Fukai YK (1990) Appl Phys Lett 57:1209
11. Bhat R, Kapon E, Hwang DM, Koza MA, Yun CP (1988) J Cryst Growth 93:850
12. Gaines JM, Petroff PM, Kroemer H, Simes RJ, Geels RS, English JH (1988) J Vac Sci Technol B6:1378
13. Fukui T, Saito H (1988) J Vac Sci Technol B6:1373
14. Nakamura Y, Koshiba S, Tsuchiya M, Kano H, Sakaki H (1991) Appl Phys Lett 59:700
15. Tsukamoto S, Nagamune Y, Nishioka M, Arakawa Y (1992) Extended Abstracts of the 1991 International Conference of Solid State Devices and Materials, Yokohama, 1991. J Appl Phys 71:533
16. Lebens JA, Tsai CS, Vahala KJ (1990) Appl Phys Lett 56:2642
17. Tanaka M, Sakaki H (1989) Appl Phys Lett 54:1326
18. Berggren KF, Thornton TJ, Newson DJ, Pepper M (1986) Phys Rev Lett 57:1769
19. Plaut AS, Lage H, Grambow P, Heitmann D, von Klitzing K, Ploog K (1991) Phys Rev Lett 67:1642
20. Tarucha S, Okamoto H, Iwasa Y, Miura N (1984) Solid State Comm 52:815
21. Tanaka T, Yamauchi T, Arakawa Y (1991) Fall meeting of Japan Society of Applied Physics, Okayama, 1991

Nonlinear Optical Properties

Time Resolved Nonlinear Optical Spectroscopy of Magnetically Confined Excitons

43

JASON B. STARK[1*], WAYNE H. KNOX[1], AND DANIEL S. CHEMLA[1**]

Summary. A perpendicular magnetic field confines the quasi-two-dimensional (2D) electronic states of a quantum well into quasi-zero-dimensional (0D) magneto-excitons. Short-pulse broadband optical studies demonstrate that at high field, a gas of magneto-excitons behaves like an ensemble of atomic two-level systems. Excitation of a nonthermal distribution of 2D carriers at zero field results in rapid relaxation to a thermal distribution; at high field, all low-energy scattering channels are eliminated, so that in 0D, relaxation rates decrease by several orders of magnitude relative to 2D. Magnetic confinement in quantum wells permits the study of the interactions and relaxation dynamics of electon-hole pairs, in materials of excellent optical quality, as their dimensionality is continuously tuned from 2D to 0D.

Introduction

Strongly confined electronic states, in materials of excellent optical and electronic quality, can be obtained by immersing a quantum well in a perpendicular magnetic field [1–6]. Optically generated electrons (e) and holes (h) are then created into so-called magneto-exciton states which experience a quadratic confining potential due to the magnetic field [7–14]. The linear absorption spectrum of these confined states consists of a sequence of isolated spectral peaks, displaying the behavior characteristic of a 0D e-h system. Moreover, the length scale of the confining potential may be varied over many orders of magnitude by controlling the magnetic field strength. The physical properties of e-h pair states may thus be studied continuously from two to zero dimensions, in a single sample, by varying the magnetic field. Since inhomogeneities in the magnetic field can easily be reduced below one part in a thousand, the size distribution of confined electronic states is determined entirely by the quality of the quantum well layers themselves. The mature technology for layered growth, therefore, determines the ultimate resolution of the spectral features.

[1] AT & T Bell Laboratories, Holmdel, NJ 07733, USA
* Present address: AT & T Bell Laboratories, Murray Hill, NJ 07974, USA
** Present address: Physics Department, University of California at Berkeley, Material Sciences Division, Lawrence Berkeley Laboratory, Berkeley, CA 94720, USA

Quantum confinement into two dimensions, in layered materials, causes qualitative physical changes relative to bulk materials. The optical absorption spectrum, which rises smoothly above the bandedge in bulk, breaks up into a series of steps in 2D. Each step in the 2D density of states corresponds to an electronic state originating from the confinement potential of the quantum well. Below each step, there are sharp features arising from the excitons. The density of states, which evolves from a smooth function in 3D to a stepped function in 2D, undergoes a further transformation as the magnetic field confines the states to 0D. The stepped 2D density of states coalesces, when the field is applied, into a series of sharp peaks, with all of the states concentrated into the macroscopically degenerate discrete allowed energies of the 0D system. The linear absorption spectrum vs magnetic field (Fig. 1), for σ_- circularly polarized light, exhibits this behavior. The smooth absorption spectrum of the zero-field scattering states, at energies above 1.59 eV, breaks up into a series of sharp magneto-exciton transitions as the magnetic field increases to 12 Tesla (T). The contour plot of the mean surface curvature of these data (Fig. 2) shows the dispersion of the main magneto-exciton peaks. The strong mixing between optically active and inactive states is apparent in the many anticrossing features seen in these spectra.

The nonlinear optical response of magneto-excitons changes from that of a 2D e-h system with strong Coulomb correlations, at low fields, to that of an ensemble of weakly interacting two-level systems in the limit of high fields [15, 16]. This occurs despite the fact that the optically generated e-h pairs in a magnetic field are

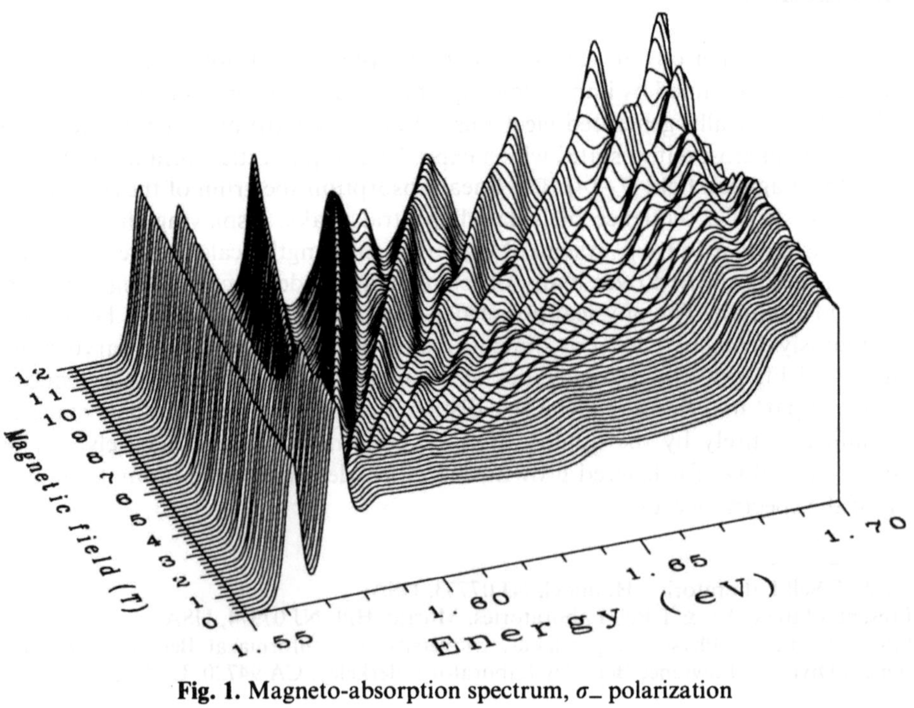

Fig. 1. Magneto-absorption spectrum, σ_- polarization

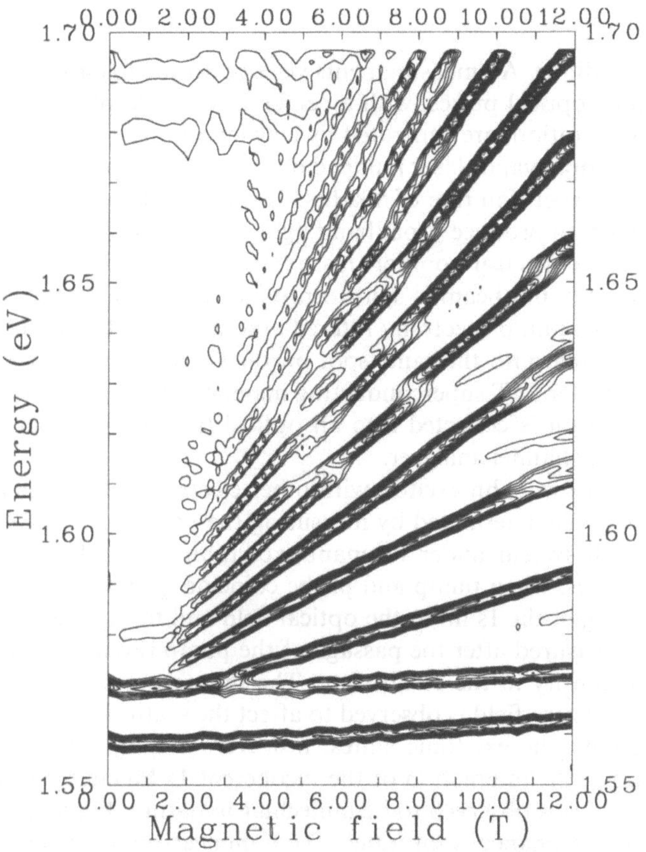

Fig. 2. Contour plot of mean surface curvature of magneto-absorption peaks, σ_- polarization

free to translate in two dimensions through the crystal. In fact, the scattering interactions of these magneto-excitons may also be studied using nonlinear optical techniques, so that the physics of magneto-excitons is richer than that of excitons trapped within a quantum dot. For these reasons, quantum wells immersed in a perpendicular magnetic field provide an ideal system in which to study the linear and nonlinear optical properties of electronic states, as they are continuously tuned from two to zero dimensions.

Experimental

Nonlinear optical measurements are performed in the excite-and-probe transmission geometry, in which a strong tunable pump pulse excites the sample, and a weak, broadband probe pulse measures the absorption of the excited material. The sample studied is an 8.4 nm multiple quantum well structure, in the intrinsic region of an unbiased *p-i-n* diode. All measurements were performed at 4 K. To facilitate these measurements, optical pulses, with a center wavelength of 805 nm and approximately 100 fs duration, are generated by a synchronously pumped dye laser, and amplified in a copper vapor laser pumped dye amplifier to produce pulses with energies of 1μJ, at a repetition rate of 8 kHz [17]. These pulses are focused into a continuum generator, to produce broadband light pulses, which retain their short pulse width and have spectral components extending from the infrared to the ultraviolet. This continuum beam is split to produce the strong pump and weak probe beams, with the pump spectrally filtered and delayed relative to the probe. Both beams are focused onto the same spot on the GaAs quantum well sample, located in the center of a 12 T superconducting magnet. After passing through the sample, the probe beam is collected into an optical fiber for parallel detection, using an optical multichannel analyzer.

The interactions of the 1s hh exciton with other electronic states, and with the scattering states, may be determined by measuring the nonlinear optical response of the quantum well system under resonant excitation of the 1s hh state. The transient nonlinearities, when pump and probe coincide, provide information on the coherent coupling of the 1s hh to the optical field and to other excitons, while the nonlinearities measured after the passage of the pump reveal the effects of an incoherent charge density in the 1s hh state. The resonant excitation of a 1s hh exciton population at zero field is observed to affect the scattering states only very weakly. The effect on the excitonic states, however, is quite strong. After the passage of the pump, the interaction of the incoherent 1s hh charge density with the other exciton states is observed to be molecular in nature. Even at high densities, when the 1s hh resonance is saturated, the nonlinear optical response can be qualitatively described by the interactions of pairs of excitons via a molecular potential [15, 18], analogous to the molecular interaction that binds two hydrogen atoms into a molecule. Transiently, the excitons respond to a combination of coherent molecular charge density effects, and a Coulomb-enhanced Rabi shift of the single particle levels [18, 19]. At high excitation densities the nonlinear response of the 1s hh excitons appears to be that of a highly correlated plasma. At the same time, the exciton states which are not resonantly excited by the pump remain spectrally intact, showing no evidence that their correlations have significantly weakened. This result indicates the very strong Coulomb correlations between electron and hole in the 2D excitonic gas.

The effect of the magnetic field on the excitonic nonlinearities is two-fold. Since saturation occurs when the excitons cover the quantum well area [20], the magnetic compression of the excitonic states results in less saturation at high field, given the same number of excitons created. Moreover, the strength of the inter-excitonic molecular potential is reduced by the strong magnetic confinement, since for a given inter-excitions separation, the charge densities of neighboring excitons cannot overlap to establish the molecular interaction as effectively when magnetically compressed, as when uncompressed at zero field. The resulting nonlinear optical response resembles that of a two-level atomic system, with saturation effects dominating over molecular interactions [15, 16, 18].

The Coulomb interaction between electron and hole in zero field yields the discrete states of the exciton, as well as the highly correlated scattering states above the bandedge. The absorption strength of the semiconductor above the bandedge is determined by the two dimensional joint density of states and the matrix element for coupling of the e-h pair state to the optical field. Coulomb correlations augment the coupling matrix element according to the Sommerfeld enhancement. At the bandedge, this results in an increase in absorption by a factor of two due to the Coulomb correlations. The Sommerfeld enhancement falls off slowly in two dimensions as a function of energy above the bandgap, so that essentially all near-bandgap pair states are highly correlated upon optical excitation. The pair states created above the bandedge, however, have excess kinetic energy, which allows the optically generated electron and hole to separate by large distances. During this separation, the particles can scatter from other charged carriers, from phonons, or from crystalline defects, so that the initial correlation between photoexcited electron and hole is rapidly lost. In addition, carrier-carrier scattering can thermalize an initially nonthermal distribution of free carriers. Short pulse pump-probe transmission experiments have the ability to measure such a non-thermal distribution as it relaxes, and so can provide information on the effects of carrier-carrier scattering on a femtosecond time scale.

The application of a high magnetic field, as observed in the experiments described above, yields a system which behaves atomically. The independent motion of electron and hole, subsequent to optical generation at zero field, is eliminated at high field, so that none of the relaxation effects associated with carrier-carrier scattering are expected in this regime. As a result, the nonthermal distribution excited by a transient pump pulse resonant with the 2s excited magneto-exciton state persists well after the initial excitation [18]. Furthermore, during this period, the binding of e-h pairs into discrete 0D pair states produces a nonlinear optical response exhibiting mainly saturation, resembling the results for resonant 1s excitation. Furthermore, excitons in 2D, at zero field, retain a high degree of correlation, even at densities for which the absorption is strongly saturated. At these high densities, the nonlinear optical response of the populated excitonic state resembles that of a highly correlated plasma, while that of the unpopulated states remains strongly excitonic. The conclusion is that the states of electrons and holes in a highly excited 2D semiconductor undergo a smooth transition from strongly saturated excitons to a highly correlated plasma, as the excitation density increases.

Conclusions

The linear absorption spectra for the quantum well magneto-exciton demonstrate the evolution of its density of states from that of a 2D system, at zero field, to 0D at high field. In this transition, all e-h pair states become bound and compressed by the quadratic effective potential of the magnetic field. At the highest field strength measured, 12 T, the 1s hh exciton is reduced in area by a factor of three. The binding and compression of the e-h pair states has a significant effect on their nonlinear optical response. At zero field, the response to resonant excitation of a population of 1s hh excitons demonstrates that 1s excitons interact through a molecular potential, much as hydrogen atoms do in molecular hydrogen. The magnetic compression of the excitons significantly weakens this molecular interaction potential. The resulting nonlinear optical response at high field expresses the physics of a gas of weakly interacting two-level atoms.

Resonant excitation of scattering states in 2D produces weak nonlinear effects on the absorption of these states, the largest effects being observed at the excitons. Relaxation of the initially nonthermal distribution occurs in times on the order of 100 fs for excitation 30 meV above the lowest energy exciton. At high field, however, resonant excitation of the excited 2s state produces predominantly saturation nonlinearities of essentially atomic states, the nonthermal distribution persisting for times on the order of 100 ps, three orders of magnitude longer than at zero field. Confinement into 0D, therefore, has a profound effect on the relaxation of excitations with large excess energy. These results indicate the wealth of information available using time-resolved nonlinear spectroscopic techniques to explore the physics of electronic systems in the transition from two to zero dimensions.

Acknowledgments. The authors would like to acknowledge valuable discussions with Stefan Schmitt-Rink and Wilfried Schäfer, as well as the expert assistance of Gail Doran and William Wiegmann for sample processing and growth. The work of one of us (JBS) was supported, in part, by NSF Grant No. EET-871841. The work of one of us (DSC) was supported by the Director, Office of Energy Research, Office of Basic Energy Sciences, Division of Materials Sciences of the US Department of Energy under contract DE-AC03-76SF00098.

References

1. Maan JC, Belle G, Fasolino A, Altarelli M, Ploog K (1984) Phys Rev B30:2253
2. Miura N, Iwasa Y, Tarucha S, Okamoto H (1985) In: Chadi JD, Harrison WA (eds) Proc 17th int Conf on the physics of semiconductors. Springer, New York
3. Ossau W, Jäkel B, Bangert E, Landwehr G, Weimann G (1986) Surf Sci 174:188
4. Rogers DC, Singleton J, Nicholas RJ, Foxon CBT, Woodbridge K (1986) Phys Rev B34:4002
5. Bugajski M, Kuszko W, Reginski K (1986) Solid State Commun 60:669
6. Ossau W, Jäkel B, Bangert E (1987). In: Landwehr G (ed) High magnetic fields in semiconductor physics. Springer, Berlin
7. Akimoto O, Hasegawa H (1967) J Phys Soc Jpn 22:181
8. Shinada M, Tanaka K (1970) J Phys Soc Jpn 29:1258
9. Shinada M, Sugano S (1966) J Phys Soc Jpn 21:1936
10. MacDonald AH, Ritchie DS (1985) Phys Rev B33:8336
11. Greene RL, Bajaj KK (1985) Phys Rev B31:6498
12. Yang SRE, Sham LJ (1987) Phys Rev Lett 58:2598
13. Bauer GEW, Ando T (1988) Phys Rev B37:3130
14. Chu H, Chang YC (1989) Phys Rev B40:5497
15. Stark JB, Knox WH, Chemla DS, Schäfer W, Schmitt-Rink S, Stafford C (1990) Phys Rev Lett 65:3033

16. Stafford C, Schmitt-Rink S, Schäfer W (1990) Phys Rev B41:10 000
17. Knox WH (1987) J Opt Soc Am B4:1771
18. Stark JB (1991) Ph D thesis, Massachusetts Institute of Technology
19. Knox WH, Chemla DS, Miller DAB, Stark JB, Schmitt-Rink S (1989) Phys Rev Lett 62:1189
20. Schmitt-Rink S, Chemla DS, Miller DAB (1989) Adv Phys 38:89

Ultrafast Behavior of Optically-Nonlinear Etalons

44

JÉRÔME PAYE AND DANIÈLE HULIN[1]

Summary. Nonlinear Fabry-Perot étalons are among the most promising candidates for all-optical switching and bistability. We have studied the temporal behavior of such a device with a GaAs/AlGaAs supperlattice as intracavity material. The use of the optical Stark effect on the excitonic resonances provides an ultrafast mechanism for intracavity nonlinearity, which allows us to obtain subpicosecond switching times.

Introduction

Optically-nonlinear Fabry-Perot étalons are interesting for their possible applications in optical signal processing, optical bistability, and logic functions. Such devices have potential applications as optical transistors, limiters, memory elements, differential amplifiers, and pulse shapers [1]. They are especially attractive for use in highly parallel optical systems; bidimensional arrays of small Fabry-Perot resonators have already been realized.

Even though parallelism is one the great advantages of optical signal processing, it is still the speed of the elementary component that controls the speed of a complex system. Therefore it is important to know the physical limits of such an optical device on an ultrashort time scale.

The steady-state characteristics of this type of device are now fairly well understood [1]. The dynamic aspect is less well documented. Although theoretical models have been developed [2, 3], little has been reported on the ultrafast behavior of nonlinear étalons. In this paper, we discuss those optical nonlinearities which have both a rapid response time and a sizeable efficiency, in order to be used for an ultrafast switching. We report experimental results demonstrating a subpicosecond speed, both for ON and OFF switching, and we raise the question of a pseudo-bistable behavior on a short time scale.

[1] Laboratoire d'Optique Appliquée, ENSTA—Ecole Polytechnique, Centre de l'Yvette, 91120 Palaiseau, France

Nonlinearities and Switching

All-optical switching can be achieved by absorptive or refractive mechanisms, or by a combination of both. The first case corresponds to an absorption change following material excitation. It may originate directly from a saturation of absorption due to the photoexcited carriers, like state filling or Coulombic interaction screening, or from the lattice heating following carrier non-radiative recombination. In the dispersive case, one takes advantage of the changes of the refractive index. There are different ways to achieve this goal, but one of the most popular is the Fabry-Perot étalon. This is partly due to the fact that, at resonance, the effective thickness of the nonlinear intracavity material is much larger than its real width.

The nonlinear Fabry-Perot étalon (NFP) consists in two mirrors with an optically-nonlinear intracavity material. It takes advantage of the change of the optical path inside the cavity following a change in the refractive index **n** of the material. We now concentrate on the case where the intracavity material is a semiconductor. This may provide modifications of the refractive index as large as 10^{-1}, so that a strongly nonlinear regime can be reached over interaction lengths as small as a few micrometers. This results in potentially very compact nonlinear optical devices.

The switching of the NFP device can be provided either by an external beam at different or identical wavelength or by the input beam itself. In this last situation, optical bistability can be encountered. The choice of the wavelength of command relies upon the mechanism which is selected for the nonlinearity. The wavelength of operation cannot be selected in the spectral region where band to band absorption is important, since this strongly perturbs the appearance of the resonances, the number of round trips being seriously reduced. It is possible to work at the absorption edge where the beam attenuation is low or in the transparency region, below the absorption threshold.

Two sets of parameters are important for the characterization of the nonlinear response of the material: the magnitude of the induced change of the refractive index and its response times. The most important times characterize the rise of the nonlinearity and its disappearance. For electronic nonlinearities, the first one may be considered as zero in most cases (carrier creation, Kerr effect, ...), although some delay can be introduced by physical effects like carrier cooling. More important is the recovery time for practical applications. In the case of carrier photoexcitation, it is determined by the carrier lifetime. It strongly varies with the material; for GaAs or GaAs multiple quantum wells (MQWS), it ranges around the nanosecond, slightly less for MQWS. Some procedures, such as proton bombardment [4] or epitaxy of the active layer on a cold substrate (low-T GaAs) [5], have been used to reduce this carrier lifetime. These improvements cannot be pushed too far, since sizeable electronic nonlinearities must be kept in the material.

The change magnitude of the refractive index **n** varies with the pump intensity and the nature of the nonlinearity. In Fig. 1, 2, and 3, we have sketched three different possibilities. The upper part of each figure represents schematic absorption spectra for a GaAs MQWS at low temperature. The continuous line is charac-

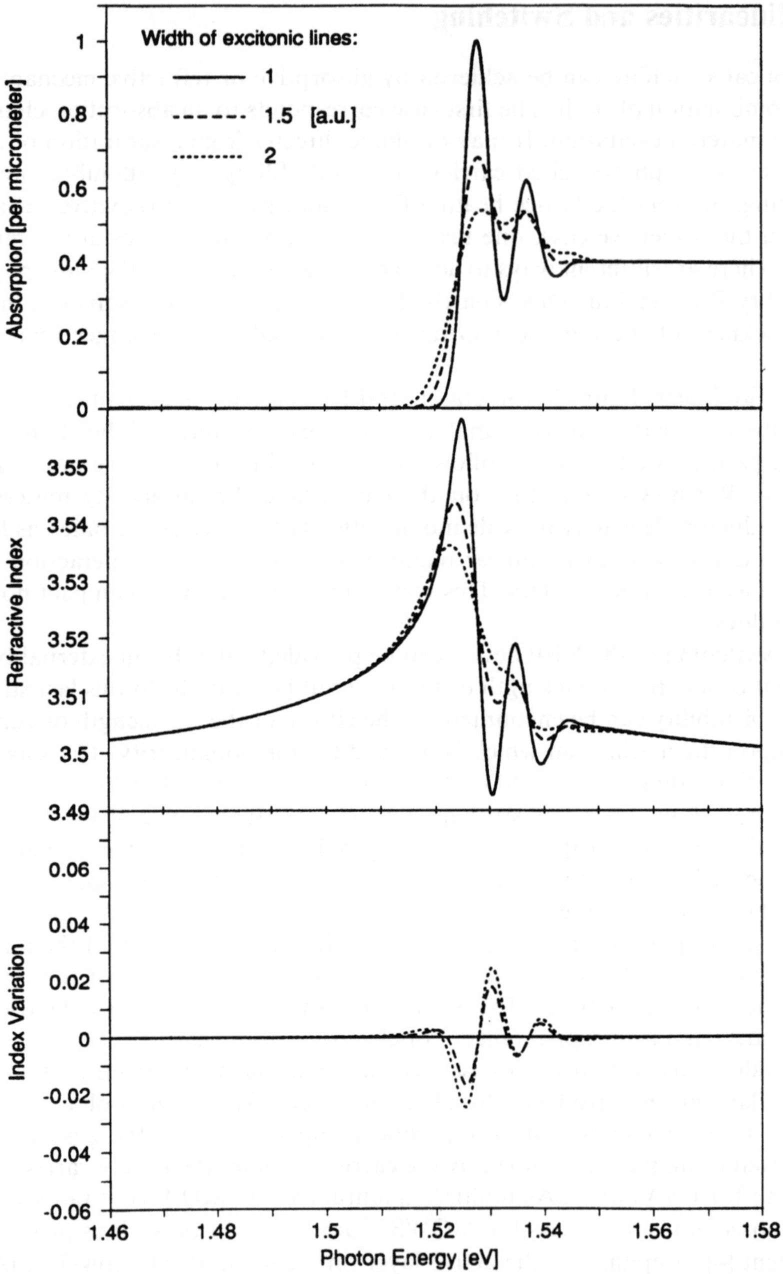

Fig. 1. Schematic representation of absorption and refractive index spectra for a GaAs/AlGaAs MQWS. The *upper part* of the Fig. shows the absorption spectra, the *middle part* the refractive index spectra calculated from absorption through the Kramers-Krönig relationships, and the *lower part* the variations of the refractive index. The *continuous lines* are representative of the unperturbed situation. The *dotted* and *dashed line* spectra correspond to a situation where the width of the excitonic absorption lines has been varied without changing the total absorption strength

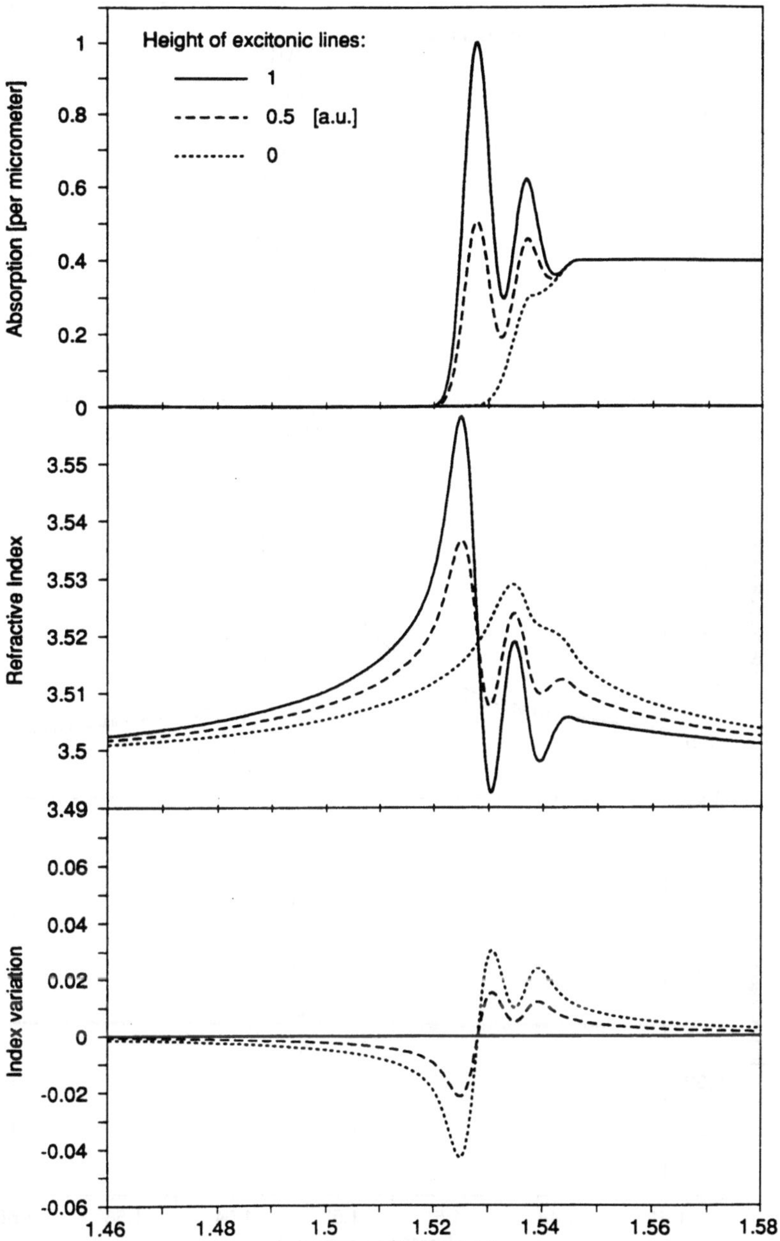

Fig. 2. Same representation as in Fig. 1, except that now the height of the excitonic lines has been changed without modification of their width

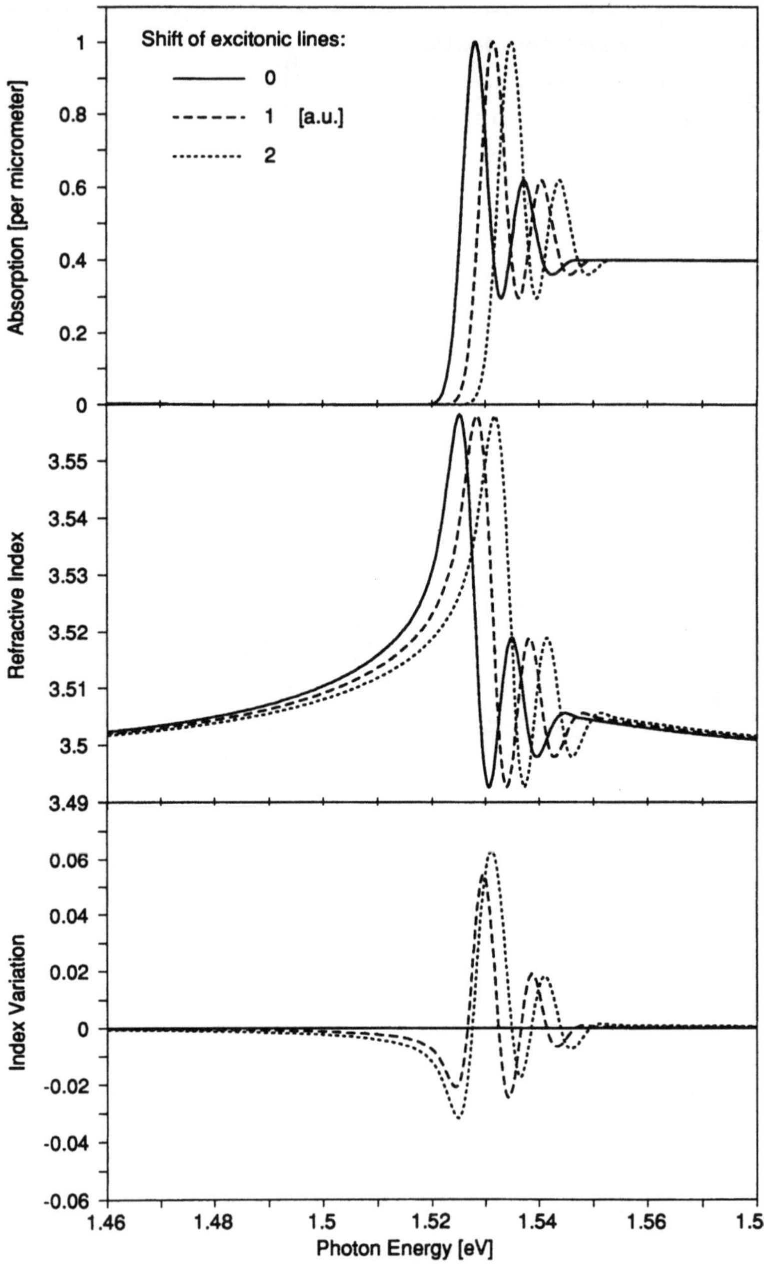

Fig. 3. Same representation as in Fig. 1, except that now the spectral position has been shifted as a whole

teristic of the unperturbed situation, with heavy- and light-exciton absorption lines modelled with gaussian shapes. In Fig. 1, we have varied the width of the excitonic absorption lines, keeping constant the total oscillator strength. The middle part of the figure represents the corresponding refractive index calculated by Kramers-Krönig transformation [6]. We have assumed that all the other optical transitions which are not in the spectral range of interest contribute as a constant background. This is not completely true, since there is always some dispersion of the refractive index even far from resonances. However, we choose this approximation for the sake of simplicity, in order to show up the main effects. It is clear that there are strong changes of n around the excitonic transitions. However little change occurs in the transparency region, below the excitonic absorption. The lower part of the Fig. shows the index modification **Δn** with respect to the unperturbed case.

In Fig. 2, we have kept constant the width of the excitonic lines and changed their heights by the same factor for heavy and light excitons. It can be seen, in the middle and lower part of the figure, that now the refractive index changes extend to a much larger spectral region, especially in the transparency region.

Figure 3 represents the simple case where the spectrum experiences a rigid blue shift without a modification of its shape. It can be seen, in the lower part of the figure, that, here also, **Δn** extends quite far below the absorption edge.

The broadening of the excitonic lines without oscillator strength modification may be linked to temperature increase or to collision broadening. The reduction of the oscillator strength may result from phase space filling and/or Coulomb interaction screening. Finally, the rigid blue shift is representative of the optical Stark effect. In reality, none of these features appears alone, but the above scheme provides an idea of the useful characteristics: large index changes in the non-absorbing spectral region are provided by the reduction of the excitonic transitions or by an energy shift.

In order to take full advantage of the highest non-linearities even at room temperature, we selected multiple quantum wells as active media in our nonlinear Fabry-Perot étalon [7]. The excitonic resonances are still present at $T = 300$ K in the absorption spectrum, providing, therefore, the possibility of large changes of the refractive index at room temperature.

Ultrafast Nonlinearities

Purely electronic optical nonlinearities have only been observed with short pulses, since it has been found that thermal effects are important on a time scale of several microseconds [8]. These effects come from the rise of temperature of the active medium, due to the heat generated by the non-radiative recombination of optically excited electron-hole pairs. In order to study the ultrafast behavior of étalons, we concentrate on electronic nonlinearities.

The compromise between a moderate absorption and a sizeable index modification through carrier creation has been achieved by pumping in the Urbach tail of the semiconductor at intermediate temperature [9]. This corresponds to the creation of an exciton with the simultaneous absorption of a phonon. The recovery of the modification is governed by the carrier lifetime. When the pump wavelength is now tuned inside the spectral transparency region of the semiconductor, there is no more absorption, but there is still a change of the refractive index which instantaneously follows the pump intensity I. The variation of **n** is described by the classical law:

$$\mathbf{n} = n_0 + n_2 I$$

where n_0 stands for the unperturbed refractive index. This is usually referred to under the general name of Kerr law and corresponds to different processes, including virtual transitions towards all the states of the semiconductor. If the pump wavelength is tuned close to the absorption edge, the virtual transition toward the excitonic level dominates: such a process is named the optical Stark effect [10, 11]. We will use it as the source of nonlinearity in our Fabry-Perot étalon.

The optical Stark effect corresponds to a coupling between the electromagnetic field of the light and the optical resonances of the material. It manifests generally as a blue shift of the excitonic lines when the pump photon energy is lower than the absorption threshold. There is no real creation of photoexcited carriers and the magnitude of the shift is proportional to the pump intensity and to the inverse detuning, $(\hbar\omega - E_{\mathbf{exciton}})^{-1}$.

Figure 4 shows an example of optical Stark effect on the absorption spectrum of a GaAs/AlGaAs superlattice at T = 20K. This has been recorded using the classical pump-probe experimental scheme with a time resolution better than 100 femtosecond for the probe. The pump duration is 1 picosecond. The curve recorded at negative delays (continuous line) is characteristic of the unperturbed absorption

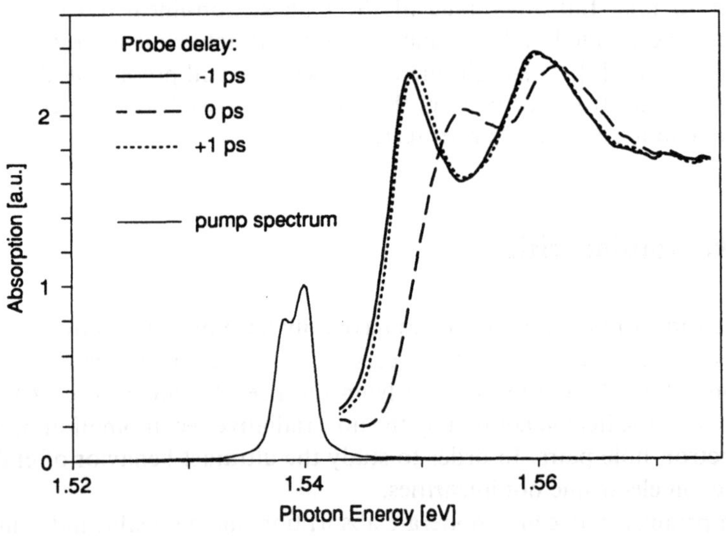

Fig. 4. Optical Stark effect on a GaAs/AlGaAs superlattice at low temperature. The heavy-hole and the light-hole excitons can be seen at 1.548 and 1.562 eV

spectrum. At zero delay between the pump and the probe, i.e., when the pump intensity is at maximum inside the sample (*dashed line*), we observe a clear shift towards higher energies of all the absorption features. Since this spectrum has been recorded by an integration over 500 shots, some fluctuations of the pump intensity from shot to shot produce an apparent broadening of the excitonic lines without physical significance. When the pump has gone through, the lines recover almost completely. For higher pump intensities, the recovery is not complete, due to carriers excited during the pump pulse. Since there is no linear absorption at the pump wavelength, the carriers originate from a two-photon absorption which occurs because of the high intensity of the pump pulse into the sample (up to 10^{10} W/cm^2).

There has recently been a strong interest in the comparison between below-gap non-linear effects and two-photon absorption (TPA). The TPA coefficient is a function of $x = 2\hbar\omega/E_g$, E_g being the gap energy, which has the form:

$$F(x) = (x - 1)^{3/2}/x^5 \quad \text{for } x > 1$$

when neglecting the Coulomb interaction and assuming a pair of isotropic and parabolic bands [12].

It has been shown by M. Sheik-Bahae et al. [13] that a universal behavior of n_2 can be deduced for a large class of semiconductors. The n_2 values, provided they are multiplied by $n_0\ E_g{}^4$, are well fitted by the same dependence on $\hbar\omega/E_g$. The values of n_2 are positive for $\hbar\omega$ not-too-close from the absorption edge and have a maximum at $\hbar\omega = E_g/2$. This clearly indicates that n_2 around the middle of the gap is "predominantly a causal consequence of two-photon absorption just as the linear index is a causal consequence of linear absorption" (Kramers-Krönig relations). However, near the one-photon absorption edge, these investigators find experimentally much larger negative values of n_2 than calculated. They attribute this to the optical Stark effect that they ignore in their calculations. This contribution, which is negative for all energies below the band gap, becomes negligible for longer wavelengths ($\hbar\omega > E_g/2$).

The fact that the two-photon absorption disappears for $\hbar\omega < E_g/2$ while n_2 does not vanish, has lead to interesting studies on nonlinear directional coupler in AlGaAs [14] and on bulk and quantum well GaAs/AlGaAs waveguides [15, 16]. The small value of n_2 is counterbalanced by the large interaction length.

In the use of waveguides, the parallelism is lost for optical signal processing. We will restrict our studies to the case of the planar Fabry-Perot and take advantage of the large non-linearities just below the absorption edge. The two-photon absorption will provide a limitation to the maximum intensity we send into the sample if we wish to minimize the amount of excited carriers.

Experimental Arrangements

Ultrashort pulses of less than 100 femtosecond duration are generated in a colliding pulse mode-locked (CPM) Rh6G dye laser. They have an energy of 100 pJ at 2 eV with a 100 MHz repetition rate. Pulses are amplified using a four-stage dye amplifier pumped by a frequency-doubled Nd:YAG laser. The amplified pulses carry up to 0.4 mJ of energy at a repetition rate of 20 Hz. After spatial filtering, the beam is focused in an ethylene glycol jet to produce a white light continuum of same pulse duration. After spectral selection by an appropriate interferential filter, the pulses are further amplified in a three-stage IR-dye amplifier up to an energy of a few μJ. The pulse duration varies with the spectral width of the interferential filter and is in the picosecond range. Care is taken to insure that the resulting beam is not inhomogeneous and has an almost constant intensity from one shot to the other. Part of the initial pulse at 620 nm is taken to generate a second continuum in a water cell and is used as a probe beam. Pump and probe beams are focused onto the sample on spots controlled by an infrared camera. The probe beam transmission is then analyzed by a spectrometer and an optical multichannel analyzer.

The device consists of 100 alternating layers of 10 nm GaAs and 2.5 nm AlGaAs grown by molecular beam epitaxy on a GaAs substrate. A 1 mm^2 area of the GaAs substrate was etched away and 20 nm gold layers were deposited on the sample surfaces [17]. These gold layers provide highly reflective interfaces yielding a measured finesse of F = 22. The very small thickness of the étalon (1.3 μm) gives a large intermode separation of 65 nm.

Experimental Results

We performed experiments in two configurations for the excitation. In both cases, we monitored the temporal evolution of the Fabry-Perot transmission peak situated just below the absorption threshold. First, we used a pump with photon energy larger than E_g, therefore directly creating electron-hole pairs by one-

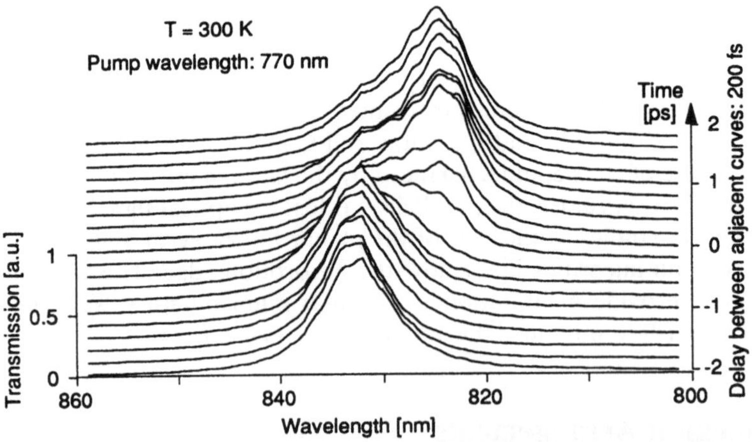

Fig. 5. Time-resolved transmission spectra of our Fabry-Perot étalon at T = 300 K for a subpicosecond pump at 770 nm. The delay between adjacent curves is 200 fs. The wavy shape of the spectra is due to some interferences in the thin glass plate on which the sample is glued

photon absorption. The refractive index changes are mainly due to phase space filling [18]. In the other configuration, the pump is tuned in the transparency region, where the optical Stark effect is the main source of nonlinearity.

Figure 5 shows the time-resolved transmission of the étalon at T = 300K. The pump has a wavelength of 770 nm and a duration of 200 fs. There is a time delay of 200 fs between adjacent curves. A significant shift of the resonance in less than a picosecond can be observed, following closely the carrier creation rate. The contrast ratio is, for example, 6/1 at 825 nm. Figure 6 shows the magnitude of the shift at a fixed delay (2 ps) in the function of pump intensity. The maximum intensity corresponds to an energy of 3 mJ/cm^2 incident on the étalon surface where the gold layer restricts the penetrating energy to 0.3 mJ/cm^2. For low excitation, the shift is linear with the excitation while it saturates at high fluence. This saturation regime corresponds to the complete disappearance of the excitonic features [18], the band gap nonlinearities then providing a minor contribution.

The above result demonstrates the subpicosecond response time of our étalon. The recovery is much longer since it is linked to the carrier recombination. Carrier diffusion [19, 20] is not important here, in view of the short time scale and the size of the spot ($150 \times 150\ \mu m^2$). The energy required for the switching depends on the desired contrast and ranges around 10 $pJ/\mu m^2$ (incident on the étalon surface). This may be obtained with different intensities following the required switching time.

We also performed experiments where the pump wavelength was tuned inside the transparency spectral region of the supperlattice. The first experiments were

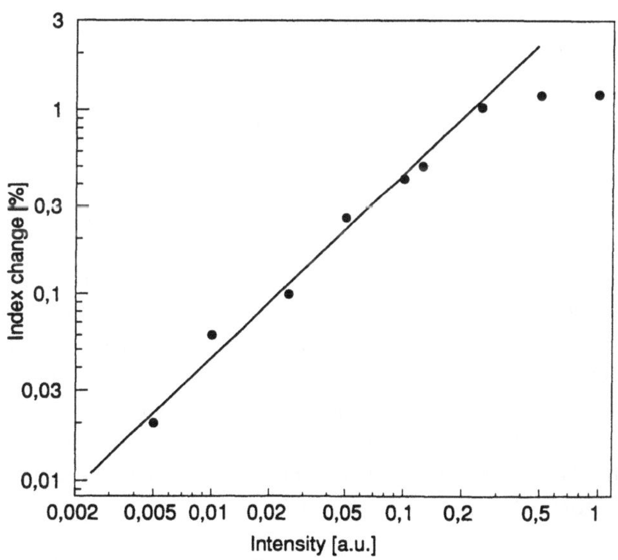

Fig. 6. Refractive index modification inside the cavity of the Fabry-Perot étalon measured through the shift of the transmission peak after excitation with a 770 nm pump pulse. The maximum intensity corresponds to an energy of 3 mJ/cm^2 incident on the étalon gold layer. The index variations are negative and n_0 is around 3.5

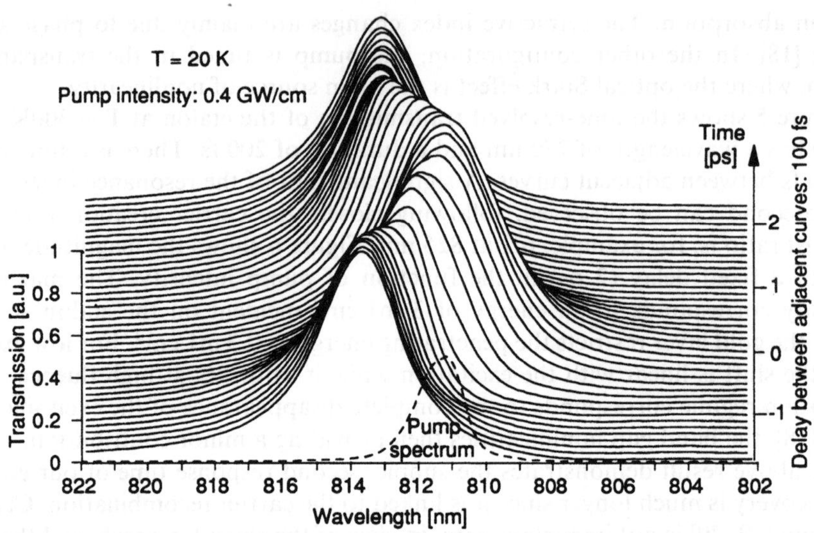

Fig. 7. Time-resolved spectra of the Fabry-Perot étalon at T = 20 K for a pump at 811 nm. The delay between adjacent curves is 100 fs

performed at room temperature. In order to benefit from the largest nonlinearities, the pump was tuned very close to the absorption edge. Consequently, excitation of carriers through absorption in the Urbach tail occurred simultaneously with the optical Stark shift, giving rise to intricate results. Therefore the étalon is now kept at T = 20K and one-photon absorption processes are avoided. Figure 7 shows the transmission spectra of the étalon for different delays between the pump and the probe. The pump spectrum is represented with a dashed line. The Fabry-Perot peak now shifts first to the blue and back to its initial position in a few picoseconds.

The temporal behavior of the étalon resonance shift seems to be mainly determined by the pump profile. In fact, it is complicated by the modification of the resonances. For example, a probe beam at 814 nm reaches the sample just before the pump pulse begins to do round trips in the cavity with an in-phase accumulation of its electromagnetic field. When the pump pulse begins to modify the intracavity refractive index, the addition is no more in phase, 814 nm is no more at resonance. Therefore the integrated transmission of the probe pulse at this wavelength is reduced, while the one at 812 nm begins to increase. This explain why, during the refractive index modification, the étalon resonance looks broader and reduced in height. We have developed a model which reproduces well the experimental features that can be seen in Fig. 7.

Discussion

Ultrafast ON and OFF switching of an optical gate based on a Fabry-Perot étalon and using the optical Stark effect has already been reported [17]. However, there is a great difference between that study and our present one: in the previous experiments, the pump spectrum was broader than the device transmission peak, so that the intracavity pump intensity did not depend on the peak spectral shift. Here now the pump spectrum is narrower than the Fabry-Perot resonance. By slightly detuning the pump spectral position with respect to the resonance, we are in the condition where optical bistability has already been observed for quasi-continuous regimes (see, for example, refs. [8] and [9]). It has also to be noticed that, since the optical non-linearity is instantaneous, the classical law $\mathbf{n} = n_0 + n_2 I$ is rigourously observed without any need to introduce a response time [2].

We discuss the experimental situation presented in Fig. 7. When the leading edge of the pump pulse reaches the sample, the pump wavelength is not exactly at resonance, so that the intracavity intensity is directly proportional to the input intensity. This leads anyway to a change of the refractive index, which brings closer the Fabry-Perot (FP) resonance and the pump wavelength. Therefore, the electromagnetic field of the pump begins to accumulate in phase, strongly increasing the intracavity intensity. The time required for this accumulation corresponds to a large number of round-trips, each round-trip time being 30fs. Our experimental conditions are such that the pulse duration is only slightly longer than the build-up time of the cavity. This leads to an interesting but not simple situation, where we can monitor the dynamics of transmitted pump intensity. A complete analysis is beyond the scope of this paper.

When the intracavity pump intensity reaches a high level, two-photon absorption adds another nonlinearity which has a long recovery time. For keeping an ultrashort response time for both ON and OFF switchings, this process has to be avoided as much as possible. This is realized by choosing a pump wavelength close to the excitonic resonance, in order to take full advantage of the optical Stark effect without increasing the pump intensity too much.

Conclusion

We have studied the behavior of an all optical gate on an ultra-short time scale. This device is realized by a Fabry-Perot étalon with a GaAs/AlGaAs supperlattice as a non-linear intracavity medium. We have shown that excitonic nonlinearities produce important changes of the refractive index in the transparency spectral region. Subpicosecond switching is observed when the pump beam is tuned in the absorption band, with a recovery time limited by the photoexcited carrier recombination. When the pump wavelength is longer than the absorption threshold, the optical Stark effect is used as an instantaneous optical nonlinearity. This leads to ultrashort ON and OFF switching times. By using a narrow pump spectrum slightly detuned with respect to the Fabry-Perot resonances, we are in the experimental conditions where optical bistability should be observed with continuous beams. In the pulsed regime, a complex analysis has to be performed in order to understand the results.

Acknowledgments. We are very grateful to W. Masselink and H. Morkoç for providing us the samples. We thank A. Mysyrowicz and A. Migus for fruitful discussions. We acknowledge the participation of N. Peyghambarian and H.M. Gibbs in the early stage of this work. Laboratoire d'Optique Appliquée is URA 1406 du CNRS. This study has been supported by DRET (grant 88/077).

References

1. Gibbs HM (1985) Optical bistability: Controlling light with light. Academic, London
2. Bishofberger T, Shen YR (1979) Phys Rev A19:1169
3. Goldstone JA, Garmire EM (1981) IEEE J Quantum Electron QE-17:366
4. Lambsdorff M, Kuhl J, Rosenzweig J, Axmann A, Schneider J (1991) Appl Phys Lett 58:1881
5. Frankel MY, Whitaker JF, Mourou GA, Smith FW, Calawa AR (1990) IEEE Trans Electron Devices 17:2493
6. Peterson CW, Knight BW (1973) JOSA B63:1238
7. Yokoyama H (1989) IEEE J Quantum Electron 25:1190
8. Kuszelewicz R, Oudar JL, Michel JC, Azoulay R (1988) Appl Phys Lett 53:2138
9. Sfez BG, Oudar JL, Michel JC, Kuszelewicz R, Azoulay R (1990) Appl Phy Lett 57:324
10. Mysyrowicz A, Hulin D, Antonetti A, Migus A, Masselink WT, Morkoç H (1986) Phys Rev Lett 56:2748
11. Von Lehmen A, Chemla DS, Zucjer JE, Heritage JP (1986) Opt Lett 11:609
12. Wherret BS (1984) J Opt Soc Am B1:67
13. Sheik-Bahae M, Hagan DJ, Van Stryland EW (1991) Phys Rev Lett 65:96
14. Aitchison JS, Kean AH, Ironside CN, Villeneuve A, Stegeman GI (1991) Paper at the Nonlinear guided-wave phenomena conference, Cambridge, UK, 2–4 September 1991
15. Islam MN, Soccolich CE, Slusher RE, Levi AFJ, Hobson WS, Young MG (1991) Paper at the Nonlinear guided-wave phenomena conference, Cambridge, UK, 2–4 September 1991
16. Tsang HK, Penty RV, White IW, Grant RS, Sibbett W (1991) Paper at the Nonlinear guided-wave phenomena conference, Cambridge, UK, 2–4 September 1991
17. Hulin D, Mysyrowicz A, Antonetti A, Migus A, Masselink WT, Morkoç H, Gibbs HM, Peyghambarian N (1986) Appl Phys Lett 49:749
18. Chemla DS, Miller DAB, Smith PW, Gossard AC, Wiegmann W (1984) IEEE J Quantum Electron QE-20:265
19. Olin U (1990) J Opt Soc Am B7:35
20. Firth WJ, Galbraith I, Wright EM (1985) J Opt Soc Am B2:1005

45

Combined Quantum Control of Electron-and Photon-Systems in Semiconductor Mesoscopic Structures: Toward an Ultimate Goal of Semiconductor Light Emitters

Masamichi Yamanishi[1]

Summary. This paper reviews two recent topics regarding physical phenomena and possible device applications which originate from combined quantum control of both electron- and photon-systems in semiconductor mesoscopic structures. The first concerns experimental results involving the systematic alteration of excitonic spontaneous emissions through continuous tuning of emission wavelength by electric fields applied to GaAs quantum wells located inside half-wavelength planar cavities sandwiched between pairs of AlGaAs distributed Bragg reflectors. The result indicates a possibility of highly efficient and extremely high speed light emitting devices, even including beam steering, which operate with the switching of the coupling efficiency of the spontaneous emission in the micro-cavities. The second is a theoretical proposal for novel semiconductor surface-emitting lasers. In the proposed lasers, all spontaneous emission is coupled into a single lasing mode. Quantum boxes biased by a *dc*-electric field separate the emission line from the absorption line by an image-charge-induced increase in the electric field, resulting in a perfect population inversion at an arbitrarily low pump rate. As a result, the threshold current of the proposed laser could be reduced down to below 100nA, which is ultimately low for a given decay rate of photons in the cavity, $\sim 10^{11}s^{-1}$.

Introduction

One of the important conclusions in cavity-quantum-electrodynamics is that the spontaneous emission of an atom is not an immutable property of the atom but can be controlled by alterations of the zero-point fluctuations of the photon fields surrounding the atom [1]. The predicted alterations of spontaneous emission have been actually confirmed by extensive studies on a variety of systems such as atoms [2–6], dye molecules [7], and semiconductors [8–13]. In all of the cases reported so far, with the exception of that in reference [4], the enhancement and inhibition of spontaneous emission has been made by modifications of the built-in structures of optical cavities involving the active materials. Obviously, it is not easy to continuously tune the resonant characteristics of the built-in cavities. As a result, an

[1] Department of Physical Electronics, Faculty of Engineering, Hiroshima University, Kagamiyama 1 chome, Higashihiroshima, 724 Japan

approach based on the modifications of the built-in structures of microcavities raises difficulties for the systematic and quantitative investigation of alteration of the spontaneous emission, since the fabrication of a number of cavities is required. On the other hand, during the last decade, the effects of an electric field on optical properties in semiconductor low dimensional systems have been extensively investigated, resulting in a number of novel optical devices (for a review, see [14]). For instance, the field effect, which is named the quantum confined Stark effect (QCSE), has offered an opportunity for substantial and continuous tuning of the emission wavelength through the control of quantum states in electron systems confined in quantum well (QW) structures, by means of static electric fields.

One natural trial is a systematic alteration of spontaneous emission by means of continuous and dynamic modification of the electron quantum states combined with the built-in modification of the photon systems. In fact, the systematic alteration of excitonic spontaneous emissions has been demonstrated with the continuous tuning of emission wavelength by electric fields applied to GaAs single QWs located inside half-wavelength planer microcavities [15, 16]. The scheme for the systematic alteration of the spontaneous emission not only enables us to perform such a systematic study in a simpler way, but also opens up an entirely new opportunity for exploratory quantum optoelectronic devices such as highly efficient, high speed and ultimately low power consumption light emitting diodes (LEDs) and lasers. The main purpose of this paper is to review our results on two recent topics regarding physical phenomena and possible device applications which originate from the combined quantum control of electron- and photon-systems in semiconductor microstructures. These are (1) experimental results on the systematic alteration of excitonic spontaneous emission in AlGaAs quantum microcavities [15, 16], together with some comments on their device application [17], and (2) the theoretical possibility of an ultimately low threshold laser with quantum confinement of photon fields in a microcavity and of single electron-hole pairs in biased quantum boxes [18].

Alterations of Excitonic Spontaneous Emission

As schematically shown in Fig. 1, for the systematic alteration of spontaneous emission, the wavelength of the excitonic spontaneous emission in GaAs QWs was tuned by means of QCSE, while the resonant characteristic of the microcavity remained essentially unchanged. We tested two kinds of devices with and without well-designed quantum microcavities [15, 16]. The first device, named the quantum microcavity (QMC) device, shown in Fig. 2a, was designed in such a way that a GaAs single QW with a thickness of 10nm was localized at an antinode position of the standing wave of the vacuum field fluctuations in a half-wavelength cavity (AlAs(67nm)/GaAs/AlAs(67nm)) sandwiched between a pair of p- and n-doped distributed Bragg reflectors (DBRs) consisting of twenty alternative quarter-wavelength layers of $Al_{0.2}Ga_{0.8}As$ and AlAs. In the second device, named the weak microcavity (WMC) device, shown in Fig. 2b, similar n-doped DBR layers were constructed prior to the growth of the AlAs(134.2nm)/GaAs(10nm)/AlAs(134.2nm) single QW layers. The purpose of the incorporation of the bottom

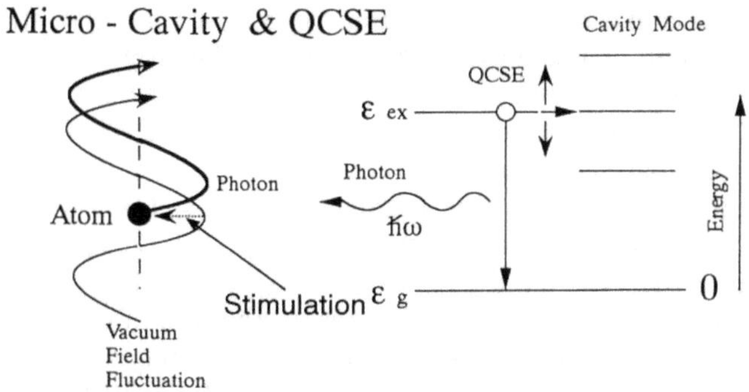

Fig. 1. Schematic illustration of coupling between controlled electron- and photon- systems. Systematic alterations of spontaneous emissions are brought about by the tuning of the emission wavelength, relative to the fixed resonant wavelength of a cavity mode if the electron system is located at an anti-node position of the vacuum field fluctuations of the cavity mode

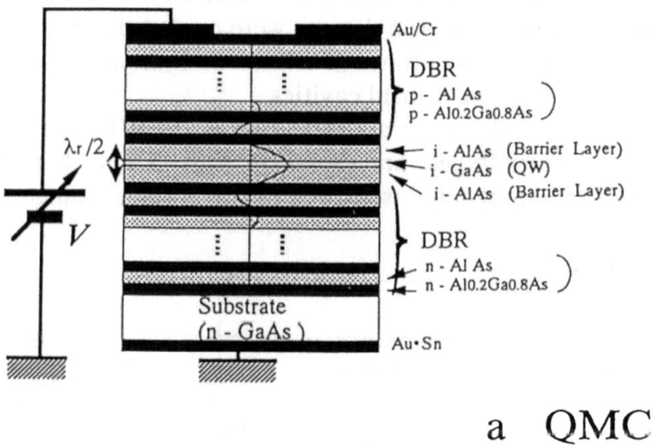

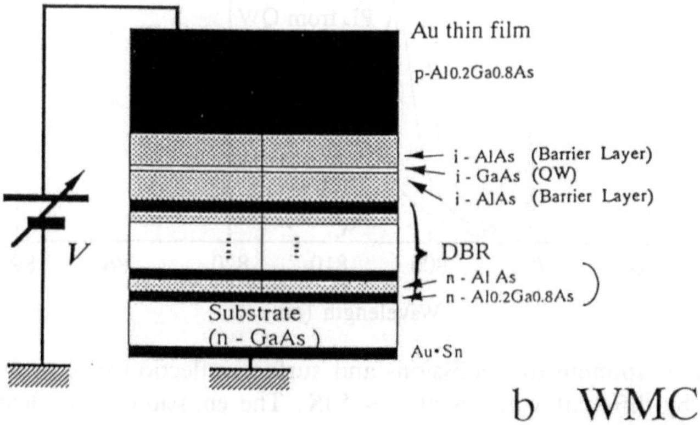

Fig. 2a, b. Structures of devices with **a** well designed quantum microcavity and **b** weak microcavity. (From [15] with permission)

DBR layers is to make the crystallographic quality of the QW structure in the WMC device as high as that in the QMC device. The stop band of the DBR of the WMC device was, however, designed to be far away from the emission wavelength of the QW, so that the spontaneous emission in the WMC device would be regarded as a free space emission.

For photoluminescence (PL) measurement, the GaAs QWs were pumped by a dye (or $Ti:Al_2O_3$) laser beam with a photon energy of 1.72eV, an energy which is lower and higher than the band gaps of the $Al_{0.2}Ga_{0.8}AS$ ($\sim$1.84eV at $\sim$50K) and of the GaAs QWs, respectively. As a result, the rate of carrier generation by the photopumping was expected to be unchanged with the changes in the applied electric field. For the maximum pumping intensity of 60mW ($\sim 5 \times 10^2 W/cm^2$), the photo-excited carrier density in the QWs was estimated to be $\sim 1 \times 10^{16} cm^{-3}$, which was far below the carrier density required for population inversion, $\sim 10^{17} cm^{-3}$. Thus, we conclude that the emissions observed in the present experiment were due to spontaneous emission (not stimulated emission). The internal electric fields perpendicular to the QW plane could be controlled by bias voltages applied to the p-i-n diodes involving the GaAs QWs and cavity structures. In the PL spectrum measurement, the radiation was detected along the optical axis normal to the device surface. Also, the devices were cooled down to $\sim$50K so as to result in spectral widths of spontaneous emissions that were narrower than the widths of the pass band of the optical cavities.

The spectra of PL and of the surface reflection in the WMC- and QMC-devices both for different bias voltages, are shown in Figs. 3 and 4, respectively. The reflection spectra are completely insensitive to the voltage change. This indicates

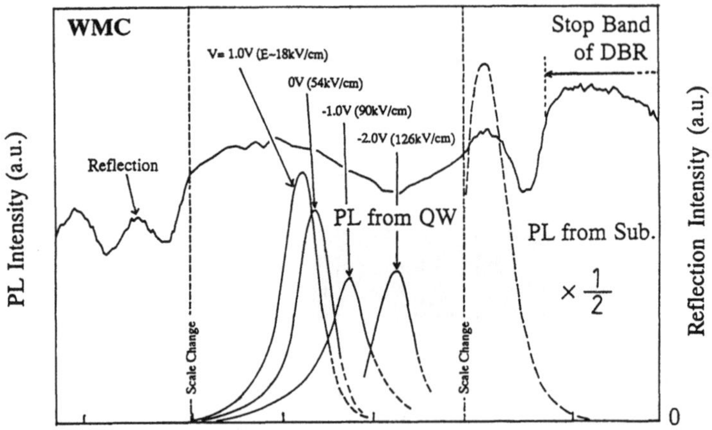

Fig. 3. Measured spontaneous emission- and surface reflection-spectra in a weak cavity device biased by different voltages at $T \sim 50K$. The emissions were detected along the direction normal to the cavity plane within a small solid angle, $< 10^{-2}\,\pi$. (From [15] with permission)

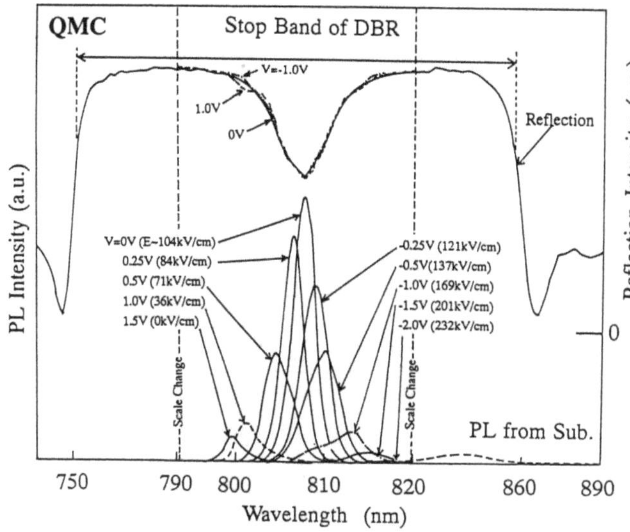

Fig. 4. Measured spontaneous emission- and surface reflection-spectra in a quantum microcavity device biased by different voltages at $T \sim 50K$. The emissions were detected along the direction normal to the cavity plane within a small solid angle, $< 10^{-2}\pi$. (From [15] with permission)

that the field-induced change in the index of refraction of the optical cavity is too small ($\Delta n/n < 0.1\%$) to affect the resonant characteristics of the cavities. In other words, the resonant wavelength of the cavities is negligibly tuned by the static electric field (~ 100kV/cm), as was anticipated. In contrast, as shown in Fig. 4, spontaneous emission into a normal direction was observed to be strongly enhanced in the $\lambda/2$-cavity (QMC-device) when the excitonic emission wavelength in the QW was tuned to the resonant wavelength of the cavity, λ_r ~ 808nm by an electric field (~ 100kV/cm) perpendicular to the QW plane. This was in marked contrast with the usual QCSE taking place in the WMC-device, as shown in Fig. 3. The PL intensity at the resonance condition in the QMC-device, detected along the axis perpendicular to the cavity plane, was confirmed to be 40 times stronger than the free space PL coming out from the WMC-device. Here, it should be noted that the generation rate of electron-hole pairs in the QWs of both the devices is kept constant with changes in the electric field and the rate of the total emission at a stationary state should be equal to the unchanged generation rate as long as the radiative recombination process is dominant over nonradiative processes, as in the present case. This means that the observed enhancement of spontaneous emission in the QMC-device, propagating into the normal direction, results from the redirection of the fixed amount of the total emission. As a result, the external quantum yield of spectrally integrated emission in the well-designed microcavity device, detected along the normal axis, was obtained to be as high as $\sim 3\%$ at the resonance condition, despite the small detection angle of the photo-detection system used in the experiment, $\sim 11°$.

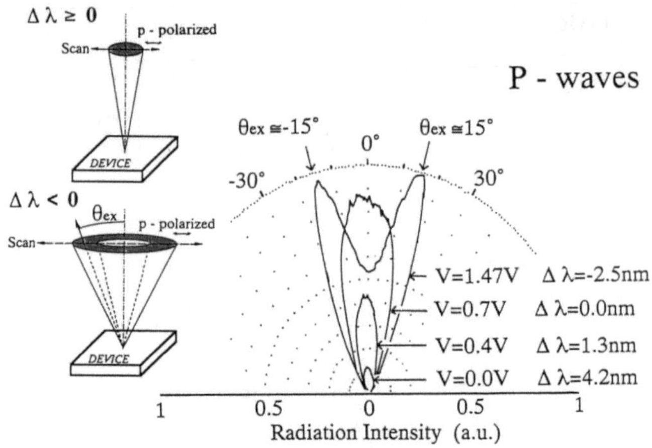

Fig. 5. Observed p-wave radiation patterns for different bias voltages, i.e., different detunings, $\Delta\lambda = \lambda_p - \lambda_r$, in another quantum microcavity device with a shorter resonant wavelength, $\lambda_r = 803$nm at $T \sim 50$K. Two-dimensional radiation patterns are schematically shown in the *inset*. The *arrows* show theoretical external emission angles. (From [15] with permission)

A particular interest lies in the beam steering of the spontaneous emission through the tuning of the emission wavelength, caused by the applied electric field. The measurement of the radiation patterns of the spontaneous emissions showed very narrow emission patterns with small divergence angles, $\sim 15°$, as shown in Fig. 5. When the peak wavelength of the spontaneous emission λ_p is shorter than the resonant wavelength of the cavity λ_r, i.e., $\Delta\lambda = \lambda_p - \lambda_r < 0$, caused by a lower electric field, the resonance condition is satisfied at an internal off-set angle, $\theta_{in} = \cos^{-1}(\lambda_p/\lambda_r)$, i.e., an external emission angle, $\theta_{ex} = \sin^{-1}\{n_{eff}\sin\theta_{in}\}$ both measured from the optical axis of the cavity. This actually resulted in conical beam patterns peaked at the emission angle θ_{ex}. In contrast, when $\lambda_p > \lambda_r$, i.e., $\Delta\lambda > 0$, caused by a higher field, the resonance condition is no longer fulfilled by any off-set angle. The radiation patterns for $\lambda_p > \lambda_r$ featured simply attenuated single peaks centered at zero degrees. This means that the coupling efficiency of the spontaneous emission into a cavity resonant mode propagating along a particular direction, such as the direction perpendicular to the cavity plane, can be substantially modulated through the tuning of emission wavelength caused by the electric fields.

The switching speed of the modulation scheme based on the switching of the coupling efficiency is expected to be extremely fast and free of carrier lifetime limitation, since the carrier population can remain fixed during a switching process which redirects a fixed amount of total spontaneous emission [17]. The high speed capability of the modulation scheme [17] can be theoretically confirmed with the rate equations for the electron population N and the photon number S in a microcavity [10],

$$\frac{dN}{dt} = P - \frac{N}{\tau_{sp}} - \frac{\beta NS}{\tau_{sp} n_{sp}}$$
$$\frac{dS}{dt} = -\gamma S + \frac{\beta N}{\tau_{sp}}\left(\frac{S}{n_{sp}} + 1\right) \tag{1}$$

where P, τ_{sp} and γ denote the pump rate, the spontaneous emission lifetime of the injected electrons, and the photon decay rate in the cavity, respectively. The population inversion parameter n_{sp} is given by,

$$n_{sp} = \frac{E_{cv}}{E_{cv} - E_{vc}} \tag{2}$$

where E_{cv} and E_{vc} are the stimulated emission and absorption rates, respectively. In this section, we shall focus on the switching of spontaneous emission, i.e., LED-mode operation. In a LED-mode operation at a pump rate below lasing threshold, the coupling coefficient β and the photon number S should be considered as the coupling efficiency of spontaneous emission into the main lobe propagating along the direction perpendicular to the cavity plane, and as the total photon number in the cavity modes producing the main lobe, respectively. Also, the term, $\beta NS/\tau_{sp}n_{sp}$ involving the population inversion parameter n_{sp} is negligible, since in LED-mode operation $S/|n_{sp}|$ is much smaller than unity. Thus, the rate equations can be approximated by,

$$\frac{dN}{dt} \sim P - \frac{N}{\tau_{sp}}$$
$$\frac{dS}{dt} \sim -\gamma S + \frac{\beta N}{\tau_{sp}} \tag{3}$$

Here, we assume that (i) the spontaneous lifetime τ_{sp} is unchanged, (ii) the photon decay rate γ is kept constant, and (iii) the coupling coefficient β can be instantaneously switched by the electric field. These assumptions would be quite reasonable in the present situation [17].

When the coupling coefficient is sinusoidally modulated as $\beta_0 + \beta_1 e^{i\omega t}$, the carrier population is kept constant, $N = P_0\tau_{sp}$ for the *dc*-pump rate P_0, whereas the *ac*-response (the power spectrum) of the photon number is evaluated by the approximated rate equations, leading to

$$\left|\frac{S_1}{\beta_1}\right|^2 = \frac{P_0^2}{\gamma^2 + \omega^2} \tag{4}$$

Apparently, the proposed modulation scheme based on the switching of the spontaneous coupling coefficient promises an extremely wide band response, together with a very high cut-off frequency, $\gamma \sim 10^{11} \sim 10^{12}$ s^{-1}, which are quite reasonable even in a half wavelength cavity if the DBRs in the cavity are properly designed [10]. Furthermore, it should be noted that the high β-value which would be possible may result in a very high external yield of light output, $\eta_{ex} = \gamma S_0/P_0 = \beta > 0.1$.

Thus, an ultrawide band response of highly efficient spontaneous emission with an extremely high cut-off frequency, $10^{11} \sim 10^{12}$ Hz is possible even at an extremely small pump current, $\sim$100nA [17]. The proposed modulation scheme may open up an entirely new opportunity for ultimately high speed and highly efficient ($\eta_{ex} > 10\%$ in principle) LEDs that have a beam steering function. In order to actually make a device which operates with the proposed modulation scheme, the device should have two functions, including carrier injection into the active region by electrical pumping, as well as the control of the electric field. One such promising device is the recently developed quantum-confined field-effect light emitting triode [19, 20]. A simple incorporation of microcavity structures into light emitting triodes may promise a successful demonstration of the modulation scheme in a device form. However, it should be stressed that the spontaneous emission line width must be narrower than the tunable range of the emission wavelength. In this view, a quantum wire or dot device is one candidate for a room temperature operation.

Ultimately Low-Threshold Laser

With regard to the second topic in this paper, we have proposed a novel surface-emitting laser where all spontaneous emission is coupled into a single lasing mode by means of a quantum microcavity, and discrete electron-hole pair emission is made free of absorption by means of a *dc*-biased quantum box (QB) [18]. To decrease the threshold pump rate for the onset of lasing, $P_{th} = \gamma n_{sp}/\beta$ [10], one can increase the coupling efficiency β of the spontaneous emission into a cavity mode to approach unity, and one can decrease the decay rate of photons in the cavity mode γ by increasing the cavity Q-value. However, a substantial population in the electron system is required to make the inversion parameter n_{sp} positive ($n_{sp} > 1$). This may set the lower limit of the threshold pump rates (currents) of semiconductor lasers. For instance, the lowest threshold current in a GaAs QW laser with a small active volume, $V = 3 \times 10^{-14}$ cm^3 (10nm-QW thickness and 2μm-diameter) and with a well-designed cavity ($\beta = 1$ and $\gamma = 10^{11}$ s^{-1}) is estimated to be 1.6μA at room temperature for a postulated spontaneous lifetime of $\tau_{sp} = 3$ns (which is determined by the condition for the population inversion). On the other hand, if a perfect population inversion, $n_{sp} = 1$, i.e., $E_{vc} = 0$ in equation (2) is possible at an arbitrarily small pump rate, the threshold current can be reduced down to 16nA for the ideal cavity with $\beta = 1$ and $\gamma = 10^{11}$ s^{-1}, which is ultimately low for the given decay rate, $\gamma = 10^{11}$s^{-1}. We have proposed a novel scheme to give rise to perfect population inversion at an arbitrarily small pump rate, no matter how small the pump current is [18]. In the proposed scheme, in the active region of a low threshold laser, an array of QBs are biased by a *dc*-electric field, as follows.

Let us consider a QB with the dimensions of $\sim$10nm cube, in which both an electron and a hole are confined in all directions. The electron and hole have discrete energies, so that the emission and absorption lines of such a QB resemble those of a two-level atom. In the case of single atom masers [21], only an excited atom can be injected into the cavity to interact with the field, resulting in an extremely low threshold pump rate for which the average number of excited Rydberg atoms inside a cavity is less than 0.1. In the semiconductor QB laser, however, there are a fixed number of QBs in the active volume. Some of these are unexcited and work as "absorptive atoms". The problem of absorption can be overcome by applying a *dc*-electric field across the QBs. Suppose that the *dc*-electric field is applied to the GaAs QB by electrodes close to the QB. (As seen in Fig 6., these electrodes are heavily-doped n^+- and p^+-$Al_xGa_{1-x}As$ layers [18].) The QB is embedded in lightly doped $Al_xGa_{1-x}As$ layers with thicknesses comparable to the size of the QB. It is assumed that the time constant for charging processes on the electrode surfaces is much shorter than the electron-hole pair recombination time $\tau_{sp} \sim 1$ns, and much longer than the quantum reaction time for electron-hole pair generation or recombination, T_2-time ~ 1ps. When a forward bias voltage of about 1.7V (slightly lower than the built-in potential of the $Al_xGa_{1-x}As$ p-n junction) is applied to the junction, an electron and a hole can be injected by tunneling through the thin barrier layer. Nevertheless, the remaining voltage drop across the p-n junction, $\sim$0.3V, results in an internal electric field, $\sim$100kV/cm. Here, the built-in potential of the $Al_xGa_{1-x}As$ p-n junction is assumed to be 2V for $x = 0.45$. Once an electron-hole pair is injected into the ground states of the QB, the junction voltage is dropped and the internal field E_{in} inside the QB is decreased by field screening due to the injected electron-hole pair, as shown in Fig. 6b. However, the initial voltage of about 0.3V across the electrodes is quickly (on the scale of the charging time) restored to the original value by the "image" charges on the electrodes, which are induced by polarized charges due to the electron-hole pair inside the biased QB. This is accomplished before the electron-hole pair is recombined to emit a photon because of the short charging time, $\ll \tau_{sp}$. The image charges on the electrodes not only restore the initial voltage, but also increase the electric field E_{out} in the i-$Al_xGa_{1-x}As$ layers, which actually acts on the excited electron-hole pair, as shown in Fig. 6b. The QCSE is thus modulated by the presence of the image charges on the electrodes. This may result in a red-shift of the emission line in an excited QB from the absorption line in an unexcited QB, $E_{1A} - E_{1E}$, as shown in Fig. 6c. It is worthwhile to note that the carrier injection into the QB can take place even in the presence of the electric field in the proposed mesoscopic diode structure.

So far, we have considered only the optical lines of completely unexcited QBs and excited QBs with single-electron-hole pairs. Some QBs may have only either electrons or holes. Such a possibility would be, however, quite low on a time-average, since the QBs tend to maintain charge neutrality. For instance, suppose that an electron is injected into a QB. As a result of the electron injection, the QB is charged up and the huge negative charge density due to the confined electron may induce a much quicker tunneling of a hole into the charged QB than that into a QB without electrons. Another problem which should be discussed is spin-state relaxation. The spin-states of the electron and hole involved in a QB may flip up and down quickly (on the scale of several tens picosecond) [22]. Therefore, the

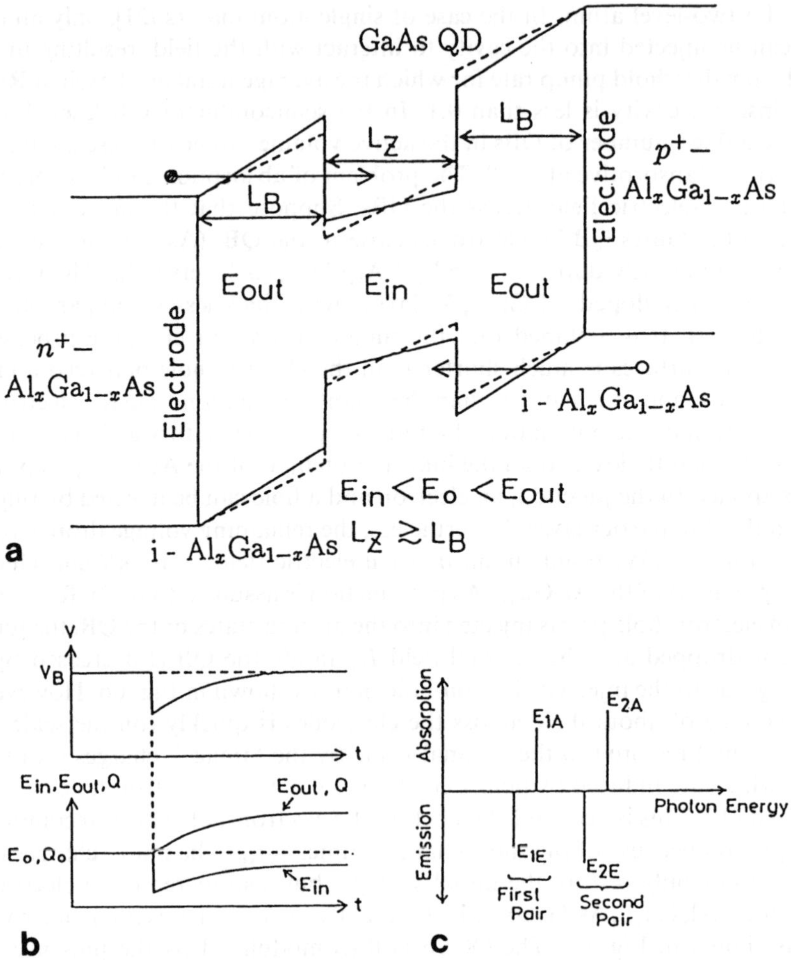

Fig. 6a–c. Proposed scheme for the active region of the novel semiconductor lasers. **a** Energy band diagrams of a quantum box structure in the direction of an applied *dc*-electric field. The *dashed* and *solid lines* show potential profiles for a test charge (not excited electron or hole) in the structure before and after the single electron-hole pair excitation in the box, respectively. **b** Time evolutions of voltage across the electrodes, *V*, electric fields inside and outside the quantum box, E_{in} and E_{out}, and electric charges on the electrodes Q. **c** Expected shift of optical lines associated with the electron-hole pair excitation. (From [18] with permission)

electron-hole pair can recombine to emit a photon, with a lifetime lengthened by a factor of two, whether the initial spin-states of the electron and hole are in "up" or "down". Note that the spin-flipping does not affect the level shifts of the quantum states, since the exchange interaction between the electron and hole is negligible, due to the orthogonality of the atomic wavefunctions in conduction and valence bands. It is, also, crucially important to recognize the following facts

[23]: Firstly, the self-Coulomb potential of the excited electron (or hole) is exactly cancelled by the self-exchange potential within the framework of the Hatree-Fock approximation. In other words, *the electron (or hole) does not really exert a Coulomb interaction on itself* (for instance, see [24]). Thus, the screened field E_{in}, which is lower than the original field E_0, acts on other electrons (or holes), not on the electron (or hole) under consideration. The field screening may result in a blue shift of absorption line E_{2A} due to a second electron-hole pair excitation in a QB which already holds the first electron-hole pair. Secondly, the above-mentioned shifts of the discrete optical lines are caused by a huge electric polarization associated with a single electron-hole pair excitation in the biased QB [23]. In fact, the surface charge density in the x-y plane, perpendicular to the direction of the electric field, produced by a single electron-hole pair attains a fairly large value, $e/L_xL_y \sim 1.6 \times 10^{-7}$ C/cm^2, for a typical size of QBs, $L_x \sim L_y \sim$ 10nm. This is because the electron and hole are confined in the QB with an extremely small volume. As a result, the amounts of the line shifts, both of the red-shift of the emission line, E_{1A}-E_{1E}, and the blue-shift of the absorption line, E_{2A}-E_{1A}, become fairly large, even though the line shifts originate from the single electron-hole pair excitation in a QB. In this view, the three dimensional confinement of the electron system is crucially important for the substantial line shifts discussed in this section.

The red shift of the emission line due to the first electron-hole pair recombination was estimated as a function of the original bias field E_0 in a GaAs QB of dimension $L_x = L_y =$ 10nm and $L_z =$ 15nm. The difference between the emission and absorption lines is 2–6meV, depending on the size of the QB, the thickness of the barrier layers L_B and the strength of the applied field E_0, as shown in Fig. 7 [18]. If the homogeneous and inhomogeneous broadening of the emission and absorption lines in GaAs QBs can be made smaller than 2–6meV, the emission originating from the single pair-excited QBs can be made free of absorption by unexcited QBs. The absorption line of an excited QB, i.e., the excitation line of the second electron-hold pair, is about 10meV higher than the absorption line of an

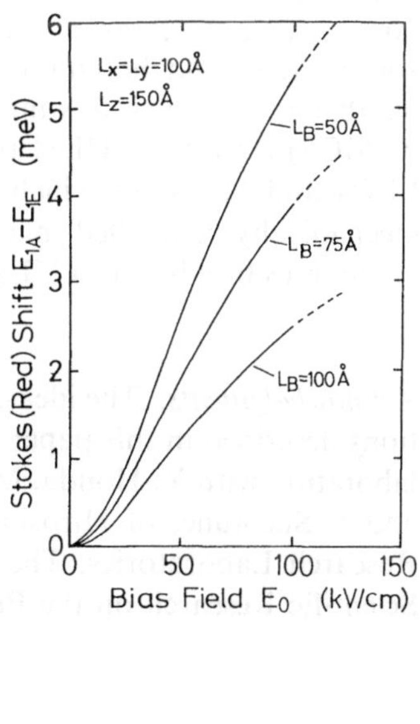

Fig. 7. Estimated red shift of the emission line due to the first-pair recombination with respect to the absorption line due to the first-pair excitation as a function of the original bias field E_0 in a GaAs quantum box with sizes, $L_x = L_y =$ 10nm and $L_z =$ 15nm, adjacent to barrier layers with thicknesses, $L_B =$ 5nm, 7.5nm and 10nm. (From [18] with permission)

unexcited QB, E_{2A}-E_{1A}, as shown in Fig. 6c [18]. Therefore, the emitted light is also free of absorption by an excited QB. In this way, the population inversion parameter n_{sp} can be made close to one, no matter how small the pump current is.

The radiation pattern of the spontaneous emission can be concentrated on a single lasing mode by the microcavity structures. For instance, suppose that a microcavity with an optical cavity layer of half wavelength thickness involving an array of biased QBs is cladded by DBRs. A DBR can have a reflection coefficient higher than 0.999 for a normally propagating cavity resonant mode, so the photon decay rate can be decreased to $10^{11} \sim 10^{12}$ s^{-1}, even though the cavity length is only a half-wavelength [10]. Moreover, if the emission linewidth is much narrower than the cavity-resonant bandwidth, and the dipole moment is oriented within the plane of the microcavity structure, the coupling efficiency β of spontaneous emission into the cavity resonant mode can be increased to more than 0.95 by using DBR materials with very different refractive indices [10, 18]. Optical lines with a narrow homogeneous broadening, <0.3meV [25] and an in-plane dipole moment would be realized by the three dimensional confined structures (QBs) subjected to a *dc*-electric field. Thus, an ultimately low-threshold current, $eP_{th} = e\gamma$, i.e., 16nA for $\gamma = 10^{11}$ s^{-1} would be possible; this value is four orders of magnitude lower than the lowest threshold current (sub-mA) in presently available QW lasers.

Conclusion

We have demonstrated the controllable alterations of excitonic spontaneous emission through the continuous tuning of emission wavelength by *dc*-electric fields applied to GaAs single quantum wells located at the center of half-wavelength planar cavities. The data indicate a possibility for highly efficient and very high speed LEDs, with the bonus of a novel function: beam steering. We also have theoretically discussed a new semiconductor surface-emitting laser, which, incorporating *dc*-biased quantum boxes for a discrete emission line free of photon absorption, and with a microcavity controlling the coupling efficiency of spontaneous emission, can attain an ultimately low threshold pump current below 100nA. It should be emphasized that the predicted current, <100nA, is comparable to currents flowing inside complementary metal-oxide-semiconductor (C-MOS) circuits of ULSI processors. In the future, such micro-semiconductor LEDs and lasers, in which both electron- and photon-systems are quantum-mechanically controlled, may find applications in new fields, such as in optical interconnection between ULSI chips and even inside chips.

Acknowledgments. The ideas, sample designs, experiments, and theoretical estimations described in this paper are the result of a collective effort performed in collaboration with Y. Honda, M. Kurosaki, Y. Lee, N. Ochi, Y. Osaka, T. Shiotani, and I. Suemune, of Hiroshima University, and Y. Yamamoto, of NTT Basic Research Laboratories. The work was partly supported by a Grant-In-Aid for Scientific Research on the Priority Area, "Electron Wave Interference Effects in

Mesoscopic Structures", from the Ministry of Education, Science, and Culture of Japan. The author would like to express his thanks to Profs. S. Datta and R.L. Gunshor, of Purdue University, for their critical reading of the manuscript.

References

1. Purcell EM (1946) Phys Rev 69:681
2. Goy P, Raimond JM, Gross M, Haroche S (1983) Phys Rev Lett 50:1903–1906
3. Gabrielse G, Dehmelt H (1985) Phys Rev Lett 55:67–70
4. Hulet RG, Hilfer ES, Kleppner D (1985) Phys Rev Lett 55:2137–2140
5. Heinzen DJ, Childs JJ, Thomas JE, Feld MS (1987) Phys Rev Lett 58:1320–1323
6. Jhe W, Anderson A, Hinds EA, Meschede D, Moi L, Haroche S (1987) Phys Rev Lett 58:666–669
7. DeMartini F, Jacobovitz GR (1988) Phys Rev Lett 60:1711–1714
8. Yablonovitch E, Gmitter TJ, Bhat R (1988) Phys Rev Lett 61:2546–2549
9. Yamamoto Y, Machida S, Horikoshi Y, Igeta K (1991) Optics Comm 80:337–342
10. Yamamoto Y, Machida S, Igeta K, Bjork G (1991) In: Yamamoto Y (ed) Coherence, amplification and quantum effects in semiconductor lasers. Wiley, New York, pp 561–615
11. Yokoyama H, Nishi K, Anan T, Yamada H, Brorson SD, Ippen E (1990) Appl Phys Lett 57:2814–2816
12. Rogers TJ, Deppe DG, Streetman BG (1990) Appl Phys Lett 57:1858–1860
13. Yamaguchi T, Arakawa Y, Nishioka M (1991) Appl Phys Lett 58:2339–2341
14. Miller DAB, Chemla DS, Schmitt-Rink S (1988) In: Haug H (ed) Optical nonlinearities and instabilities in semiconductors. Academic San Diego, pp 325–359
15. Ochi N, Shiotani T, Yamanishi M, Honda Y, Suemune I (1991) Appl Phys Lett 58:2735–2737
16. Honda Y, Lee Y, Yamanishi M, Ochi N, Shiotani T, Suemune I (1992) Surf Sci 267:612
17. Yamanishi M, Yamamoto Y, Shiotani T (1991) IEEE Photonics Technology Lett 3:888
18. Yamanishi M, Yamamoto Y (1991) Jpn J Appl Phys 30:L60–L63
19. Kan Y, Okuda M, Yamanishi M, Ohnishi T, Mukaiyama K, Suemune I (1990) Appl Phys Lett 56:2059–2061
20. Mukaiyama K, Yamanishi M, Kan Y, Ohnishi T, Okuda M, Suemune I (1990) Jpn J Appl Phys 29:L967–L970
21. Meschede D, Walther H, Muller G (1985) Phys Rev Lett 54:551–554
22. Takeuchi A, Muto S, Inata T, Fujii T (1990) Appl Phys Lett 56:2213–2215
23. Yamanishi M, Osaka Y, Kurosaki M (1990) Jpn J Appl Phys 29:L308–L311
24. Slater JC (1968) McGraw-Hill, New York, pp 326–335
25. Ohtoshi T, Yamanishi M (1991) Jpn J Appl Phys 30:L1406–L1408

Superradiance of Excitons in Mesoscopic Systems

46

EIICHI HANAMURA, TETSUJI TOKIHIRO, AND YOSHIHIRO MANABE[1]

Summary. Dicke's [1] superradiance from a gas system of atoms and molecules was described in terms of a two-level system, i.e., as an assembly of half-spin operators. Then the magnitude of the total spin, which is called a cooperation number, is conserved at the maximum for the whole radiative process. We will answer in this paper how this superradiance is modified in a crystal in which an excitation can propagate from atoms to atoms through dipolar interactions. As a crystal prototype, we choose a linear chain of two-level atoms. The superradiance master-equation of this system can be solved in terms of the exact solution of Lieb, Shultz, and Mattis [2] for the spin system. The first effect is chirping, i.e., the emitted frequency shows opposite shifts of red and blue, respectively, for the first and second halves of the superradiance pulse, or vice versa, depending upon the sign of the transfer matrix element of the excitation. The second effect is the appearance of the slow component of the emission tail, which originates from mixture of the states with smaller cooperation numbers or smaller oscillator strength under the transfer of the excitations. These results are demonstrated for a linear chain mesoscopic system.

Introduction

Dicke [1] introduced the concept of superradiance for the coherent spontaneous emission of radiation from highly excited two-level atom systems. Each two-level atom is described by spin 1/2 operator s_j. The Hamiltonian of the whole electronic system is represented by z-component S_z of the total spin $\boldsymbol{S} = \sum_{j=1}^{N} s_j$ as

$$H_0 = \hbar\omega_0 S_z = \sum_{j=1}^{N} \hbar\omega_0 s_j^z \tag{1}$$

where $\hbar\omega_0$ is the excitation energy of the two-level atom. As long as the dimension of the gas cell which contains N atoms is small compared with the relevant wavelength, the interaction Hamiltonian H_1 is approximated by

$$H_1 = -A(0)\cdot(e_x S_x + e_y S_y) \tag{2}$$

[1] University of Tokyo, Department of Applied Physics, 7-3-1 Hongo, Bunkyo-ku, Tokyo, 113 Japan

where e_x and e_y are unit polarization vectors, which are the same for all atoms, and $A(0)$ denotes the product of the vector potential and the momentum operator of the atomic electron multiplied by the constant (e/mc). Let us consider n atoms from N are excited. Then the eigen-value of the Hamiltonian H_0 in eq. (1) is $\hbar\omega_0(n - \frac{N}{2})$ so that the eigen-value $M = (n - \frac{N}{2})$ of the operator S_z measures the degree of population inversion.

The square of the total spin $\boldsymbol{S}^2 \equiv S_x^2 + S_y^2 + S_z^2$ commutes with the total Hamiltonian $H = H_0 + H_1$;

$$[\boldsymbol{S}^2, H_0 + H_1] = 0 \tag{3}$$

The operator $\boldsymbol{S}^2$ has eigen-values $S(S + 1)$, S is integral or half-integral and positive, such that

$$|M| \leqslant S \leqslant \frac{1}{2}N \tag{4}$$

Equation (3) means that the quantum number S of the total spin is conserved in the whole radiation process. The eigen-value S is called the cooperation number of the gas system. The spontaneous radiation probability I from the gas state (S, M) to the state $(S, M - 1)$ is given by

$$I = I_0(S + M)(S - M + 1) \tag{5}$$

where I_0 is the radiation rate of a gas composed of one atom or molecule in its excited state. When all N atoms or molecules are excited, i.e., $M = S = \frac{1}{2}N$, $I = NI_0$, then $I = \frac{1}{2}N(\frac{1}{2}N + 1)I_0$ for $S = \frac{1}{2}N$ and $M = 0$. This is the greatest rate at which a gas with an even number of atoms ("atoms" is used instead of "atoms or molecules" here and hereafter) can radiate spontaneously. It should be noted that the emission rate is proportional to the square of the number of atoms for large N.

The question arises of how this superradiance is modified in a crystal which is a dense and regular array of atoms. Here the excitation can propagate from atom to atom. We will analyze new aspects of superradiance from a mesoscopic crystallite, which are aspects missing in that from a dilute atom gas.

The superradiance master equation of Frenkel excitons is derived in the second section. In the third section, this master equation is solved in terms of the exact solution of the spin system in a linear chain given by Lieb, Shultz, and Mattis [2]. Here we will demonstrate two effects of excitation transfer on the superradiance of fully excited Frenkel excitons in a mesoscopic system. The first effect is the chirping in the superradiance pulse. The peak frequency in the time-analyzed emission spectra shows the blue shift in the first half of the superradiance pulse and the red shift in the second half, or vice versa, depending upon the sign of the transfer matrix element. The second effect of excitation transfer is the non-conservation of the cooperation number S in the radiation process, so that the states of the whole system with the smaller oscillator strength or cooperation number are populated. These states bring about the slow components of the emission spectrum.

While only macroscopic physical quantities such as magnetization have been obtained and discussed in magnetic systems, we can evaluate and compare all the microscopic physical quantities with observations, such as level spacings in multi-excited states and their relaxation rates. Therefore the time-analyzed emission spectrum from fully excited Frenkel excitons will open new fields to study interacting two-level systems.

The exciton is the elementary excitation coherent over a mesoscopic size in a crystal. This mesoscopic size is determined by the boundary condition in systems consisting of semiconductor microcrystallites embedded in insulators and glasses and in large molecules systems. On the other hand, the coherent length of excitons is limited by the scattering of the exciton with lattice imperfections and lattice vibrations in bulk or two-dimensional crystals.

First, the Wannier excitons, e.g., in CuCl microcrystallites embedded in insulators or glasses, were proposed [3] to show superradiative decay $2\gamma = 64\pi(R_0/a_B)^3\gamma_s$, where $\gamma_s \equiv 4|\mu_{cv}|^2/3\hbar\lambda^3$ is the band-to-band radiative decay rate and R_0 and a_B are the radius of the spherical microcrystallite and the exciton Bohr radius 6.7Å for CuCl, respectively. The factor $64\pi(R_0/a_B)^3$ is the mesoscopic enhancement of exciton decay rate. The cooperation number of this excitation is at maximum, although this is a single excitation per a microcrystallite so that it superradiatively decays with the mesoscopically enhanced decay rate. This characteristic superradiance was also experimentally confirmed [4]. The superradiance from more highly excited Wannier excitons indicates future theoretical, as well as experimental, problems. These may be solved with the help of the present concept of the superradiance of highly excited Frenkel excitons.

Master Equation of Superradiance

We start from the Hamiltonian of atoms and radiation field as follows:

$$H = H_{atom} + H_{rad} + V \tag{6}$$

$$H_{atom} = \hbar\omega_0 \sum_i s_i^z \tag{6a}$$

$$H_{rad} = \sum_{\varepsilon k} \hbar\omega_k \left(a_{\varepsilon k}{}^{\dagger} a_{\varepsilon k} + \frac{1}{2} \right) \tag{6b}$$

$$V = -\sum_i \{E^+(r_i) + E^-(r_i)\} \cdot D_{ai} \tag{6c}$$

where D_{ai} is the atomic transition dipolemoment operator and ε denotes the polarization of the radiation. We confine ourselves to the superradiance configuration in which the escape time of the emitted photon from the gas cell is the shortest among the characteristic time constants. Then the radiation field can follow adiabatically the equation of motion for the atomic system. Therefore we can eliminate the variables describing the radiation field and we derive the equations of motion for the atomic density operator $\rho(t)$, i.e., the superradiance master equation. In the second Born and Markovian approximations, the atomic density operator obeys the following equation of motion [5]:

$$\frac{d\rho}{dt} = -\frac{\Gamma_0}{2} \sum_{ij} [s_i^+ s_j^-, \rho]_+ + \Gamma_0 \sum_{ij} s_j^- \rho s_i^+ - i[\Omega_{XY}, \rho]_- \tag{7}$$

where Γ_0 is the spontaneous emission rate of a single atom and is given with a transition dipolemoment d by

$$\Gamma_0 = \frac{8\pi^2 d^2}{3\hbar\varepsilon_0\lambda^3} \tag{8}$$

and $\hbar\Omega_{XY} = H_{XY}$ denotes the Hamiltonian of Frenkel excitons in a linear chain. We accept here the following Hamiltonian in an XY model [2]:

$$H_{XY} = \hbar\omega_0 \sum_l s_l^z + \hbar T \sum_l [s_l^+ s_{l+1}^- + s_{l+1}^+ s_l^- + \gamma(s_l^+ s_{l+1}^+ e^{-2i\omega_0 t} + s_{l+1}^- s_l^- e^{2i\omega_0 t})] \tag{9}$$

where $s_l^+ \equiv s_l^x + is_l^y$ and $s_l^- \equiv s_l^x - is_l^y$. Here $\hbar\omega_0$ is the excitation energy of a two-level atom without any transfer, $2\hbar T$ is the dipolar interaction between the nearest neighbor (ij), and others are neglected:

$$2\hbar T = \frac{d^2}{4\pi\varepsilon_0 r_{ij}^3}\left[1 - \frac{3(\varepsilon_a \cdot r_{ij})^2}{r_{ij}^2}\right] \tag{10}$$

and $\hbar T\gamma$ comes from (1) the anti-resonant excitation or deexcitation processes, and (2) the higher-order Born approximation for the atom-radiation interaction. Although the γ-term causes nonlinear optical phenomena, such as higher harmonic generation, and the superradiance process will be made more colorful and complicated for the case of $\gamma \neq 0$, in this paper we focus our attention on the superradiance process itself for the case $\gamma = 0$. The sign of the transfer matrix element T depends on the atomic polarization ε_a: when ε_a is parallel to the chain, $\hbar T = -d^2/4\pi\varepsilon_0 r_{ij}^3$, while $\hbar T = d^2/8\pi\varepsilon_0 r_{ij}^3$ for ε_a perpendicular to the chain.

Superradiance of Frenkel Excitons

When we solve the master equation of superradiance eq. (7) by putting the excitation transfer $\hbar T$ in H_{XY} of eq. (10) to be zero, we have Dicke's superradiance from an atom gas. This is justified only when $\Gamma_0 \gg |T|$. Unfortunately, however, $|T| \gg \Gamma_0$ in solids, since the distance r_{ij} between the nearest neighbor atoms is much smaller than the wavelength λ of emitted radiation, so that $|T| \gg \Gamma_0$, as realized from eqs. (8) and (10). In this section, we solve the master equation (7) for the superradiance, fully taking into account the propagation effect of excitations.

The energy is measured with the unit of $d^2/4\pi\varepsilon_0 r_{ij}^3$ and our Hamiltonian is written as

$$H_{XY} = \omega_0 \sum_l s_l^z - \sum_l (s_l^+ s_{l+1}^- + s_{l+1}^+ s_l^-) \tag{11}$$

Here and hereafter $\hbar = 1$, so that $H_{XY} = \Omega_{XY}$. Because $\hbar\Gamma_0 = 8\pi^2 d^2/3\varepsilon_0\lambda^3$ is much smaller than the present unit energy $d^2/4\pi\varepsilon_0 r_{ij}^3$, it is very natural to solve the superradiance master equation (7) using the eigen-function of H_{XY} as the basis functions.

The eigen-values and eigen-functions of Hamiltonian H_{XY} (11) are obtained under Jordan-Wigner transformation [2]:

$$c_l \equiv \exp\left[i\pi \sum_{j=1}^{l-1} s_j^+ s_j^-\right] s_l^- \tag{12a}$$

$$c_l^\dagger \equiv s_l^\dagger \exp\left[-i\pi \sum_{j=1}^{l-1} s_j^+ s_j^-\right] \tag{12b}$$

Then the Hamiltonian (11) is rewritten as follows:

$$H_{XY} = \omega_0 \hat{n} - \frac{N}{2} - \sum_l^N (c_l^\dagger c_{l+1} + c_{l+1}^\dagger c_l) + [\exp(i\pi\hat{n}) + 1](c_N^\dagger c_1 + c_1^\dagger c_N) \tag{13}$$

where

$$\hat{n} \equiv \sum_l^N c_l^\dagger c_l = \sum_j^N \left(s_j^z + \frac{1}{2}\right) \tag{14}$$

and the last term of eq. (13) describes the periodic boundary condition. The eigen-vectors of H_{XY} are grouped according to the eigen values of $\hat{n}$, i.e., the number of excitations $n = 0, 1, 2, \cdots, N$. For even n, H_{XY} is diagonalized by the tight binding model with the anti-periodic boundary condition, while for odd n, this is done with the periodic boundary condition. The number of states with n_0 excitations is $N!/(N - n_0)!n_0!$. When we introduce the wave number $k = 2m\pi/N$ for the periodic boundary condition, and $k = (2m + 1)\pi/N$ for the anti-periodic boundary condition, the Hamiltonian H_{XY} is diagonalized in each subspace as

$$H_{XY} = \sum_k [\omega_0 - 2\cos k] c_k^\dagger c_k \tag{15}$$

where

$$c_k^\dagger = \frac{1}{\sqrt{N}} \sum_{l=1}^N e^{ikl} c_l^\dagger \tag{16a}$$

$$c_k = \frac{1}{\sqrt{N}} \sum_{l=1}^N e^{-ikl} c_l \tag{16b}$$

Note that the distance between the nearest neighboring atoms is chosen as a unit of length.

Finally, the eigen-states and eigen-values are given as follows: for $n = 2m + 1$ with the periodic boundary condition,

$$K = (k_1, k_2, \ldots, k_n), \quad k_j = \frac{2j\pi}{N}$$

$$\Psi(K) \equiv c_{k_1}^\dagger c_{k_2}^\dagger \cdots c_{k_n}^\dagger |0\rangle \tag{17}$$

and

$$E(K) = n\omega_0 - 2\sum_{j=1}^n \cos k_j \tag{18}$$

for $n = 2m$ with the anti-periodic boundary condition and $k'_j = (2j+1)\pi/N$,

$$\Psi(K') \equiv c^\dagger_{k'_1} c^\dagger_{k'_2} \cdots c^\dagger_{k'_n} |0\rangle \tag{19}$$

and

$$E(K') = n\omega_0 - 2\sum_{j=1}^{n} \cos k'_j \tag{20}$$

Here $|0\rangle$ is the vacuum state of the quasi-particle $(c_k, c^\dagger_k)$, i.e., $c_k|0\rangle = 0$, corresponding to the crystal ground state.

We can reduce the Laplace transform of the superradiance master equation to the coupled linear algebraic equations by taking the matrix elements between the eigen-states of $\Psi(K)$ [eq. (17)] and $\Psi(K')$ [eq. (19)], e.g., using the following relation

$$\langle K'|S^+|K\rangle \equiv \langle \Psi(K')|S^+|\Psi(K)\rangle$$
$$= 2^n N^{-n+(1/2)} \delta[\exp\{i(\textstyle\sum k'_i - \sum k_i)\}, 1] \frac{\prod_{i>i'} (e^{ik_i} - e^{ik_{i'}}) \prod_{j>j'} (e^{-ik'_j} - e^{-ik'_{j'}})}{\prod_{i=1}^{n}\prod_{j=1}^{n+1} (1 - e^{i(k'_j - k_i)})} \tag{21}$$

The four-atom linear chain is the smallest system that can show two transfer effects on the superradiance. The energy diagram and the superradiant channels are drawn in Fig. 1.

We define the time analyzed emission spectrum by the following equation:

$$I(t,\omega) = \Gamma_0 \mathrm{Im}\left[i\int_0^{\Delta T} d\tau e^{i\omega\tau} \langle S^+(t) S^-(t+\tau)\rangle \right] \tag{22a}$$

$$\doteq \Gamma_0 \sum_{n=0}^{N-1} \sum_{K_n, K'_{n+1}, K''_{n+1}}$$
$$\mathrm{Im}\left[\frac{\langle K'_{n+1}|\rho(t)|K''_{n+1}\rangle \langle K''_{n+1}|S^+|K_n\rangle \langle K_n|S^-|K'_{n+1}\rangle}{E(K_n) - E(K'_{n+1}) + \omega + i[\Gamma(K_n) + \Gamma(K'_{n+1})]/2} \right] \tag{22b}$$

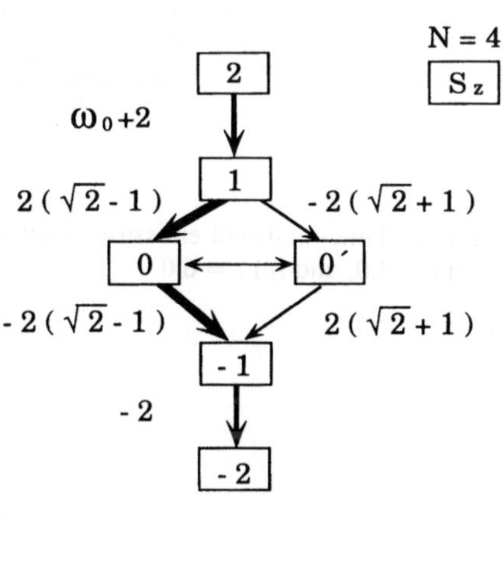

Fig. 1. Energy diagram and radiative decay channels of a 4 atom system. The *number in the box* denotes the eigen value of S_z, i.e., the degree of population inversion. The *thickness of the arrow* denotes the oscillator strength, and the *number beside the arrow* denotes the transition frequency with the energy unit of dipole-dipole interaction of Frenkel excitons

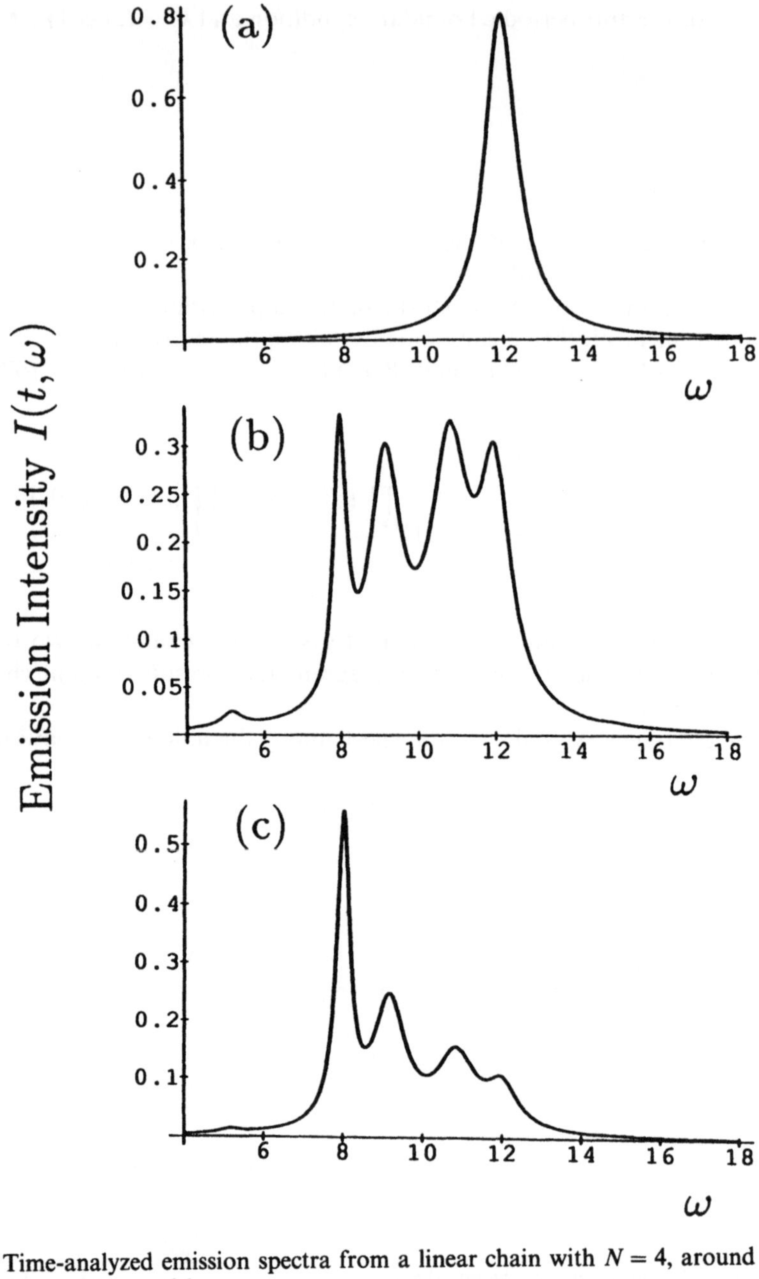

Fig. 2. Time-analyzed emission spectra from a linear chain with $N = 4$, around (*a*) $t = 0$, (*b*) $t = 3.0$, and (*c*) $t = 6.0$

where

$$\Gamma(K_n) = \Gamma_0 \langle K_n | S^+ S^- | K_n \rangle$$

and $\langle S^+(t) S^-(t+\tau) \rangle$ was expanded in Γ_0 and only the lowest term was kept in obtaining eq. (22b), which is well justified for the numerical results presented in this paper. Here the observation time ΔT for the spectroscopy was chosen so as to satisfy the condition $\{\Gamma(K_n) + \Gamma(K'_{n+1})\}\Delta T > 1$. The time-analyzed emission spectra from the fully inverted $N = 4$ system are shown in Fig. 2.

Here we have chosen $\omega_0 = 10$, $T = 1.0$ and $\Gamma_0 = 0.1$ so that whole transition lines are well resolved. The characteristic features are summarized as follows:

(1) At the beginning, only the transition from the fully inverted m $= 2(n_0 = 4)$ to $m = 1(n_0 = 3)$ is observed at $\omega_0 + 2T = 12$ (see Fig. 2a).
(2) In the next stage, there are four strong transitions, from $m = 2$ to $m = 1$ at $\omega_0 + 2T = 12$, from $m = 1$ to $m = 0$ at $\omega_0 + 2(\sqrt{2} - 1) \doteq 11$, from $m = 0$ to $m = -1$ at $\omega_0 + 2(\sqrt{2} - 1) \doteq 9$, and from $m = -1$ to $m = -2$ at $\omega_0 - 2 = 8$. The first effect is observed as the blue-shifted emission line at the beginning and the red-shifted emission line at the final stage (see Fig. 2 b).
(3) The transition from $m = 1$ to $m = 0'$ at $\omega_0 - 2(\sqrt{2} + 1) \doteq 5$ is observed at the intermediate time $t \doteq 3$ to 12, while that from $m = 0'$ to $m = -1$ at $\omega_0 + 2(\sqrt{2} + 1) \doteq 15$ is observed only at the final stage $t = 12$ to 15. The state $|0'\rangle$ has a smaller "cooperation number" than the state $|0\rangle$, so that hybridyzation of the states with a smaller cooperation number results in the appearance of the slow component.
(4) The most important feature of the superradiance is the rapid size-dependent radiative decay. In this case the main emission process is completed within $t = 3$, less than the decay time 10 of a single atom Γ_0^{-1} with $\Gamma_0 = 0.1$.

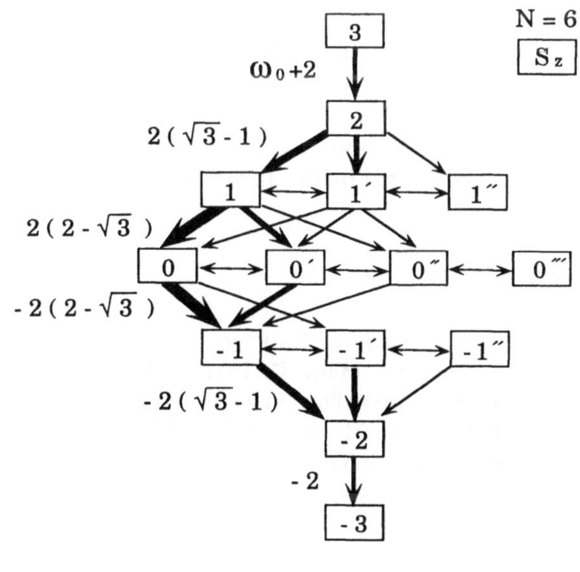

Fig. 3. Energy diagram and radiative decay channels of a 6 atom system. Numbers in the figure have the same meaning as in Fig. 1. Note that the number of decay branchings increases rapidly as N increases

When we increase the number N of atoms, the number of multi-excited states increases very rapidly. For example, see the energy diagram and the superradiance channels for $N = 6$ drawn in Fig. 3. The oscillator strength is represented schematically by the degree of thickness of arrows. The main structure of the emission pulses is determined by the decay channels shown with the thickest arrows. This shows the blue shift in the first half of the emission pulse and the red shift in the second half. On the other hand, the many states connected by weak arrows are effective in inducing the slow component of the emission tail. As the number of atoms N increases further, the energy levels are highly distributed over the same energy region between $\omega_0 - 2T$ and $\omega_0 + 2T$, and the superradiant decay rate also increases rapidly so that the line structures of the emission spectra are easily smeared out. In conclusion, we have clear chirping in the superradiant pulse and the appearance of the slow component in a larger mesoscopic system.

Future Problems

There are a number of properties of exciton superradiance that we have not discussed. In particular, we have not discussed the interesting features of higher dimensional systems. The equivalence between interacting two-level atoms and quantum XY spins is preserved in two and three spatial dimensions, and we can expect that superradiant pulses reflect the physical properties of higher dimensional quantum spin systems. Actually, the expression of the time analyzed emission spectrum is quite similar to that of the spin-spin correlation function. We have not considered the photon statistics of the emitted pulse, nor have we presented a careful analysis of higher harmonic generation from the system with finite γ in eq. (9). These and other questions are the subjects of ongoing research.

Acknowledgments. This work has been financially supported by Grants-in-Aid for Scientific Research on the Priority Areas, "Electron Wave Interference Effects in Mesoscopic Structures" and "Electronic Properties of Mesoscopic Particles", from the Ministry of Education, Science, and Culture of Japan. It has also been partially supported by cooperative research between Japan and the USA sponsored by the Japan Society for the Promotion of Science.

References

1. Dicke RH (1954) Phys Rev 93:99–110
2. Lieb E, Schultz T, Mattis D (1961) Ann Phys 16:407–466
3. Hanamura E (1988) Phys Rev B38:1228–1234
4. Itoh T, Furumiya M, Ikehara T (1990) Solid State Commun 73:271–274
5. Gross M, Haroche S (1982) Phys Reports 93:301–396 (and references therein)

Hole-State Mixing and Nonlinear Optical Properties of Semiconductor Quantum Dots

47

NASSER PEYGHAMBARIAN, STEPHAN W. KOCH, BRIAN MCGINNIS, KOO KANG, SANDALPHON, YUAN Z. HU[1], SUBHASH H. RISBUD, LI-CHI LIU[2], ANDRE MYSYROWICZ, AND DANIELE HULIN[3]

Summary. Our recent experimental and theoretical results on optical nonlinearities of semiconductor microcrystallites in glass will be reviewed. One-photon and two-photon absorption spectroscopy are used to show the importance of confinement-induced valence-band mixing for small semiconductor particles. Deviations from the parabolic band approximation are demonstrated.

The confinement of electron-hole pairs in semiconductor microcrystallites can result in discrete optical transitions if all three dimensions are sufficiently small [1–12]. From a fundamental point of view, these microcrystallites, often called quantum dots, represent an intriguing system comprising both the discrete nature of isolated elements and the band nature of large-ordered systems. At present, quantum dots are still not well understood, due to the complexity of the system with a large uncharacterized distribution of sizes, shapes, stoichiometry, defects, and surface interactions. To date, most quantum dots exhibit broad absorption resonances that become sharper with decreasing dot radius. These resonances are usually assigned to quantum-confined states based on a model which assumes simple parabolic valence and conduction bands. This model has been successful in describing the optical nonlinearities of quantum dots measured in pump-probe experiments. For example, the dashed curve in Fig. 1 shows the measured linear absorption spectrum of the sample, exhibiting the lowest quantum-confined transition [13] (the peak labeled A). Pumping inside this transition with a short optical pulse results in a nonlinear spectrum ($-\Delta\alpha L$), shown by the dotted curve in Fig. 1. In this experiment, a weak broad band pulse was used as a probe, and its transmission was measured with and without the pump. The positive peak in the ($-\Delta\alpha L$) spectrum around the pump frequency specifies bleaching of the lowest quantum-confined transition (1s-1s), while the negative region on the high-energy side of the bleached component is a result of induced absorption. The model calculation with parabolic bands [9] gives a qualitatively similar ($-\Delta\alpha L$) spectrum, as shown in Fig. 2. The calculations were conducted for dots with $R/a_0 = 1.0$, where R and a_0 are the radius of the dot and the exciton Bohr radius, respectively, and

[1] Optical Sciences Center, University of Arizona, Tucson, AZ 85721, USA
[2] Mechanical, Aeronautical, and Material Engineering, University of California, Davis, CA 95616, USA
[3] Ecole Polytechnique, ENSTA, Laboratoire d'Optique Appliqueé, Palaiseau, France

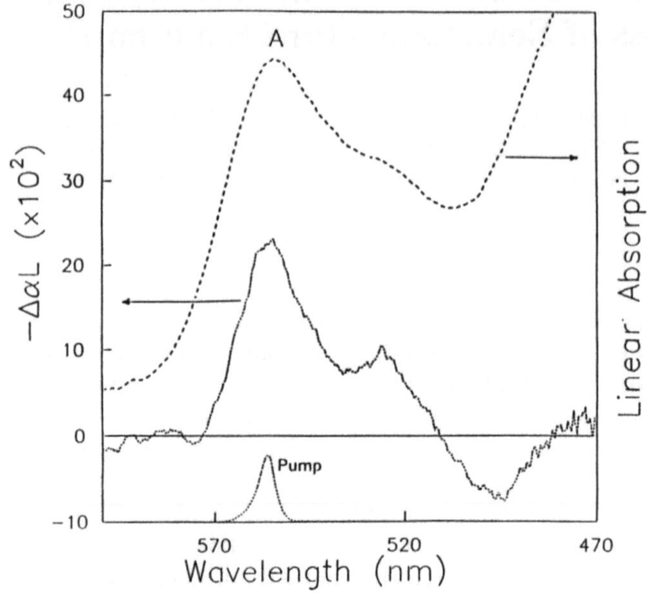

Fig. 1. The linear absorption spectrum of the 26-Å-CdSe quantum dot sample is plotted together with the absorption changes obtained using 100 fs pulses

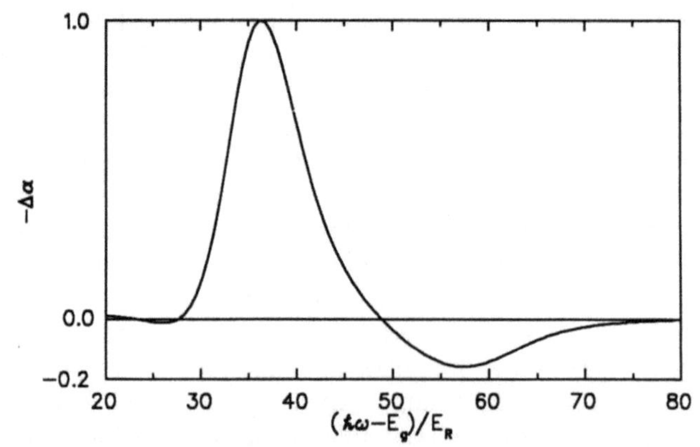

Fig. 2. Computed absorption change, $-\Delta\alpha$, for pump-probe excitation assuming pumping into the energetically lowest one-pair state. The parameters are $m_e/m_h = 0.24$, $R/a_0 = 0.5$, and $\hbar\gamma/E_R = 10$

for a broadening of $\hbar\gamma/E_R = 10$, with E_R being the bulk exciton binding energy. This spectrum was calculated for the case of resonant excitation of the energetically lowest exciton resonance, which in this case amounts to $(\hbar\omega - E_g)/E_R = 36$, where E_g is the bulk bandgap energy. The dominant feature in Fig. 2 is the bleaching of the exciton state, giving rise to the positive peak in the spectrum. In addition to this exciton saturation, we see a region of increasing absorption (negative

$-\Delta\alpha L$) in the region around $(\hbar\omega - E_g)E_R \simeq 58$, which is due to the transition to excited two-pair (biexciton) states. The dominant contribution to these transitions comes from two-pair states where the two electrons are in the 1s state and the two holes are either both in the 1p state, or one hole is in the 1s and the other in the 2s state, respectively. Such transitions would be dipole forbidden in the case without Coulomb interaction, and they are dipole-allowed only because the Coulomb interaction breaks the symmetry of the quantum dots.

Even though simple parabolic valence and conduction band models have been successful in describing the optical nonlinearities of quantum dots measured in pump-probe experiments, recent theoretical publications [8, 10] indicate that this model needs to be modified to account for mixing of the valence bands caused by the spherical confining potential of small quantum dots. To obtain the experimental verification of such confinement-induced valence-band mixing, we measured and compared one- and two-photon absorption spectra (see Fig. 3a) of CdS quantum dots in glass. Because of the large number of poorly known system parameters, one-photon absorption measurements alone are not sufficient to unambiguously determine the energies between the quantum-confined states. However, the combination of one- and two-photon absorption spectra allows a direct comparison of electron-hole pair states with total angular momentum of zero and one, thus eliminating many of the experimental uncertainties. The results show significant deviations from calculations based on the parabolic valence-band approximation.

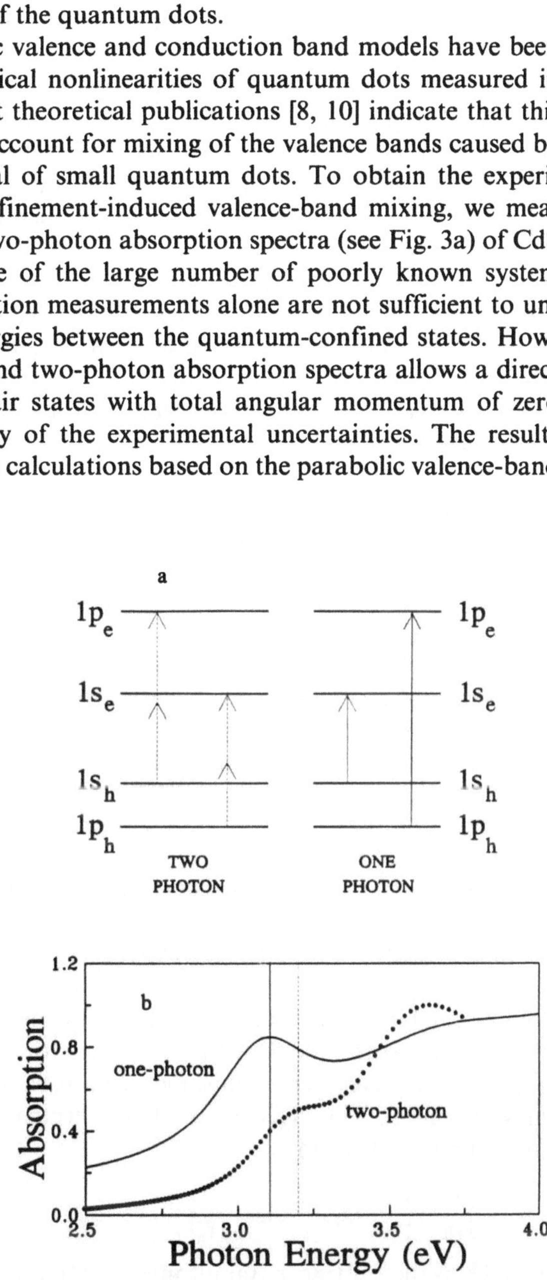

Fig. 3. a One- and two-photon allowed transitions between the discrete states in the spherical potential well. **b** Theoretical calculation of one- and two-photon absorption based on a parabolic band model

To demonstrate the significance of the confinement-induced valence-band mixing effects, we first calculated the one- and two-photon absorption spectra in the simple parabolic band model. Figure 3b shows the results of such a calculation where effective masses appropriate for the A band in bulk wurtzite CdS, $m_e \simeq 0.2\ m_0$ and $m_h \simeq 1.35\ m_0$ ($m_\perp = 0.7\ m_0, m_\parallel = 5\ m_0$, and thus, $m_h = \sqrt[3]{(m_\perp^2 m_\parallel)} = 1.35\ m_0$) have been used. A phenomenological broadening parameter has been introduced to account for the homogeneous linewidth and the inhomogeneous contribution from the distribution of dot sizes. The first two-photon resonances are expected to be situated between the first two one-photon transitions, as seen in Fig. 3b. The splitting between the one- and two-photon absorption peaks is a result of the nondegeneracy of the valence-band states. By experimentally measuring both the one- and two-photon absorption spectra, we can check the predictions of this model, which assumes a single uncoupled parabolic valence band for these quantum dots.

For our experiments, we used samples containing CdS microcrystallites in a glass matrix. The sample with the smallest microcrystallites had average radii of 0.9 ± 0.7 nm, while the sample with the largest microcrystallites had average radii of 8.0 ± 0.7 nm.

Optical characterizations consisted of both one- and two-photon absorption measurements for each sample. One-photon spectra were obtained through direct transmission measurements or by excitation spectroscopy. Two-photon spectra were obtained through two-photon excitation spectroscopy, where the photoluminescence (PL) intensity is detected as the excitation frequency is varied at constant input intensity excitation. Most of the PL from our samples was emitted in a broad band, peaking near 800 nm. These PL spectra were observed to be independent of the excitation photon energy, allowing us to monitor a constant band of PL while tuning the excitation source to obtain the two-photon absorption spectrum. The validity of this technique was checked by obtaining one-photon absorption spectra through PL excitation, comparing it to absorption spectra obtained from transmission measurements, and verifying that the two techniques gave identical results (see Fig. 4).

Examples of one- and two-photon spectra for two samples with different average quantum dot radii between 1 and 2 nm are shown in Fig. 5a, b. It is apparent that the one- and two-photon absorption peaks occur at the same energies. Similar spectra were obtained for other quantum dot samples with different average sizes. The poor agreement between the experimental spectra in Fig. 5a, b and the theoretical spectra in Fig. 3b shows that the independent parabolic valence-band approximation is insufficient to explain the optical transitions of CdS quantum dots.

To improve the model, we performed a theoretical analysis, which included mixing of the heavy- and light-hole valence bands induced by the spherical confining potential. The eigenstates of the electron-hole pairs in the quantum dot satisfy the Hamiltonian

$$H = -\frac{\hbar^2}{2m_e}\nabla_e^2 + H_L - \frac{e^2}{\varepsilon|\vec{r}_e - \vec{r}_h|} \tag{1}$$

where H_L is the Luttinger Hamiltonian for the valence bands in a spherical coordi-

Fig. 4. The linear absorption spectrum of the CdS quantum dot sample measured using transmission (*solid curve*) and luminescence excitation (*dotted curve*) spectroscopies

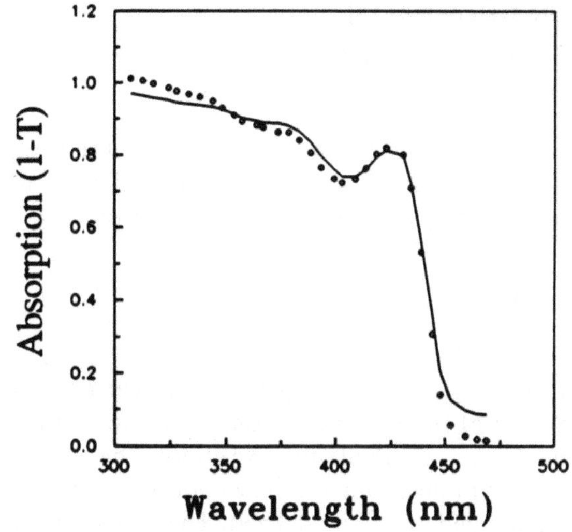

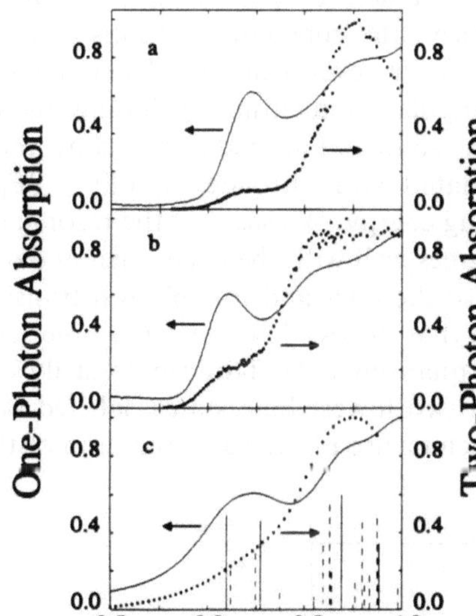

Fig. 5. a Experimental results of one-photon (*solid line*) and two-photon (*dots*) absorption spectra for a quantum dot sample heat-treated at 640°C for 1 h. No observable differences are seen in the transition energies between the one- and two-photon spectra. **b** Similar experimental results for a quantum dot sample heat-treated at 640°C for 3 h. **c** Calculations of one-photon (*solid line*) and two-photon (*dots*) absorption spectra based on the Luttinger Hamiltonian. The *vertical lines* are calculated with no broadening, while the *dashed and solid curves* represent spectra for a broadening of $\gamma = 8E_R$, where E_R is the bulk exciton binding energy of 27 meV. The magnitude of a vertical line represents the absorption strength. The other parameters used in the calculations are $R/a_B = 0.5$, with $\gamma_1 = 2.97$ and $\mu = 0.75$. The experimental data of Fig. 5a, b should be compared with Fig. 5c

nate system [14, 15],

$$H_L = \frac{\gamma_1}{2m_0}\left(p^2 - \frac{\mu}{9}(p^{(2)} \times J^{(2)})\right) \tag{2}$$

with $\vec{p}$ denoting the hole linear momentum operator, $\vec{J}$ the angular momentum operator, and $p^{(2)}$ and $J^{(2)}$ the second rank irreducible tensor operators of the momentum p and the angular momentum $J = 3/2$, respectively. The coupling constant μ is

$$\mu = \frac{4\gamma_2 + 6\gamma_3}{5\gamma_1} \tag{3}$$

with γ_1, γ_2, and γ_3 being the Luttinger parameters. Equation (2) is a simplified version of the Luttinger-Kohn Hamiltonian for the case of $(\gamma_3 - \gamma_2) \ll \gamma_1$ [14]. In the numerical calculation, we choose the eigenstates of the square of the total pair angular momentum and its z component as the basis into which the Hamiltonian is expanded and diagonalized. The Coulomb energy is also included. The numerically computed one- and two-photon absorption spectra from this model, using parameters appropriate for cubic CdS ($\mu = 0.75$), are shown in Fig. 5c. The vertical lines in this figure represent the energetic positions of the transitions and their oscillator strengths. There are several predicted new resonances that satisfy the selection rules of total angular momentum, including quantum confinement and Coulomb interaction. The solid and dashed curves show the one- and two-photon absorption spectra calculated with a broadening of $\gamma = 8E_R$, where $E_R = 27$ meV is the bulk exciton binding energy. We see that the resonances merge and form an absorption curve that agrees well with the experimental data.

The near degeneracy of the one- and two-photon transition energies observed here results from the fact that the two lowest energy-hole states are roughly degenerate for the case of CdS quantum dots. This can be easily seen in Fig. 6, where the calculated energy of the two lowest hole states, labeled $s_{3/2}$ and $p_{3/2}$ to indicate their s- and p- like characters, are plotted as a function of the coupling constant μ

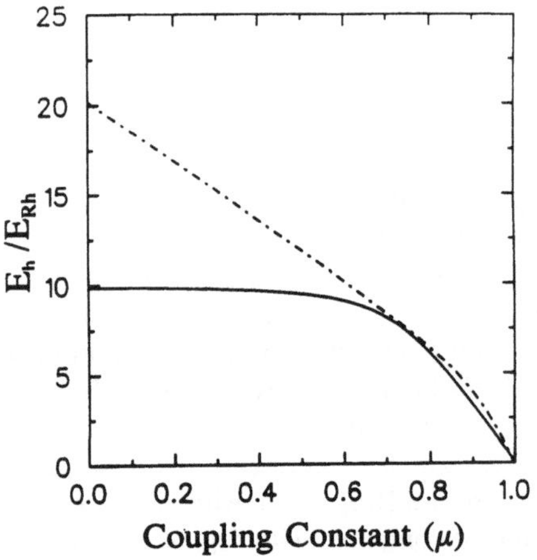

Fig. 6. Calculated normalized energies of the two lowest valence-band states, $s_{3/2}$ and $p_{3/2}$, in the quantum dot with crystallite size of $R = a_{hB}$. Here, E_{Rh} and a_{hB} are the Rydberg energy and the Bohr radius computed for the hole. The *solid line* (*dashed line*) represents the $s_{3/2}$ ($p_{3/2}$) states

for a quantum dot radius equal to half of the bulk exciton radius. In the vicinity of $\mu = 0.75$, which is appropriate for CdS, we see that the two curves nearly touch, suggesting near degeneracy caused by the confinement-induced modification of the original heavy- and light-hole states. Figure 6 indicates that for materials with smaller coupling constant μ values, these states are nondegenerate and the one- and two-photon transition energies are not expected to coincide. It is noted that the results of Fig. 6 are not strongly size-dependent for the range of quantum dot sizes analyzed in this study. For very large dot sizes the assumptions made in the calculations are no longer valid and the treatment does not apply.

In conclusion, even though the qualitative features of linear and nonlinear absorption spectra of quantum dots can be understood on the basis of simple parabolic band theory, more detailed spectroscopic analysis of the one- and two-photon absorption spectra of CdS quantum dots requires the analysis of confinement-induced valence-band mixing. Our one- and two-photon spectroscopy results can be well explained on the basis of Luttinger theory for the hole states in spherical quantum dots. The observation of better resolved one- and two-photon resonances requires quantum dot samples with reduced size distribution and inhomogeneities.

Acknowledgments. The authors would like to acknowledge support from NSF, Joint Services Optical Program, AFOSR, ARO, SDI/ONR, the Optical Circuitry Cooperative of the University of Arizona and the NATO travel grant.

References

1. Efros Al L, Efros AL (1982) (in Russian). Soviet Phys Semicond 16:772
2. Ekimov AI, Onushchenko AA (1982) (in Russian). Sov Phys Semicond 16:775
3. Brus LE (1983) J Chem Phys 79:5566; Bawendi MG, Carroll PJ, Wilson WL, Brus LE (1992) J Chem Phys 92:946
4. Bawendi MG, Wilson WL, Rothberg PL, Carroll PJ, Jedjin TM, Steigerwald ML, Brus LE (1990) Phys Rev Lett 65:1623; Brus L (1991) Quantum crystallites and nonlinear optics. Appl Phys A 53:465
5. Hanamura E (1988) Phys Rev B37:1273
6. Roussignol P, Richard D, Flytzanis C, Neuroth N (1989) Phys Rev Lett 62:312
7. Takagahara T (1989) Phys Rev B39:10206
8. Xia J-B (1989) Phys Rev B40:8500
9. Hu YZ, Koch SW, Lindberg M, Peyghambarian N, Pollock EL, Abraham FF (1990) Phys Rev Lett 64:1805
10. Vahala KJ, Sercel PC (1990) Phys Rev Lett 65:239
11. Kang KI, McGinnis BP, Sandalphon, Hu YZ, Koch SW, Peyghambarian N, Mysyrowicz A, Liu LC, Risbud SH (1992) Phys Rev B45:3465
12. Rama Krishna MV, Friesner RA (1991) Phys Rev Lett 67:629
13. Peyghambarian N, Fluegel B, Hulin D, Migus A, Joffre M, Antonetti A, Koch SW, Lindberg M (1989) IEEE J Quant Elect 25:2516
14. Baldereschi A, Lipari NO (1973) Phys Rev B8:2697
15. Luttinger JM (1956) Phys Rev 102:1030

Polarization-Dependent Optical Nonlinearities in GaAs/AlGaAs Fractional Layer Superlattices

48

ARTURO CHAVEZ-PIRSON, JUNJI YUMOTO, HIROAKI ANDO, TAKASHI FUKUI, AND HIROSHI KANBE[1]

Summary. $(AlAs)_m(GaAs)_n$ sub-monolayer superlattices (m,n < 1), also called fractional layer superlattices (FLS), exhibit nanometer-scale compositional corrugation parallel to the grown surface. This attractive feature makes it possible to obtain structures exhibiting quantum confinement effects in the direction perpendicular to, as well as along, the growth direction. One important consequence of lateral confinement is the appearance of anisotropies in the linear and nonlinear optical properties. The study of the optical anisotropies not only reveals the underlying physical properties but also may offer new optical device functionalities. We measure the room temperature polarization-dependent nonlinear absorption and refractive index spectra of a $(Al_{0.5}Ga_{0.5}As)_{1/2}(GaAs)_{1/2}$ fractional-layer superlattice (FLS) structure grown by metalorganic chemical vapor deposition. The size of the room temperature optical nonlinearity expressed by σ_{eh} ($=\Delta\alpha/N$) is 6×10^{-15} cm^2. The microscopic nonlinear mechanism in this FLS is band-filling and the switch-on time of the nonlinearity we measure to be less than 4 ps. We directly measure the anisotropic nonlinear effects between the directions parallel and perpendicular to the superlattice steps that give rise to a nonlinear optical birefringence in the plane of the growth surface. The nonlinear birefringence makes the FLS attractive as a nonlinear material for arrays of surface-normal, polarization-based semiconductor optical switching devices.

Introduction

The study of mesoscopic effects in semiconductors is interesting because of the possibility of exploring new physical phenomena that can improve present optical and electronic device performance, and because this study may lead to the introduction of new device functionalities not available with present structures. Strongly motivated by the goal to realize quantum wires in semiconductors, investigators have recently developed crystal growth techniques that make it possible to control, on a nanometer scale, the compositional modulation perpendicular to the direction of crystal growth. $(AlAs)_m(GaAs)_n$ sub-monolayer superlattices (m,n < 1), also called fractional-layer superlattices (FLS), exhibit compositional corrugation parallel to the grown surface [1, 2] and are a strong candidate for the realization of

[1] NTT Basic Research Laboratories, 3-9-11 Midori-cho, Musashino, Tokyo, 180 Japan

quantum wire arrays. The ability to obtain growth-induced quantum confinement effects in two orthogonal spatial dimensions makes the FLS structure promising for the study of mesoscopic effects in GaAs. One important consequence of the growth-induced lateral confinement of carriers is the appearance of anisotropies in the optical properties of the structure. Photoluminescence (PL) and photoluminescence excitation (PLE) spectroscopy are typically used to characterize the quality and size of the optical anisotropy [3]. Recently, direct measurements of the anisotropy in the linear absorption and refractive index spectra have clearly confirmed the compositional modulation perpendicular to the direction of crystal growth [4, 5]. We report on our measurements of the room temperature polarization-dependent nonlinear absorption and refractive index spectra in a FLS [6]. The polarization-dependent optical nonlinearities are a direct consequence of the growth-induced lateral quantum confinement.

Sample Characteristics

The fractional layer superlattice sample is grown by metalorganic chemical vapor deposition (MOCVD) on a (001) vicinally oriented semi-insulating GaAs substrate tilted towards the $[\bar{1}10]$ crystal orientation. Figure 1 shows a schematic representation of the FLS. To grow a $(Al_{0.5}Ga_{0.5}As)_{1/2}(GaAs)_{1/2}$ FLS, first TEGa only, and then a mixture of TEGa and TEAl, are alternately supplied. The superlattice is formed in the $[\bar{1}10]$ direction. Details of the crystal growth system and growth procedure are reported elsewhere [7]. Surface view transmission electron microscope (TEM) bright field images of the FLS, exhibiting clear contrast of the GaAs and AlAs regions and no fluctuation of the superlattice period, confirm the compositional corrugations parallel to the grown surface [8]. The $(Al_{0.5}Ga_{0.5}As)_{1/2}(GaAs)_{1/2}$ FLS layers used in these experiments have a lateral period of 80Å, defined by the monolayer steps on the tilted GaAs substrate. The nominal widths of the GaAs and $Al_{0.5}Ga_{0.5}As$ layers in the FLS are 40Å, and the thickness in the growth direction is 2540Å. On each side of the FLS there are 300Å-thick $Al_{0.5}Ga_{0.5}As$ alloy cap layers which serve both to confine photo-generated car-

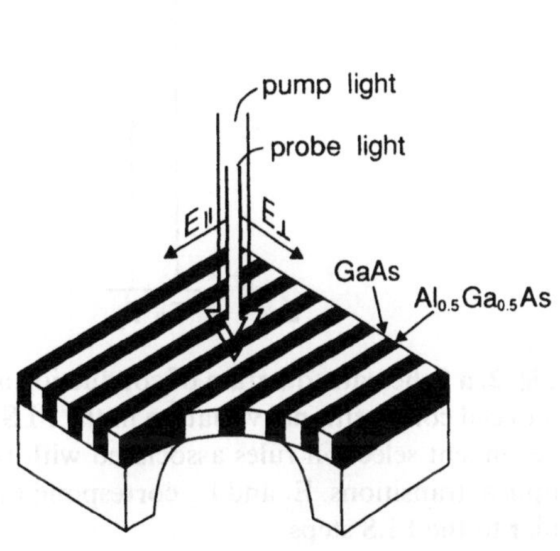

Fig. 1. Schematic drawing of the fractional layer superlattice (FLS) structure showing the orientation of the FLS steps. The input light beams are perpendicular to the sample surface

riers in the FLS regions and also to provide a chemical etch stop layer for the GaAs substrate removal process. For transmission experiments, we remove the GaAs substrate by a wet chemical etching technique [9].

Anisotropy in Linear Optical Properties

The reason for the anisotropy in the FLS absorption originates from the polarization-dependent selection rules associated with the heavy- and light-hole related optical transitions [4, 10, 11]. Figure 2 depicts the conduction and valence band of the FLS and schematically indicates the allowed optical transitions. It is useful to note that the optical absorption characteristics for light propagating through the FLS are formally analogous to the case of light propagating in the plane of a quantum well structure. Because of the carrier confinement effects, the ground state energy levels, which are degenerate in bulk material, separate into individual levels in the FLS structure. The band edge energy for probe light with an electric field parallel to the confined layers is determined by the heavy-hole related transitions, while that for an electric field perpendicular to the layers is determined by the light-hole related transitions.

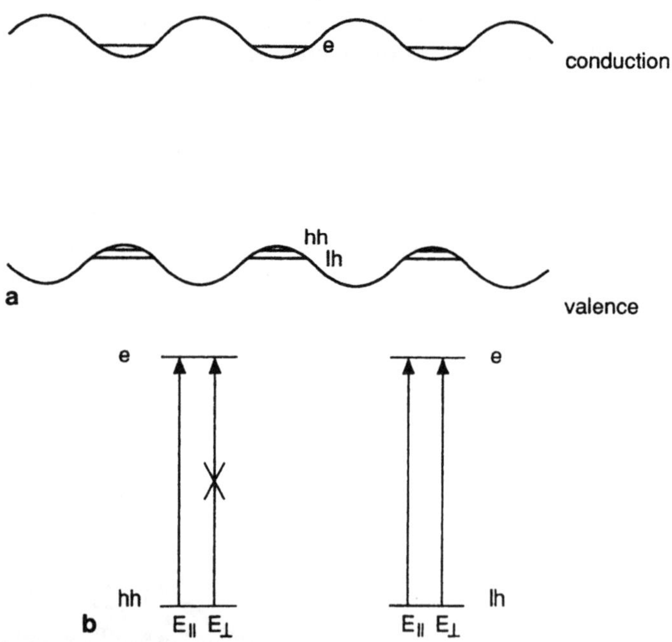

Fig. 2. a Schematic diagram of conduction and valence band modulation due to the growth-induced compositional variation in the FLS. **b** Schematic representation of the polarization-dependent selection rules associated with the electron to heavy- and light-hole (*e-hh*, *e-lh*) optical transitions. $E_{\parallel}$ and $E_{\perp}$ correspond to the electric field vectors parallel and perpendicular to the FLS steps

Nonlinear Optical Properties

We use the output from the combination of a colliding-pulse mode-locked dye laser and copper vapor laser pumped amplifier to do the room temperature nonlinear pump and probe measurements. The pump beam ($\lambda = 622$nm) produces photo-generated electrons and holes which are responsible for the nonlinear optical effects. We use a polarizer to select the probe beam polarizations parallel and perpendicular to the FLS steps. The polarization of the pump beam is oriented 45 degrees off the FLS step direction. By varying the delay time of the probe relative to the pump we confirmed that a 4 ps delay was longer than the intraband relaxation times. Using a similar experimental arrangement, we determined the electron-hole recombination time to be 860 ps.

The transmitted sub-picosecond continuum probe beam, with and without pumping, is collected by an optical spectroscopic multichannel analyzer. Figure 3a displays the nonlinear absorption changes for probes parallel $\Delta\alpha_{\parallel}$ and perpendicular $\Delta\alpha_{\perp}$ to the FLS. From Fig. 3a we can see that the maximum anisotropy in the nonlinear absorption principally occurs at the band edge. At higher photon energies the anisotropy is reduced. The number of electrons and holes populating the various energy bands determines the size of and the anisotropy in the nonlinear absorption. To calculate the relative populations precisely requires knowledge of the transition probabilities, the underlying density of states, and the Fermi-Dirac distribution functions for all the quantum confined energy bands in the FLS. Since the difference in the lowest energy heavy- and light-hole energy bands is small (3 meV) relative to the thermal energy, we can assume that within 4 ps the photogenerated carriers distribute themselves almost equally between the two energy bands, with the result that the magnitude of the nonlinear absorption effects for the two polarizations is comparable. Since the size of the absorption saturation is approximately the same for both polarizations, the anisotropy in the nonlinear absorption at the band edge is caused mainly by the energy separation of the heavy- and light-hole bands.

The room temperature absorption saturation of the FLS absorption band edge is consistent with a band-filling nonlinear mechanism. We determine the magnitude of the optical nonlinearity by evaluating σ_{eh}, where σ_{eh} is the ratio of the absorption change $\Delta\alpha$ per photo-generated electron-hole pair. The value of σ_{eh} in this FLS at room temperature is 6×10^{-15}cm^2. This value is of the same order as the band-filling nonlinearity in bulk GaAs. In an FLS structure with rectangular compositional modulation, we expect that excitonic saturation will be the dominant nonlinearity, and therefore, we expect at least one order of magnitude increase in the σ_{eh} value.

We perform a Kramers-Kronig transformation on the nonlinear absorption spectra to obtain nonlinear refractive index spectra, as shown in Fig. 3b. As in the nonlinear absorption spectra, the anisotropy in the nonlinear refractive index principally occurs at the band edge. For the pumping energy of 1.25 nJ in a 100-μm diameter spot, we subtract the nonlinear refractive index values for parallel and perpendicular probe polarizations to obtain the nonlinear birefringence as a function of photon energy, as shown in Figure 4. We define the nonlinear birefringence

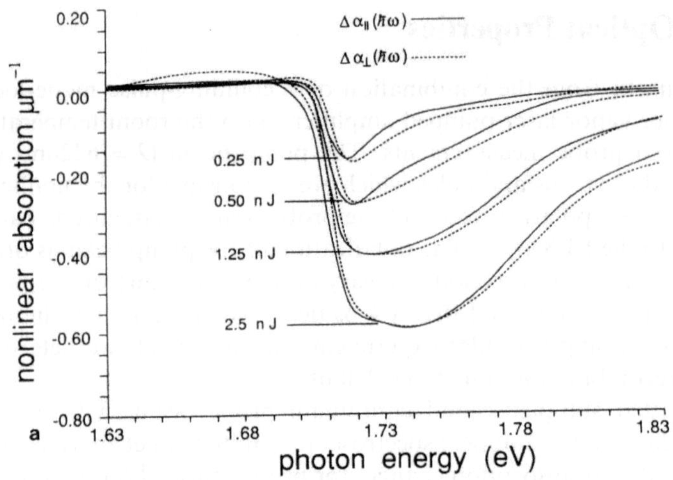

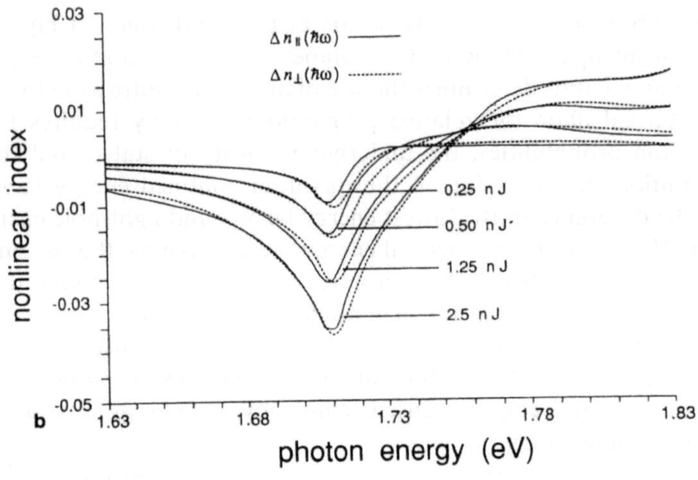

Fig. 3a, b. Room temperature **a** nonlinear absorption spectra $\Delta\alpha_{\parallel,\perp}(\hbar\omega)$ and **b** refractive index spectra $\Delta n_{\parallel,\perp}(\hbar\omega)$ in the $(Al_{0.5}Ga_{0.5}As)_{1/2}(GaAs)_{1/2}$ FLS for various pump pulse energies in a 100-μm diameter spot. The pump pulse energies absorbed by the FLS are indicated

as $\Delta n_{\parallel-\perp}(\hbar\omega) = \Delta n_{\parallel}(\hbar\omega) - \Delta n_{\perp}(\hbar\omega)$. In an FLS with rectangular compositional modulation, we expect at least an order of magnitude increase in the maximum nonlinear birefringence value.

Conclusions

We suggest that the FLS structure offers new opportunities for arrays of polarization-based semiconductor switching devices. The nonlinear birefringence observed here is the essential component in realizing an optically controlled polarization

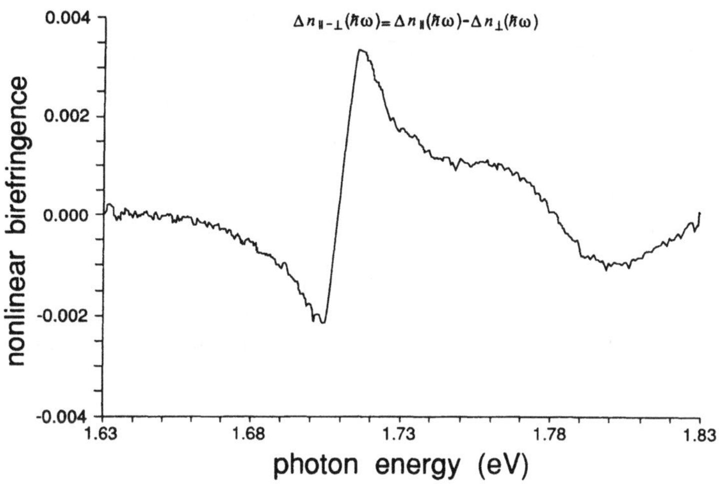

Fig. 4. Room temperature nonlinear optical birefringence, $\Delta n_{\parallel-\perp}(\hbar\omega)$, in the $(Al_{0.5}Ga_{0.5}As)_{1/2}(GaAs)_{1/2}$ FLS. $\Delta n_{\parallel-\perp}(\hbar\omega) = \Delta n_{\parallel}(\hbar\omega) - \Delta n_{\perp}(\hbar\omega)$ for the absorbed pump pulse energy of 1.25 nJ in a 100-μm diameter spot

rotation switch. In this case, an optical pump beam controls the amount of birefringence, and therefore controls the polarization state of a probe beam passing through the FLS. Moreover, since the nonlinear birefringence in the FLS is observed with light propagating in a direction parallel to the growth direction, one can couple into and out of many devices simultaneously, making array operation possible. The speed of the polarization switching is determined by the switch-on time of the nonlinear action, which we measure to be less than 4 ps. The switch-off time is determined by the electron-hole recombination time, which in this FLS is 860 ps, but which, in general, can be controlled by adjusting the surface recombination conditions [12].

In conclusion, we measured the polarization-dependent nonlinear absorption and refractive index spectra for a $(Al_{0.5}Ga_{0.5}As)_{1/2}(GaAs)_{1/2}$ fractional layer superlattice. We characterized the spectral dependence of the nonlinear optical birefringence and determined the size of the optical nonlinearity to be $\sigma_{eh} = 6 \times 10^{-15} cm^2$. Because of its anisotropic nonlinear behavior, the FLS structure offers the possibility of arrays of polarization-based semiconductor optical devices of high speed and great contrast.

Acknowledgments. The authors would like to thank Dr. H. Saito for sample preparation and Dr. Y. Horikoshi for his encouragement throughout this work.

References

1. Petroff PM, Gossard AC, Wiegmann W (1984) Appl Phys Lett 45:620
2. Fukui T, Saito H (1987) Appl Phys Lett 50:824
3. Kasu M, Ando H, Saito H, Fukui T (1991) Appl Phys Lett 59:301

4. Ando H, Fukui T, Saito H (1990) In: Extended abstract of the 22nd Conference on solid state devices and materials, B-4-7. Business Center for Academic Societies Japan, Tokyo, pp 123–126
5. Kanbe H, Chavez-Pirson A, Ando H, Saito H, Fukui T (1991) Appl Phys Lett 58:2969
6. Chavez-Pirson A, Yumoto J, Ando H, Fukui T, Kanbe H (1991) Appl Phys Lett 59:2654
7. Fukui T, Saito H (1988) J Vac Sci Technol B6:1373
8. Fukui T, Saito H (1990) Jpn J Appl Phys 29:L731
9. LePore JJ (1980) J Appl Phys 51:6441
10. Yamanishi M, Suemune I (1984) Jpn J Appl Phys 23:L35
11. Asada M, Kameyama A, Suematsu Y (1984) IEEE J Quant Electron QE-20:745
12. Lee YH, Gibbs H, Jewell J, Duffy J, Venkatesan T, Gossard A, Wiegmann W, English J (1986) Appl Phys Lett 49:486

Nonlocal Formulation of Optical Response of an Assembly of Fine Particles

49

YASUSHI OHFUTI AND KIKUO CHO[1]

Summary. A nonlocal optical response formulation developed recently is applied to an assembly of fine particles with a resonant level. In the theory, the field produced by electron excitation is fed back selfconsistently to the motion of the electrons. The selfconsistency between Maxwell and Schrödinger equations is kept within the linear response for polarization and it gives rise to the radiative lifetime and the Lamb shift within the semi-classical treatment. From the obtained response spectra, the radiative width and the Lamb shift are evaluated as functions of the number of the particles. It is observed that the radiative width deviates from the linear dependence on the system size even for a system size of about one-tenth of the light wavelength.

Introduction

It seems that, although our knowledge has been expanded by many studies, many important problems still remain to be resolved with regard to the response of matter on a mesoscopic scale to light. Even in the linear response, we cannot answer the question of how the radiative decay width (or the oscillator strength) is determined for a general system size, although it is qualitatively expected to increase and become saturated as the system size grows. Such behavior has been investigated only for slab geometry [1]. A nonlocal formulation is needed to investigate such behavior, since deviation from linear dependence on the system size limits the validity of local theories (or long wavelength approximation).

Recently the size enhancement of third order nonlinear susceptibility $\chi^{(3)}$ in mesoscopic systems has been attracting much attention. The size dependence of the oscillator strength has a more significant meaning in nonlinear processes, since the nonlinear susceptibilities themselves depend on the system size, and the enhancement of $\chi^{(3)}$ has been discussed in terms of oscillator strength in the long wavelength approximation [2]. Because a careful treatment of $\chi^{(3)}$ by which $\chi^{(3)}$ is saturated in large systems is restricted to a periodic chain model [3] at present, nonlocal investigations of the linear response of other systems may help us to gain a more general understanding of size-dependent responses and susceptibilities.

[1] Faculty of Engineering Science, Osaka University, Machikaneyama 1-1, Toyonaka, Osaka, 560 Japan

From the application viewpoint, materials with large oscillator strength and fast switching time are being investigated extensively for optimization of nonlinear optical devices. The theoretical investigations have been, however, confined mainly to isolated units of various structures. As they are integrated, the optimized properties may be changed, since the optical response would strongly depend on the size and the shape of systems in the mesoscopic region. Alternatively, it may be possible to get optimum properties by integrating fine unit structures (quantum dots), each of which has well separated (size quantized), but not necessarily optimized, excitation levels.

In this paper, we numerically study the linear optical response of an assembly of fine particles, applying a nonlocal formulation, recently developed by one of the authors (Cho) [4], to a simple chain model. From the response we evaluate the half width corresponding to the radiative decay width as a function of the number of fine particles. While the radiative decay width corresponds to the imaginary part of the self-energy correction of the electronic levels coupled to light, the real part correspond to the Lamb shift, which we also estimate.

In the following, we first summarize our formulation. The derivation of the formulation without assuming a background susceptibility is given in detail in Ref. [4]. The extension to the case with the introduction of the background susceptibility is straightforward. Next we introduce a simple model and show the results of numerical calculation.

Formulation

We consider a system of electrons with fixed positions of nuclei. We choose the Coulomb gauge and treat the vector potential $\boldsymbol{A}$ as an external perturbation. By applying the linear response theory [5] (at $T = 0$) to the Schrödinger equation under the electro-magnetic field $\boldsymbol{A}$, the Fourier component of the polarization with frequency ω is given by

$$\boldsymbol{P}(\boldsymbol{r}, \omega) = \int \mathrm{d}\boldsymbol{r}' \chi(\boldsymbol{r}, \boldsymbol{r}', \omega) \boldsymbol{E}_s(\boldsymbol{r}', \omega) \tag{1}$$

where

$$\boldsymbol{E}_s(\boldsymbol{r}, \omega) = \frac{i\omega}{c} \boldsymbol{A}(\boldsymbol{r}, \omega) = \boldsymbol{E}(\boldsymbol{r}, \omega) - \boldsymbol{E}_{dep}(\boldsymbol{r}, \omega) \tag{2}$$

$$\boldsymbol{E}_{dep}(\boldsymbol{r}, \omega) = \text{grad div} \int \mathrm{d}\boldsymbol{r}' \frac{1}{|\boldsymbol{r} - \boldsymbol{r}'|} \boldsymbol{P}(\boldsymbol{r}', \omega) \tag{3}$$

and

$$\chi(\boldsymbol{r}, \boldsymbol{r}', \omega) = \sum_{\lambda} \left\{ \frac{\boldsymbol{\rho}_\lambda(\boldsymbol{r})^* \boldsymbol{\rho}_\lambda(\boldsymbol{r}')}{E_\lambda - \hbar\omega - i\gamma_\lambda} + \frac{\boldsymbol{\rho}_\lambda(\boldsymbol{r}) \boldsymbol{\rho}_\lambda(\boldsymbol{r}')^*}{E_\lambda + \hbar\omega + i\gamma_\lambda} \right\} \tag{4}$$

The $\boldsymbol{\rho}_\lambda(\boldsymbol{r})$ is the transition dipole density with excitation energy E_λ, and $\boldsymbol{E}$ is the Maxwell electric field obeying

$$\text{rot rot}\boldsymbol{E}(\boldsymbol{r},\omega) - q^2\boldsymbol{E}(\boldsymbol{r},\omega) = 4\pi q^2\boldsymbol{P}(\boldsymbol{r},\omega), \qquad \left(q \equiv \frac{\omega}{c}\right) \tag{5}$$

In the definition of the source field, $\boldsymbol{E}_s$, the depolarization field, $\boldsymbol{E}_{dep}$, is subtracted from $\boldsymbol{E}$ to ensure that $\boldsymbol{E}_s$ is transverse. It should be noted that γ_λ comes from the adiabatic switching-on and may include the effect of non-radiative decay, but not that of radiative decay, since χ is expressed in terms of unperturbed quantities.

For practical use, the sum of λ in eq. (4) is restricted to resonant states. Instead we add the background susceptibility, $\chi_b\delta(\boldsymbol{r}-\boldsymbol{r}')\theta(\boldsymbol{r})\theta(\boldsymbol{r}')$, to χ, where θ is unity in the region which the matter fills and zero elsewhere. The χ_b may also include the contribution of the motion of nuclei. The important property of χ to make use of is that χ is separable with respect to the position variables. Even if the background term is added, this property holds by virtue of the expansion of the delta function with respect to any complete set;

$$\delta(\boldsymbol{r}-\boldsymbol{r}') = \sum_\nu \varphi_\nu^*(\boldsymbol{r})\varphi_\nu(\boldsymbol{r}') \tag{6}$$

By using the separability of χ mentioned above, $\boldsymbol{E}$ is expressed as

$$\boldsymbol{E} = \boldsymbol{E}_0 + \boldsymbol{E}_r \tag{7}$$

where $\boldsymbol{E}_0$ is the incident field and $\boldsymbol{E}_r$ is given by

$$\boldsymbol{E}_r = \tilde{\boldsymbol{E}}_r - \frac{1}{q^2}\text{grad div}\tilde{\boldsymbol{E}}_r \tag{8}$$

with

$$\tilde{\boldsymbol{E}}_r(\boldsymbol{r},\omega) = q^2\int d\boldsymbol{r}'\frac{e^{iq|\boldsymbol{r}-\boldsymbol{r}'|}}{|\boldsymbol{r}-\boldsymbol{r}'|}\left\{\sum_\lambda (F_\lambda\boldsymbol{\rho}_\lambda^*(\boldsymbol{r}') + \tilde{F}_\lambda\boldsymbol{\rho}_\lambda(\boldsymbol{r}')) + \sum_\nu \boldsymbol{G}_\nu\tilde{\varphi}_\nu(\boldsymbol{r}')\right\} \tag{9}$$

where

$$\tilde{\varphi}_\nu(\boldsymbol{r}) \equiv \theta(\boldsymbol{r})\varphi_\nu(\boldsymbol{r}) \tag{10}$$

and the coefficients F's, $\tilde{F}$'s, and $\boldsymbol{G}$'s are determined by a set of linear equations

$$(E_\lambda - \hbar\omega - i\gamma_\lambda)F_\lambda - \sum_\mu [A_{\lambda\mu}F_\mu + \tilde{A}_{\lambda\mu}\tilde{F}_\mu] - \sum_\nu \boldsymbol{B}_{\lambda\nu}\cdot\boldsymbol{G}_\nu = \int \boldsymbol{\rho}_\lambda(\boldsymbol{r})\cdot\boldsymbol{E}_0(\boldsymbol{r})d\boldsymbol{r} \tag{11}$$

$$(E_\lambda + \hbar\omega + i\gamma_\lambda)\tilde{F}_\lambda - \sum_\mu [A'_{\lambda\mu}\tilde{F}_\mu + \tilde{A}'_{\lambda\mu}F_\mu] - \sum_\nu \tilde{\boldsymbol{B}}_{\lambda\nu}\cdot\boldsymbol{G}_\nu = \int \boldsymbol{\rho}_\lambda^*(\boldsymbol{r})\cdot\boldsymbol{E}_0(\boldsymbol{r})d\boldsymbol{r} \tag{12}$$

$$\frac{1}{\chi_b}\boldsymbol{G}_\nu - \sum_\sigma \boldsymbol{C}_{\nu\sigma}\boldsymbol{G}_\sigma - \sum_\mu (\tilde{\boldsymbol{B}}'_{\nu\mu}F_\mu + \boldsymbol{B}'_{\nu\mu}\tilde{F}_\mu) = \int \tilde{\varphi}_\nu(\boldsymbol{r})\boldsymbol{E}_0(\boldsymbol{r})d\boldsymbol{r} \tag{13}$$

with

$$A_{\lambda\mu} = \iint d\boldsymbol{r}d\boldsymbol{r}'\left\{\boldsymbol{\rho}_\lambda(\boldsymbol{r})\cdot\boldsymbol{\rho}_\mu^*(\boldsymbol{r}')\frac{q^2e^{iq|\boldsymbol{r}-\boldsymbol{r}'|}}{|\boldsymbol{r}-\boldsymbol{r}'|} - \text{div}\boldsymbol{\rho}_\lambda(\boldsymbol{r})\text{div}\boldsymbol{\rho}_\mu^*(\boldsymbol{r}')\frac{e^{iq|\boldsymbol{r}-\boldsymbol{r}'|}-1}{|\boldsymbol{r}-\boldsymbol{r}'|}\right\} \tag{14}$$

$$(B_{\lambda\mu})_\xi = \iint \mathrm{d}\boldsymbol{r}\mathrm{d}\boldsymbol{r}' \left\{ \rho_{\lambda\xi}(\boldsymbol{r})\tilde{\varphi}_\nu(\boldsymbol{r}') \frac{q^2 e^{iq|\boldsymbol{r}-\boldsymbol{r}'|}}{|\boldsymbol{r}-\boldsymbol{r}'|} - \mathrm{div}\boldsymbol{\rho}_\lambda(\boldsymbol{r}) \mathrm{grad}_\xi \tilde{\varphi}(\boldsymbol{r}') \frac{e^{iq|\boldsymbol{r}-\boldsymbol{r}'|}-1}{|\boldsymbol{r}-\boldsymbol{r}'|} \right\} \tag{15}$$

$$(C_{\nu\sigma})_{\xi\eta} = \iint \mathrm{d}\boldsymbol{r}\mathrm{d}\boldsymbol{r}' \left\{ \tilde{\varphi}_\nu^*(\boldsymbol{r})\tilde{\varphi}_\sigma(\boldsymbol{r}')\delta_{\xi\eta} \frac{q^2 e^{iq|\boldsymbol{r}-\boldsymbol{r}'|}}{|\boldsymbol{r}-\boldsymbol{r}'|} - \mathrm{grad}_\xi \tilde{\varphi}_\nu^*(\boldsymbol{r}) \mathrm{grad}_\eta \tilde{\varphi}_\sigma(\boldsymbol{r}') \frac{e^{iq|\boldsymbol{r}-\boldsymbol{r}'|}-1}{|\boldsymbol{r}-\boldsymbol{r}'|} \right\} \tag{16}$$

and the other coefficients with ' and ~ are given similarly.

The most important point of the present formulation is that $\boldsymbol{A}$ contains the radiative decay width and the "Lamb shift" due to the interaction of electrons and light [4]. Our interest is how they are modified as the system size grows. Since the directions of the propagation and the polarization of the incident field are contained only in the right hand sides of the linear equations (11)–(13), the directions just select the modes which respond to the incident field. However, internal interaction through the emission and absorption of light is induced by the incident field, and thereby a relevant wavelength is introduced into the system.

It should also be noted that the boundary conditions for electro-magnetic fields were already considered when the Green's function of the Maxwell equation was chosen. This is the reason why we expect the present formulation to be suitable for systems with complex boundary conditions.

Model

We consider a linear chain along z-axis of N equivalent fine spherical particles with equal spacing b. It is assumed that each particle is so small that the level separation due to size quantization is quite large, and thereby a degenerate two-level model may be adopted to describe the resonant structure of each particle. By applying the long wave approximation to each particle (but not to the assembly), the transition dipole density is expressed by a uniform spherical density;

$$\boldsymbol{\rho}_{m\nu}^0(\boldsymbol{r}) = \boldsymbol{\mu}_\nu \theta(R - |\boldsymbol{r}-\boldsymbol{r}_m|)/\sqrt{V}, \qquad (\nu = x, y, z) \tag{17}$$

where R and V are the radius and the volume of the particle, respectively, $\boldsymbol{r}_m$ is the the position of the m-th particle, θ is the Heaviside function, and ν refers to the degenerate excitation levels with the transition dipole densities aligned to the x-, y-, and z-axes. It is also assumed that charge transfer does not occur between the particles but that the excitation itself can hop like a Frenkel exciton due to the dipole-dipole interaction. Because of the geometry we have chosen, the transition dipole densities with different alignments are decoupled. The transition dipole density in the chain is then written as a linear combination of $\boldsymbol{\rho}_{m\nu}^0$;

$$\boldsymbol{\rho}_{\lambda\nu}(\boldsymbol{r}) = \sum_m c_{\lambda m\nu} \boldsymbol{\rho}_{m\nu}^0(\boldsymbol{r}) \tag{18}$$

where $c_{\lambda m\nu}$ and the corresponding excitation energy are determined by

$$\sum_n t_{mn\nu} c_{\lambda n\nu} = E_{\lambda\nu} c_{\lambda m\nu} \tag{19}$$

with

$$t_{mnx} = t_{mny} = \frac{\mu^2}{b^3 |m - n|^3}, \qquad (m \neq n) \tag{20}$$

$$t_{mnz} = \frac{-2\mu^2}{b^3 |m - n|^3}, \qquad (m \neq n) \tag{21}$$

$$t_{mm\nu} = E_p, \qquad (\nu = x, y, z) \tag{22}$$

and μ is the amplitude of $\boldsymbol{\mu}_\nu$ and E_p the resonant energy of an isolated particle. The μ is related to the longitudinal and transverse splitting Δ_{LT} as

$$\mu^2 = V\frac{\Delta_{LT}}{4\pi}\varepsilon_b \tag{23}$$

where ε_b is the background dielectric constant.

Final approximation is concerned with $\{\varphi_\nu\}$. We replace it by $\{\theta(R - |\boldsymbol{r} - \boldsymbol{r}_m|)\}$. Such coarse graining is considered valid when the particles are small and the distance between them is not very small.

In the numerical calculation, we choose $R = 15$Å and $b = 50$Å as the parameters of the system size and use $\Delta_{LT} = 5.7$meV, $E_p = 3.2041$eV, and $\varepsilon_b = 5.59$ for the particle's internal parameters.

Results

We calculate the spectrum of the intensity of the radiated field at a position far from the chain. We make γ_λ equal to zero to see the net radiative decay width. As the incident field, we choose a linearly polarized light to separate the responses of the decoupled modes polarized to different directions. We have calculated for three geometries in which the directions of propagation and the electric field of incident field are different. To simplify the description in the following, we will refer to them as p_z, p_y and s. In p_z geometry, the incident field comes along the x-axis and $\boldsymbol{E}_0$ is parallel to the z-axis. In p_y geometry, the incident field propagates in the same direction, but $\boldsymbol{E}_0$ is parallel to the y-axis. The direction of the induced polarization is different and thereby the hopping energy of the activated exciton differs by factor 2 between them. In s geometry, the incident field propagates along the chain axis (z-axis) and therefore the direction of $\boldsymbol{E}_0$ is perpendicular to the chain. A wavelength is explicitly introduced along the chain by the incident field in this geometry.

Figure 1 shows the spectra for a $N = 1$ and b $N = 2$ for the p_z geometry. It is realized that the radiative width grows linearly with N while the peak height is unchanged. As N becomes large, other peaks with narrow widths (that is, with small oscillator strengths) appear, due to the effect of the nonlocality. In what follows, we will concentrate only on the main peak. The radiative width is determined from the half width of the main peak in the obtained spectra. The Lamb shift is evaluated by the shift of the main peak from $E_{\lambda\nu}$ of the mode with the

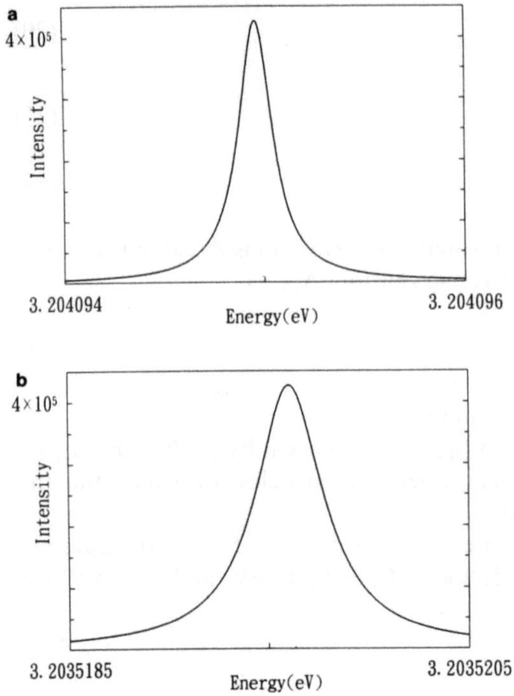

Fig. 1. Spectra for **a** $N = 1$ and **b** $N = 2$ for p_z geometry

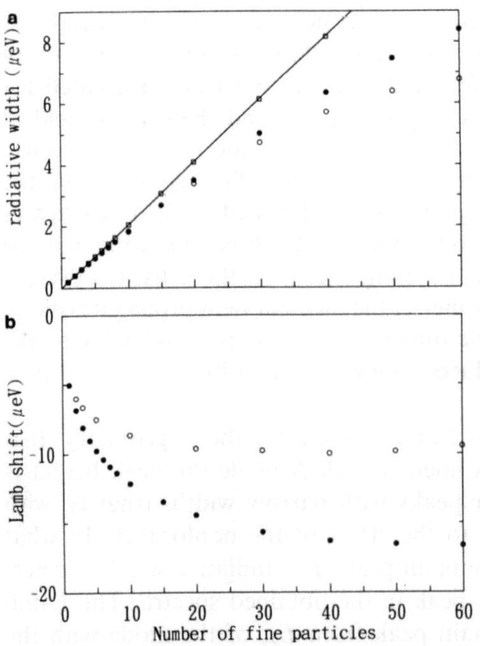

Fig. 2. a Radiative width and **b** Lamb shift vs N for p_z (*closed circles*) and p_y (*open circles*) geometries. The *open squares* show the radiative width estimated by means of the long wave approximation

main contribution to the response, which is always the mode without node in the calculated range.

In Fig. 2, the radiative width and the Lamb shift are plotted as functions of N for p_z (closed circles) and p_y (open circles) geometries. The initial linear growth of the radiative width is understood as the giant oscillator strength [6, 7]. The deviation from the linear dependence is obvious even for $N = 10$, for which the system size is about 500Å, and the radiative width seems to be saturated for a larger size. For comparison, the radiative width estimated by means of $\Gamma = (4/3)N\mu^2 q^3$, which is obtained by the long wave approximation, is also plotted (open squares) by using the values of q at the peak for p_z geometry. Though q at the peak changes as N is changed, mainly due to the effect of the depolarization, the deviation from the linear dependence is negligibly small. The results for s geometry are almost the same as those for p_y. This justifies the statement in the last part of the paragraph below eq. (16).

The difference between the two widths for p_z and p_y comes from the fact that the hopping energy of the activated exciton and the strength of the interaction through light differ by a factor 2. The saturation values of the widths seem to differ (presumably by a factor 2). An intrinsic "coherent length", defined as the system size for which the saturation is achieved, may be also different for the two cases, since the tendency to saturation at large N is more obvious for p_y. It seems that the coherent length is the same order as the wavelength at the resonant frequency and the factor is determined by parameters in the system. The Lamb shift also has interesting behavior, seeming to reach saturation more rapidly. This would also help us to understand the mechanism through which the coherent length is determined by the system parameters.

Acknowledgments. This work was supported, in part, by Grants-in-Aid (a) for Scientific Research on the Priority Area, "Electron Wave Interference Effects in Mesoscopic Structures" and (b) for a New Project, "Nanoscale Structure", from the Ministry of Education, Science, and Culture of Japan.

References

1. Cho K, Ishihara H, Okada T (1992) In: D'Andrea A, Del Sole R, Girlanda R, Quattropani A (eds) Proceedings of the international meeting on the optics of excitons in confined systems. The Institute of Physics Bristol, Philadelphia and New York, Conference Series 123:17–24
2. Hanamura E (1988) Phys Rev B37:1273–1279
3. Ishihara H, Cho K (1990) Phys Rev B42:1724–1730
4. Cho K (1991) Prog Theor Phys Suppl 106:225–233
5. Kubo R (1957) J Phys Soc Japan 12:570–586
6. Rashba ÉI, Gurgenishvili GÉ (1962) Sov Phys Solid State 4:759–760
7. Henry CH, Nassau K (1970) Phys Rev B1:1628–1634

Resonant Enhancement of Internal Field as a Function of Mesoscopic System Size: An Unnoticed Source of Large Optical Nonlinearity in Nonlocal Response

50

HAJIME ISHIHARA[1] AND KIKUO CHO[2]

Summary. By means of the nonlocal theory of linear response, it is shown that the internal field in resonance with excitons in a mesoscopic system has a remarkable size dependence, which plays an important role in the size dependence of nonlinear response of the excitonic system. In the energy region of the exciton resonance, the internal field is enhanced resonantly with system size due to the interference of the exciton polaritons. This is shown by calculating the polariton modes in a thin film with, using an ABC-free theory for the linear response in a nonlocal system. The important size region for the resonant enhancement of the internal field coincides with that for the size-enhancement of third-order nonlinear susceptibility $\chi^{(3)}$, therefore, a proper use of the former effect could bring about quite a large nonlinearity of exciton resonance.

Introduction

For ultra fast photonic signal processing and switching, nonlinear optical devices are key requisites in optelectronics systems. To realize efficient devices for practical use, materials which have large nonlinearity and quick response are being sought.

Excitonic systems are thought to be likely candidates for such materials because of their large oscillator strength. Confined excitonic systems in particular, such as thin films, microcrystals, quantum wells (wires, dots), and so on are attracting much attention, since they show a remarkable size dependence and a more effective performance than bulk materials. Therefore, the size dependence of the nonlinearity in mesoscopic systems is now a primary problem in the establishment of a guiding principle in the search for efficient nonlinear material, as well as being a problem of fundamental interest.

The size dependence of the nonlinear response of an excitonic system is caused by the nonlocal nature of the media. In nonlocal media, the optical response on a certain site is not determined by the field on the same site alone, namely, the

[1] Solid State Quantum Electronics Dept., Central Research Laboratory, Mitsubishi Electric Corporation, 1-1, Tsukaguchi Honmachi 8-Chome, Amagasaki, Hyogo, 661 Japan
[2] Department of Material Physics, Faculty of Engineering Science, Osaka University, 1-1, Machikaneyamacho, Toyonaka, Osaka, 560 Japan

polarization at a point is written as a correlated contribution of the field at various sites. This is due to the coherent extension of the exciton wavefunction.

The nonlocal effect appears in two factors determining the magnitude of nonlinearity. One is the third-order nonlinear susceptibility $\chi^{(3)}$, which shows size enhancement in the mesoscopic region. This effect has been discussed enthusiastically [1–7] and now we can describe it in a consistent way from micro to macroscopic systems [6, 7], namely $\chi^{(3)}$ is enhanced in a small size region, and is then saturated to a constant value.

Another factor is the internal field. In mesoscopic systems, the interference effect of exciton polaritons appears in various quantities near the energy region of the exciton resonance [8]. This is because, in the resonant region, the wavelength of the polariton changes more rapidly than its damping, as a function of sample size, which enables us to see its coherent character in a mesoscopic range of sample size. Since the interference is sensitive to sample size, we can expect resonant behavior of the internal field accompanying a polariton. In contrast to the size-enhancement of $\chi^{(3)}$, there has been little attention paid to this effect of the internal field.

In this paper, to reveal the essential role of the internal field in the size dependence of the nonlinear response by exciton resonance, we calculate the field in a thin film within the framework of the linear response, and we show its remarkable size dependence due to the resonant enhancement. We then point out the possibility of obtaining very large nonlinearity by the use of this effect.

Internal Field in Exciton Resonant Region

In this section, we calculate the internal field in resonance with an exciton in a slab, along the lines of the ABC-free theory of Cho [9], which was developed for the nonlocal treatment of the linear response. The model consists of a discrete lattice with N layers which has Frenkel exciton states as resonant excited levels. We consider the case of normal incidence. Then, the Hamiltonian for calculating the resonant part of susceptibility can be written as

$$\mathcal{H} = \sum_{l=1}^{N} \varepsilon_0 b_l^\dagger b_l - t \sum_{l=1}^{N} (b_{l+1}^\dagger b_l + b_l^\dagger b_{l+1}) \tag{1}$$

where ε_0 and t are the site and transfer energy, respectively, and $b_l^\dagger$ and b_l the creation and annihilation operators of the excitons localized in the l'th layer, respectively. The eigenvalues and eigenstates of this system are

$$E(K) = \varepsilon_0 - 2t \cos K \tag{2}$$

and

$$|K\rangle = \sqrt{\frac{2}{N+1}} \sum_l \sin Kl\, b_l^\dagger |0\rangle \tag{3}$$

respectively. The lattice constant is taken to be unity. The allowed values of K are

$$K = \frac{n\pi}{N+1}, \qquad \{n = 1, 2, \ldots, N\} \tag{4}$$

Using these eigenvalues and eigenfunctions, we can perform the remaining part of the ABC-free theory calculations, i.e., the calculation of nonlocal susceptibility and the solving of the Maxwell equation, in almost the same way as in ref. [9]. The main difference is the discreteness of the lattice in this case in contrast to the continuum model in ref. [9].

After the calculation, we get the expression of the internal field at site j as

$$\begin{aligned} E_j = {} & (2\cos k_1 - 2\cos q_0)\{\mathscr{E}_1 \sin k_1 j + \mathscr{E}_2 \sin k_1(N+1-j)\}/\sin k_1(N+1) \\ & - (2\cos k_2 - 2\cos q_0)\{\mathscr{E}_1 \sin k_2 j + \mathscr{E}_2 \sin k_2(N+1-j)\}/\sin k_2(N+1) \end{aligned} \tag{5}$$

where $\mathscr{E}_1$ and $\mathscr{E}_2$ are arbitrary constants, q_0 is defined as

$$\cos q_0 = (\varepsilon_0 + 2t - \hbar\omega - i\hbar\Gamma)/2t \tag{6}$$

and k_1 and k_2 are the wavenumbers of the polaritons which are the roots of the following dispersion relation.

$$(2\cos k - 2\cos q)(2\cos k - 2\cos q_0) - B = 0 \tag{7}$$

In the above expressions (6) and (7), $\hbar\Gamma$ is the phenomenological damping constant of the exciton, and

$$q = \frac{\omega a_0}{c}\sqrt{\varepsilon_b} \tag{8}$$

$$B = \frac{\Delta_{LT} a_0^2 \omega^2 \varepsilon_b}{tc^2} \tag{9}$$

where

$$\Delta_{LT} = \frac{4\pi |M|^2}{a_0^3} \tag{10}$$

ε_b is the background dielectric constant, a_0 is the lattice constant, and M is the matrix element of the transition dipole moment of the atom. Then, giving the amplitude of the incident field and using the Maxwell boundary conditions, we can determine the amplitudes of the internal field.

We show some examples of the spectrum of the internal field in the exciton resonant region. The parameters we use here are those of CuCl, namely

$$\begin{aligned} &\hbar\omega_T = 3.2022 \text{ meV}, \qquad t = 57 \text{ meV}, \qquad \Gamma = 0.06 \text{ meV} \\ &\Delta_{LT} = 5.6 \text{ meV}, \qquad \varepsilon_b = 5.7 \end{aligned} \tag{11}$$

where ω_T is the transverse exciton energy which is defined as

$$\hbar\omega_T = \varepsilon_0 - 2t \tag{12}$$

and ε_b is the background dielectric constant. Figure 1 shows the spectra of the

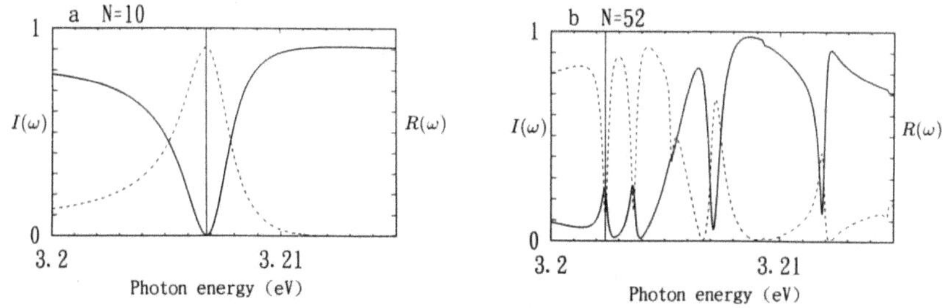

Fig. 1a, b. The reflectance (*broken line*) and the intensity of the internal field averaged over the layers (*solid line*). The intensity of the incident field is taken to be unity. The *vertical line* indicates the energy position of the lowest exciton energy; **a** and **b** are for $N = 10$ and $N = 52$, respectively. The parameters are shown in the text

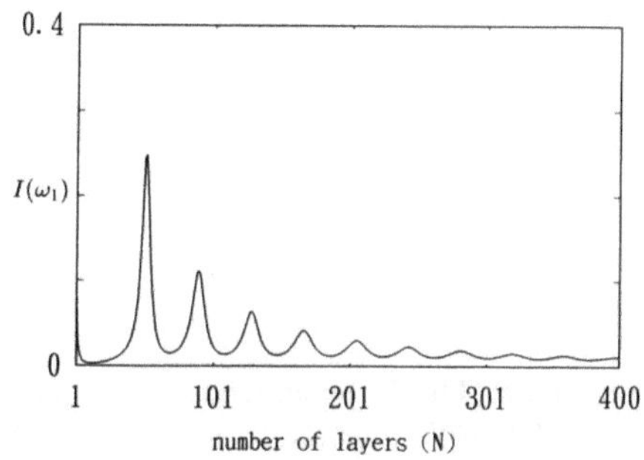

Fig. 2. The intensity of the internal field as a function of the size N. The parameters are the same as in Fig 1

reflectance $R(\omega)$ and the intensity of the internal field $I(\omega)$ averaged over the layers. The intensity of the incident field is taken to be unity. The vertical line indicates the energy position of the lowest exciton state. In the case of $N = 10$ (Fig. 1a), the intensity of the internal field is much depressed just around the lowest exciton state, which is easily understood from the large reflectance and absorption in this energy region. On the other hand, in the case of $n = 52$ (Fig. 1b), a peak due to the polariton interference can be seen near the lowest exciton region. Notice that the energy position of the lowest exciton state and the peak of $I(\omega)$ (or dip of $R(\omega)$) coincide with one another. This means that the internal field which is tuned to the lowest exciton energy is enhanced resonantly at this size. Figure 2 shows $I(\omega_1)$ as a function of the size N, where $\hbar\omega_1$ is the lowest exciton energy. In the small size region ($N = 1 \sim 12$), $I(\omega_1)$ is reduced as size increases, and is then enhanced resonantly around $N = 30 \sim 60$. Repeating the patterns of reduction and enhancement, $I(\omega_1)$ decreases as the size increases. Figures 3 and 4 show the change of the size dependence with the change of Δ_{LT} and Γ. From these figures, we understand that the contrast of reduction and enhancement is strengthened by making Δ_{LT} large and Γ small.

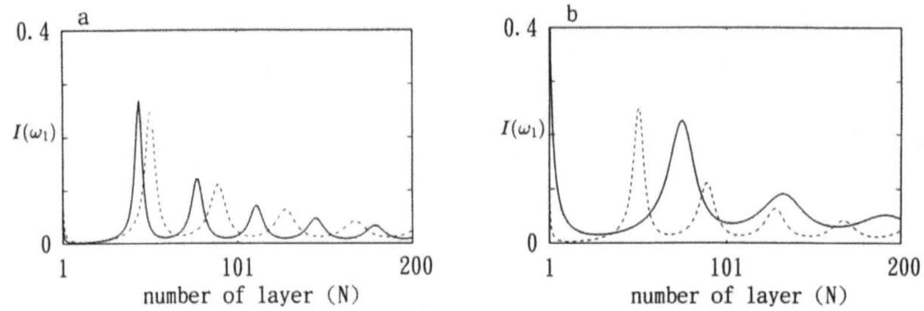

Fig. 3a, b. The Δ_{LT}-dependence of the intensity of the internal field; **a** shows the curves for $\Delta_{LT} = 10.0$ meV (*solid line*) and for $\Delta_{LT} = 5.6$ meV (*broken line*), **b** shows the curves for $\Delta_{LT} = 1.0$ meV (*solid line*) and for $\Delta_{LT} = 5.6$ meV (*broken line*)

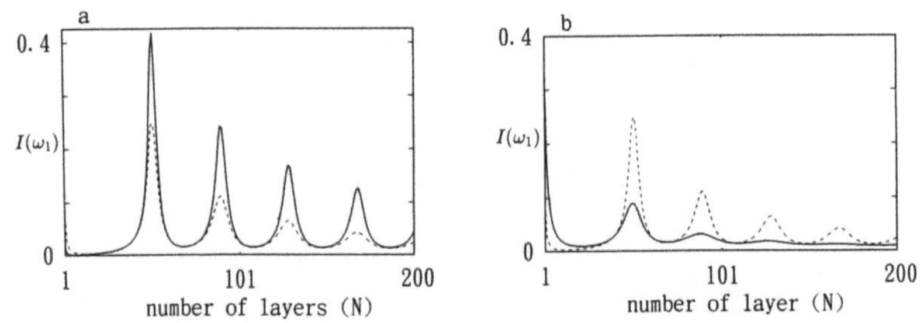

Fig. 4a, b. The Γ-dependence of the intensity of the internal field; **a** shows the curves for $\Gamma = 0.02$ meV (*solid line*) and for $\Gamma = 0.06$ meV, which is the original value (*broken line*), **b** shows the curves for $\Gamma = 0.2$ meV (*solid line*) and for $\Gamma = 0.06$ (*broken line*)

Discussion

In the previous section, we showed the notable size dependence of the internal field in resonance with excitons. It should be noted that the change of the internal field is most remarkable in the size region where $\chi^{(3)}$ shows size enhancement. For example, in the case of CuCl, $\chi^{(3)}$ (evaluated in the long-wave approximation) continues to be enhanced in proportion to the size until $N \cong 60$ [7], and below this size ($N = 52$), the internal field takes a peak value due to resonant enhancement. As we understand from this, the size dependence of nonlinearity in the observed physical quantities of a mesoscopic system must include the effect of not only the size enhancement of $\chi^{(3)}$, but also the change in the internal field. Therefore, the study of susceptibility is not sufficient for understanding the nonlinearity in signal intensity, which is relevant for realistic applications.

As we showed in the previous section, the larger the dipole moment and the smaller the damping, the clearer was the contrast of reduction and enhancement. Since $\chi^{(3)}$ is proportional to the square of $|M|^2$ and inverse to the square of the phase relaxation constant (Γ in this text), this means that in the case of a material (or structure) of large $\chi^{(3)}$, the influence of the change of the internal field appears more clearly, and a prominent difference of nonlinearity arises depending on the size.

From the viewpoint of development of large nonlinear materials, the large enhancement of the internal field is quite attractive. Though a full nonlocal calculation of nonlinear response is necessary for a quantitative treatment (Ishihara H, Cho K, Nonlocal theory of the third-order nonlinear response of confined excitons, manuscript in preparation), we can expect that a very large nonlinearity should be obtained by making the best use of the size-linear enhancement of $\chi^{(3)}$ and the resonant enhancement of the internal field. If we choose a size where the internal field is reduced, the large nonlinearity due to the enhanced $\chi^{(3)}$ does not emerge in the response field.

Since the size-resonant enhancement of the internal field which we have shown here with a slab model is a general phenomenon based on polariton interference, a similar effect can be expected in the case of other structures for exciton confinement, such as fine particles, wires, boxes, and so on.

Acknowledgments. This work was supported, in part, by a Grant-in-Aid for Scientific Research on the Priority Area, "Electron Wave Interference Effects in Mesoscopic Structures," from the Ministry of Education, Science, and Culture of Japan

References

1. Hanamura E (1988) Phys Rev B37:1273–1279
2. Takagahara T (1989) Phys Rev B39:10206–10231
3. Banyai L, Hu YZ, Lindberg M, Koch SW (1988) Phys Rev B37:8142–8153
4. Spano FC, Mukamel S (1989) Phys Rev A40:5783–5801
5. Spano FC, Mukamel S (1989) Phys Rev Lett 66:1197–1200
6. Ishihara H, Cho K (1990) Phys Rev B42:1724–1730
7. Ishihara H, Cho K (1992) J Nonlinear Opt Phys 1:287–309
8. Cho K, D'Andrea A, Del Sol R, Ishihara H (1990) J Phys Soc Jpn 54:1853–1867
9. Cho K (1986) J Phys Soc Jpn 55:4113–4121

MIX
Papier aus verantwortungsvollen Quellen
Paper from responsible sources
FSC® C105338

If you have any concerns about our products,
you can contact us on
ProductSafety@springernature.com

In case Publisher is established outside the EU,
the EU authorized representative is:
Springer Nature Customer Service Center GmbH
Europaplatz 3, 69115 Heidelberg, Germany

Printed by Libri Plureos GmbH
in Hamburg, Germany